FRAMINGHAM STATE COLLEGE

3 3014 00058 3379

D0849164

Defining
the
Laboratory
Animal

IV Symposium, International Committee on Laboratory Animals
organized by the
INTERNATIONAL COMMITTEE ON LABORATORY ANIMALS
and the
INSTITUTE OF LABORATORY ANIMAL RESOURCES
NATIONAL RESEARCH COUNCIL

NATIONAL ACADEMY OF SCIENCES   Washington, D.C.   1971

WHITTEMORE LIBRARY
Framingham State College
Framingham, Massachusetts

These *Proceedings* were directly supported by Contract PH 43–64–44, Task Order Number 12, with the Animal Resources Branch, National Institutes of Health, United States Public Health Service; Grant #R-24 from the American Cancer Society, Inc.; Contract 277–E–ARS–68 with the United States Department of Agriculture, Agricultural Research Service; Contract C–310, Task Order Number 144, with the National Science Foundation; and supported in part by Contract AT(49–7)643 with the Atomic Energy Commission; Grant RC–1K from the American Cancer Society, Inc.; and Contract Nonr–2300 (24) with the Office of Naval Research, United States Army Medical Research and Development Command, and United States Air Force.

ISBN 0–309–01862–5

Available from:

Printing and Publishing Office
National Academy of Sciences
2101 Constitution Avenue
Washington, D.C. 20418

Library of Congress Catalog Card Number 72–609048

Printed in the United States of America

QL
55
I5
1969

FOREWORD

This volume is a record of the program of the IVth International Symposium on Laboratory Animals. As such it can be regarded as a bench mark of the state of the science and art of laboratory animals in 1969. Those familiar with the scientific process will recognize at once that this is only temporary; one can confidently predict further progress by the time of the Fifth International Symposium.

The symposium, held in Washington, D.C., April 8–11, 1969, was organized under the joint auspices of the International Committee on Laboratory Animals (ICLA) and the Institute of Laboratory Animal Resources of the National Research Council of the United States of America.

The organizers of this symposium felt that the time was ripe to specify, as completely as possible, the parameters that give dimension to the world of animals in the laboratory. The theme, then, of the meeting, and of this volume, is "Defining the Laboratory Animal in the Search for Health." The papers published here embrace the variables in the animal itself—of genetics, associated microflora, antigenic experience, husbandry, nutrition, and emotional history; and the variables that surround the animal—the physical and social environments and the never-ceasing pressure of microbial agents that seek admittance to the tidy world of animals in the laboratory.

Additional luster was conferred on the symposium by the thoughtful perspectives provided by the three keynote speakers, George W. Beadle, John C. Eccles, and James A. Shannon.

HOWARD A. SCHNEIDER, Ph.D.
Chairman
Institute of Laboratory Animal Resources

iii

ORGANIZING COMMITTEE FOR THE IVth INTERNATIONAL SYMPOSIUM OF THE INTERNATIONAL COMMITTEE ON LABORATORY ANIMALS

Berton F. Hill, General Chairman
Roger D. Estep, Assistant General Chairman
Charles G. Durbin
William I. Gay
Samuel M. Poiley
M. M. Rabstein
Marie Woodard

Program Committee

William I. Gay, Chairman, U.S.A.
B. K. Batra, India
Thomas B. Clarkson, U.S.A.
Stian Erichsen, Norway
Henry L. Foster, U.S.A.
Willi Heine, West Germany
William Lane-Petter, England
Tatsuji Nomura, Japan

CONTENTS

I

Defining
the Laboratory
Animal

TOWARD STANDARDIZED
LABORATORY RODENTS:
THE MANIPULATION OF
RAT AND MOUSE LITTERS

W. Lane-Petter
M. E. Lane-Petter

For the biological assay of protamine zinc insulin injection, which must be carried out against a standard preparation of insulin, the *British Pharmacopoeia* (1968 edition) specifies that no fewer than 96 mice, all within a 5-g weight range, be used; the *United States Pharmacopoeia* of 1965 specifies no fewer than 24 rabbits, each weighing at least 1.8 kg; and the *British Pharmacopoeia* of 1963 specifies 96 mice within a 5-g weight range or 12 rabbits, each weighing from 1.8 to 3.0 kg.

If for this assay 12 (or 24) rabbits suffice, why are 8 (or 4) times that number of mice required? Moreover, if the rabbits may weigh anything upwards of 1,800 g, why must the mice all be within a 5-g range, approximately half the variance allowed to rabbits? I would not have the effrontery to suggest that the smaller numbers and greater variance permitted for rabbits had anything to do with their greater cost, for science recognizes no such venal considerations. Therefore, the reason for this difference is presumably that the sort of mice customarily used for insulin assay, as well as for countless other purposes, are expected to be so variable that to get meaningful results, significantly greater numbers of them have to be used. For this reason, efforts are being made to eliminate some of the variables in laboratory rodents by producing uniform animals of high quality. In this paper we discuss some of the variables and the steps that are being taken to avoid their recurrence.

Factors Influencing Variability

VARIATION IN LITTER SIZES

It has been said for many years that to produce animals of uniform quality and characteristics requires a strain that possesses a measure of genetic uniformity and an environment that is also uniform. It has been further emphasized that the environment must not only be uniform but optimal, for a poor environment, however uniform, will of itself contribute to undue variation in the animals.

One of the commonest causes of lack of uniformity in the environment, especially the early environment, is variation in litter size and preweaning influences. It seemed a logical step, therefore, to attempt to contribute to a standardized preweaning environment by manipulating litter size.

The factors influencing litter size at birth are both genetic and environmental. In 100 consecutive litters of CFY rats, we found variations in litter sizes as born from 6 to 21, with an average of 13.85. Even in inbred strains, litter sizes vary as much as they do in random-bred strains, and the variation in litter size follows a normal distribution. Individual weights of pups vary inversely to litter size, both at birth and at weaning (Table 1). Festing[2] has reported that

TABLE 1 Weights of Three Litters of CFY Rats as Born and at Weaning

Litter Size	Average Weight (g)	
	Birth	Weaning (21 Days)
5	8.6	65.4
10	8.1	57.9
15	7.3	54.8

in inbred mice the average weaning weight goes down by 0.13 ± 0.03 g for every extra pup in the litter. Our figures for noninbred mice show a much greater variation (Table 2).

Weight gain from birth to weaning is not the only factor that is grossly affected by litter size. The onset of sexual maturity is also strongly influenced.

4

W. LANE-PETTER
AND
M. E. LANE-PETTER

TABLE 2   Weights of Three Litters of CFLP (Noninbred) Mice at Weaning

Litter Size	Average Weight (g) at Weaning (18 Days)
5	13.0
10	9.5
12[a]	8.3

[a]This litter started with 15, but only 12 survived.

AVERAGE AND OPTIMAL LITTER SIZES

In any breeding colony the average litter size at birth depends on the strain and on the system of management. There is no particular reason why the average litter size born should be the same as the optimal litter size reared. A heavily lactating strain might well be characterized by the production of small litters, or a colony in which large litters were born might possess mothers that were poor lactators. If the average litter size born is substantially greater than the optimal size for lactation, then the dams will be overmilked and the pups will not get enough nourishment. On the other hand, if the average litter size born is smaller than the dams can nurse in comfort, there will be an excess of milk, which may be associated with engorgement mastitis in the dams.

POLYGYNOUS GROUPS

In polygynous groups of one male and a number of females (often called harems), the females may be allowed to produce their young in the group and in the presence of the male so that advantage may be taken of mating at the postpartum estrus.

We have not found this to be a practical way of breeding rats because it leads to a very high rate of infant mortality. It is, however, a common way of breeding mice. From one point of view, it is an economical breeding method, and in some hands it leads to the maximum production of animals per foot of shelf and per pair of hands employed. However, we have observed that in permanently mated polygynous groups of mice, the infant mortality rate may be high and there is a considerable variability in the pups at weaning. Moreover, when the ages of pups in the nest differ by as much as 10 days, the mortality rate among young pups rises steeply because the older ones strip

5

the milk that they should be receiving. Also, the variability of weight for age, which is the natural result of differences in litter sizes, is increased when the semistarvation factor is introduced.

Methods of Standardization

ADJUSTMENT OF LITTERS

Consideration of the variability factors involved led us to investigate the possibility of standardizing conditions in the nest during the first 3 weeks of life.

Our rats are mated in harems of one male and six females, and the females are removed and put into maternity cages a day or two before they bear their litters. There is no possibility of postpartum mating, and under these conditions and with a special diet, we have found that the optimal litter size for lactation is 14 pups. Larger litters grow markedly less well, and when the litters are smaller, the dams often develop engorgement mastitis. Therefore, we have standardized at 14 pups.

Our mice are mated in harems of one male and four females, and the females are allowed to have their litters in the mating cage and thus be exposed to postpartum mating. We have found that under these conditions, 10 pups per mouse is the optimal litter size for rearing.

CROSSFOSTERING

Because 14 seems to be the optimal litter size for rats, we adjust the litters to this figure at 2 days of age. At this age, the young of each harem are sorted into litters of 14 and fostered onto one of the dams that contributed to the group. The rejection rate by dams is zero. For greater convenience, 14 members of each artifically adjusted litter are of the same sex. The average litter size of our rats as born may be less than 14, and since we may also adjust the sex ratio of the young at this age, it follows that there will not be quite enough pups to go round, and some of the dams will have no litter to raise. These dams are put straight back to the male and will come into estrus within a few days.[1, 4]

Any poor-looking pups are discarded at 2 days of age, when they are cross-fostered, and we have found that the subsequent infant mortality rate is negligible.

The mouse litters are adjusted at 1 day of age. The pups born into harems are sorted out into groups of 40, and each group is put into a new cage with 4 lactating dams. As in the case of rats, the rejection rate is zero. We have found a group of 4 dams and 40 pups to give rather better results than 3 and 30 or 2 and 20 and to be very much better than 1 and 10.

At two days of age, rats are comparatively easy to sex, so there is little, if any, difficulty in making the artificial litters unisexual. The demand for the animals when they come to be used is by no means always for equal numbers of each sex, but the disparity of demanded sex ratio is to some extent predictable. Therefore, at the time of crossfostering it is a simple matter to adjust the number of male and female litters according to the expected demand. The demand for male rats nearly always exceeds that for females, so there are usually some female pups at 2 days of age that are not going to be needed. It is much more economical to dispose of the pups at 2 days of age than to grow them on to weaning, only to destroy them then.

Sexing mice at 2 days is rather more troublesome, and we have not found it practicable to do so. Moreover, our experience has been that approximately equal numbers of mice of each sex are used.

Crossfostering by skilled workers is a quick and accurate procedure that depends more than most on the technique by which it is accomplished. We are fortunate in having a team of highly skilled animal technicians. We have not found it difficult to train staff to sex and crossfoster rats at 2 days of age. Our technicians can count and sex a group of 14 2-day-old rats in about as many seconds. When it comes to weaning, the time saved is considerable. The litters are, or should be, all of the same sex, so it is necessary only to verify that no mistake was made in sexing at 2 days of age. In our experience, such mistakes are rare. (For a discussion of the problem of selecting good staff for work in animal houses see reference 3.)

Results of Litter Manipulation

The results of manipulating litters have been very encouraging. Figures 1 and 2 show the weights of 100 rats, 50 of each sex, at 24 and at 29 days of age. Figure 3 shows the weights at 17 days of 200 mice, 100 of each sex.

It has been possible to predict with great confidence the weight for age of both rats and mice up to 6 or more weeks of age, and we have developed a simple calculator for this purpose. In the case of rats 6 weeks of age or younger, it is not necessary to do a lot of weighing and reclassifying, because the weights predicted on the calculator, ± 5 percent, will be achieved by at least 80 percent of the animals, and often by 95 percent. Since every cage is marked with the birth date of the rat in it, a single check weighing of each animal at the time of shipment is enough. The very few animals that fail to come within a few grams of the predicted weight are usually discarded.

Much the same is true for mice, but there is one variable that affects them that is not applicable to rats. Although all the female mice are necessarily ex-

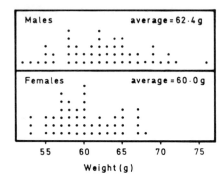

FIGURE 1 Weights of 50 male and 50 female CFHB rats at 24 days of age.

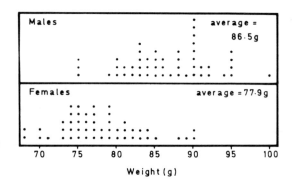

FIGURE 2 Weights of 50 male and 50 female CFHB rats at 29 days of age.

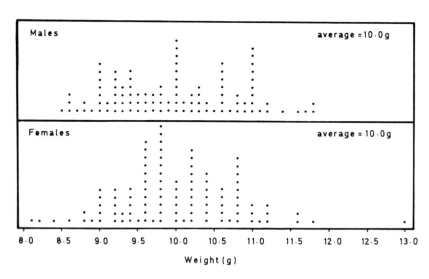

FIGURE 3 Weights of 100 male and 100 female CFLP mice at 17 days of age.

8

posed to postpartum mating, not more than 80 percent of them become pregnant at the first estrus. This means that in any crossfostered nursing group of 4 dams and 40 young, one or more of the dams may not be pregnant.

This possibly contributes to the fact that, despite the standardization of the litter groups of 4 dams and 40 young, the weights at weaning cover as much as a 2-g range, with a few outliers. It is our custom to classify the mice at weaning largely by their appearance, weighing only those that seem to be outside the range. The heavier outliers will be promoted to a group one day older, and most of the lighter ones will be regarded as runts and discarded. Of the remainder, the vast majority, nearly all will make the predicted weight for age when they come to be machine-weighed immediately before packing and shipment.

Thus, in the case of both rats and mice, there is no machine-weighing at weaning, except for occasional spot checking. It is not necessary to weigh the mice until the moment of packing for shipment, when it can be expected that at least 90 percent of any given age will fall within a 4-g weight range—the range that most users specify. This saves a lot of weighing time, but it has the added advantage of not mixing the communities. In our experience, mixing a community, especially of male mice, disturbs the normal pattern of weight gain, and the animals take a few days to settle down.

When they are sold, all our animals are weighed at the point of shipment. The rats that are not sold are weighed at 5 weeks of age, that is, at 120–140 g. At this stage, the demand for older animals is reviewed, contract orders are reserved, and future breeding stock replacements are put aside allowing also a small reserve for contingenices. Any remaining unallocated animals are discarded. From 5 weeks of age to maturity we still find that some 80 percent of rats of the same age fall within a 20-g weight range.

By these systems we find we can make the most efficient use of our space, our labor and our animals. In a room of approximately 960 ft^2, two technicians can select, breed, rear, and pack for shipment some 1,600 rats per week, coming from some 1,350 breeding females and about 150 breeding males. In a room of about 1,700 ft^2, four or five animal technicians can select, breed, rear, and pack for dispatch 8,000–10,000 mice per week, coming from 4,000 breeding females and 660 breeding males. These animal technicians carry out all the work that has to be done in the rooms, including feeding and watering and the changing of cages. They do not have any cage cleaning duties. They work a 40-hour week.

Animals coming from such a colony are all of known age because birth dates are marked on every cage. If littermates are wanted, suitable litters can be held back from crossfostering, provided reasonable notice is given. However, the degree of uniformity produced by crossfostering is such that work

9

for which littermates have been required can, in most cases, be done as well on crossfostered litters.

Conclusion

So far we have achieved a high degree of uniformity of weight for age at weaning by means of crossfostering and standardizing litter and group sizes. The process has, of course, to be continued after weaning. This is done by adjusting the density of animals per cage. The optimal density is that which permits optimal growth, without checks. We have found that too high or too low a density will result in a falling away from the optimal growth curve. The actual number of animals per cage will have to be decreased as the animals grow, and the age at which densities are reduced is fairly critical. Thus, in our colonies, a cage that accommodates 20 rats at weaning will take only 10 rats 10 days later.

The optimal density depends on the animals, their age, the size and type of cage, the standard of ventilation, and the ambient temperature. If the temperature falls a few degrees, the result is likely to be not a significant failure to gain weight, but a compensatory consumption of extra food. If the temperature falls further, then there is a failure to gain weight, which can cause a breeding colony to fail to achieve its calculated weight at age.

Nature is diverse in her manifestations, not least in the fecundity she confers on her creatures. But that unnatural being, the scientist, likes to have all his materials ordered to size, and therein lies the conflict. By improving on nature in the manipulation of litters, the rat and mouse breeder can go some way to satisfying the scientist.

References

1. Bowtell, C. W., and M. Lane-Petter. 1968. Breeding and cross-fostering technique of CFE rats. J. Inst. Anim. Tech. 19:184.
2. Festing, M. 1969. Research: (e) Genetic Studies. Lab. Anim. Centre Newsletter 37:9.
3. Lane-Petter, W. 1967. Selection, training and control of staff. In M. L. Conalty [ed.] Husbandry of laboratory animals. Academic Press, London.
4. Lane-Petter, W., M. Lane-Petter, and C. W. Bowtell. 1968. Intensive breeding of rats. I. Crossfostering. Lab. Anim. 2:35.

DISCUSSION

DR. GAY: Are you confident that unisexual litters can develop normally? Intersexual experiences early in life may contribute significantly to later behavior.

DR. LANE-PETTER: We have done no experiments to check this, but we find no evidence among the animals developed in this way that they are deprived of their normal sexual proclivities or abilities. We have a very, very low infertility rate. It is really negligible. That is the best answer I can give to that question.

DR. GAY: How do you select future breeders with the crossfostering method to assure uniform randomization?

DR. LANE-PETTER: To begin with, we keep a very careful check on the performance of all our breeders, both the males and the females. Every breeding female is marked on the tail, and she is not given a chance to overstay her time in the colony unless she is working as she should be. This is to say that the less efficient breeding females are eliminated and therefore do not contribute to the next generation of breeders.

Secondly, we eliminate any animals that have any black marks against them. If they have small litters, or if they fail to raise most of the young that are given to them, they are marked down and eliminated. So, by negative selection, we get rid of the bad ones. So it happens that our breeding stock is taken in a completely random way—an unplanned random way, I should say—from the remainder of the colony, which is the majority.

There is only one thing about which we are rather careful. We do not take male and female breeders from the same harem groups. This makes it almost, but not quite, impossible that there should ever be any sib mating. I think it is impossible in the case of rats. It is almost impossible in the case of mice. The odds are very heavily against mating between closely related animals, and repeated mating of this sort is impossible.

DR. GAY: I have three more questions that seem to me to be related. First, are you selecting future breeding stock with the aim of producing an even narrower weight range? Also, we have found that in RFM mice weight at weaning is only slightly correlated to weight at 10 weeks of age. Is the uniformity of weight you observe at weaning continued at later ages? Third, is that uniformity continued in animals up to 600 days old?

DR. LANE-PETTER: In answer to the third question, I must say that we have no animals in our colony at the present time who are 600 days old. Perhaps we should, but we don't.

On the question about selecting future breeding stock with the aim of producing an even narrower weight range: Yes, we do intend to select to produce a narrower range. The outliers will certainly not find their way into future breeding stock. This is not a very intensive selection, but it is a selection that eliminates the ones that are not within the main group.

Our experience is that weight at weaning is correlated with weight some weeks later. If the animals get off to a good start, they maintain a good growth rate. For example, in the illustrations of the weights of rats at 24 and 29 days of age (Figures 1 and 2), the male outlier that weighed over 75 g at 24 days is the same animal that weighed 100 g at 29 days. It is our experience that these big ones continue to go ahead.

DR. GAY: What microbiological studies are done on your animals and what studies are done on your animal handlers?

DR. LANE-PETTER: We do not do enough studies on our animals. There can never be enough. We check the microflora at infrequent intervals, but in my view not nearly as often as we should.

We have done very few checks on the handlers. There have been a few throat swabs and skin cultures, and we have never cultivated anything from the skin cultures within a half hour of washing up. We do have the handlers rinse their hands with alcohol after washing up. I think the answer to both of those questions is "not enough."

DR. GAY: One more questioner asks two questions. Is crossfostering practical in nervous DBA? What stain did you use on the animals' tails?

DR. LANE-PETTER: The first is not for me to answer because we do not breed DBA and have never tried to. However, I cannot see any reason why it should not be practical to crossfoster this strain. We have done crossfostering on another inbred strain, and it seems to be no more difficult than on the outbred strains.

On the question of the marking of the tail, we use a waterproof felt marker pen. There are some that do much better than others, and we did quite a bit of investigating to discover, first, the best marker for use on the animals, for which we needed one that would not come off for at least two or three weeks, and second, the best one for marking the cages, where we wanted one that would come off. We spoiled a number of cages because the manufacturer of the pen that we were using changed to waterproof ink without telling us.

DEFINED MICE IN
A RADIOBIOLOGICAL
EXPERIMENT

L. J. Serrano

Introduction

The literature records that laboratory animals have
been of transcendent importance in the advancement of experimental biology.
It also records the experiences of many investigators whose animals have died
in mid-experiment or who have found an excessively large variability in the
results of an experiment presumed to be carefully controlled. Some of the
investigators subjected to these frustrations converted their experiments into
studies of the cause of their experimental difficulties, and these studies have
given us much of our knowledge about the diseases of laboratory animals and
about the interaction of the animals with various facets of their environment.
Presently this knowledge is being expanded by many approaches, such as those
included in this volume.

All of us recognize that the complexity of today's problems in experimental
biology and medicine requires us to continue improving both the quality of
animals used in experiments and the quality of methods for their maintenance.
Therefore, we must assiduously apply the knowledge we have gained about
methods for controlling every extraneous variable that can influence the results
of an experiment whether the variable manifests itself as a rampaging disease
or as a subtle change that can be detected only as an inordinate variability in
the results.

The objectives of this report are (1) to give a brief description of an experiment that illustrates an exceptional need for preventing changes in the environment of the experimental animal; (2) to describe the facility and methods of

13

operation, showing how we achieved environmental stability; and (3) to describe our experiences in conducting the experiment under those environmental conditions.

The Low-Level Experiment

One of the problems of modern society to which experimental biologists must apply their skills is defining the effect of low levels of ionizing radiation on health. The chance of being exposed increases with the increased use of radioactive sources in power production, science, industry, and medicine, and with the advent of flight above the protective mantle of the atmosphere.

The immediate effects of exposing mammals to a dose of less than 50 rads from an external source are usually negligible; however, it is difficult to estimate with confidence the effects that occur later in life and long after exposure. For example, the data from various experiments testing the effect on life-span suggest at least three hypotheses, each with different implications (Figure 1). We need information to define curves of this type if we are to estimate the hazard of low-level radiation.

We have therefore undertaken a large, long-term experiment to determine the relation of small and medium doses of radiation to induction of neoplasms and other late-occurring degenerative changes and to shortening of life-span in mice. Other studies in the experiment, including biochemical, physiological,

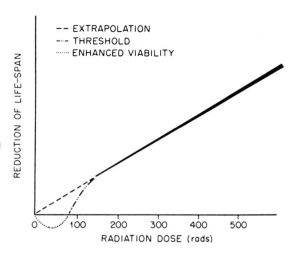

FIGURE 1 Three hypothetical extensions of the dose–response curve into the low dose region. The linear extrapolation suggests that the smallest dose has an effect. The threshold hypothesis suggests that no effect is seen until a minimum dose is given. The third hypothesis suggests that the low doses of radiation may have some beneficial effect.

and microbiological investigations, may aid in clarifying the fundamental mechanism of such radiation-induced changes. Table 1 shows how the experiment is set up.

TABLE 1   Radiation Doses and Sample Sizes

Dose (rads)[a]	Number of Female Mice To Be Irradiated	Number of Mice To Be Irradiated Each Week
0	4,000	32
10	3,000	24
25	1,000	8
50	1,000	8
100	1,000	8
150	1,000	8
300	4,000	32
Total	15,000	120

[a]Whole-body ^{137}Cs gamma rays, about 45 rads per min.

To have acceptable confidence limits in an experiment of this kind, a large number of animals must be exposed. Using a large number of animals creates many problems, including the costs of handling, housing, and maintaining the mice, and the costs of collecting, analyzing, and interpreting the concomitantly large amounts of data. This requires a considerable financial investment for which some valuable results will be expected.

A second difficulty is the length of time required to complete the experiment. Even if it were possible to put all the mice into the experiment in one day, the length of the experiment could be no shorter than the maximum life-span of the mice, about 3 years, which is a long time to expose experimental animals to the hazards commonly found in the animal facility environment. The logistics of putting mice into the experiment further extends its duration. The length of the present phase of our experiment will be about 5 years. Not only is this a long time, it is also an appreciable part of the productive career of an experimental biologist. If we are to invest significant parts of our careers in the experiment, we must be protected from the multitude of variables that could render the experiment valueless.

A third difficulty is that the differences in the results among the various groups are very likely to be small. Hence, any uncontrolled variable, even though the changes it produces are minor, could mask, cancel, or amplify the effect of the experimental variable—radiation. Also, we cannot assume that simply because control groups and experimental groups are in the same room a given environmental variable will affect all animals equally and, therefore, that its effect will be of no consequence.

There are many other difficulties in an experiment such as this, but those stated above are sufficiently compelling to require us rigorously to minimize variables and maximize control of those we cannot eliminate.

Animal Facility

There are few obvious characteristics of an animal facility in which the environment and the transmission of microbes within its walls are rigorously controlled. Also, many of the necessary controls may be no more than subtle extensions of the same type of controls seen in other animal facilities not specifically designed or operated with a view to maximizing control. We shall try to elucidate those differences, without which we could not successfully perform the experiment, in the following description of the animal facility.

DESIGN

The general design of the facility is illustrated in Figure 2, a diagrammatic floor plan. Some construction features are shown in the photographs of the clean corridor (Figure 3) and of a mouse room (Figure 4). The design is similar to other modern animal facilities planned around a clean-to-dirty corridor system. We can say, after more than 4 years of successful operation, that the design has been adequate to meet our requirements and that many of its features have been applied to more recently constructed facilities.[1,3,12]

We can further illustrate many of the features of design and operation by describing the traffic patterns of some of the many objects that are continually moved into and through the facility.

TRAFFIC PATTERNS

Mice

Mice are transferred biweekly into the facility directly from a germfree plastic isolator through a sterilized porthole in the door near the north elevator (Figure 2, "EL"). They are immediately placed in a transfer cart that is pressurized with filtered air and taken to an empty

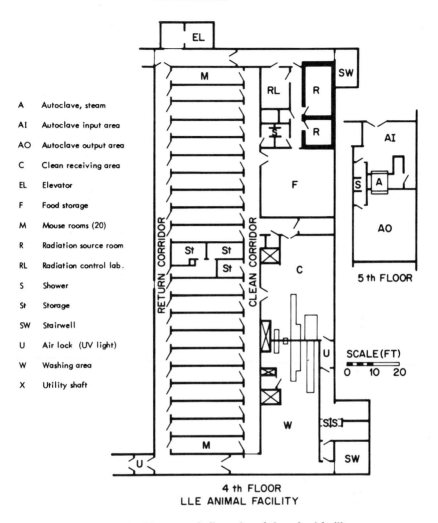

A   Autoclave, steam

AI   Autoclave input area

AO   Autoclave output area

C   Clean receiving area

EL   Elevator

F   Food storage

M   Mouse rooms (20)

R   Radiation source room

RL   Radiation control lab.

S   Shower

St   Storage

SW   Stairwell

U   Air lock (UV light)

W   Washing area

X   Utility shaft

FIGURE 2 Diagrammatic floor plan of the animal facility.

mouse room for quarantine. There they receive food pellets soaked with a broth containing six selected species of bacteria. Ten days later, their feces are tested. If no contaminants are found and the six bacteria are present in the fecal samples, the mice are transferred to breeder rooms on the fourteenth day. Mice produced in the breeder rooms for the experiment are weaned at 20-23 days of age and their feces are then tested the same way. If the results are satisfactory, the experimental mice are transferred to another room and

FIGURE 3 Clean corridor. The exit from the clean receiving area is on the left. The doors at the ends of the corridor are for emergency use only and are sealed with tape. The boxes to the right of each are clock-controlled switches for the lights in the room. The protrusion on the conduit leading from these boxes is a condulet that allows the insertion of an expanding mastic to seal the lumen of the conduit.

kept there until they are 9½ to 10½ weeks of age. One week before they reach this age, their feces are tested again. If the results are satisfactory, the mice are transferred to the radiation source room, irradiated, and then placed in a room where they will remain until death or sacrifice. Their feces are retested about once every 20 weeks. All transfers from room, to cart, to room are made in the return corridor after it has been thoroughly cleaned.

People

People entering the animal facility must first re-move all their clothes and shoes in a room outside the facility. Then they shower and don clean clothes and shoes. In the entrance air lock (Figure 2) they remove their shoes, don knee-high boots, step into a disinfecting tank and scrub their boots, and then enter the facility. If they are to enter any room on the "clean side" (any room with a door onto the clean corridor), they must again disrobe, shower, and don sterile clothing. They place their boots in disinfectant tanks while showering and again scrub them with disinfectant in the air lock leading into the clean receiving area (Figure 2, "C"). When a person enters a mouse room, the entrance door locks behind him, thus preventing accidental return to the clean corridor. When he leaves the room, the door locks behind him. He must again shower and change clothes if he wants to return to the clean side. No one is exempt from this procedure. Entrance is limited, of course, to those whose activities are necessary to the operation.

Air

Air Supply About 14,000 ft^3 of air passes through the animal facility every minute. All the air is drawn in from the outside through vents 70 ft above the ground. The air is then filtered through high-

FIGURE 4 Mouse room. Cages on pipe racks line each wall. The hopper on the cage-changing cart holds mouse food. The uniform worn by the animal attendant is required for everyone entering the facility. The method for handling mice while transferring them to clean cages is demonstrated.

19

efficiency particulate air (HEPA) filters, which are tested with dioctylphthalate particles after installation. The filtered air passes through a curtain of water and through moist sinuous channels in a cooling coil. Although this water scrubbing was designed for cooling and humidifying the air, it also helps to trap particles that might pass through the filters and to remove soluble gases with which the water is not saturated. The water, however, can become heavily contaminated with bacteria if it is not disinfected periodically. The filtered, humidified, and cooled air then enters the animal facility.

Airflow   Air that enters the facility, although clean, as defined above, immediately mixes with whatever microbes are present in the facility and thus can be a way of transmitting organisms from an undetected focus of contamination to uncontaminated areas. Therefore, airflow through the facility is as carefully controlled as other traffic through the facility. The following features of the facility help to minimize back-flow of air from less clean to cleaner areas:

(1) Ultraviolet-irradiated air locks (Figure 2, "U"). A bactericidal curtain of ultraviolet light[9] with a minimum intensity of $35 \mu W$ at the floor helps to free the air in the lock of viable particles.

(2) Interlocked doors. Opening one door of the air lock activates an electrically controlled lock on the opposite door. Thus, it is impossible to open both doors at the same time, and there is always a solid barrier between contaminated and clean areas.

(3) Door-closers. Automatic door-closers are installed on every door in the facility to ensure that the door is closed when not in use and to minimize the time it remains open.

(4) Weather stripping. All doors are weather-stripped to minimize the flow of air around the doors when they are closed.

(5) Pressure gradient. The highest air pressure is in the clean receiving area (Figure 2, "C"). The lowest pressure is in the wash area (Figure 2, "W"). Through the other rooms, the pressure decreases in increments of 0.05 in. of water. Pressure changes beyond set limits activate an alarm on a remote pressure monitoring panel. Control of air pressure is "fail-safe"; i.e., if either the air-supply system or the exhaust system and the standby equipment should fail, then the air-pressure differential is still maintained. The air pressure in the clean receiving area and in the clean corridor cannot be lowered below that of adjacent areas because there are no exhaust vents in either of these clean areas. Pressure in the wash area, which during operating hours has the highest concentration of airborne particles, cannot be increased above that of the clean areas because there are no connections between the wash area and the air supply system. The wash area is ventilated by air exhausted from other parts

20

of the facility. For purposes of airflow, the wash area can be considered as a large protrusion on the main exhaust duct.

Air Changes   The ventilation rate provides enough air to change the room air twelve times per hour. This rate is sufficient to clear the room of suspended particles in 20 to 30 min or, where particles are continuously being generated, to greatly reduce their concentration in the air.[10]

Food
Mouse food (Purina 5010-C) is delivered weekly to the floor above the animal rooms. It is removed from the bags and placed in 1-in.-deep trays (about three pellets deep). Several racks of trays (up to 1,000 lb of food) are placed in the high-pressure, high-vacuum steam autoclave. The chamber is evacuated to 14 mm Hg absolute, filled with steam until the temperature of the food reaches 270° F for 30 sec, and then evacuated again. The food is then removed and stored in a sealed metal tank connecting the autoclave area with the clean side of the animal facility (Figure 2, "AO" and "F"). After bacteriological confirmation of the sterility of the food and of *Bacillus stearothermophilus* spore strips that accompanied the food, the tank is opened and food is carried to each room as needed.

Bedding
Chipped corn cob (San-I-Cel) used as bedding (litter) is received in porous cotton bags containing 50 lb each. Twenty bags at a time are placed in the steam autoclave. The chamber is evacuated to 14 mm Hg absolute, filled with steam until the bedding reaches 270° F for 25 min, and then evacuated again. The bedding is then removed and stored in the autoclave output area. After bacteriological confirmation of the sterility of the bedding and spore strip, it is placed on a conveyor that carries it to a hopper in the clean receiving area (Figure 2, "C"). The bedding is then automatically dispensed into cleaned cages as they emerge from the cage-washer.

Water
The drinking water for the mice is highly chlorinated (14–18 ppm) before it comes to the facility. The water used for cleaning comes from the same source but is not highly chlorinated (0.5 ppm). Although bacteriological tests have detected bacteria in the cleaning water, it is seldom used without the addition of disinfectant–detergents.

Cages
Cages are removed twice a week from the animal rooms and taken to the wash area where the soiled bedding is dumped into a

hopper that pneumatically forces the bedding out of the building to a trash container. The cages are then placed on a conveyor in the cage-washer that carries them through the sections for prewash (75° F), wash (170° F), recirculated rinse (192° F), rinse (200° F), and dry. The cage emerges in the clean receiving area where it is automatically filled with about 0.5 lb of bedding. Twice each wash day, a cage is randomly selected for microbiological testing to confirm that established washing and handling procedures remain effective. Cages are then transferred either to a mouse room or to a storage area.

Bottles

Drinking units—consisting of a glass pint bottle, rubber stopper, and stainless steel tube—are removed from the mouse rooms three times a week. They are transferred to the wash area where the bottles are emptied and placed in the bottle washer. After 6 min of washing and rinsing, they are passed into the clean receiving area where they are filled with highly chlorinated water. The stoppers and tubes are soaked for 30 min in disinfectant–detergent solution, boiled for 30 min, soaked in a 1,000 ppm chlorine solution, and then inserted in the bottle top. Each wash day, three drinking units are randomly selected for bacteriological testing.

Miscellaneous Materials

There are many other materials that must pass into the clean side of the facility, e.g., tools, cleaning equipment, disinfectants, data cards, papers, the wheeled dollies used to carry the cages and bottles, replacement parts, and light bulbs. The usual procedure is to sterilize them in steam or ethylene oxide autoclaves, but where this is not efficient, other decontamination procedures (and bacteriological tests of their effectiveness) have been developed. We have duplicated much of our cleaning equipment and other labor-saving devices rather than subject them to frequent sterilization, as would be necessary if they were passed from one room to another.

Vermin

Since vermin can find entry routes not easily detectable by humans during careful inspection, a wandering insect might bypass our physical and operational barriers and set up housekeeping in the animal facility. We did not want to introduce insecticides or other poisons into the facility, so an additional line of defense was established outside the facility. We began by introducing pest-control measures on the floor above and the floor below the animal facility, and then expanded, by logical extension, to the other floors in the building, adjoining buildings, and the warehouse where our food and bedding are stored. We have seen no signs of vermin infestation in the facility since the initiation of this program.

ENVIRONMENTAL CONTROLS

Achieving a stable environment that is identical for each experimental mouse is particularly difficult in an experiment of this size and duration. Housing mice in several rooms (3,000 per room) requires a geographic stability; i.e., we must maintain the same environmental conditions in all mouse rooms, regardless of the distance separating them. Since the experiment will last about 5 years, we must also achieve a temporal stability; i.e., provide the same environment for mice entering the experiment next year as for those entering last year and then prevent change for the duration of the experiment.

Accumulating reports suggest that there are numerous complex interactions between the laboratory animal and the many components of its environment and that not all the components of an environment capable of affecting experimental results have yet been identified.[5] Those that we have identified as being important in our experiment are considered individually in the following description of the control methods.

Air Conditioning

Five large air-conditioning units serve the building. If one fails or is turned off for routine preventive maintenance, the others can be quickly adjusted to compensate for it. Various protective measures have been taken to reduce the probability of total failure of the air-conditioning and exhaust systems to less than once in 10 years.

Temperature Each room has a thermostat for controlling electric "strip heaters" in the duct supplying air to the room. The thermostat is activated by a 1° F change from set temperature. However, daily measurements taken in each room with thermometers that mark the highest and lowest temperatures since the last periodic adjustment of the thermometer demonstrate that minimum and maximum temperatures of 72 and 78° F are recorded at least once each week. A second thermostat in each room is activated by higher or lower temperatures to sound alarms on remote panels that are monitored around the clock. The temperature inside cages is not measured, but the cage tops do not impede convection of heat generated in the cage.

Humidity Relative humidity is controlled at 45 to 60 percent. Only one mouse room has a high–low monitor for activating an alarm, because humidity is centrally controlled, and loss of control would be detected in any one of the rooms.

Light

Fifteen 40-W, 48-in. long, cool-white fluorescent tubes (GEF-40CW) provide light in each mouse room. The intensity of the light varies with distance from the source as follows: The cages on the top row are 41 in. from the lights, and they receive about 150 fc (as measured with a General Electric 213 light meter) on the front of the cage and about 70 fc in the rear of the cage. The cages on the bottom row are about 8 ft from the lights, and they receive 70 fc at the front of the cage and heavy shadow at the rear. Most of the mice in cages on the top row sleep under the food well of the cage cover. In other rows, most of the mice sleep near the front of the cage. Room illumination may seem weaker to mice than to humans, since the highest energy output from the phosphor in the tubes is in the wavelengths of yellow and orange.[13] Lights in the rooms are automatically turned on at 7:00 A.M. and off at 6:00 P.M. EST. In the adjoining corridors, each alternate light goes off 30 min later, and the corridors are darkened 1 hr later. Half of the corridor lights go on at 6:00 A.M., and the remainder, 30 min later. The incremental reduction and increase of light intensity simulate dusk and dawn in the mouse rooms.

Noise

A survey of sound-pressure level in the range of 20 to 10,000 Hz was performed in an empty mouse room. The over-all relative intensity was between 60 and 70 dB. Most of the sound seemed to be coming from the ventilation system. Cages in the mouse rooms probably absorb some of the sound waves, but the activity of the mice and the activities of the attendants servicing the rooms add to the noise level. To provide background noise that might mask startling noises, such as may occur during routine servicing of the rooms and when announcements are made over the public address system, music from the Muzak Co. is played in the room 24 hr a day.

Caging

The cages are of the "shoe-box" type and are constructed of transparent polycarbonate plastic (Lexan); they measure approximately 13 × 7½ × 5 in. The cage covers are made of stainless steel sheet, with ventilation openings of 54 in.^{2} covered with stainless steel mesh 20 × 20 × 0.011 in. (61 percent air space); each has wells for food and a water bottle. The mice are transferred to clean cages twice a week. About one half pound of sterile chipped corn cob (San-I-Cel) is placed on the bottom of each cage, in contact with the mice, to absorb urine and feces. No nesting materials are used.

Cage Position

It is obvious that light itensity is different for each cage shelf in a mouse room. Because particles that are too heavy to remain suspended in calm air for more than a few minutes settle on the floor and rise only a few feet each time they are disturbed by someone walking by, the concentration of airborne particles differs from shelf to shelf. The intensity of other environmental components may also vary at different shelf levels. For these reasons, each time mice are transferred to clean cages, the clean cage is placed on the shelf above the one from which the soiled cage was taken, and the mice from the top shelf are moved to the bottom shelf. We feel that this rotation will expose all mice equally to whatever conditions may exist at different levels of the room. When cages of new mice are brought into the room, they are placed on the shelves in vertical columns next to the previous week's columns of cages.

Food

The mice are fed commercially prepared pellets (Purina 5010-C) that contain increased levels of certain vitamins to compensate for loss during sterilization of the food. Although there is considerable variability in composition among various lots of food, the alternative of a diet prepared from high-purity chemicals introduces other problems of greater magnitude. Therefore, we standardize food with regard only to its manufacturer and its microbial content and are dependent on the manufacturer for limiting other variables. The autoclaving procedure is standardized, and food is discarded if exposure to heat exceeds certain limits of time or temperature, if bacteria are found in samples of the sterilized food, or if spore strips are positive. Frequent analyses for thiamine are also performed to monitor sterilization procedures, which reduce thiamine from the concentration in untreated food of 59 ppm ± 10 (SD) to a poststerilization level of 35 ± 7 ppm. An excess of food is kept in the food well of the cage top.

Drinking Water

Drinking water for the mice comes from a filtered water system that is also the source of water used for human consumption. Analysis reveals that it is fluoridated, has a hardness of 116 ppm $(CaCO_3)$, a pH between 7.3 and 7.8, and alkalinity of 105 ppm. Before it is given to the mice, chlorine gas is added to a concentration of 14-18 ppm. Spectrographic analysis of the water for 40 elements reveals considerable variation. We have not attempted to control this variation since the alternative—demineralized water—could deprive the mice of necessary trace elements. Water bottles are left on the cages no longer than 72 hr to avoid reduction of the chlorine concentration below 2 ppm, which would allow bacterial growth in the water. Each bottle contains about 400 ml—more than twice the amount the mice would consume in 72 hr; the water level in each bottle is checked twice a day.

Cage Population

At weaning age (20–23 days), usually eight and no less than six mice of the fourth generation (after entry into the facility) are placed in a cage and assigned for experimental use. The age span of the mice in one cage does not exceed 1 week and is usually less. No more than two mice from the same litter are placed in one cage. No mice are added after the cage group is established.

Room Population

Each week the same number of cages of experimental mice are placed in a room until the maximum of about 400 cages is reached. Although the first groups of mice are placed in an almost empty room, the last groups are placed in one that is almost full. We have not been able to eliminate this difference, but our record system will allow us to determine if it influenced the experimental results.

People

The continued existence of the mice and much of the control of their environment depend on people. Yet people are, understandably, the most difficult component of the environment to subject to unvarying behavior. To help people perform as the experiment requires, we try to instill in each individual an understanding of the importance of his performance to the success of a complex experiment. For those working in the facility, there is an extensive on-the-job training program combined with occasional formal lectures. During this training, we try to impart not only the skills involved in caring for the animals, but also an understanding of why it is necessary to have standardized procedures that must be performed precisely each time and to avoid the creeping modifications that can occur because of minor "improvements" in established procedures. We provide close supervision, of course; but since supervision that is excessively close can have the opposite effect of that desired, the experiment is largely dependent upon the understanding, the capability, and the attitude of each attendant.

Handling

Mice are held in the human hand only at weaning while their sex is being determined and, in the case of mice being assigned to the experiment, while one toe is being amputated as a means of identifying the mouse in its cage. At these times the animal attendant wears rubber gloves that are disinfected each time mice from a different cage are handled. At other times, such as during transfer to clean cages, the mice are picked up with 12-inch-long forceps (Figure 4).

Time

When they are 9½ to 10½ weeks old, all mice in a cage receive the same dose of radiation at the same time. Control groups are not irradiated. Mice are transferred to the radiation source room in weekly replicates and are all irradiated on the same day of the week, Monday, between 9:00 A.M. and 12:00 noon. All other procedures, especially those involving handling or moving the mice, are similarly scheduled to occur on a specific day and at about the same time of day each occurrence. We feel that this variable is particularly important, since its effect on radiation studies has been demonstrated.[8]

Preweaning Experience

Although our mice are as genetically similar as possible for so large a group, they have all experienced different environments by the time they are born into the facility.[2] For example, they have shared a uterus with various numbers of fetuses and their mothers are of different parities and varying physiological preparation for motherhood. Such differences continue after birth. For example, the litters are of different sizes with different ratios of sex and are composed of mice of different weights or rates of growth. Therefore, by the time mice are weaned, they have received differing challenges to their homeostatic abilities. Dr. Lane-Petter discussed this in the preceding paper (page 3) and has described a system of crossfostering to reduce these differences. We cannot adopt such a system because of our need for genealogical identification of each mouse, but we do record such differences for future analysis of their influence on the experiment. In addition, we try to prevent adding other differences by keeping a constant environment around the breeder cage and by sorting the mice into experimental groups as soon as they are weaned.

Pheromones

It is obvious that the natural odors of outside air vary considerably with seasons of the year and with weather conditions. Fortuitously, most of these odors are removed by filtering and by water scrubbing of air supplied to the animal facility.

Odors generated within the facility but outside of the mouse rooms are controlled by the airflow system, which minimizes exchange of air between mouse rooms and any other part of the facility. Also, we avoid any activities, such as painting, that would produce unusual odors.

A lack of possibly necessary odors became a problem when the male groups of mice were eliminated from the experiment. Since there is evidence that a female mouse must have contact with male odors (pheromones) or her ovaries function aberrantly (as evidenced by an irregular estrous cycle),[15] and since we intend to measure the incidence of ovarian tumors in the mice, we felt that we had to add some male pheromones in the mouse room. We did this by placing in each room one cage of eight unirradiated males for every 100 cages of females.

Microbial Control

Most of the environmental components we have described are relatively easy to control in a well-designed facility with the aid of appropriate mechanical devices, constant attention to detail, precise performance of standardized procedures, ample feedback of information, and careful supervision. There is, however, one component of the environment that cannot be controlled easily even in small or short-term experiments: namely, microbes—ubiquitous microscopic organisms such as viruses, bacteria, other protists, and metazoan parasites that are of microscopic size during their transmissible stage.

Microbes must be controlled because of (1) the variable degree of their effect, which ranges from occult infection, through mild morbidity, to furious disease; (2) the continuing variation of the quantity and types present in the environment; (3) the ease with which they are transmitted; (4) the difficulty of detecting them before they have done their damage; and (5) their possible interaction, in unrecognized ways, with radiation.

The desired degree of control of transmission of microbes is a major consideration in an experiment because it can influence the design of the facility, the operating methods, the protocol of the experiment, the costs, and the probability of loss of experimental data. In determining the degree of control of microbes needed for our low-level radiation experiment, we unanimously agreed that so large a colony should not be exposed to contagious pathogenic organisms, to those organisms causing occult (latent) infection, or to those opportunistic microbes commonly found in healthy mice but causing disease only in stressed mice. There were two points, however, on which we had varying opinions: (1) the definition of pathogenic organism and (2) the importance of the indigenous microbes of the mouse as sources of variation. Since there are pathogenic members in almost every class of microorganisms, whatever

methods used to prevent contact between pathogen and mouse would necessarily be similar to methods used to prevent contact with any other micro-
organism. In either case, we would apply the same methods of epidemiology,
preventive medicine, public health, and sanitation used to prevent transmission
of contagion. We therefore decided to try to control, within reasonable limits,
the transmission of all organisms.

METHODS

Our methods for controlling the transmission of
microbes are guided by several basic principles of preventive medicine and
epidemiology: (1) minimize contact between the susceptible population and
an undesirable microbe, (2) quickly detect failures of the preventive program,
(3) isolate infected and exposed individuals, (4) sacrifice individuals known to
be infected, and (5) review the history of a contamination with the purpose
of improving prevention or detection.

Two lines of prevention or protection against either contamination of the
mice or an unacceptable change in the microbial environment are the design of
the animal facility and its operating procedures, which were described in the
section on traffic patterns. A third line of defense is the mouse's own resistance
to infection.

Each of these three defenses is subject to continuous challenge by the dynamic process of sustaining a large colony of mice, yet none of them is an
absolute barrier, as suggested in Figure 5. We can thus expect that some microbes will occasionally slip through our defenses. If we are to avoid a consequent gradual increase in the number of species of microbes that associate
themselves with the mice, we must be able to detect them, preferably before
they contact the mice, and then identify their sources of entry or modes of
transmission. To do this requires a microbiological surveillance program. Such
a program must be adequate to detect contaminants in the facility rapidly,
thus allowing quick application of isolation and eradication procedures. If the
surveillance program is subliminal, then its only value is to record the approximate times that specific organisms contaminated the mice. An adequate microbiological surveillance program is, therefore, the *sine qua non* for maintaining
microbially defined mice. In biological terms, such a program provides the
necessary sensory feedback to activate the defensive mechanisms to restore
homeostasis.

The major components of our microbiological program are described below.

Laboratory Tests

The part of our surveillance program that utilizes
standard laboratory diagnostic procedures is summarized in Table 2. The number and type of samples, the frequency of collecting them, and the type of

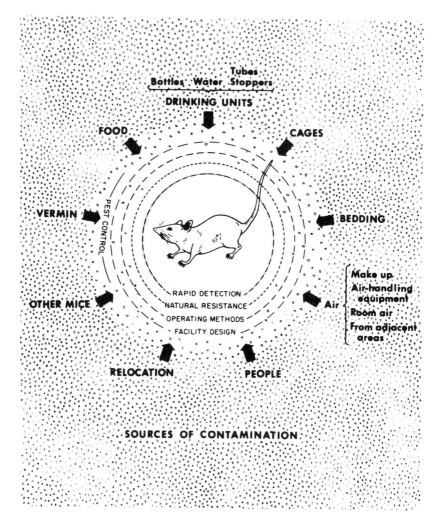

FIGURE 5 Schematic representation of our systems for control of transmission of microbes. The mouse is shown in a sea of microbes that is being held back by four imperfect barriers. The items named are a partial listing of the things that must enter the animal facility and be cleaned of their load of microbes before they contact the mice.

test applied to each sample has changed often since the program began in December, 1963, but the major objective continues to be the detection of a contaminant in the facility before it can contact the mice (area and object samples) or in the mice before it can spread to a large number of other mice (fecal and organ samples).

30

TABLE 2 Surveillance Program for Microbes

Contaminant	Sample Size and Frequency	Total Samples Tested to 2/69
PARASITES		
Intestine	5 mice per week	4,380
Skin and hair	5 mice per week	420
Blood	5 mice per month	65
MYCOPLASMA		
Lung and ear	5 mice per week	410
VIRUSES		
Serum	40 mice per month	450
BACTERIA		
Feces		
Aerobic	50 samples per week	8,210
Anaerobic	5 samples per week	820
Organs	4 mice per week	822
Area		
Swab	12 per week	4,325
Rodac	12 per week	4,325
Objects	20 per week	5,654
People	Sporadic	432

We selected tests that would detect "indicator organisms" and some overt pathogens. The indicator organisms were those that would be most likely to contaminate the mice, and thus indicate the effectiveness of contamination control. The pathogens were those that could produce an epizootic and might be present in the laboratory environment—based on a review of the literature on mouse diseases. The tests for these included (1) flotation techniques for ova or embryos of intestinal parasites, such as helminths; (2) soaking carcasses in detergent solution to remove and concentrate parasites of skin and hair, such as lice and mites; (3) splenectomizing mice and then examining blood from the tail for *Eperythrozoöan coccoides*; (4) examining lung and tympanic bullae for mycoplasma; (5) serological tests for antibodies[7] to viruses commonly associated with mice, such as pneumonia virus of mice (PVM), reovirus 3, encephalomyelitis virus (GDVII), newborn pneumonitis virus (K), polyoma (parotid tumor virus), Sendai (pneumonitis), mouse adenovirus, and mouse hepatitis, with occasional tests for lymphocytic choriomeningitis, ectromelia, and minute virus of mice (MVM); and (6) bacteriological tests for a wide range of bacteria, especially those that are contagious or that are commonly found in mouse feces.

Mice used for these tests are usually selected randomly from retired breeders from each breeder room in the colony. For bacteriological study of organs and tissues, retired breeders are selected only if mice in breeder rooms that have shown signs of illness or abnormal behavior or that are cage mates of mice that died of unknown causes are not available. Sera for virological studies are also collected from experimental mice being used for age-dependent serial studies.

Periodic Inspections

Attendants inspect each mouse twice a week. If any mouse appears ill or acts abnormally, a report is made and the cage is marked for careful daily inspection. The animal is examined more closely or removed for diagnostic studies if the report suggests anything suspicious. When an attendant predicts that a mouse will die within 24 hr, it is removed to the autopsy laboratory. A less rigorous daily inspection of all cages for dead mice is also performed. Because of this system of inspection, most mice that would otherwise not be found until after they had died can be removed from the facility when death is imminent. The pathologists are then able to examine the mice and to collect tissues for microscopic examination before autolysis begins.

Necropsy Data

A contagious infection in a population of animals is usually revealed by a pattern of lesions, morbidity, or mortality. No such pattern has been seen, although every experimental mouse that dies or is killed when death is imminent, and every mouse in the breeder colony that shows signs of distress or other abnormalities or dies from unknown causes is necropsied and its tissues are examined microscopically by pathologists.

Vital Data

Our semiautomatic system for recording and analyzing data[11] from the mice in the breeder and experimental colonies allows us frequently to plot and review various periodic measurements, such as reproductivity, that may reveal sudden changes or prolonged trends indicating the presence of some deleterious influence. No such changes have been noted to date.

Germfree Monitors

We have also used germfree mice in open cages placed in areas of heavy traffic in the facility as an aid to rapid detection of contaminants. Surprisingly, we found that bacteriological culturing of swab samples collected periodically from the floor of the facility was a faster and more sensitive test for detecting contaminants of the environment.

Enhanced Resistance

To evoke the mouse's own defenses against infection (as well as to ensure that over a long period of time all mice would start off with the same exposure to the same bacteria), we inoculate our first-generation breeders with six selected species of bacteria. They, in turn, pass them on to their progeny. The bacteria we selected were chosen to achieve the following objectives: (1) to reproduce a typical bacterial population of the intestinal tract of the mouse, thus filling the ecological niches of the gut, and thereby placing contaminants at a competitive disadvantage; (2) to educe normal structure and tonus of the gut wall; and (3) to stimulate the non-specific immune system so that the mice would not be unusually sensitive to ordinarily nonpathogenic contaminants.[6] An additional requirement governing the selection of bacteria was that they not greatly interfere with bacteriological methods for detecting contaminants. To achieve these objectives, we selected the following bacteria: *Lactobacillus acidophilus, Clostridium bifermentans, Staphylococcus albus, Streptococcus faecalis* var. *liquifaciens, Escherichia coli* 1324, and *Bacterioides theta iota omicron.* In addition, an anaerobic oxygen-sensitive fusiform rod, resembling the one described by Dr. Savage (this volume, page 60), is found in the feces of all mice in the facility. In the germfree mice, it appears within 2 weeks of their entry into the facility.

We have some evidence that these organisms do, in fact, provide some protection. For example, when we place germfree weanling mice in a room with "conventional" mice, about half the young mice die within 2 weeks, and most of the remainder appear runted. If we place weanlings that have experienced only the selected bacteria in the same room, some develop a mild diarrhea from which they recover, and, at a later age, most of them retain their comparatively healthy appearance.

RESULTS OF THE SURVEILLANCE PROGRAM

When we compile the data and find many negative results, we are tempted to reduce the size of the testing program, even though we have learned through experience that our present program is minimal for rapid detection of contaminants. We have made reductions in the laboratory tests, but only after careful review and with full cognizance that a long history of minimal contamination only gives evidence that today's defenses are adequate to meet yesterday's challenges. The fact that we have not been able to demonstrate the presence in mice of protozoan or metazoan parasites, 11 common murine viruses, mycoplasma, or any pathogenic bacteria suggests that we have maintained a stable microbial environment. However, as expected, our colony has not been entirely free of contamination. Our experience with specific contaminants found in the mice, as presented below, is summarized in Table 3.

TABLE 3 Bacterial Contaminations Found in Mice in LLE Animal Facility

Contaminant	Date Detected	Type of Room	Number of Cages Positive[a]	Number of Cages in Room	Number of Cages in Facility
Staphylococcus and *streptococcus*	February 17, 1964	Germfree	2	2	2
Fusiform	February 24, 1964	Germfree	2	2	2
Proteus	September 29, 1964	Breeder	99	99	130
Coliforms	January 24, 1966	Breeder and test	80	80	340
		Mixed	3	116	
Citrobacter	May 16, 1966	Diet test	25	110	360
β Hemolytic *Streptococcus*	August 29, 1966	Breeder	3	190	600
		Quarantine	28	28	
	September 6, 1966	Breeder	1	276	730
		Quarantine	10	10	
Gram-negative	September 13, 1966	Quarantine	1	16	760
Staphylococcus aureus	July 3, 1967	Postradiation	~150	197	1,800
	September 18, 1967	Mixed	1	180	2,100
Enterobacter-Klebsiella	October 16, 1967	Mixed	20	196	2,300
	January 23, 1968	Quarantine	18	18	2,400
	January 29, 1968	Breeder	8	384	2,400
	September 9, 1968	Breeder	20	371	3,300
Aerococcus	January 1, 1968	All rooms	80[b]		2,400

[a]Out of 32,300 cages tested.
[b]Percent of samples positive (see page 37).

34

Streptococcus faecalis and *Staphylococcus albus*

As an initial test, we placed several groups of germ-free mice in the animal facility after it was decontaminated with beta propio-lactone vapors. The mice were maintained under the same conditions that we planned to use for the experimental mice. Within a few days, streptococci and staphylococci were found in their feces. Because of this and because these two organisms are commonly found in mouse feces, these bacteria were selected for initial inoculation of all other mice brought into the facility.

Fusiform Rods

In the same group of mice, about a week later, another organism, fusiform rods, could be seen in Gram-stained smears of feces, but it could not be isolated by bacteriological methods. This organism is present now in the feces of all mice in the facility, and it can usually be seen in the feces of germfree mice before they leave the quarantine room, 2 weeks after they have entered the facility. The organism comprises about 90 percent of the total bacterial population in the feces of our mice. We have not, how-ever, determined its source or how it is transmitted. The organism is a contam-inant in the sense that it has bypassed our defenses, but we do not consider it to be one because it is not an undesirable organism. In fact, since it so closely resembles the organism described by Dubose (see Savage, this volume, page 60) as being autochthonous, we would probably include it in our group of selected bacteria if we could culture it and if it did not distribute itself so quickly through all the mice.

Proteus mirabilis and *Enterobacter*

Our first group of breeders was established in the barrier in May 1964 to produce mice for a pilot experiment that would allow us to test the facility and operating methods, train personnel, test the protocol of the experiment, and determine if these mice would respond differently from conventional mice. Four months later *Enterobacter* and *Proteus mirabilis* were found in a fecal sample from these mice. By testing every mouse over a period of 3 weeks we were able to identify, isolate, and kill all mice contami-nated with the *Enterobacter*, but the *Proteus* spread to all mice in the room faster than we could isolate known positives. We accepted the *Proteus* as a temporary resident because the alternative would have been to kill all the mice and delay gaining the benefits of the pilot experiment. The source of the *Enterobacter* was found to be a contaminated stock culture of the selected bacteria used to inoculate the germfree mice. The source of the concomitant

Proteus infection was not determined. We sacrificed the breeders in February 1966. The experimental mice produced by these breeders, and thus contaminated with the *Proteus*, were kept in another room for 42 months. During this period seven other rooms were filled with mice not contaminated with the *Proteus*. The *Proteus* room, as we called it, was a constant test and confirmation of our ability to control room-to-room transmission within the facility, because that room was serviced in the same way as all other rooms; e.g., cages and drinking units from the room were washed by routine methods and used in other rooms.

Coliforms

A small group of breeders in a room with other breeders was used to test a liquid diet sterilized by filtration. Two or three coliforms, two *Streptococci*, an *Enterobacter*, and *Proteus* were found in their feces. All mice in the room were killed after confirmatory tests. Three cages of mice transferred to another room before the coliforms were detected were also killed before they transmitted the contaminants. The source of contamination was traced to faulty sterilization of the glass tubes used to contain and dispense the liquid diet.

Citrobacter

We set up groups of breeders in a room by themselves to test three diets including the filter-sterilized liquid diet. Three months later, a *Citrobacter* was found in the feces of all mice receiving the liquid diet. They were all killed. The other mice in the room were tested, and none of the results were positive.

Streptococcus

Occasionally, a colony of β hemolytic streptococci is seen in media plates of fecal samples and area samples. They are biochemically similar to the *Streptococcus faecalis* we give the mice.

Staphylococcus aureus

Experimental mice in another preliminary experiment had been housed in a room by themselves for 13 months when *Staphylococcus aureus* was found in their feces. Positive mice were isolated, but all fecal samples were positive for the organism within a few weeks after initial detection. No further action was taken to eliminate it. The source was not determined. Three months later *Staphylococcus aureus* was found in mice in one cage in a breeder room. The mice were killed, and the organism has not been detected in any other room.

Enterobacter-Klebsiella

We have had three separate contaminations with Gram-negative rods of the *Enterobacter-Klebsiella* group. Each time we were able to eliminate it, but this required killing mice in about 70 cages. We identified the last contaminant as *Enterobacter cloacae.*

Aerococcus

In January 1968 an organism identified as *Aerococcus viridans* was found in 80 percent of the fecal samples collected during 1 week. It was not present in the samples the week before. Samples from mice in every room were positive. A review of the literature suggested that this organism was usually found in dust and might be associated as an opportunist with some infections of humans.[4] Within 3 months the incidence was down to about 25 percent of the fecal samples and has remained there. The incidence in mice (not samples) is lower since a routine fecal sample is a composite of feces collected from five cages. Retesting samples collected individually from each cage represented in a composite sample that was positive suggests that the actual incidence in mice was closer to 5 percent since mice in more than one of the five cages were seldom positive on retest. Also, the organism appears to establish only a transient relationship with most mice, for it frequently cannot be found in the feces of previously positive mice after three or four weekly tests. We kill any mice scheduled to be placed in the experiment if *aerococci* are found in their feces. Hence, all mice entering the experiment still have the same bacteria in their guts at the time they are irradiated.

Other Bacteria

Bacteria different from those we give the mice have been found rarely (0.1 percent) in samples of organs and tissues from 800 mice. The organisms are those usually considered to be laboratory contaminants. Other bacteria, such as *Bacillus, Pseudomonas*, and molds, are occasionally detected in the facility by area and object samples, but they have not been detected in mice there.

Viruses

Antibodies to 11 viruses (listed above) commonly found in mice have not been detected in our mice. Signs of other viruses are present, however. Leukemia that is presumably virus-induced still occurs in the mice in about the same incidence as in their conventional progenitors, and virus-like particles resembling the Gross virus have been seen in electron-microscopic studies.

Other Microbes

Signs of contagious infection or of contamination with organisms other than those described above have not been revealed by clinical examinations, diagnostic tests, evaluations of demographic data, or necropsy data.

Genetic Control

Mice of the RFM strain were selected for the experiment because of their sensitivity to induction of myeloid leukemia by ionizing radiation and because of the extensive information derived from other radiation studies using this strain.[14]

A litter from an RFM/Un female (generation 54) was removed by hysterectomy, transferred aseptically into a germfree isolator, and foster nursed by an ICR female. The litter was expanded by brother X sister mating in the isolators. This subline is now in the fourteenth generation and has been designated RFMf/Wg. All sublines of this germfree foundation colony are related to a common ancestor no more than five generations removed. Each week, about 10 pairs of germfree weanling mice of this strain are transferred into the animal facility where they are expanded by brother–sister mating for three more generations. To avoid accidental crosses, two cages of breeders may not be open at the same time in one room, and no other strain of mouse may be placed in a room producing mice for the experiment. Only mice of the third generation born in the animal facility are used in the experiment. Within a given cage each experimental mouse is uniquely identified by a toe mark, and its genealogy is recorded, thus allowing the relationship of any mouse to any other mouse to be traced to a common ancestor.

Comparison to Conventional Mice

To determine if mice maintained under our conditions of control would be satisfactory for the experiment, we compared them to mice of the same strain born and maintained in a conventional facility. Groups in each facility were given the same dose of radiation at the same age and then followed for their life-span. The results indicated that the defined mice would be satisfactory for the experiment. These mice showed a higher incidence of radiation-induced lymphomas and a lower incidence of granulocytic leukemias than their conventional counterparts, but the combined incidence was similar to that seen previously in conventional mice of the same strain.[14] In nonirradiated mice the mean survival time was about 668 days for males and 603 days for females maintained in our animal facility, but it was 572 days in males and 581 in females maintained in the conventional animal facility. Another difference was the number of abscesses in these two groups. In 1,036 conventional mice, abscesses were found in 6 percent of lungs, 6.5 percent of livers, and 3 percent of kidneys; but in 602 defined mice, the incidence was 0.3 percent, 0.7 percent, and 1.8 percent, respectively. These differences were significant ($P = 0.05$) and suggested that each group is subject to different challenges from its microbial environment.

While setting up the experiment, some other comparative but less objective observations were made. For example, in selecting mice for the experiment, many of the conventional mice were discarded as unfit, whereas few of the defined mice were discarded because they appeared to be larger, healthier, and cleaner.

Conclusions

The evidence strongly suggests that we have been able to maintain a stable environment for the experimental animals during the past 4 years. Although the worth of our efforts can be evaluated only when the experiment ends, our history of apparent success augurs well for the future.

The animal facility and its methods of operation are adequate (1) to produce and maintain mice suitable for the experiment, (2) to minimize variation in the environment, and (3) to prevent the spread of infection.

We feel that there is good evidence that the mice in this experiment have experienced a microbial environment that is stable and free of pathogens. However, because of the limitations of detection techniques, we recognize that we cannot say with absolute certainty that all microbes other than those discussed above are absent from the mice.

We feel that it is feasible to maintain a large number of microbially defined animals without using isolators.

Summary

We have undertaken a study to determine the relation of small doses of ionizing radiation to shortening of life-span and induction of neoplasms and other late-occurring degenerative changes in mice. Because this experiment requires that a large number of mice be maintained for their life-span, and because the anticipated differences among control and experimental groups are small, we stringently minimize variables that can endanger the health of the mice or otherwise diminish the quality of the experimental results.

To aid in achieving these objectives, a 15,000 ft^2 animal-isolation facility ("barrier") was constructed to enable us to produce and maintain genetically and microbially defined mice in a controlled environment. Since control depends on prevention and rapid detection of changes, we developed surveillance procedures (such as frequent periodic tests of mice and materials for the presence of microbes) and monitoring equipment (such as temperature, humidity, and air-pressure alarms). This facility has been operating for 4 years, thus allowing adequate evaluation of its design and function. It is basically a two-corridor facility with a one-way traffic cycle, in which physical and operational barriers against the transmission of microbes are imposed in the path of mice, air, supplies, personnel, and equipment moving into and within the facility. To aid in controlling the microbial component of the environment, mice in the breeder colony were replaced periodically by germfree mice that were inoculated orally with six species of bacteria. These mice and their descendants remained free of all protozoan and metazoan parasites, 11 common murine viruses, PPLO, and, in the main, all bacteria other than those with which they were inoculated. To minimize genetic variation, we used an inbred strain of mice, RFM, that had been brother X sister mated for more than 65 generations. Each mouse in the experiment was separated by no more than one generation from a brother X sister breeder pair and was uniquely identified so that if a mutation were detected, it could be traced genealogically to its origin.

Results of an experiment comparing barrier mice with conventionally maintained mice, both of the same strain, revealed that the barrier mice were larger, more active, appeared healthier, had a longer life-span, and differed in the types of leukemia induced by irradiation and in the incidence of other induced and spontaneous lesions.

Acknowledgments

This experiment, under the direction of Dr. A. C. Upton, was conducted as a team effort, and each of the following has made significant contributions to the development and conduct of the experiment: Drs. R. C. Allen, R. C. Brown, N. K. Clapp, G. E. Cosgrove, E. B. Darden, Jr., T. T. Odell, Jr., R. L. Tyndall, and H. E. Walburg, Jr. We could not have undertaken this experiment without the assistance of many others, including all those who have participated in the evolution of laboratory animal care. We are especially grateful for the assistance and advice of the following: J. F. Hacker III and N. L. Ensor, supervisors of the animal facility; H. J. Hicks, Sr., B. H. Hale, L. E. Lebo, and more than a score of other engineers from the Y-12 Plant at Oak Ridge; and Dr. C. B. Richter for assistance with parasitological testing.

The research reported in this paper was sponsored by the U.S. Atomic Energy Commission under contract with the Union Carbide Corporation.

References

1. Allen, R. C. 1969. Bidirectional containment marks facility design. Lab. Manage. 7(5):38–44.
2. Albert, S., P. L. Wolf, C. O'Mara, W. Barany, and I. Pryjma. 1965. Influence of maternal age and parity on development of lymphoreticular organs of offspring in mice. J. Gerontol. 20:530–535.
3. Brick, J. O., R. F. Newell, and D. G. Doherty. 1969. A barrier system for a breeding and experimental rodent colony: Description and operation. Lab. Anim. Care 19:92–97.
4. Kerbaugh, M. A., and J. B. Evans. 1968. *Aerococcus viridans* in the hospital environment. Appl. Microbiol. 16:519–523.
5. Laroche, M. J. 1965. Influence of environment on drug activity in laboratory animals. Food Cosmet. Toxicol. 3:177–191.
6. Lev, M. 1963. Studies of bacterial associations in germfree animals and animals with defined floras, p. 325–336 *In* P. S. Nutman and B. Mosse [ed.] Symbiotic associations. Cambridge University Press, Cambridge, England.
7. Parker, J. C., R. W. Tennant, T. G. Ward, and W. P. Rowe. 1965. Preparation of serologic diagnostic reagents and survey of germfree and monocontaminated mice for indigenous murine viruses. J. Nat. Cancer Inst. 34:371–380.
8. Pizzarello, D. J., D. Isaak, K. E. Chua, and A. L. Rhyne. 1964. Circadian rhythmicity in the sensitivity of two strains of mice to whole-body radiation. Science 145:286–291.
9. Phillips, G. B. 1965. Microbiological contamination control. A state of the art report. J. Amer. Ass. Contam. Control 4(11):16–25.
10. Riley, R. L., and F. O'Grady. 1961. Airborne infection. Macmillan, New York. 180 p.
11. Serrano, L. J., and C. C. Amsbury. 1967. Semiautomatic recording and computer processing of mouse breeding data. Lab. Anim. Care 17:330–341.
12. Simmons, M. L., L. P. Wynns, and E. E. Choat. 1967. A facility design for production of pathogen-free inbred mice. Proc. Amer. Ass. Contam. Control 6:98–102.
13. Spalding, J. F., R. F. Archuleta, and L. M. Holland. 1969. Influence of the visible color spectrum on activity in mice. Lab. Anim. Care 19:50–54.
14. Upton, A. C., R. C. Allen, R. C. Brown, N. K. Clapp, J. W. Conklin, G. E. Cosgrove, E. B. Darden, Jr., M. A. Kastenbaum, T. T. Odell, Jr., L. J. Serrano, R. L. Tyndall, and H. E. Walburg, Jr. 1969. Quantitative experimental study of low-level radiation carcinogenesis, p. 425–438 *In* Radiation-induced cancer, International Atomic Energy Agency.
15. Whitten, W. K., F. H. Bronson, and J. A. Greenstein. 1968. Estrus-inducing pheromone of male mice: Transport by movement of air. Science 161:584–585.

DISCUSSION

DR. GAY: The first questions come from your colleagues interested in photobiology. (1) What is your lighting cycle within the mouse rooms? (2) Besides the intensity of light, are you controlling the quality of light? (3) Do you think the spectrum is important, particularly in the ultraviolet range?

DR. SERRANO: As I stated, light is one of the things we try to control. The answer to the first question is given in my paper. In answer to the second question: Controlling the spectrum of light is far more difficult than controlling the intensity of duration of light. The only practical control we can exert is to use the same type of fluorescent tube in each room and to replace them with new tubes of the same type before the spectrum of the old tubes changes greatly. A graph of the spectrum and the average longevity of each type of tube is available from the manufacturers.

In answer to the third question: The presence or absence of specific wavelengths in "white" light can have biological effects, as suggested by the observations of John Ott. Also, a recent report by Spalding[13] demonstrated that activity of mice was related to the color of light in their environment. Reports such as these illustrate the importance of the quality of light. As far as the ultraviolet range is concerned, as you know, there is a relatively high output of light energy around the 400-nanometer wavelength from all fluorescent tubes. Cages on the top shelves are exposed much more than those on lower shelves.

DR. GAY: Why don't you use filtered cages, and in a program of this magnitude, would an automatic watering system reduce the cost as well as the sources of contamination and variability?

DR. SERRANO: We have not found the need for replacing filter covers on the cages because other methods of preventing transmission of microbes have proved to be sufficient. Also, the influence of filter covers on the environment in a cage is still uncertain. We are now doing some studies on the effect of protective covers, such as filter covers, on the levels of carbon dioxide and ammonia in covered cages. The preliminary results show that gas levels in the cages are surprisingly higher than the gas levels in the air surrounding them.

We have investigated the use of automatic watering systems. We did want to use them, but, as I mentioned, we transfer cages from one shelf level to another, and we would be transferring cages, then, from one dirty automatic watering valve to the next. We could not transfer the valve with the cage, even though there are some quick-release ones, because this would require human handling. This was one consideration.

42

DR. GAY: I have a question about the diet. What are the methods of sterilization of the food and water?

DR. SERRANO: Food is sterilized in a high-vacuum, high-pressure autoclave. We use the Purina 5010-C highly fortified food. We have done some thiamin-level studies, and Purina has assisted us with other nutritional-level studies. The levels remain satisfactorily high. The water is chlorinated to 16–18 ppm. We consistently test this and have never found bacteria in it, even though the original supply does have some. In fact, we have used *Pseudomonas aeruginosa* as a test organism and found that they will not grow in water that is chlorinated above 2 ppm. We change the water bottles three times a week because the chlorine concentration in many bottles will be as low as 2 ppm within 72 hours.

DR. GAY: Dr. Sabourdy has a question. When you replace mice in the breeder colony with inoculated germfree animals, how do you proceed, a room at a time after sterilization, or by regular infusion of new breeders with controlled flora?

DR. SERRANO: We receive about ten pairs of germfree mice a week or twenty pairs every other week, depending on various considerations. After inoculation and quarantine, they replace other first-generation breeders in the colony. As I mentioned, we expand four generations from the germfree origin before we get the experimental animals. We have five criteria for retiring the females. Our computer system is programmed to tell us when a breeder has met one of these criteria, and then we replace it with a mouse of the same generation.

DEFINING
THE SPONTANEOUSLY
DIABETIC
LABORATORY RODENT

Herbert K. Strasser

History and Introduction

More than 200 years have elapsed since the first mention by Gibson in 1751[8] of a spontaneously diabetic condition in an animal—the horse. As in man, a more detailed examination of *Diabetes mellitus* in animals was not instituted until after the discovery of insulin by Banting and Best.[1] The rapid development of pharmacologic, biochemical, immunological, and genetic methods during the intervening decades has made it possible to gain deeper insight into the functional interrelations of this disease than had been feasible during the preceding 150 years. Nevertheless, even in human diabetes, which has been studied intensively, there are still many uncertainties in connection with the etiology and the course of the disease.

Investigations of genetic problems are hard to conduct in man because of the slow succession of generations and because of the ethical questions involved in using human subjects in research. To be able to carry out extensive experimental studies on all aspects of diabetes mellitus, biological models other than man are required. The spontaneously diabetic animal makes the best model for such studies.

The species commonly thought of as domestic animals and those species that are kept in zoological gardens are unsuitable for such investigations because occurrence of spontaneous diabetic disturbance is limited to isolated cases, so the percentage incidence is inadequate for systematic work. Further complications arise relative to financing extensive research with large experimental animals.

Thus it became especially important for experimental diabetes research that, first by chance as a result of secondary findings made in other studies, and later systematically by means of directed investigations, a large number of small species of experimental animals with a spontaneous diabetic disturbance of the metabolism were discovered. These animals constitute useful models for a wide variety of investigations because they are characterized by (1) a high percentage incidence of the disease, (2) the possibility of economic breeding in large numbers, and (3) a rapid succession of generations. In this paper I intend to deal only with rodents and spontaneously occurring forms of diabetes, although experimentally induced diabetes in laboratory animals can be as valuable.

Specification of Rodents with Spontaneous Diabetes

The rodent species in which the occurrence of spontaneous *Diabetes mellitus* has been observed can be divided into (1) inbred strains, mutants, and hybrids of *Mus musculus* and (2) various species of recently domesticated rodents that exhibit, under laboratory conditions, a high incidence of diabetic breakdown of the metabolism.

Tables 1–3 show the known species and strains in which there is a high percentage incidence of spontaneous *Diabetes mellitus* and their systematic grouping. Tables 4–6 present the studied criteria of diabetes in these animals as reflected in the literature.

Criteria of Diabetic Status

The data presented in Tables 4–6, which are the reported results of investigations conducted to determine species and strains of spontaneously diabetic rodents, are based largely on the current summary by Brunk, "Spontaneous Diabetes in Animals."[3]

The paucity of research findings in the field is evidenced by the blank spots in those three tables. Many criteria that may be of some importance in determining whether a strain is characterized by spontaneously occurring *Diabetes mellitus* have been investigated in only one strain; others, however, have been studied in many or all strains. From the results of those analyses, we may conclude that positive determination of certain criteria is requisite to characterization of a strain as diabetes-prone. Again referring to Tables 4–6, for each of the three groups of findings reported, the criterion for which positive determination has been made in all or most strains is labeled "Leading Symptom."

TABLE 1a   Types of Inherited Diabetes for Single Gene Mutations of the Species _Mus musculus_[a]

Gene Symbol[b]	Gene Names	Existing Stocks[c]	Synonyms Previously Used[d]
A^y	Yellow or lethal yellow	Many	Obese yellow, yellow obese
A^{vy}	Viable yellow	$C_{57}BL/6J\text{-}A^{vy}$	—
A^{iy}	Intermediate yellow	$C_{57}BL/6J\text{-}A^{iy}$	—
ob	Obese	$C_{57}BL/6J\text{-}ob$	AO, obese hyperglycemic, North American obese hyperglycemic
ad	Adipose	—	Adipose–Edinburgh
db	Diabetes	$C_{57}BL/KsJ\text{-}db$	—

TABLE 1b   Types of Inherited Diabetes for Inbred Strains and F_1 Hybrids of the Species _Mus musculus_[a]

Recommended Name[c]	Synonym	Synonym Previously Used[d]
NZO	New Zealand obese	—
KK	KK mouse	Japanese obese
$C_3Hf \times I\,F_1$	$C_3fl\,F_1$	Wellesley mouse

[a] Data from Renold and Dulin.[22]
[b] Data from _Mouse News Letter_, International Committee on Laboratory Animals and Laboratory Animals Centre, MRC Laboratories, Carshalton, Surrey, England.
[c] Data from J. Staats, Standardized Nomenclature for Inbred Strains of Mice. Third Listing, _Cancer Res._ 24:147 (1964). In many instances the investigator should check his source of animals to get the correct name for the strain or mutant he is using.
[d] Not now recommended.

TABLE 2 Types of Inherited Diabetes in Small Laboratory Rodents Other Than *Mus musculus*[a]

Recommended Name	Synonyms	Synonym Previously Used[b]
Acomys cahirinus	Spiny mouse	*Acomys dimidiatus*
Psammomys obesus	Sand rat, desert rat	–
Cricetulus griseus	Chinese hamster	–
"Fatty"; single mutant gene in the rat	Fatties	–

[a]Data from Renold and Dulin.[22]
[b]Not now recommended.

There are, of course, other symptoms that, more-or-less distinctively, define the conditions of the disease, and in this sense they can be described as facultative or characterizing criteria. For lack of space, however, the discussion of symptoms presented here is confined to the leading symptoms.

The results of clinical research indicate adiposity, which is more-or-less pronounced in all species and strains (except the Chinese hamster), to be the leading symptom. It does occur independently of the various types of diabetes, a fact that is discussed later in this report. Of the pathological–histological findings, degranulation of the β-cells of the islets of Langerhans is the most marked characteristic; this can be understood to be the necessary consequence of functional activity. In the pathological–biochemical findings, hyperglycemia is the leading symptom. However, this characteristic of *Diabetes mellitus* is pronounced to a different degree.

TABLE 3 The Genera in Which Certain Species Developed Spontaneous *Diabetes mellitus*[a]

Classification	Genus

Order: Rodentia
  Superfamily: Muroidae
    Family: Muridae
      Subfamily: Murinae
	Mus
	Acomys
	Rattus

Subfamily: Gerbillinae	*Psammomys*
Subfamily: Cricetinae	*Cricetulus*

[a]From Strasser.[23]

47

TABLE 4 Clinical Findings in Spontaneous *Diabetes mellitus* in Laboratory Rodents Described in the Literature

Gene Symbol	Recommended Name	Leading Symptom	Other Symptoms				
		Adiposity	Polyphagia	Polydipsia	Polyuria	Cataracts	Retinopathy
A^y	Yellow or lethal yellow	+					
A^{vy}	Viable yellow	+					
A^{iy}	Intermediate yellow	+					
ob	Obese	+	+				
ad	Adipose	+					
db	Diabetes	+	+	+	+		
	NZO	+		+	+		
	KK	+	+		+	+	+
	$C_3Hf \times I\ F_1$	+					
	Acomys cahirinus	+					
	Psammomys obesus	+				+	
	Cricetulus griseus[a]	+	+	+	+		+
	"Fatty" (rat)	+					

[a]Chinese hamster. Adiposity is considerably less pronounced in this species.

TABLE 5 Pathological–Histological Findings in Spontaneous *Diabetes mellitus* in Laboratory Rodents Described in the Literature

Gene Symbol	Recommended Name	Leading Symptom	Other Symptoms					
		Degranulation of β-Cells	Hypertrophy of the Islets of Langerhans	Hyperplasia of the Islets of Langerhans	Glycogenosis of β-Cells	α-β-Cell Relation	Arterio-sclerosis	Decrease of Spermatogenesis
A^y	Yellow or lethal yellow	+	+	+				
A^{vy}	Viable yellow	+	+	+				
A^{iy}	Intermediate yellow	+	+	+				
ob	Obese	+	+	+		+		+
ad	Adipose							
db	Diabetes	+						
	NZO	+	+	+				
	KK	+	+	+				
	$C_3Hf \times I F_1$	+	+	+		+		
	Acomys cahirinus	+	+	+	+	+	+	
	Psammomys obesus	+						
	Cricetulus griseus	+			+	+	+	+
	"Fatty" (rat)							+

49

TABLE 6a Pathological–Biochemical Findings in Spontaneous *Diabetes mellitus* in Laboratory Rodents Described in the Literature (Part I)

Gene Symbol	Recommended Name	Leading Symptom	Other Symptoms						
		Hyperglycemia	Glycosuria	Glucose-Tolerance Test	Insulin-Tolerance Test	Insulin-Like Activity (ILA)	Insulin Contents of the Pancreas	Immuno-reactive Insulin (IRI)	Liver Glycogen Contents
A^y	Yellow or lethal yellow	+	+		+				+
A^{vy}	Viable yellow	+	+		+				+
A^{iy}	Intermediate yellow	+	+		+				+
ob	Obese	+	+	+	+	+	+		+
ad	Adipose		+						
db	Diabetes	+	+						
	NZO	+	+	+	+	+	+		+
	KK	+	+	+			+	+	+
	$C_3Hf \times I F_1$	+	+				+	+	+
	Acomys cahirinus	+	+			+	+		+
	Psammomys obesus	+	+	+		+	+	+	+
	Cricetulus griseus	+	+	+			+	+	+
	"Fatty" (rat)								

50

TABLE 6b Pathological-Biochemical Findings in Spontaneous *Diabetes mellitus* in Laboratory Rodents Described in the Literature (Part II)

Gene Symbol	Recommended Name	Ketonemia	Ketonuria	Free Fatty Acids (FFA)	Nonprotein Nitrogen (NPN)	Biochemical Adiposity	Lipoidemia	Plasma-Insulin Contents	Paradoxical Blood Glucose Reaction
A^y	Yellow or lethal yellow						+		
A^{vy}	Viable yellow						+		
A^{iy}	Intermediate yellow						+		
ob	Obese						+		
ad	Adipose							+	
db	Diabetes		+						
	NZO					+			+
	KK								
	$C_3Hf \times I F_1$		+						
	Acomys cahirinus		+						
	Psammomys obesus	+	+				+	+	
	Cricetulus griseus	+	+	+	+			+	
	"Fatty" (rat)						+		

51

WHITTEMORE LIBRARY
Framingham State College
Framingham, Massachusetts

ADIPOSITY

According to Mayer,[15,16] adiposity can be thought of as occurring in two forms, metabolic and regulatory. The metabolic form occurs in conjunction with hyperglycemia and, as a rule, is connected with reduced glucose tolerance, increased free fatty acids in the plasma, increased serum insulin activity, and hyperplasia of the islets of Langerhans. It is manifested in connection with a disturbance of the carbohydrate metabolism—in this case, impaired glucose utilization. The regulatory type of adiposity is triggered by external influences, such as nutrition, that cause the carbohydrate metabolism disturbance. Utilization of this twofold approach for defining the forms of diabetes in spontaneously diabetic rodents appears to me to be of limited value, because in many cases, the two forms merge. In the case of the sand rat, in which the regulatory type is the more pronounced, it is exogenic influences such as feed and environment that have a decisive effect on the genesis of adiposity and diabetes. This is true for the spiny mouse as well.[11] According to investigations conducted by Hefti and Flückiger,[11] diabetes also occurs in nonadipose spiny mice, but less frequently than in adipose animals.

DEGRANULATION OF THE β-CELLS

Degranulation of the β-cells is a manifestation of an increase in the burden on these cells. If the amount of insulin required is greater than the supply, then the β-cells that continue to function react, first, by enlarging, then by dividing. Degranulation as a result of long-term overloading can develop into glycogen infiltration and hydropic degeneration, with the eventual destruction of the β-cells. This results in decompensation and an absolute insulin deficiency. This metabolic situation can be countered only by direct administration of insulin. At this advanced stage of degeneration, the application of sulphonyl ureas is ineffective because of the weakened target organ.

HYPERGLYCEMIA

Hyperglycemia is a manifestation of various disturbances in the complicated mechanism that serves to maintain the blood-glucose concentration. On the one hand, hyperglycemia is caused by an absolute or relative deficiency of insulin, the only blood-sugar-lowering hormone; when it is lacking, both glucose retention and glucose utilization are reduced or annulled. On the other hand, the cause of hyperglycemia may stem from an excess of hormones that, as antagonists of insulin, prevent an excessive lowering of the blood-sugar level. Among the hormones that act as antagonists of insulin are suprarenal gland steroids, ACTH, growth hormone, adrenalin or noradrenalin, an excess of thyroid secretion, and the glucagon formed in the islets of Langerhans. The details of the points of attack and modes of action of these different substances are too complex to be treated here.

Technical Problems

Because spontaneously diabetic rodents are so small, all continuous experiments, e.g., prolonged observations and long-term loading tests, are very difficult or impossible to conduct. Repeated withdrawals of minimal amounts of blood constitute a relatively large strain on the blood balance and thus on the organism as a whole. According to Weihe,[24] in the course of a glucose-tolerance test on the hemoglobin and PCV of Chinese hamsters, the constitution of the blood after four withdrawals was shown to be normalized only after four weeks from the time of the last withdrawal.

Moreover, the withdrawal of a blood volume of 0.05 to 0.1 ml, which is necessary for many tests, is possible with sufficient safety only by puncture of the retrobulbar venous plexus. I feel that application of this technique without shallow ether anesthesia is unreasonable for both man and the animal. The metabolic–physiological loading caused by the anesthesia and the influence this has on the results is certainly no greater than that occasioned by stress due to carrying out the procedure without anesthesia. In any case, such a withdrawal of blood at short intervals can be carried out, at most, two or three times.

The blood volume and number of samples withdrawn are therefore generally not sufficient for obtaining a glucose-loading curve that is complete in the conventional sense. This, among other things, indicates why results of glucose-loading tests, which might be very important for characterizing *Diabetes mellitus*, are available on only a few of the spontaneously diabetic rodents. The glucose-assimilation coefficient, which is important in detecting latent diabetes, cannot be calculated for small experimental animals because the number of values obtained is so low. Also, the simultaneous testing of several parameters from one blood sample is often not possible because of the blood-volume problem (e.g., the behavior of free fatty acids and blood glucose after glucose loading).

The Prediabetic Condition

The diagnosis and definition of the prediabetic condition have varying significance in man and animals. In man, early recognition of this state serves to institute measures that prevent or retard a manifestation of diabetes. In the spontaneously diabetic animal, an early recognition of pre-

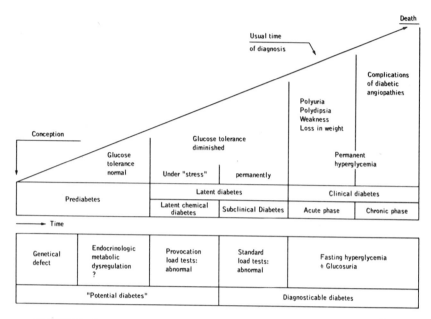

FIGURE 1   Course and diagnosis of diabetes. (From Jahnke et al., reference 12.)

diabetes could make it possible to segregate those animals that probably will not develop a diabetic disturbance of the metabolism. The unnecessary maintenance of large groups of animals over prolonged periods can be avoided in this way.

There has not yet been much consideration of the prediabetic condition of small rodents. The early detection of biochemical adiposity in NZO mice[5] may be very interesting scientifically, but because the detection method is complex, it is virtually impossible to use it in large-scale experimentation. Another fact that can be demonstrated in NZO mice is that glucose tolerance is already reduced before the onset of clinical adiposity, which permits predicting the course of development of the metabolism.[5] (Strasser, 1968 unpublished observation.) The efficacy of using the level of the a_2-protein concentration in the blood for diagnosing the prediabetic condition[10] has not been confirmed, although it has been tried in investigations by Gerritsen and Dulin.[6,7]

Histological findings in newborn animals with a striking hyperplasia of the islets of Langerhans, which can be made, for example, in spiny mice,[9] are, with regard to the detection of the prediabetic status, mainly of theoretical scientific interest because they can be made only in the dead animal. In this connection, the concept of a dynamic course of the development of diabetes is of importance as Jahnke and co-workers have demonstrated[12] (Figure 1). This study makes clear the dynamic development of diabetes and its connection with a genetic defect, an aspect of the process that is discussed later in this paper.

Comparison of Rodent and Human Forms of Diabetes

For the purpose of comparing the spontaneous diabetes forms of rodents, it is helpful to recall the definition of the forms of diabetes in man. According to Creutzfeldt,[4] genuine diabetes of man is characterized as follows:

Juvenile type lean, insulin-sensitive, tendency to ketosis, sulphonyl ureas have no effect
Maturity-onset diabetes tendency to obesity, relatively insensitive to insulin, no tendency to ketosis, reacts to sulphonyl ureas

Comparison of the forms of diabetes found in rodents with those found in humans gives evidence that there is no direct, complete comparability in any case.

If, however, we consider the disease as it occurs in rodents and in man relative to insulin, certain similarities can be seen. Spontaneous diabetes in the Chinese hamster bears the closest resemblance to juvenile insulin-deficiency diabetes in man. In the hamster, the disease is characterized by pronounced hyperglycemia, frequently connected with ketosis; the number of β-cells in the islets of Langerhans is reduced, and the remaining cells tend toward degranulation. In diabetic hamster colonies it is possible, by means of insulin therapy, to prolong the lives of individual animals and to effect an increase in the number of pregnancies, with a reduction of the mortality rate among young animals.[17,18]

Less marked similarities have been noted in comparisons of human diabetes with the form of the disease, coupled with adiposity, as it occurs in various strains of mice and spiny mice. In the rodents, a considerable increase in the number of β-cells results in a concomitant increase in the insulin content of the pancreas and serum. This form of the disease, known as insulin-resistance diabetes, resembles maturity-onset diabetes in man. The disease as it occurs in man, however, is not characterized by β-cell degranulation and the resultant exhaustion of the remaining β-cells that causes insulin-deficiency diabetes in the rodents. On the contrary, plasma and serum insulin levels in man are generally normal, and the course of maturity-onset diabetes can be influenced by sulphonyl ureas.

Spontaneous diabetes of the sand rat, which in the acute stage is characterized by pronounced hyperglycemia and ketosis, exhibits parallels with juvenile diabetes in man. However, the coincidence of these manifestations with diet-dependent adiposity indicates connections with maturity-onset diabetes in man as well. Research must be conducted to determine the nature of those suspected parallels.

The phenomenon that adiposity and diabetes can occur in the presence of high reserves of insulin in a number of animal strains is viewed by Renold[21] as an especially interesting subject for future diabetes research. By clarifying the question of the creation of relative insulin resistance and the coupling of diabetes with adiposity, it would be possible to make a contribution toward a better understanding of maturity-onset diabetes in man.

Thus we can see that spontaneously diabetic laboratory animals can be of great interest for experimental diabetes research, even if the course of the disease in the individual case is not fully comparable with one of the forms of human diabetes.

The Genetic Defect

In discussing the genetic aspect of spontaneous diabetes of rodents I shall not consider the mutants of the individual genes because they will hardly correspond to the conditions of determination of human diabetes.

Of the remaining species, spiny mice and sand rats have not yet been investigated with regard to diabetes. The species on which extensive genetic investigations were carried out is the Chinese hamster (*Cricetulus griseus*), which is of special interest because of its relatively low chromosome number ($2n = 22$) and the suitability of its cheek pouches for accepting tissue transplants.

While diabetes of *Cricetulus griseus* was initially considered to have come about through mutation, subsequent investigations by Yerganian[25,26] indicate that a complicated plurigenetic determination of diabetes might be present. Based on this hypothesis, analogies with the current views on human genetics in connection with *Diabetes mellitus* have been created. Yerganian found that the incidence of diabetes may not only be increased by a rising in-breeding coefficient but that it can be reduced again in the offspring of brother X sister pairings from strains with an almost 100 percent diabetes rate.

In NZO mice[2] and in KK mice, too,[19] the findings to date indicate that a plurigenetic disposition toward diabetes has to be reckoned with. Both strains exhibit relatively low adiposity.

The $C_3Hf \times I F_1$ hybrids (Wellesley mice) are of special interest genetically because they stem from crossing two adiposity-free and diabetes-free inbred strains.[4] These hybrids are distinguished by a pronounced hyperplasia of the islets of Langerhans.

Discussion

Thus we can speak of a genetic definition of spontaneously diabetic rodents only in the case of mutations of individual genes of a number of strains of mice. However, they are of little interest genetically as compared with human diabetes, since diabetes in man is probably based on a plurigenetic disposition.[13,20] However, those laboratory animal strains in which a plurigenetic cause of diabetes is probably or possibly present have not been tested enough genetically in connection with diabetes to allow precise statements.

The metabolic–physiological definition on the basis of clinical, pathological-histological, and pathological-biochemical test results is sufficient in the case of the animal strains described to determine the diabetic condition of individual animals. A further characterization of the varying forms of diabetes in the different species and strains is possible only incompletely.

The difficulties described as inherent in the test methods undoubtedly contribute to this situation. A premise for progress in the experimental work with spontaneously diabetic animals would appear to be a systematic improvement of long-term clinicochemical experiments.

Parameters that are still lacking for defining the metabolic situation of the animal strains concerned should be ascertained simultaneously with the further investigation of diabetes itself. Through a more detailed knowledge of the animals used for research, it will be possible to continue working in this field with greater ease and more precise direction, and with higher promise of success.

References

1. Banting, F. G., and C. H. Best. 1922. The internal secretion of the pancreas. J. Lab. Clin. Med. 7:251–266.
2. Bielschowsky, M., and F. Bielschowsky. 1953. A new strain of mice hereditary obesity. Univ. Otago School 31:29–31.
3. Brunk, R. 1971. Spontandiabetes bei Tieren. Heffter-Heubner, Handbuch der experimentellen Pharmakologie, Band Insulin. Springer, im Druck, Berlin.
4. Creutzfeldt, W. 1963. Zur Theorie des *Diabetes mellitus*. Med. Klin. 58:41–46.

5. Crofford, O. B., and Ch. K. Davis. 1965. Growth characteristics, glucose tolerance and insulin sensitivity of New Zealand Obese mice. Metabolism 14:271–280.

6. Gerritsen, G. C., and W. E. Dulin. 1965. Studies on the diabetic Chinese hamster. Diabetes 14:448–449.

7. Gerritsen, G. C., and W. E. Dulin. 1966. Serum proteins of Chinese hamsters and response of diabetics to tolbutamide and insulin. Diabetes 15:331–335.

8. Gibson: 1751. Diseases of horses. London: Millar. (Quoted by R. Brunk, 1969.)

9. Gonet, A. E., W. Stauffacher, W. Pictet, and A. E. Renold. 1965. Obesity and *Diabetes mellitus* with striking congenital hyperplasia of the islets of Langerhans in Spiny Mice (*Acomys cahirinus*). Diabetologia 1:162–171.

10. Green, M. N., G. Yerganian, and H. Meier. 1960. Elevated a-2 serum proteins as a possible genetic marker in spontaneous hereditary *Diabetes mellitus* of the Chinese hamster. Experientia 16:503–504.

11. Hefti, F., and E. Flückiger. 1967. Obesitas und *Diabetes mellitus* bei *Acomys cahirinus*. Rev. Suisse Zool. 74:562–566.

12. Jahnke, K., H. Daweke, W. Schilling, R. Rüenauver, and K. Oberdisse. 1967. Der potentielle Diabetes (sog. Frühdiabetes). 12. Symp. Deutsch. Ges. Endokrin., Wiesbaden 1966. Springer-Verlag, Berlin, p. 57–74.

13. Jörgensen, G. 1967. Vergleichende Pharmakogenetik des Menschen und der Säugetiere. Med. Welt 18:32–36; 84–91.

14. Jones, E. E. 1964. Spontaneous hyperplasia of the pancreatic islets associated with glucosuria in hybrid mice. Proc. 3rd Intern. Symp. "The Structure and Metabolism of the Pancreatic Islets," Stockholm 1963. Pergamon Press, Oxford, p. 189–191.

15. Mayer, J. 1955. Mechanism of regulation of food intake and multiple etiology of obesity. J. Proc. 3rd Int. Nutr. Congr. Voeding 16:62–88.

16. Mayer, J. 1960. The obese hyperglycemic syndrome of mice as an example of metabolic obesity. Amer. J. Clin. Nutr. 8:712–718.

17. Meier, H., and G. A. Yerganian. 1961. Spontaneous hereditary *Diabetes mellitus* in the Chinese hamster (*Cricetulus griseus*). II. Findings in the offspring of diabetic parents. Diabetes 10:12–18.

18. Meier, H., and G. A. Yerganian. 1961. Spontaneous hereditary *Diabetes mellitus* in the Chinese hamster (*Cricetulus griseus*). III. Maintenance of a diabetic hamster colony with the aid of hypoglycemic therapy. Diabetes 10:19–21.

19. Nakamura, M. 1962. A diabetic strain of the mouse. Proc. Jap. Acad. 38:348–352.

20. Neel, J. V. 1962. *Diabetes mellitus*: A "thrifty" genotype rendered detrimental by "progress"? Amer. J. Hum. Genetics 14:353–362.

21. Renold, A. E. 1967. Zur Pathogenese des *Diabetes mellitus*. 12. Symp. Deutsch. Ges. Endokrin., Wiesbaden 1966. Springer, Berlin, p. 45–56.

22. Renold, A. E., and W. E. Dulin. 1967. Spontaneous diabetes in laboratory animals. Diabetologia 3:63–64.

23. Strasser, H. 1968. A breeding program for spontaneously diabetic experimental animals: *Psammomys obesus* (sand rat) and *Acomys cahirinus* (spiny mouse). Lab. Anim. Care 18:328–338.

24. Weihe, W. H. 1968. Glukosebelastungsteste bei gesunden und diabetischen chinesischen Hamstern. Paper presented at the symposium on "The Use of the Chinese Hamster in Research," 29th and 30th of Nov. 1968, Zürich, Switzerland.

25. Yerganian, G. 1964. Spontaneous *Diabetes mellitus* in the Chinese hamster, *Cricetulus griseus*. IV Genetic aspects. Ciba Foundation Colloquia on Endocrinology 15:25–48.

26. Yerganian, G. 1965. Spontaneous *Diabetes mellitus* in the Chinese hamster (*Cricetulus griseus*). p. 612–626 *In* B. S. Liebel and G. A. Wrenshall [ed.] Current trends and projected views on the nature and treatment of diabetes. Excerpta Med. Foundation, Amsterdam.

DISCUSSION

DR. FESTING: Can you tell me where the "fatty" rats can be obtained?

DR. STRASSER: No, I am sorry I cannot, but perhaps someone here can tell you. I cannot, because we never have been interested in that animal.

DR. LANE-PETTER: Are there diabetogenic diets that affect the appearance of diabetes in susceptible animals, and are they useful in this connection?

DR. STRASSER: Yes, there are diets that produce a higher incidence of diabetes through provocation of fatty animals. The diet has a very important influence on bringing the diabetic condition into appearance. Especially well known is the case of the sand rat, in which diabetes can be produced merely by change in the diet. The paper of Hefti and Flückiger, mentioned in my paper, shows that this is also possible in spiny mice.

ESTHER CHOLNOKY: Do the parental strains C_3H and I, respectively, show symptoms of diabetes, or can these be detected in the F_1 hybrids only?

DR. STRASSER: The parent strains of these hybrids do not show any diabetic disturbance. Both of them are normal in this respect. Only the F_1 hybrids show this metabolic disturbance.

DR. GEORGE WOLFF: Can spontaneous diabetes be characterized as general metabolic dysregulation due to numerous and varied basic causes?

DR. STRASSER: I have a little difficulty in understanding this question, but I think I could say the following. Diabetes as a whole, that is, in human medical research as well, is not as well defined as it should be, so I am not able to characterize these animals by comparing this disease in animals with the human forms of diabetes. If we do not exactly know what diabetes is, and that is the case today, we must have difficulties in describing these animal strains as diabetic or not. I think it might be possible that some of these strains, which are acknowledged today as spontaneously diabetic, might, from another viewpoint, not be spontaneously diabetic.

DEFINING THE GASTROINTESTINAL MICROFLORA OF LABORATORY MICE

Dwayne C. Savage

The Indigenous, Normal, and Autochthonous Microbiota

The words "indigenous microbiota" are commonly used to describe the microbial population resident in animals. The indigenous microbiota resident in the gastrointestinal canal of mice has been studied extensively in recent years.[1-11] The results of these systematic studies have led to the proposal that the indigenous microbiota is composed of the normal and the autochthonous biotas.[8]

The "normal biota" is thought to be composed of micro-organisms that can be found only in an individual community of a certain type of mammal. Such micro-organisms are ubiquitous in the community and establish themselves in all of its members. The normal biota can vary from community to community.

In contrast, the "autochthonous biota" is thought to be uniform in all communities of a certain type of mammal. Autochthonous micro-organisms seem to be involved in a special relationship with their mammalian hosts. The micro-organisms and the animal appear to have evolved together.

The criteria for classifying certain micro-organisms as autochthonous members of the indigenous microbiota cannot be established firmly at this time. However, certain parameters may be drawn from retrospective analysis of the results of the extensive experimentation. *These parameters are not hard-and-fast criteria and should be regarded only as guidelines for research.* The parameters may be summarized as follows: Autochthonous micro-organisms are

60

always found in mice possessing an indigenous microflora. They colonize the body early in life. Their populations increase to high levels soon after initial colonization and remain at those levels throughout the lives of healthy animals. Autochthonous micro-organisms may be associated in some intimate way with an area of the epithelium on the body surface.

In the following paragraphs, I will summarize the results of efforts to define the autoch+honous bacterial flora of the gastrointestinal tracts of laboratory mice. In addition, I will describe some of what is known about the physiological interactions between mice and the indigenous bacterial populations of their alimentary canals. Where possible, I will indicate where such interactions may involve autochthonous bacteria.

The parameters of autochthony described above were drawn primarily from results of observations on the mode of bacterial colonization of the stomachs and guts of infant mice. Accordingly, the discussion of the autochthonous bacterial flora must begin with the birth of mice.

Development of the Gastrointestinal Microbiota

BACTERIAL COLONIZATION OF THE INFANT ALIMENTARY CANAL

One of the more important characteristics of autochthonous micro-organisms may be that they colonize an animal's body very early in life. Evidence to support this contention has been derived primarily from studies with specific-pathogen-free (SPF) mice. The NCS[1] and NCS-D[12] mice from the SPF colony at the Rockefeller University in New York have been particularly useful for such experimentation.

Under normal circumstances the mammalian fetus is sterile bacteriologically, but during and after birth it is contaminated with a variety of different microbial types. Remarkably, however, indigenous micro-organisms do not establish in the alimentary canal in a random way. Bacterial colonization of the guts of infant mice takes place in a constant sequence that suggests a complicated interplay among the host, the microbiota, and nutritional factors.[7,10] Figure 1 shows the bacterial colonization of the gastrointestinal tracts of infant NCS mice.

First to appear in significant numbers are lactobacilli and Group N streptococci. These bacteria can be cultured from all areas of the tract within the first 24 to 48 hr after birth. Their populations rise to high levels within the first week after birth and then remain at those levels throughout the lives of healthy mice maintained under good environmental conditions.[7,10] Histological sections show these particular lactic acid bacteria in a layer on the kera-

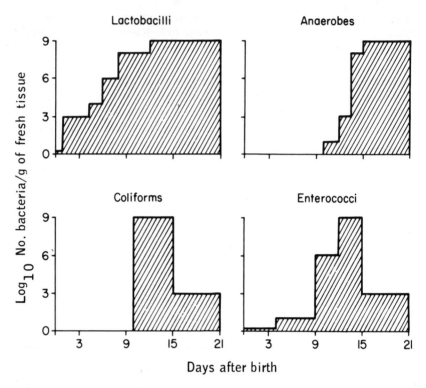

FIGURE 1   Bacterial colonization of the large bowel of infant NCS mice. The graph for lactobacilli also illustrates colonization by Group N streptococci. The Anaerobes are mostly bacteroides. (This summary plot was prepared from data in Table III, reference 5 and Table I, reference 7. The media and methods used to culture the bacteria are described in References 3, 7, and 10.)

tinized surface of the stratified squamous epithelium of the nonglandular portion of the stomach (Figure 2).[8,10]

Cultures made from mice in the second week after birth yield enterococci and coliforms from all areas of the tract.[7,10] The populations of these bacteria reach high levels very quickly but decline rather abruptly to comparatively low levels during the third week. Once their populations have dropped to low levels, enterococci and coliforms can be cultured only from the cecum and other areas of the large bowel. Low population levels of these facultative anaerobes are characteristic of the microflora of healthy adults housed in protected environments.[7,10]

The enterococci and coliforms are not quantitatively significant components of the microflora of adults, but certain types of these facultative anaerobes may play an important role during the colonization of the infant tract. Enormous numbers of particular coliforms and enterococci can be cultured during most of the second week after birth. Histological sections made during this period show microcolonies of Gram-negative and Gram-positive bacteria in the mucus on the mucosal epithelium of the cecum and colon.[10] No other bacteria appear in microcolonies in the mucin, although large masses of Gram-positive rods (probably lactobacilli) can be seen in the center of the gut lumen. The bacteria in the microcolonies are morphologically identical to enterococci and coliforms. Moreover, the microcolonies disappear from the mucus when the population levels of enterococci and coliforms drop during the third week after birth. It seems very likely, therefore, that the microcolonies are made up of enterococci and coliforms. If so, the localization of the microcolonies in the mucin layer suggests a more than passive role in the colonization sequence for certain enterococci and coliforms. This possibility is discussed in a later section of this paper.

Anaerobic bacteria first appear on appropriate cultures of the infant mouse gut at the end of the second week after birth.[7,10] This anaerobic population

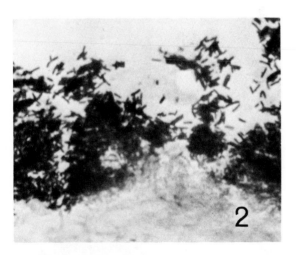

FIGURE 2 Gram-stained histological section of the stomach of a normal adult NCS mouse. The dark mass is composed of Gram-positive lactobacilli and Group N streptococci in a layer on the stratified squamous epithelium of the nonsecreting portion of the stomach ($\times 900$). (The method for preparing the section is described in reference 10.)

can be cultured only from the cecum and other regions of the large bowel; it quickly increases to adult levels early in the third week. The bacteria on the anaerobic plates are almost all bacteroides, although occasionally a few clostridia also appear.

Anaerobes other than bacteroides can be found in significant numbers in the guts of NCS and NCS-D mice; however, these are fusiform-shaped rods that can be seen in histological sections of the cecum and colon that are made at the end of the second week after birth.[10] These tapered rods first can be seen in the mucin on the mucosal epithelium scattered among the microcolonies of Gram-negative and Gram-positive bacteria. Three to 5 days later the Gram-positive micro-organisms cannot be seen in the mucin. The fusiform rods then

FIGURE 3 Gram-stained histological section of a cecum from a normal adult NCS mouse. The mucosa is to the left. The long cigar-shaped rods are oxygen-sensitive anaerobes that form a layer in the mucin on the epithelium of the mucosa (×958). (The method for preparing the section is described in reference 10.)

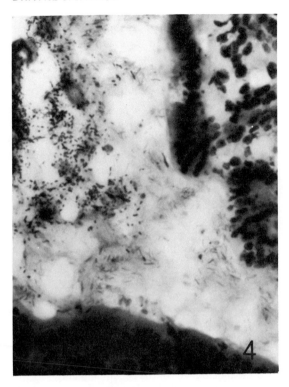

FIGURE 4　Gram-stained histological section of a colon from a normal adult NCS mouse. The mucosa is to the right at the bottom of the photograph; the remainder shows the lumen containing digesta and various micro-organisms. A lightly colored layer separates the mucosa and the darkly stained material in the center of the lumen. This lightly stained layer is composed of massive numbers of Gram-negative oxygen-sensitive tapered rods in the mucin on the mucosa ($\times$ 465). (The method for preparing the section is described in reference 10.)

form thick layers on the epithelium and seem to occupy the entire mucin layer (Figures 3 and 4). These bacterial layers can be seen in histological sections of the cecums and colons of all mice over about 2 weeks of age. The tapered rods in the layers appear to outnumber by a significant margin all other bacteria in the alimentary microflora and therefore deserve recognition as important components of that flora. In spite of their apparent superiority of numbers in the gut, these fusiform-shaped anaerobes were only recently cultivated quantitatively *in vitro*.[11]

The successful quantitative cultures of these bacteria were made on Petri plates of agar media, and the entire process of cultivation from gut to surface colony was carried out in an anaerobic environment.[11] These tapered rods are exceedingly sensitive to oxygen and cannot be cultivated after even brief exposure to an atmosphere containing that gas. The cultures made without exposing the gut material or media to oxygen yielded over 1×10^{10} of these fusiform-shaped anaerobes per gram of cecum.[11] These oxygen-sensitive anaerobes outnumber the other anaerobic bacteria in the normal flora by as much as ten to one and all other bacteria by many thousands to one. They are undoubtedly the dominant bacterial population in the cecum and colon.

THE AUTOCHTHONOUS BACTERIAL FLORA OF THE MOUSE ALIMENTARY CANAL

Of the several types of bacteria that colonize the alimentary canal of infant mice, only certain ones may qualify as autochthonous within the limits of the parameters. Lactobacilli, Group N streptococci, bacteroides, and the fusiform-shaped bacteria certainly qualify, but coliforms and enterococci cannot be regarded as autochthonous at this time. The adult population levels of the latter types of bacteria are very low. However, it has been noted that certain enterococci and coliforms may serve an important function during bacterial colonization of the infant gut. It can be speculated that certain types of these facultative anaerobes in some way prepare the mucin layer in the cecum and colon for the oxygen-sensitive anaerobes that appear later in the colonization sequence. If this speculation receives experimental support, it would mean that particular coliforms and enterococci occupy a position of considerable importance in the indigenous microflora. Then perhaps the parameters for autochthony would have to be expanded so that certain coliforms and enterococci could be included, in spite of their low population levels in adult animals.

In any case, of the bacteria that qualify as autochthonous under the terms described above, the vast majority are anaerobic. Even the lactobacilli, although they can grow with oxygen present, have no aerobic metabolism and must have reduced oxygen tension to multiply efficiently.[13] Thus anaerobic bacteria so predominate in the adult alimentary canal that the population of bacteria with aerobic metabolism can be considered insignificant. It seems logical to conclude, therefore, that the gut lumen, even on the immediate surfaces of the mucosal epithelium, provides an anaerobic microworld. This conclusion seems paradoxical; the gut mucosa is extremely well supplied with blood in the normal animal.[14] This apparent contradiction may be explained only by systematic studies of the intimate physiological interactions between indigenous bacteria and the alimentary mucosa.

66

DWAYNE C. SAVAGE

Physiological Interactions between Microbiota and Alimentary Tissue

ALTERATIONS IN THE MOUSE CECUM AND ITS BIOTA INDUCED BY ANTIBACTERIAL DRUGS

Germfree animals are a very powerful tool in the study of physiological interrelationships of animals and their indigenous microbiota. In fact, studies with germfree rodents provided the first solid clues to the extent of such interactions.[15] The large cecums of these animals are of particular interest. These organs will reduce to the size of cecums of conventional rodents when the germfree animals are contaminated with microorganisms from gastrointestinal tracts of the conventional animals.[9,15,16] Thus, the germfree cecums are large because the normal microflora is absent. This phenomenon makes the germfree animals well suited for experiments on the interactions of the indigenous microflora and alimentary tissue, but the difficulty and expense of maintaining germfree colonies still limit the breadth and scope of most such experimentation.

Recently, enlarged cecums were observed in NCS and NCS-D mice given antibacterial drugs in their drinking water.[17] The enlarged cecums of the mice treated with the drugs seemed similar in many respects to the cecums of germfree mice. Therefore, the changes in cecal weights and microflora were examined in detail in mice given several different antibacterial drugs. It was hoped that the large cecums of drug-treated mice would prove to be as useful as cecums of germfree animals in experiments on the physiological interactions of the gut and its indigenous microbiota. The findings are summarized below.

Figure 5 illustrates cecal enlargement in mice given penicillin or Terramycin in the drinking water. Kanamycin also induces enlarged cecums when given in the drinking water.[17] The cecums enlarge within 12 to 24 hr after the ordinary drinking water is replaced with an aqueous solution of one of the drugs. The organs remain larger than normal as long as the antibacterial drug is administered. Penicillin induces larger cecums than either of the two broad-spectrum antibacterials. None of the drugs affected growth as determined by increases in body weight in these experiments.[17]

Penicillin also is more effective than the other two drugs in its early[17] and long-term[4,17] effects on the bacterial microflora. Cecal bacteria disappear within 12 to 24 hr after penicillin is first given (Figure 6) and cannot be cultured from the cecums or observed in histological sections of those organs up to 48 hr thereafter. However, 48 to 72 hr after the drug treatment is begun, the cecums are repopulated by a characteristic bacterial flora in mice given penicillin solution to drink for prolonged periods.[4,17] The cecums remain large in spite of this drug-induced bacterial population.[17]

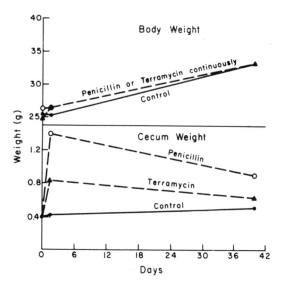

FIGURE 5   Changes in cecal weights and body weight in 6- to 8-week-old NCS mice given aqueous penicillin solution (0.3 g/liter) or aqueous Terramycin solution (1.7 g/liter) in the place of drinking water. (This summary plot was drawn from data taken in part from Tables III, VII, and VIII in reference 17.)

The two broad-spectrum antibacterial agents also rapidly eliminate certain elements of the indigenous microflora, but they do not eliminate the entire bacterial population as does penicillin. Mice given Terramycin retain their populations of bacteroides. Mice given kanamycin still possess essentially normal numbers of lactobacilli. But the cecums enlarge even though those bacteria remain. Similarly to penicillin, both Terramycin and kanamycin induce a characteristic microflora in the cecums of mice continuously given the drugs for more than 48 hr. Again, however, the organs continue to be larger than normal as long as the drugs are given. The cecums and their microfloras return to normal only when the drug treatment is discontinued. They return to normal most rapidly after the drug treatment is stopped in mice contaminated with cecal contents from untreated mice.[17]

As noted, each antibacterial drug given for over 48 hr will induce a cecal microflora characteristic of the drug given. These drug-induced microfloras consist of various combinations of coliforms, enterococci, lactobacilli, bacteroides, or clostridia.[17] Importantly, of the various types of bacteria in the indigenous microflora of untreated mice, only the oxygen-sensitive anaerobes (fusiform-shaped rods) do not appear in the microflora of mice treated with one or another of the antibacterial drugs.[17] It seems likely, therefore, that these oxygen-sensitive anaerobes are involved in maintaining normal cecal size. Admittedly, the evidence for this hypothesis is indirect at the present time. Moreover, until recently it has been difficult to test the idea because the spindle-shaped anaerobes have been so difficult to culture *in vitro*. They are now being cultured,[11] however, and it should be possible to test the hypothesis directly in the near future.

THE MECHANISMS OF CECAL ENLARGEMENT

There are striking similarities between the large cecums of germfree mice and the enlarged cecums of mice given antibacterial drugs. In particular, mice given aqueous penicillin solution to drink for about

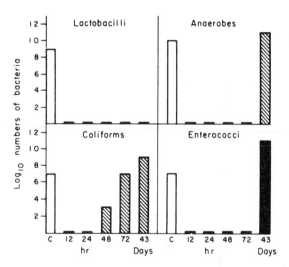

FIGURE 6 Changes in the microbial flora that can be cultured from the cecums of 6- to 8-week-old NCS mice given aqueous penicillin solution (0.3 g/liter) in the place of drinking water. The numbers of bacteria are per gram of fresh tissue. The "C" indicates values from untreated controls. *Bars labeled "43 days" indicate condition from 4 to 43 days.* The anaerobes are mostly bacteroides in untreated mice, but they are mostly clostridia in the treated animals. The types of coliforms and enterococci from the treated mice also differ from the types of these facultative anaerobes cultured from untreated mice. These differences in microbial types are indicated by stipling or blackening the bars. (This summary plot was drawn from data taken in part from Table IV of reference 17.)

24 hr have cecums that markedly resemble the germfree organs. The cecums in mice given penicillin are 3 to 4 times normal size and are dark in color. They appear distended by internal pressure, and if ruptured, they yield dark fluid contents that pour out as if under pressure.[17] Cecums in germfree mice could be described in much the same way.[15] Moreover, as noted previously, contamination of germfree mice with micro-organisms of the indigenous microflora will reduce their cecums to conventional size.[9,15,16] Similar treatment will accelerate the return to normal of the cecum size and microflora in mice first treated with penicillin.[17] Thus it seems highly likely that the cecums enlarge for the same reasons in both germfree mice and mice treated with antibacterial drugs.

The cecal contents of germfree mice yield some pharmacologically active substances that induce relaxation of smooth muscle.[18] One of these substances has been identified as a fecal kallakrein[19,20] and could induce enlarged cecums by preventing muscular contraction, thus inducing stasis in the organs. These substances are not found in ordinary animals with indigenous microfloras. Presumably, they are produced in the small intestine of all mice but are destroyed by micro-organisms in the cecums of normal mice.[18] Fecal kallakrein should be detectable in the cecums of mice treated with antibacterial drugs. Unfortunately, due to technical difficulties, the tests for the substance have been unsatisfactory (D. C. Savage, unpublished observations).

The studies of the cecums of mice treated with antibacterial drugs suggest that a factor other than fecal kallakrein may also influence cecum size. Water accumulates in the lumens of the cecums of mice given solutions of antibacterial drugs in place of drinking water.[17] The increased weight of the cecums with contents intact is due almost entirely to water in mice given penicillin solution for 24 to 48 hr (Figure 7). The dry weights of the cecums with their contents intact do not increase relative to normal values as much as do the wet weights. This finding indicated[17] that the water-transport mechanisms were deranged in the cecums of the treated mice. Subsequent findings have reinforced that impression.

These findings, yet to be published, can be summarized briefly as follows:

1. The wet cecal mucosa washed free of contents weighs the same whether from untreated control mice or from mice given aqueous penicillin solution to drink for 24 hr.

2. The dry cecal mucosa weighs the same whether from control mice or from mice given the penicillin solution for 24 hr.

3. The dry cecal contents from mice given penicillin solution for 24 hr weigh slightly less than the contents from cecums of control mice.

4. The packed cell volume is greater per unit volume of blood in mice treated with penicillin for 24 to 48 hr than in untreated control mice.

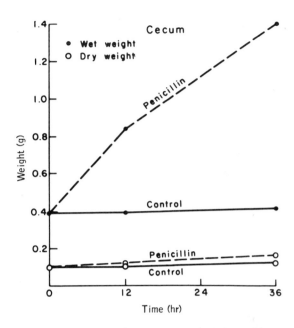

FIGURE 7 Changes in the wet weights and dry weights of cecums weighed with their contents intact from 6- to 8-week-old NCS mice given aqueous penicillin solution in the place of drinking water. (This summary plot was drawn from data taken in part from Table VII in reference 17.)

5. The treated mice consume penicillin solution at a faster rate than control mice drink water.

Similar observations have been made in germfree mice.[15] Moreover, pharmacologically active substances that interfere with water transport have recently been found in the guts of germfree mice but not in conventional animals.[20]

Apparently, the cecums enlarge in germfree mice or mice treated with penicillin, at least in part, because water accumulates in the lumen. Apparently, the water accumulates because it is not transported effectively from the cecal lumen into the blood compartment.

It seems clear that the mechanisms underlying cecal enlargement in the absence of indigenous microflora are complex and yet to be understood. It is just as clear, however, that the indigenous microflora does interact physiologically with the gut mucosa. The truly important interactions may be those between the mucosa and the autochthonous bacteria. This and related ideas are under active investigation.

Summary

The microbiota indigenous to mammalian body surfaces is now believed to consist of the normal and the autochthonous biotas. Autochthonous micro-organisms seem to be involved in special physiological relationships with their mammalian hosts. The autochthonous bacterial flora of the mouse gastrointestinal canal has been partially defined. Some of these bacteria may be involved in maintaining the tone of the musculature and the water-transport mechanisms of certain areas of the alimentary canal.

The recent investigation discussed in this review was supported by Public Health Service research grant AI-08254-01 from the Institute of Allergy and Infectious Diseases.

References

1. Dubos, R. J., and R. W. Schaedler. 1960. The effect of the intestinal flora on the growth rate of mice and on their susceptibility to experimental infections. J. Exp. Med. 111:407–417.
2. Dubos, R. J., and R. W. Schaedler. 1962. The effect of diet on the fecal bacterial flora of mice and on their resistance to infection. J. Exp. Med. 115:1161–1172.
3. Schaedler, R. W., and R. J. Dubos. 1962. The fecal flora of various strains of mice. Its bearing on their susceptibility to endotoxin. J. Exp. Med. 115:1149–1160.
4. Dubos, R. J., R. W. Schaedler, and M. Stephens. 1963. The effect of antibacterial drugs on the fecal flora of mice. J. Exp. Med. 117:231–243.
5. Dubos, R. J., R. W. Schaedler, and R. Costello. 1963. The effect of antibacterial drugs on the body weight of mice. J. Exp. Med. 117:245–254.
6. Dubos, R. J., R. W. Schaedler, and R. Costello. 1963. Composition, alteration, and effects of the intestinal flora. Federation Proc. 22:1322.
7. Schaedler, R. W., R. J. Dubos, and R. Costello. 1965. The development of the bacterial flora in the gastrointestinal tract of mice. J. Exp. Med. 122:59–66.
8. Dubos, R. J., R. W. Schaedler, R. Costello, and P. Hoet. 1965. Indigenous, normal, and autochthonous flora of the gastrointestinal tract. J. Exp. Med. 122:67–76.
9. Schaedler, R. W., R. J. Dubos, and R. Costello. 1965. Association of germfree mice with bacteria isolated from normal mice. J. Exp. Med. 122:77–82.
10. Savage, D. C., R. J. Dubos, and R. W. Schaedler. 1968. The gastrointestinal epithelium and its autochthonous bacterial flora. J. Exp. Med. 127:67–76.
11. Lee, A., J. Gordon, and R. J. Dubos. 1968. Enumeration of the oxygen-sensitive bacteria usually present in the intestine of healthy mice. Nature 220:1137–1139.
12. Mushin, R., and R. J. Dubos. 1965. Colonization of the mouse intestine with *Escherichia coli*. J. Exp. Med. 122:745–757.
13. Stanier, R. Y., M. Doudoroff, and E. A. Adelberg. 1963. The microbial world, 2nd ed., p. 432–436. Prentice-Hall, Englewood Cliffs, New Jersey.
14. Hummel, K. P., F. L. Richardson, and E. Fekete. 1966. Anatomy, p. 247–307. *In* E. L. Green [ed.] Biology of the laboratory mouse. McGraw-Hill, New York.
15. Gordon, H. A. 1965. Germ-free animals in research: An extension of the pure culture concept. Triangle 7:108–121.

16. Skelly, B. J., P. C. Trexler, and J. Tanami. 1962. Effect of a *Clostridium* species upon cecal size of gnotobiotic mice. Proc. Soc. Exp. Biol. Med. 110:455–458.
17. Savage, D. C., and R. J. Dubos. 1968. Alterations in the mouse cecum and its flora produced by antibacterial drugs. J. Exp. Med. 128:97–110.
18. Gordon, H. A. 1965. Demonstration of a bioactive substance in caecal contents of germfree animals. Nature 205:571–572.
19. Gordon, H. A. 1968. The role of the intestinal microflora in maintaining normal function of the lower bowel, p. 193–195. *In* M. Miyakawa and T. D. Luckey [ed.] Advances in germfree research and gnotobiology. CRC Press, Cleveland, Ohio.
20. Csaky, T. Z. 1968. Intestinal water permeability regulation involving the microbial flora, p. 151–159. *In* M. E. Coates, H. A. Gordon, and B. S. Wostman [ed.] The germ-free animal in research. Academic Press, New York.

DISCUSSION

DR. GAY: Have you performed electron-microscope studies to demonstrate morphologic relationships between bacteria and mucosal cells?

DR. SAVAGE: We have such a study under way at the present time, and I have looked at many of the plates. Sometimes we see some evidences of connections between the bacteria and the tissue, but I begin to disbelieve that it is a real phenomenon. I do not think we are going to find it at the electron-microscope level. I believe this is a micromolecular phenomenon that may require more elegant electron microscopy and perhaps some elegant biochemistry, which is now being started.

DR. GAY: The next two questions are related: First, is the difficulty in culturing the anaerobic fusiform rod due to oxygen sensitivity or to a failure to provide an adequate nutritional culture? The antibiotic effect suggests this.

Second, please give the references on cultivating the fusiform bacteria.

DR. SAVAGE: The problem with the fusiform bacteria does, indeed, seem to be one of oxygen sensitivity. If you have achieved a sufficiently anaerobic environment, you can culture these bacteria on blood agar or the usual anaerobic medium that we use for bacteroides. They are being cultured now by a number of people, not only by Dr. Lee. I am sorry I do not know this reference offhand; it is to an article that appeared in *Nature* early in 1969. There is a doctor in Virginia, Dr. Moore, who is culturing fusiform-shaped bacteria, and of course, R. E. Hungate at Davis, California, has been culturing bacteria from the rumen for a long time. The trick is not to expose the bacteria to oxygen at any step from the time you take them from the rumen—in fact you can't even expose them in the lumen. You must do all of the procedures in an anaerobic environment, in nitrogen or hydrogen. Again, I am sorry I can't give you the reference right now.

II

Defining the Condition of the Laboratory Animal

PROBLEMS IN CHECKING INAPPARENT INFECTIONS IN LABORATORY MOUSE COLONIES: AN ATTEMPT AT SEROLOGICAL CHECKING BY ANAMNESTIC RESPONSE

Kōsaku Fujiwara

Introduction

Serology is considered to be a very useful method of checking for most infectious diseases in mouse breeder colonies, but the evaluation of this method remains incomplete.

Since March 1967 we have examined serologically many retired breeder mouse sera to check for corynebacteriosis, salmonellosis, Tyzzer's disease, murine viral hepatitis, and Sendai virus infections, which are commonly observed in conventional mouse colonies in Japan. With regard to corynebacteriosis, salmonellosis, and Sendai virus infections, the serologic results were confirmed by the incidence of corresponding lesions at autopsy as well as by the results of cortisone provocation tests on young mice.[1,2] There is a problem, however, in detecting Tyzzer's disease and viral hepatitis within the breeder colonies. Although only a very limited number of retired breeder mouse sera reacted positively to serologic tests for these diseases, the offspring from the same colonies were suspected of having these infections because of their reactions to cortisone tests and their unexpected resistance to inoculation tests.

It seems likely that cortisone tests or some other provocation tests might be more effective for checking a silent contamination among breeder colonies than serology and that such tests should be combined with the serological checking method. The provocation tests, however, give rise to some problems, especially the propagation of infective agents, which results in further contamination of the breeder colonies as well as difficulty in identifying the

77

specific agents causing any detected lesions. To eliminate these problems, more sensitive and specific serological methods are needed to check for silent infections in breeder colonies.

This report deals with the possibility of devising a more effective serological checking method for revealing inapparent infections by anamnestic response.

Materials and Methods

ANIMALS AND TEST SERA

Mice of ddY or ICR were obtained from either conventional or barriered breeder colonies. Seven colonies, A, B, C, D, E, F, and G were branches of one group located in one city, and the mice were fed commercial pellets and water in a conventional environment without air conditioning, although some degree of hygienic precaution was exercised. Colonies H and J were also conventional but institutional. On the other hand, colonies P, Q, S, T, and U were sustained in a strict barrier system and kept on sterilized pellets and water. From these colonies were obtained retired breeder mice of age 15 to 25 weeks weighing 25 to 40 g as well as 4-week-old weanlings weighing 14 to 19 g. Most of them were females.

Animals were bled immediately on arrival at the laboratory or after a certain maintenance according to needs in a conventional environment. Each individual serum was diluted 1:8 with phosphate buffer saline, (PBS) pH 7.2, and heated at 56°C for 30 min before testing.

SEROLOGY
Complement Fixation (CF) Test

Antigens of Tyzzer agent and mouse hepatitis virus (MHV) were prepared from severely affected livers of mice previously inoculated with each infective material. Tyzzer agent used was of SK strain[4] detected at this laboratory and MHV was provided by Dr. S. Suzuki of the Institute of Virus Research, Kyoto University. A 1:10 homogenate of infected livers in distilled water was stored at 4°C overnight and then treated with an equal volume of ethyl ether also overnight at 4°C. Then, water phase of the mixture was separated by centrifugation and evaporation and used as CF antigen. With 4 units of each homologous mouse antiserum, the Tyzzer and MHV antigens showed titers of 1:1,280 and 1:64, respectively.

Sendai virus antigen was a water phase of ethyl-ether-treated chorioallantoic fluid of 12-day-old chick embryos infected with MN strain of the virus. Mouse adenovirus antigen was a supernatant of mouse kidney cell culture inoculated with FL strain of the virus, which was supplied by Dr. K. Hashimoto, Keio University Medical School. Before the latter was used for testing, it was heated at 56°C for 30 min. The Sendai and mouse adenovirus antigens showed titers of 1:16 and 1:1,280 with four units of each homologous mouse antiserum, respectively.

The CF test for screening was made at a 1:8 dilution of sera with four units of antigen, according to the method described by Fujiwara,[3] except that the system was reduced in amount so as to consist of 0.15 ml each of antigen and test sera and 0.25 ml each of complement and hemolytic systems. Titration was made using a microsystem consisting of 0.025 ml each of the four elements.

Agglutination Test

Serum agar culture of *Corynebacterium kutscheri* isolated from liver abscesses in mice was suspended in PBS, and the suspension was incubated at 38°C for 72 hr after adding formalin at 0.5 percent. Then, washed repeatedly with PBS, the killed organisms were suspended in PBS at a concentration of 10 mg per ml and used as the antigen (Coryne). Also, from nutrient agar culture of *Salmonella enteritidis* G14 the antigen (Sal) was prepared in a similar way, but in this case, the organisms were killed by heating at 100°C for 30 min.

To 0.15 ml of a 1:8 dilution of test sera was added 0.05 ml of each bacterial antigen. The mixture was allowed to stand overnight at room temperature, and agglutination was examined. Titration was made using a microsystem consisting of 0.025 ml each of antigen and serum dilution. As diluent of serum for titration, a 1:8 dilution of pooled negative mouse serum was used to inhibit nonspecific agglutination.

INDUCTION OF ANAMNESTIC RESPONSE
Preparation of Antigens

To form the Tyzzer antigen, formalin-treated SK-infected mouse liver homogenate was prepared by the method described in a previous report[4] and diluted with PBS to contain 10^6 organisms per ml. In the same manner, MHV antigen was prepared from infected liver homogenate showing an infective titer of $10^7 \times MLD_{100}$ per 0.1 ml for ICR weanlings[5] before adding formalin.

Corynebacterial antigen, and Sendai and mouse adenovirus antigens, prepared for CF and agglutination tests, were used for induction of anamnestic response.

Injection of Antigens

Two kinds of mixed antigens, bivalent and pentavalent, were used throughout the experiments. The bivalent (TM) antigen consisted of one part each of Tyzzer and MHV antigens and three parts of PBS,

TABLE 1 Routine Serology of Retired Breeder Mouse Sera Collected from March 1967 to July 1968

Colony	Strain or Stock	Number of Sera Tested	Percent Positivity (Range of Each Month)[a]				
			Coryne	Sal	Tyzzer	MHV	Sendai
Conventional							
A	ddY	803	23.7(0–47.1)	0	0.1(0–2.2)	2.5(0–8.5)	0.5(0–4.1)
B	ddY	822	2.8(0–8.2)	0	0.0	0.6(0–6.1)	0.6(0–8.0)
C	ddY	790	13.3(2.1–21.7)	0	0.1(0–2.1)	2.5(0–7.7)	57.5(2.0–92.1)
D	ddY	779	18.5(0–48.9)	0	1.0(0–9.8)	2.2(0–7.3)	77.4(30.0–100.0)[b]
E	ddY	696	24.6(14.3–30.2)	0	0.4(0–3.8)	2.3(0–6.0)	0.3(0–2.4)
F	ddY	839	26.0(6.1–39.6)	0	0.7(0–4.2)	1.0(0–6.4)	72.9(18.0–100.0)[b]
G	ddY	800	21.6(14.0–41.3)	0	0.0	1.5(0–8.7)	0.4(0–5.6)
H	DDD	574	3.8(0–16.7)	0	0.0	1.2(0–3.6)	77.0(39.3–100.0)
J	BALB/C	198	0.0	0	0.0	3.5(0–22.2)	25.8(0–100.0)
Total		6,301	16.6	0	0.3	1.8	20.5[c]
Barriered							
P	ICR	2,867	0.0	0	0.3(0–2.1)	0.2(0–1.1)	1.2(0–11.3)
Q	ICR	174	0.0	0	2.9(2.6–3.1)	0.0	0.6(0–1.0)
T	ICR	88	0.0	0	0.0	1.1(1.1)	4.4(4.4)
U	ICR	82	0.0	0	0.0	3.7(3.7)	3.7(3.7)
Total		3,211	0.0	0	0.4	0.3	1.3

[a]Percent of positive sera at a 1:8 dilution.
[b]Partly vaccinated with inactivated Sendai virus adjuvant vaccine.
[c]Calculated not including D and F colonies.

while the pentavalent (CTAMS) one consisted of equal parts of all five antigens described above. For induction of anamnestic response, 0.5 ml of either the bivalent or pentavalent antigen was injected intraperitoneally.

Results and Discussion

Routine Serology of Retired Breeder Mice

From March 1967 to July 1968 retired breeder mice were examined for corynebacteriosis, salmonellosis, Tyzzer's disease, murine viral hepatitis, and Sendai virus infections at a 1:8 dilution of their sera.

As presented in Table 1, none of 9,512 sera from either conventional or barriered colonies reacted with salmonella antigen. With corynebacterial infection, however, 8 of 9 conventional colonies were shown to be contaminated at levels of 2.8 to 26.0 percent, while all four barriered colonies were found to be free of the infection. Tests for Tyzzer antigen revealed that only a few positive cases could be checked in repeated samplings, although 5 of 9 conventional and 2 of 4 barriered colonies gave positive cases. With MHV, only one barriered colony was perfectly negative, but the positivities of other colonies remained very low. Sendai virus infection was detected in all the colonies examined, but the degree of contamination varied remarkably. In some colonies the positivities remained at lower levels than 1 percent, suggesting incidental invasions of the virus without spreading, while in conventional C, H, and J colonies, considerably high levels of contamination were consistently observed.

When this routine serology was compared with either clinical observations or incidence of specific lesions encountered at autopsy, it seemed useful to check salmonellosis, corynebacteriosis, and Sendai virus infection. However, for Tyzzer's disease and murine viral hepatitis, the routine serology was not considered to be sensitive enough to check the inapparent infection by examination of a limited number of sera. From the results of cortisone tests[2] and unexpected resistance to inoculation tests, the degree of contamination seemed to be much higher than that observed in the routine serology. Moreover, such low positivities could not enable us to rule out nonspecific reactions. In these respects, sera of young and old animals treated with cortisone were tested, but the results obtained were not satisfactory.

Since many Tyzzer-free or MHV-free animals are needed for use in experimental work, a more effective checking method is required. Although cortisone treatment of young animals seems to be more sensitive to check these infections than serology, it has considerable disadvantages. It is not only dangerous in provoking propagation of pathogens within colonies, but identification of induced lesions is always difficult. In addition, if young animals have received

81

TABLE 2 Response of Sensitized and Nonsensitized Retired Breeder Mice to Reinjection of Antigens

Days after Reinjection[a]	Number Tested	Percent Positivity[b]					
		Coryne	Tyzzer	Adeno	MHV	Sendai
Sensitized[c]						
0	20	85.0	5.0	0	85.0	35.0
3	20	90.0	10.0	50.0	100.0	85.0
6	20	100.0	55.0	65.0	100.0	100.0
9	20	95.0	60.0	80.0	100.0	95.0
12	20	100.0	70.0	60.0	100.0	100.0
16	20	95.0	60.0	85.0	100.0	90.0
19	20	100.0	40.0	55.0	100.0	95.0
Nonsensitized						
0	20	0.0	0.0	0.0	40.0	0.0
3	19	0.0	0.0	0.0	97.4	0.0
6	20	0.0	0.0	0.0	100.0	0.0
9	20	10.0	0.0	0.0	95.0	5.0
12	19	26.3	0.0	0.0	100.0	52.6
16	20	30.0	0.0	0.0	100.0	60.0
19	19	47.4	0.0	0.0	100.0	42.1

[a] 0.5 ml of pentavalent (CTAMS) antigen was injected intraperitoneally 4 weeks after sensitization.
[b] Percent of sera reacting positively at a 1:8 dilution.
[c] 0.5 ml of pentavalent (CTAMS) antigen was injected intraperitoneally.

82

maternal antibodies, it seems not to be applicable.[1] On the contrary, serology is the most specific method; it is not dangerous and can be applied even within breeder colonies. Moreover, it is less expensive, since retired breeder mice are used.

Then, an attempt was made to find a more sensitive serological test and the possibility of checking a faint sensitization due to natural infection by anamnestic response was examined.

EXPERIMENTAL BASES FOR THE PROPOSED METHOD

The first experiment was used to study whether an anamnestic response induced in sensitized animals is distinguishable from the primary response induced by specific antigen injection.

From barriered colony P, which had not given any cases that were positive to any antigens by routine serology since early in 1968, 280 female ICR retired breeder mice were obtained and about a half of them were injected intraperitoneally with 0.5 ml of pentavalent (CTAMS) antigen. Four weeks later, 20 animals from both the pretreated and the untreated groups were bled, and the remaining ones were given the same injection they had received 4 weeks earlier. After injection, 20 mice from each group were bled at intervals of 3 to 4 days, and all the sera were tested for five antigens at a 1:8 dilution.

As the data in Table 2 indicate, 4 weeks after the first injection, many of the sensitized animals were shown to have antibodies to corynebacterial and Sendai virus antigens, while none of the nonsensitized ones had that response. Only one positive case of Tyzzer antigen was detected in the sensitized group.

However, with MHV, many of the nonsensitized animals were found to be contaminated. This might have been due to natural infection because virulent viruses are frequently used in the same laboratory.

After reinjection of antigens, in sensitized animals, an early response was observed at remarkably high positivities for each antigen 3 or 6 days after the injection. None of the nonsensitized animals responded at this time to any of the antigens except MHV. Only a few of them reacted 9 days after the injection with corynebacterial and Sendai virus antigens, but no positive cases with Tyzzer and adenovirus antigens were observed even after 19 days following antigen injection. With MHV antigen, however, sensitized and nonsensitized animals reacted showing 100 percent positivity 3 and 6 days, respectively, after the antigen introduction. A similar experiment was made using 5-week-old mice from the same colony, and perfectly analogous results were obtained, including a similar contamination with MHV.

Sera obtained 6 and 9 days after reinjection of antigens in the experiment described above were titrated; the results are shown in Table 3. Except for MHV, titers of the sera from previously sensitized mice were much higher

83

TABLE 3 Antibody Titer of Sensitized and Nonsensitized Retired Breeder Mice after Reinjection of Antigens

Number Tested		Antibody	Titer of 6 and 9 Days after Reinjection[a]					Mean (-log$_2$)
			<1:8	1:8–16	1:32–64	1:128–256	1:512≤	
Sensitized[b]	40	Coryne	1[c]	7	11	19	2	<6.23≤
		Tyzzer	17	13	10			<3.88
		Adeno	11	2	22	5		<5.10
		MHV		1	14	24	1	6.78≤
		Sendai	1	2	20	16	1	<6.18≤
Nonsensitized	40	Coryne	39	1				<3.00
		Tyzzer	40					<3.00
		Adeno	40					<3.00
		MHV	1	3	20	16		<6.10
		Sendai	39	1				<3.03

[a] 0.5 ml of pentavalent (CTAMS) antigen was injected intraperitoneally 4 weeks after sensitization.
[b] 0.5 ml of pentavalent (CTAMS) antigen was injected intraperitoneally.
[c] Number of sera showing the titer indicated at the top column.

84

than those from unsensitized mice, indicating that the earlier and higher positivities observed with a 1:8 dilution would be of an anamnestic response.

The next experiment was made to confirm specificity of such anamnestic response. Among three groups of about 50 female retired breeder mice from colony P, one was injected intraperitoneally with 0.5 ml of bivalent (TM) antigen, while another was injected with 0.5 ml of pentavalent (CTAMS) antigen. The third group served as a nonsensitized control.

Four weeks later, 10 animals of each group were bled and the sera were tested for five antigens. Then, each half of the remaining 40 animals was injected intraperitoneally with 0.5 ml of either the bivalent (TM) or the pentavalent (CTAMS) antigen, and bled 6 days later. All the sera were tested at a 1:8 dilution. The results are presented in Table 4.

When the sensitization was bivalent (TM), only the response to Tyzzer and MHV antigens was remarkable after injection of either bivalent or pentavalent antigen, although an unexpectedly high degree of contamination of MHV was also inevitable in this experiment, as observed previously. Some of the bivalent-sensitized and pentavalent-challenged animals showed positive reaction to corynebacterial antigen. In many animals sensitized with pentavalent (CTAMS) antigen, anticorynebacterial, MHV, and Sendai antibodies were checked before reinjection, but after reinjection of antigen, elevation of positivity was much more remarkable with the reinjected antigen.

Interpretation of these experiments suggests that the early response after antigenic introduction is anamnestic due to the previous exposure to the homologous antigen. It may also be possible to check by such response a faint sensitization caused by natural infection that is undetectable by routine serology.

It should be noted, however, that some animals showed comparatively early response to corynebacterial and Sendai virus antigen. Such early response might be due either to characteristics of antigenic properties or to different kinetics of antibody production with these antigens. On the other hand, corynebacterial and Sendai virus infection could be checked by routine serology, and there seems to be no need to expect anamnestic response for checking the infections.

CHECKING OF BREEDER COLONIES BY ANAMNESTIC RESPONSE

Based upon the experimental results described above, checking of conventional and barriered breeder colonies was attempted by the anamnestic response of retired breeder mice after antigen injection.

From October 1968 to January 1969 several lots of retired breeder mice were obtained from 8 conventional and 4 barriered colonies and a part of each lot was injected intraperitoneally with 0.5 ml of bivalent (TM) antigen on arrival. The remaining animals served as noninjected controls. Six days

TABLE 4 Response to Reinjection of Retired Breeder Mice Sensitized with Bivalent or Pentavalent Antigen

Sensitization[a]	Reinjection[b]	No. Tested	Percent Positivity[c] 6 Days after Reinjection				
			Coryne	Tyzzer	Adeno	MHV	Sendai
Nonsensitized	Before Reinjection	10	0	0	0	50.0	0
	Bivalent	21	0	0	0	100.0	0
	Pentavalent	20	15.0	0	0	95.0	0
Bivalent (TM)	Before Reinjection	10	0	0	0	60.0	0
	Bivalent	20	0	65.0	0	100.0	0
	Pentavalent	19	10.5	63.2	0	100.0	0
Pentavalent (CTAMS)	Before Reinjection	10	80.0	0	0	80.0	50.0
	Bivalent	20	90.0	60.0	5.0	100.0	40.0
	Pentavalent	22	95.5	40.9	81.8	100.0	100.0

[a] 0.5 ml of each antigen was injected intraperitoneally.
[b] 0.5 ml of each antigen was injected 4 weeks after sensitization.
[c] Percent of sera reacting positively at a 1:8 dilution.

later, all the animals were bled and the sera were tested for five antigens.

The results are summarized in Table 5. Seven of eight conventional colonies that had given no positive case by routine serology were shown to be contaminated with Tyzzer's disease. Above all, colonies D, E, and F showed high positivities of 21.1 to 42.1 percent after antigen injection. On the other hand, all the tested barriered colonies proved to be free from contamination by Tyzzer's disease when checked by serology of anamnestic response.

With MHV, most of conventional colonies were shown to be contaminated at low positivities less than 7.1 percent without antigen injection, but positivities after injection of the specific antigen were five to ten times higher than those observed in the routine serology. Except for one that was highly contaminated, the barriered colonies were shown to be negative with MHV.

In contrast to such specific response to Tyzzer and MHV antigens, reactions to heterologous antigens seem to remain unaffected as shown in Table 5.

The experiments described above were made using retired breeder mice, since it was considered that older animals would be exposed to natural infection more frequently than young ones. In another experiment, response to bivalent (TM) antigen injection was compared for retired breeder mice and for weanlings from E colony, which seemed to be highly contaminated with Tyzzer's disease and viral hepatitis (see Table 5).

As presented in Table 6, young animals did not react to either Tyzzer or MHV antigens even after injection of the same antigens, although high positivities were obtained with retired breeders.

It remains unknown whether such negative results with weanlings occur because the animals did not have a chance to be exposed to the pathogens or because antibodies transferred from the mother interfered with the production of antibodies in weanlings.

Animals should be maintained at least 6 days after antigen injection for checking by anamnestic response, and contamination is possible during this period. In the experiments presented in Tables 2 and 4, severe contamination with MHV was observed after maintenance at the laboratory for several weeks. However, if the maintenance after antigen injection is confined to a period of less than 7 days, involvement of natural infection during required maintenance can be ruled out because of the negative results with noninjected animals maintained under the same conditions as antigen-treated animals.

From these results it might be concluded that the method presented here is applicable in breeder colonies as well as in laboratories for detecting a faint sensitization due to natural infection, especially of Tyzzer's disease and MHV infection, both of which are undetectable by routine serology. By this method it may be possible to obtain a positivity high enough to detect the contaminations, checking a limited number of test animals.

TABLE 5 Serological Checking of Retired Breeder Mice by Anamnestic Response, from October 1968–January 1969

Colony	Strain or Stock	Bivalent (TM) Antigen[a]	Number Tested	Percent Positivity[b] 6 Days after Injection				
				Coryne	Tyzzer	Adeno	MHV	Sendai
Conventional								
A	ddY	Injected	94	21.3	5.3	0	28.7	1.5
		Noninjected	86	17.4	0.0	0	1.2	0
B	ddY	Injected	99	0.0	1.0	0	27.3	5.7
		Noninjected	91	1.1	0.0	0	2.2	0.0
C	ddY	Injected	97	5.2	5.2	0	43.3	52.9
		Noninjected	88	3.4	0.0	0	3.4	55.9
D	ddY	Injected	76	9.2	42.1	0	14.8	36.8
		Noninjected	90	10.0	0.0	0	1.1	42.2

G	ddY	Injected	97	12.4	3.1	0	41.2	1.4
		Noninjected	84	15.5	0.0	0	7.1	0.0
H	DDD	Injected	170	0.0	0.0	0	12.4	45.3
		Noninjected	141	1.4	0.0	0	0.0	38.2
Barriered								
P	ICR	Injected	51	0	0	0	0.0	0.0
		Noninjected	48	0	0	0	0.0	0.0
S	ddY	Injected	173	0	0	0	97.1	0.0
		Noninjected	186	0	0	0	81.7	0.0
T	ICR	Injected	53	0	0	0	0.0	1.9
		Noninjected	48	0	0	0	0.0	0.0
U	ICR	Injected	88	0	0	0	0.0	0.0
		Noninjected	100	0	0	0	0.0	0.0

[a] 0.5 ml intraperitoneally.
[b] Percent of sera reacting positively at a 1:8 dilution.

TABLE 6 Response of Weanlings and Retired Breeder Mice from Colony E to Injection of Bivalent Antigen

Mouse	Antigen[a]	Number Tested	Percent Positivity[b] 7 Days after Injection					
			Coryne	Sal	Tyzzer	MHV	Sendai
Retired Breeder 15–20-week-old	Injected	40	10.0	0	37.5	32.5	0.0
Female: 34–38 g Male: 28–34 g	Noninjected	41	9.8	0	4.9	2.4	4.9
Weanling 4-week-old	Injected	44	2.3	0	0	0	0
Male: 14–19 g	Noninjected	37	5.4	0	0	0	0

[a]Bivalent (TM); 0.5 ml injected intraperitoneally.
[b]Percent of sera reacting positively at a 1:8 dilution.

References

1. Fujiwara, K., Y. Takagaki, M. Naiki, K. Maejima, and Y. Tajima. 1964. Tyzzer's disease in mice. Effect of cortico steroids on the formation of liver lesions and the level of blood transaminases in experimentally infected animals. Japan. J. Exp. Med. 34:59–75.
2. Takagaki, Y., M. Naiki, M. Ito, G. Noguchi, and K. Fujiwara. 1967. Checking of corynebacteriosis and Tyzzer's disease in mice by cortisone treatment. Exp. Anim. 16:12–19.
3. Fujiwara, K. 1967. Complement fixation reaction and agar gel double diffusion test in Tyzzer's disease of mice. Japan. J. Microbiol. 11:103–117.
4. Fujiwara, K., H. Kurashina, K. Maejima, Y. Tajima, Y. Takagaki, and M. Naiki. 1965. Actively induced immune resistance to experimental Tyzzer's disease of mice. Japan. J. Exp. Med. 35:259–275.
5. Naiki, M., and K. Fujiwara. 1968. Some factors affecting host susceptibility to murine hepatitis virus. Exp. Anim. 17:53–58.

DISCUSSION

DR. GRAFTON: Have you tried to test serologically for mycoplasma or other organisms? Which others can be demonstrated serologically and which cannot?

DR. FUJIWARA: I have done nothing on mycoplasma infection, only the infections we have presented here.

DR. FLYNN: In using cortisone, what dosage was used, and what form of cortisone?

DR. FUJIWARA: That was cortisone acetate. We injected mice that were 4 or 5 weeks old with 5 mg in a single dose and observed them for 7 days.

DR. HOLMES: Have you isolated and cultured the Tyzzer organism, and if so, by what technique?

DR. FUJIWARA: We do not have any culture of Tyzzer organisms. We maintained the Tyzzer organisms by mouse passage.

DR. ARMIGER: Do you have a special method of collecting sera? Assuming that you bleed individuals to death, how many tests could you carry out on the volume of serum obtained?

DR. FUJIWARA: For either serum tests or agglutination tests, we make them with 0.15 ml of mouse serum. We bleed the serum of mice by cutting the necks of the mice and collect the whole blood in a Petri dish. We harvest the serum by adding PBS, so as to obtain a 1:8 dilution. The serum test is made by standard admixture with only 0.15 ml of serum.

DR. GREENWOOD: What was the animal used and what was the method of preparation of the monospecific serum? How did you make the monospecific serum? In the case of murine viruses, what method of isolation was used?

DR. FUJIWARA: Preparing monospecific serum is very different for the different kinds of antigens. For example, for Tyzzer's antigen, we inject pathogenic agents into mice, and then we harvest the liver. For MHV antigen, we use the supernatant of infected liver, and for Tyzzer we use the supernatant of the homogenate of infected liver in water, because Tyzzer organisms rapidly show a lysis in distilled water at 4° C.

DR. GREENWOOD: In the case of murine viruses, what method of isolation was used?

DR. FUJIWARA: I have made no attempt to isolate the virus. I used virus imported from the United States. In the case of mouse adenovirus, we imported this too from the United States. I have never seen a natural case of mouse adenovirus infection in Japan.

IMMUNOFLUORESCENT DETECTION OF MURINE VIRUS ANTIGENS

Roger E. Wilsnack

The taxonomically diverse viruses capable of asymtomatic infection in mice, a seemingly ubiquitous group of agents, have frequently obscured virologic and immunologic data derived from mouse-based experimentation.[1,2] Exhaustive serologic surveys have attested to the prevalence of colony-wide infections and have contributed significantly to the ecological comprehension of indigenous murine viruses.[1,3]

Despite their value in epizootiologic studies, these serologic methods cannot be applied as antigen-detection tests unless they are carefully manipulated as a mouse antibody production (MAP) test.[4] The utility of the MAP test definitely exceeds the level of expediency attainable with conventional virus-isolation methods; however, its sensitivity varies from agent to agent, it requires a supply of mice free from the agent and antibody in quest, careful control of cross contamination is essential during the holding period, a 21-day holding period is inherent in the technique, and a thorough knowledge of the pertinent serological technique is essential in assessing the antibody profile of the recipient mice. Direct and indirect immunofluorescent (IF) staining can serve as a valuable adjunct to serological and virus-isolation tests. Standardization and control must be relentlessly applied to immunofluorescent tests to achieve the desired objectives of specificity, speed, and sensitivity. The principal advantages of the IF method are the ability to visualize the antigen directly on a cellular level and the speed with which the antigen identification can be performed. This paper will deal with the application of IF techniques to the detection of some commonly encountered murine virus antigens.

93

Materials and Methods

ANIMALS

Random-bred mice (ICR) used in experiments with lymphocytic choriomeningitis virus (LCM) were purchased from Charles River Mouse Farms, Norwalk, Massachusetts. Germfree mice (CFW) utilized in experiments with pneumonia virus of mice (PVM) and reagent preparation for the virus of epizootic diarrhea of infant mice (EDIM) were procured from Carworth Lab Cages Inc., New City, New York. All other titrations and reagent preparations were conducted with random-bred mice (NLW) from National Laboratory Animal Co., Creve Coeur, Missouri. The mice, including those of germfree origin, were housed in filter-top cages.

Young mixed-breed goats were purchased from a livestock dealer and were maintained in a loose housing pen barn. Adult guinea pigs (Hartley) were procured from various sources and were housed in drawer-type cage batteries.

VIRUS STOCKS

Lymphocytic choriomeningitis virus (strains CA 1371 and WCP) were obtained from Dr. W. P. Rowe, National Institutes of Health, Bethesda, Maryland. Other viruses, epizootic diarrhea of infant mice (EDIM), pneumonia virus of mice (PVM), mouse encephalomyelitis virus (GDVII strain), newborn mouse pneumonitis virus (K virus), and mouse hepatitis viruses (MHV) were obtained from Dr. J. C. Parker, Microbiological Associates, Bethesda, Maryland. Preparation and propagation of these viruses was as previously described.[5]

VIRUS TITRATIONS

LCM and GDVII viruses were quantitated by intracerebral inoculation (0.03 ml) of weanling mice, and end points were computed on the basis of LD_{50}. K virus was titered by the intracerebral inoculation (0.02 ml) of mice less than 5 days of age, with end points determined by LD_{50}. MHV was inoculated by the intracerebral (0.02 ml) and intraperitoneal (0.05 ml) routes, simultaneously, into mice less than 5 days of age. Virus activity was determined by death of the animals. Quantitation of PVM virus was effected by the intranasal instillation (0.05 ml) of weanling, germfree mice, and viral activity was ascertained by the presence of haemagglutinins (mouse cells) in the lungs of recipient mice. Mice less than 5 days of age were inoculated *per os* (0.05 ml) in the determinations of EDIM virus activity; units of infectivity were determined by clinical evidence of diarrhea at 5 days postinoculation.

CELL CULTURES

Primary mouse lung cell cultures were prepared by the trypsinization of minced lungs harvested from mice less than 5 days of age. These cultures were maintained on medium 199 (Earle's base) + 10 percent fetal bovine serum with 200 units and 200 micrograms of penicillin G and streptomycin sulfate per ml, respectively. A mouse liver cell line (NCTC 1469) was maintained on medium NCTC 109 + 20 percent equine serum with an antibiotic component similar to the one described above. For purposes of immunization, NCTC 1469 cells were maintained on medium NCTC 109 + 20 percent autologous caprine serum. BHK-21 cells were maintained on MEM medium (Earle's base) with the above antibiotic component. All cells intended for staining were propagated on 9 X 35-mm coverslips.

IMMUNE SERUM PREPARATION

Antiserum to LCM and EDIM viruses was supplied by Dr. J. C. Parker of Microbiological Associates, Bethesda, Maryland. All antigens were clarified (5,000 X G, 10 minutes) prior to injection with complete Freunds adjuvant. Goats were injected by the intramuscular route, mice by the intraperitoneal route, and all antigens were administered at approximately 7–10-day intervals. Table 1 denotes specifics of antigen preparation, potency, and characteristics of resultant antisera. Goats were bled from the external jugular vein, and mice were exsanguinated (anaesthetized with ether) by incising the thoracic inlet. Ascitic fluid, when present, was aspirated by hypodermic syringe.

SEROLOGY

Serologic assessments of immune sera were conducted by the microtiter technique according to protocol previously described by Parker et al.[5,6] Antigens and control sera to K virus, MHV (A-59, MHV-1, MHV-S, and JHM), Reo 3, GDVII, Sendai virus, mouse adenovirus, polyoma, and PVM, were supplied by Dr. J. C. Parker of Microbiological Associates. Caprine sera frequently demonstrated anticomplementary activity as well as autoagglutination of human, cavian, and mouse blood cells. As a result, caprine sera were sorbed 3 times with 50 percent concentrations of appropriate blood cells, were mixed with equal volumes of undiluted guinea pig complement (held 24 hr at $5°C$), and finally, were heated at $56°C$ for 30 min.

SERUM PROTEIN FRACTIONATION

Serum proteins intended for conjugation were fractionated by one of the following methods: (1) 50 percent saturation with

TABLE 1 Preparation of Antisera to Murine Virus Antigens

| Virus Antigen | | | Antiserum | | Serological Response | |
Type	Prepared In	Titer[a]	Species	Immunization Schedule (weeks)	Homologous Titer[a]	Heterologous Titer
LCM	Guinea pig brain, spleen, and liver	$10^{-5.4}LD_{50}$	Guinea pig	4	1:512 CF	ND[b]
LCM	African green monkey kidney cell culture	$10^{-6.0}LD_{50}$	Goat	12	1:128 Neut.	Reo 3 1:64 GDVII 1:28
K	Mouse lung and livers 10 percent (W/V)	$10^{-5.5}LD_{50}$	Mouse	6	1:128 HI	ND
EDIM	Mouse intestinal tract—10 percent (W/V)	$10^{-5.0}LD_{50}$	Mouse	16	1:256 CF	ND
PVM	Mouse lung 20 percent (W/V)	1:128 HA	Goat	22	1:128 HI	Reo 3 1:64
PVM	Mouse lung 20 percent (W/V)	1:128 HA	Mouse	12	1:512 HI	ND
GDVII	Mouse brain 20 percent (W/V)	$10^{-6.0}LD_{50}$	Goat	24	1:512 HI	Reo 3 1:32
MHV	NCTC 1469 cells	$10^{-5.3}TCD_{50}$	Goat	18	1:32	1:128 GDVII
MHV	NCTC 1469	$10^{-5.0}TCD_{50}$	Mouse	12	1:512	ND

[a]HI = haemagglutination inhibition titer; Neut. = neutralization titer; CF = complement fixation titer; TCD_{50} = tissue culture infectious dose.
[b]ND = none detected.

$(NH_4)_2 SO_4$; (2) ion-exchange chromatography with DEAE-Sephadex, A-50 coarse (Pharmacia Fine Chemicals, Piscataway, N.J.); (3) 50 percent saturation with $(NH_4)_2SO_4$, followed by digestion with trypsin (1:300). Method (1) involved the sedimentation (12,000 X G, 20 min) of those proteins rendered insoluble by 50 percent saturation with $(NH_4)_2 SO_4$. This sediment was washed in a solution of $(NH_4)_2 SO_4$ (50 percent saturation) equivalent to the original serum volume and was resedimented. These proteins were dissolved in 0.85 percent NaCl solution equivalent in volume to one half the original serum volume. The resulting globulin solution was dialyzed against 0.85 percent NaCl until it was free of sulfate, immediately after this, the solution was conjugated or stored at $-20°C$.

Sera fractionated by Method (2) were initially subjected to $(NH_4)_2 SO_4$ fractionation and were dialyzed for 24 hr against Tris-HCl buffer (0.1 M, pH 8.1). If necessary, sediment was removed from the dialyzed globulins by centrifugation (12,000 X G, 20 min). Dry DEAE-Sephadex (A-50 Coarse) was mixed with 30 volumes of the same Tris-HCl buffer, was held overnight at 5°C, and was washed with the Tris-HCl buffer until the effluent from the Büchner funnel recorded pH 8.1 ± 0.1. The equilibrated gel was then mixed with the dialyzed globulins in a beaker at a ratio of 2 parts gel to 1 part globulin. The mixture was stirred and held at 5°C for 4 hr. After 4 hr the mixture was poured into a Büchner funnel and slowly eluted with Tris-HCl buffer equal in volume to 6 times the original serum volume. At this point the eluate containing the IgG was placed in a dialysis casing and was concentrated to one fourth the original serum volume with Carbowax 20,000 (Union Carbide Co.) at 5°C. The concentrated eluate was then dialyzed against 0.01 M phosphate buffer saline (pH 7.2) overnight and was immediately conjugated or stored at $-20°C$.

Method 3, like method 2, utilized sera previously fractionated by the $(NH_4)_2 SO_4$ method. Globulin solutions were dialyzed against 0.01 M phosphate buffered saline (pH 8.3) at 5°C for 48 hr, with one change of buffer solution. If a precipitate formed, it was removed by centrifugation (12,000 X G, 20 min). Dry Trypsin (1:300) was added to the dialyzed globulins at the rate of 1 mg of trypsin to each mg of serum protein. The mixture was held at 5°C, with constant mixing on a magnetic stirrer, for 24 hr. The digested protein solution was then clarified by centrifugation and was separated from trypsin residue by molecular filtration on G-75 Sepadex (Pharmacia Fine Chemicals, Piscataway, N.J.) equilibrated with 0.01 M phosphate buffered saline (pH 7.2).

The efficacy of all three fractionation methods was monitored by immunoelectrophoresis against antisera (rabbit) with precipitating activity against the whole serum protein of the species in question. A 1.5 percent solution of Agarose (Bausch and Lomb) in barbital-HCl buffer (pH 8.3, ionic strength 0.05) served as support in a system utilizing a 160 volt potential at 30 milliamperes. Specimens were usually subjected to the electrical field for 2.5–3 hr.

The precipitation arcs were developed by holding the Agarose slides for 15-18 hr (at room temperature) after the application of antisera.

SERUM PROTEIN QUANTITATION

Three methods were utilized in the quantitation of serum proteins for conjugation. A Protein meter (Bausch and Lomb) was used to measure serum proteins by refraction; its use was confined to those fractions resulting from 50 percent saturation with $(NH_4)_2$ SO_4. The Biruet and Lowry[7] methods of protein determination were applied interchangeably on proteins resulting from ion-exchange and trypsin-digestion methods of fractionation. All the protein measuring tests were standardized against a whole-serum standard, Lab-Trol (Dade Reagents, Inc., Miami, Florida).

CONJUGATE PREPARATION

All conjugates were prepared with the fluorochrome fluorescein isothiocyanate, isomer I (BBL, Cockeysville, Maryland). The relatively crude fractions resulting from (NH_4) SO_4 50 percent saturation were coupled at fluorescein to protein ratios (F/P) of 1:60. The IgG fractions resulting from ion-exchange chromatography and trypsin digestion were coupled at F/P ratios of 1:80 and 1:100, respectively. Prior to the addition of the dye, the protein solution was adjusted to pH 9.0 by the addition of 0.5 M $NaHCO_3$-Na_2CO_3 buffer. The required amount of dye was added to the solution, and the conjugation procedure was allowed to proceed for 18 hours in a sealed flask at 5°C, with constant mixing by a magnetic stirrer.

Uncoupled fluorescein and its breakdown products were removed from the conjugate by molecular filtration on G-25 or G-75 Sephadex (Pharmacia Fine Chemicals, Piscataway, New Jersey). The G-25 porosity gel was used for conjugates incorporating proteins from $(NH_4)_2$ SO_4 and ion-exchange fractionation; G-75 porosity was applied specifically to those reagents containing trypsin digested serum proteins. Both Sephadex gels were equilibrated and eluted with 0.01 M phosphate-buffered saline (pH 7.2). The conjugate eluates containing the immunoglobulins were collected from G-25 Sephadex columns by visual observation; however, the trypsin digests conjugated did not offer a readily observable separation and were collected by volume, a volume equal to that applied to the top of the column. Acetone-extracted tissue powders were not employed in conjugate preparations. The finished conjugates were adjusted to contain 2 percent bovine albumin (W/V) and were either stored at -20°C or lyophilized. In addition, trypsin digest conjugates received 2 mg of ovomucoid (Mann Research Laboratories, New York, New York) per mg of trypsin originally added. The ovomucoid effectively inhibited any residual proteolytic activity.

TISSUE PREPARATION AND FIXATION

Tissues were prepared for staining primarily by touch impressions on microscope slides (0.8–1.0-mm thickness). Impressions were taken from a cut surface of parenchymous organs or a minced segment of the intestinal tract. The tissues were conveniently handled by placement on a tongue depressor, and care was taken to minimize the amount of tissue adhering to the slide. Tissues intended for frozen sections were immersed in O.C.T. mounting media (Lab.-Tek, Westmont, Illinois) and frozen on a block of dry ice. Tissues frozen and mounted in O.C.T. media at a later date were less satisfactory for sectioning. Sections were cut at 4 μm on a Minot rotary microtome (International Equipment Co.) with a Jung microtome blade (profile A, biconcave) at $-20°$C. Cell cultures were handled on 9- $\times$ 35-mm coverslips.

All slides, with the exception of EDIM antigen, were fixed in acetone (spectro grade) at $-20°$C for 4–18 hr. EDIM antigens were fixed under the same conditions, but 95 percent ethyl alcohol was substituted for acetone. At $-20°$C, fixation times were not critical and could have been varied for expediency. After fixation, slides and coverslips were stored at $-20°$C. All antigens, with the exception of LCM and MHV, were stable for at least 6 months under these conditions.

IMMUNOFLUORESCENT STAINING PROCEDURES

Conjugates were diluted for staining in 10–20 percent (W/V), uncentrifuged tissue suspensions prepared from germfree or specific-virus-free mice. To attain specific inhibition of staining, virus-infected tissues were used in the preparation of suspensions. A phosphate-buffered saline (0.1 M, pH 7.2) served as effective diluent for tissue suspensions in contrast to cell culture media, which frequently resulted in acidic (pH 6.5) suspensions after prolonged storage at $-20°$C. After the application of conjugate, slides were incubated (in a humid atmosphere) for 30 min at $35°$C and were washed for 20 min in 0.01 M phosphate-buffered saline on a mechanical rotator. Finally, slides were rinsed with distilled water, and coverslips (Corning No. 1, 22 $\times$ 40 mm) were mounted with glycerine buffered to pH 8.5 with 0.1 M Tris-HCl.

TESTS OF STAINING SPECIFICITY

Five staining controls were employed to evaluate the specificity of the conjugates: (1) Staining was not observed when specific conjugates were incubated with sections or impressions of normal tissue (specific-virus-free). (2) Staining was effectively inhibited when conjugates were diluted in homologous, *unlabeled* immune sera. (3) Staining was not observed when control conjugates (from antisera prepared against normal tissue components) were applied to virus-infected tissues. (4) Staining was inhibited by the dilution of conjugates in homologous antigen. Live virus preparations, with the exception of LCM, were employed for inhibition. The preparation of an inactivated LCM antigen has been described elsewhere.[8] The antigen inhibition of staining was more efficient, particularly in the GDVII and MHV systems, if the antigens were sonified at 1.6 amperes for 1 min per 10 ml volume (MSE Sonicator). (5) Staining was not observed on tissue sections and impressions infected with heterologous antigens.

MICROSCOPE AND LIGHT SOURCE

Two binocular microscopes were employed, a Richert "Biozet" and an American Optical "Microstar." Both microscopes (with darkfield condensers) were used in conjunction with light sources incorporating Osram HBO-200 mercury vapor burners. Exciter filter systems consisting of Corning #8079 or #5970 (both ½ thickness) were employed interchangeably; however, a Wratten 2A filter was used exclusively as a barrier filter. The intensity of ultraviolet light was monitored with a SM-600 meter (Westinghouse); a mercury vapor burner was discarded if light intensity at the inlet aperture of the condenser did not exceed 25 microwatts per cm^2.

Results

DETECTION OF LCM VIRUS ANTIGENS

LCM antigens were readily detected in the tissues of acute and chronically infected mice and in African green monkey kidney cell cultures.[8] Cellular localization, both *in vivo* and *in vitro*, was exclusively cytoplasmic. Cytoplasmic localization is in agreement with the observation of Benda *et al.*[9] as well as Pedersen and Volkert.[10]

Following intracerebral inoculation with strain CA 1371 LCM virus, antigen was observed largely in the cytoplasm of cells in the leptomeninges, chorioid plexus, and ependyma.[8] Mice chronically infected with LCM or acutely infected with the WCP strain exhibited a wide distribution of antigen in the lungs, kidneys, spleen, and brain. In both acute visceratropic and chronic LCM infections, the liver was the most consistently and uniformly involved organ, making it the logical selection as a sentinel organ in the immunofluorescent diagnosis of LCM.

Although the various strains of LCM exhibited marked differences in tissue trophism and mortality rate (by various routes of inoculation), all fluoresced with equal intensity. The WCP stain appeared as the agent of choice in the preparation of a staining inhibitor.[8] In addition to the universal staining of various strains of LCM virus, the immunofluorescent method would appear to be equal in sensitivity to the detection by intracerebral mouse inoculation.

DETECTION OF EDIM VIRUS ANTIGENS

EDIM virus proved especially amenable to immunofluorescent detection because the antigen is confined to the intestinal tract and conventional methods of detection are tedious and frequently difficult to interpret.[11,12] Orally infected mice exhibited stainable antigen in the cytoplasm of intestinal epithelium from the duodenum to the colon (from section preparations). Antigen was detected within 48 hr of inoculation, an interval that agreed favorably with virus isolation by mouse inoculation (*per os*). EDIM antigen distribution in naturally infected mice was similar to that in mice inoculated in the laboratory. Antigen could not be demonstrated in the stomachs or the livers of mice infected orally or in naturally infected mice. Of possible epizootiologic significance was the observation that stainable EDIM virus could be viewed in the intestines of suckling and weanling mice, neither of which exhibited clinically discernible diarrhea at the time of sacrifice.

In deference to the ubiquitous nature of this agent, antisera and virus were prepared in germfree mice. Antisera prepared in rabbits were unsatisfactory for conjugate preparation. Since EDIM antigen was confined to the intestinal tract, selection of tissue for immunofluorescent diagnosis was greatly simplified.

DETECTION OF K VIRUS ANTIGENS

In vitro and *in vivo* immunofluorescent studies on the kinetics of K virus replication yielded strikingly dissimilar observations. K virus antigen was not recognized in the nucleus of any *in vivo* system examined, an observation seemingly irreconcilable with the electron microscopy reports of Dalton *et al.*[13] on nuclear localization of K virus in mouse lung endothelial cells. In marked contrast, intranuclear K virus antigen was regularly observed *in vitro*, beginning approximately 3 days postinoculation.

Antigen was first detected at 3–4 days postinoculation (IP) in the liver and lungs of suckling mice (Table 2). In the lungs, antigen was noted in the cytoplasm of alveolar and septal cells; the unique lesions of endothelial nuclei described by Fischer and Kinam[14] were not discernible on stained frozen sections. Virus antigen was also demonstrable, by immunofluorescent stain, in spleen, kidney, and to the least extent in the brain (Table 2). Antigen in the liver was initially discernible in sinusoidal cells and later in parenchymal cells. Spleen-associated antigen was randomly distributed in cells of the red pulp, and K virus antigen in the kidneys was frequently associated with glomeruli.

101

TABLE 2 Sequential Appearance of K Virus Antigen in Suckling Mice as Detected by Immunofluorescence and Mouse Inoculation

Time[a] Postinoculation (hr)	Quantitation of K Virus Antigen In									
	Liver		Spleen		Lung		Brain		Kidney	
	Titer[b]	IF[c]	Titer	IF[c]	Titer	IF[c]	Titer	IF[c]	Titer	IF[c]
24	<1	0	<1	0	<1	0	<1	0	<1	0
48	1	0	1	0	1	0	<1	0	1	0
72	1	±	1	0	ND	+	1	0	1	0
96	3	+	2	±	3	++	1	0	ND	ND
120	ND[d]	++	2	+	4	+++	1	±	2	±
144	4	++	2	+	4	+++	2	±	3	+
168	4	+++	4	++	4	+++	4	+	4	+
192	5	+++	6	++	6	+++	4	+	6	+
216	6	+++	6	++	6	+++	5	+	6	+
240	8	+++	8	++	7	+++	7	+	6	+

[a]Intraperitoneal inoculation, approximately 1,000 LD$_{50}$ contained in 0.1 ml.
[b]Titer to the nearest reciprocal Log 10 of centrifuged organ suspensions (10 percent W/V), inoculated intracerebrally (0.02 ml) into suckling mice.
[c]IF = immunofluorescent staining; 0 = no visible staining; ± = few positive cells; + = less than 10 percent of cells stain; ++ = 10–25 percent of cells stain; +++ = 25 percent or more of cells stain.
[d]ND = not done.

102

Antigen stained in liver, spleen, kidney, lung, and brain was exclusively cyto-plasmic, and in late stages of infection the immunofluorescent material coalesced to an inclusion-like, amorphous mass.

Mouse embryo and suckling mouse lung cells infected with K virus *in vitro* (10^4–10^5 LD_{50}) elicited nuclear and cytoplasmic staining beginning 3 days postinoculation. The total number of antigen-containing cells usually did not exceed 10 percent at 10 days postinoculation. In positive-staining nuclei the entire nucleus, with the exception of nucleoli, fluoresced. Infected cells under-going mitosis exhibited antigen in the nuclei of both daughter cells. Nuclear lobes or satellite nuclei, as described by Fraser and Crawford[15] for polyoma-infected BHK-21 cells, were noted in infected suckling mouse lung cell cul-tures. The cytoplasmic staining was of a granular nature, the fineness of which was inversely proportional to time postinoculation. The relative proportion of cells exhibiting nuclear and cytoplasmic infection appeared independent of multiplicity of infection and elapsed time. Suckling mouse lung cell cultures prepared from suckling mice infected with K virus *in vivo* exhibited exlusively cytoplasmic staining in 80 percent of cells, and nuclear staining did not appear until the sixth day after the cells were plated.

Natural K virus infection was detected in lung impressions by immuno-fluorescent stain in one out of four AKR Mice (4 months of age) sampled from a colony that had exhibited serologic evidence of K virus infection (mice sup-plied by Dr. J. Parker, Microbiological Associates, Bethesda, Maryland). Inter-estingly, the mouse yielding K virus antigen had shown a 1:10 serum HI titer to K virus prior to sacrifice.

DETECTION OF MHV ANTIGENS

Anti-MHV conjugates prepared from MHV-immune mouse and caprine sera effectively stained MHV antigens *in vivo* and *in vitro* (Table 3). Anti-MHV conjugates prepared from mouse immune serum or ascitic fluid exhibited higher staining titers than those prepared in goats. Although the caprine MHV-immune serum elicited 1:128 HI titer to GDVII virus, the conjugates prepared from this serum did not stain GDVII virus antigen.

Four stains of MHV (MHV-1, MHV-S, JHM, and A-59) were stained in the cytoplasm of MHV-inoculated NCTC-1469 cells. Antigen could be detected as early as 24 hr postinoculation, and the percentage of positive cells increased with the interval postinoculation. With our reagents, the fluorescence associated with MHV-1 and MHV-S was less intense than that of A-59 and JHM. Staining assumed an inclusion-like configuration.

Weanling mice injected (IP) with 10^4 LD_{50} of A-59 virus yielded staining in liver sections at 24 hr postinoculation. Antigen was initially discernible in sinusoidal cells, and later staining was confined to the cytoplasm of parenchymal cells in the foci of necrosis or contiguous to it. This sequence of antigen forma-

TABLE 3 Conjugate Parameters

Antibody To		Species of Serum	Protein Fractionation Method	Final Protein Concentration (mg/ml)	Diluent Tissue Suspension 10–20% (W/V)[a]	Highest Effective Staining Dilution
LCM	RC-9	Guinea pig	$(NH_4)_2SO_4$	14	Mouse brain and liver	1:50
K virus	RC-85	Mouse	DEAE-Sephadex	30	Mouse brain and liver	1:40
	RC-11	Caprine	DEAE-Sephadex	28	Mouse brain and liver	1:5
PVM	5C-2	Mouse	$(NH_4)_2SO_4$	20	Mouse lung	1:40
	RC-131	Caprine	DEAE-Sephadex	33	Mouse lung	1:5
GDVII	RC-101	Mouse	DEAE-Sephadex	6	Mouse brain	1:5
	5C-32	Caprine	$(NH_4)_2SO_4$ and Trypsin	47	Mouse brain	1:20
EDIM	RC-104	Mouse	$(NH_4)_2SO_4$	6	Germfree mouse intestine	1:30
	RC-50	Rabbit	$(NH_4) SO_4$	13	Germfree mouse intestine	1:5
MHV	5C-22	Caprine	$(NH_4)_2SO_4$	20	Mouse liver	1:5
	5C-48	Mouse	$(NH_4)_2SO_4$ ascitic fluid	9	Mouse liver	1:20

[a]In 0.1 M phosphate buffered saline pH 7.2.

tion is consistent with the observations of Boss and Jones.[16] Antigen was readily detected in livers of mice infected by cage contact, and detection of this antigen by impression slides was accomplished with ease.

In an application of MHV staining to natural infections, livers were harvested and sectioned from 20 weanling Balb/C mice from a colony exhibiting serological evidence of MHV infection. Three of the livers exhibited stainable MHV antigen, which was confirmed by suckling mouse inoculation (characteristic liver lesions). The livers exhibiting natural MHV infection differed histologically from experimental infections in that the natural disease exhibited minimal necrosis.

Since in its normal ecology MHV is an enteric agent, the detection of MHV in intestinal impressions was studied. Weanling mice infected *per os* with 10^3 LD_{50} of A-59 virus elicited stainable antigen in intestinal impressions within 24 hours postinoculation. The slides viewed were impressions, and cell types were not established. MHV could be stained in intestinal impressions of mice infected by cage contact, but with a longer lag period.

Inhibition of staining by antigen dilution was most difficult to attain, and inhibition was frequently enhanced by sonication as previously described earlier in this paper.

DETECTION OF PVM ANTIGEN

PVM possesses inherent characteristics unique among murine viruses: It cannot be recovered from organs other than lung, and successful infection is predicated on intranasal inoculation.[17] Based on these observations, detection of PVM antigen by immunofluorescent stain was confined to mouse lung and BHK-21 cells.[18]

Conjugates prepared from PVM-immune goat serum were unsatisfactory although HI antibody was demonstrable (Table 1). Conjugates prepared from mouse immune sera effectively stained PVM antigen *in vivo* and *in vitro*; cytoplasmic localization occurred in all instances. When observed in PVM-infected mouse lung sections, PVM antigen was usually localized in the cytoplasm of alveolar and septal cells within 48 hours of intranasal inoculation. On rare occasions, antigen is present in the cytoplasm of bronchial epithelium. The percentage of positive cells was variable; however, 30–60 percent was the normal range of cells containing antigen (uniformly distributed) (Table 4).

PVM-infected cell culture (BHK-21) demonstrated cytoplasmic antigen in 5-60 percent of cells, depending on multiplicity of infection and interval postinoculation. Staining was originally noted at 48 hr postinoculation and increased quantitatively until it reached a peak at 5 days postinoculation. These kinetics of replication are slower than those reported by Harter and Choppin.[19]

Staining inhibition by antigen dilution could not be produced with prepara-

TABLE 4   Appearance of PVM Virus Antigen in
Mouse Lung After Intranasal Inoculation

Time Postinoculation (Days)[a]	Virus Detection in Lungs	
	H A Titer[b]	IF[c]
1	1:4	0
2	1:4	±
3	1:8	++
4	1:32	+++
5	1:32	+++
6	1:64	+++
7	1:128	+++

[a]Inoculated with 0.05 ml of PVM virus with an HA titer 1:32.
[b]Centrifuged, 10 percent (W/V) lung suspensions heated 70° C for 30 min.
[c]IF = immunofluorescence; 0 = staining not observed; + = 15 percent of cells stained; ± = few cells stained; ++ = 15–30 percent of cells stained; +++ = 3 percent of cells stained.

tions other than sonicated suspensions of PVM-infected mouse lung (uncentrifuged).

DETECTION OF GDVII VIRUS ANTIGEN

GDVII was the only system in which a highly satisfactory immunofluorescent reagent was prepared from caprine serum. However, this success was attained only if the tryptic digestion method of serum fractionation was employed (Table 3).

Since GDVII is an enteric murine virus capable of inducing infrequent encephalomyelitis, efforts to stain antigen were confined to mouse brain and mouse intestinal tract. Antigen was readily stainable in the cytoplasm of neurons at 25 hr postinoculation, with maximal involvement attained at 5 days postinoculation (1,000 LD_{50} IC). A detailed study of GDVII antigen formation by Liu et al.[20] reported antigen stainable in mouse brain to be at near maximal involvement at 2 days postinoculation (IC).

Detection of GDVII virus in intestinal tract impressions, after inoculation per os (10,000 LD_{50}), proved erratic. Virus was consistently observable the first day after inoculation; however during the period between 1 and 12 days postinoculation, invalidating inconsistency was noted between mice sampled. At 2-day intervals the number of positive mice never exceeded two out of five.

Commentary

The primary objective of the study reported here was to ascertain the validity and practicality of the immunofluorescent detection of the common murine virus antigens and to chronicle the methods by which such detection was achieved. Specific sensitive detection was attained with the agents of LCM, MHV, EDIM, PVM, GDVII, and K viruses. In those systems where tissues from natural infections were available, the patterns and degree of antigen deposition were similar to that noted in induced infections.

In most instances immune sera prepared in mice were the most satisfactory for conjugate preparation, with LCM and GDVII systems proving the exception. Highly suitable reagents were prepared from caprine serum in the GDVII system and from cavian serum in the LCM system. When murine sera were utilized, the 50 percent $(NH_4)_2 SO_4$ method of serum protein fractionation proved satisfactory, a phenomenon conceivably attributable to the use of homologous antigens. The tryptic degradation of serum globulins imparted a drastic reduction in nonspecific staining in those systems involving heterogenetic antigens. With the concentration of trypsin and ambient conditions described, the staining activity of the immunoglobulins was selectively resistant.

Acetone fixation proved suitable to all antigens except EDIM, for which 95 percent ethyl alcohol proved superior. Although most of the antigens can be acetone-fixed at room temperature, fixation at $-20°C$ offers the advantage of minimizing the importance of the duration of fixation. Most antigens fixed in acetone for 4–24 hr at $-20°C$ were stable after 6 months storage at $-20°C$. LCM and MHV antigens were less stable and were more durable at $-80°C$. A factor of importance to be considered in the storage of fixed tissue sections or impressions is the phenomenon of increased nonspecific staining as the duration of storage lengthens.

The substitution of tissue impressions for frozen sections sacrifices histologic conformation but not sensitivity of detection. Most satisfactory results are attained if the quantity of tissue adhering to the slide is minimized. The antigens described are nonreactive to immunofluorescent stain when tissues are subjected to formalin fixation.

Immunofluorescent microscopy of viral antigens was most conveniently and validly conducted with a colorless barrier filter and an exciter filter transmitting a portion of the visible spectrum. Orange barrier filters complicate the visual perception of the characteristic green of FITC against the blue-gray of normal tissue autofluorescence. Contrast stains prepared with Lissamine rhodamine are helpful in accentuating the specific green of FITC.[21]

Summary

The direct immunofluorescent technique was applied to the detection of indigenous murine virus antigens, K virus, pneumonia virus of mice (PVM), Theilers mouse encephalomyelitis (GDVII), epizootic diarrhea of infant mice, mouse hepatitis, and lymphocytic choriomeningitis viruses. Virus antigens were discernible, with adequate specificity and sensitivity, in both experimental and natural infections. With the exception of lymphocytic choriomeningitis and mouse encephalomyelitis, the most satisfactory staining reagents were obtained from hyperimmune sera prepared in germfree or specific-virus-free mice. Tryptic digestion of the immunoglobulins prior to conjugation proved to be a useful procedure in minimizing nonspecific staining.

The work was supported by the National Cancer Institute under Contract 43-63-1161.

References

1. Rowe, W. P., J. W. Hartley, and R. J. Huebner. 1962. Polyoma and other indigenous mouse viruses. p. 131–142. *In* R. J. C. Harris [ed.] The problems of laboratory animal disease. Academic Press, New York.
2. Tennant, R. W. 1966. Taxonomy of murine viruses. p. 47–53. *In* R. Holdenried [ed.] Viruses of laboratory rodents. National Cancer Institute Monograph 20.
3. Parker, J. C., R. W. Tennant, and T. G. Ward. 1966. Prevalence of viruses in mouse colonies. p. 47–53. *In* R. Holdenried [ed.] Viruses of laboratory rodents. National Cancer Institute Monograph 20.
4. Rowe, W. P., J. W. Hartley, J. D. Estes, and R. J. Huebner. 1959. Studies of mouse polyoma virus infection. I. Procedure for quantitation and detection of virus. J. Exp. Med. 109:379–391.
5. Parker, J. C., R. W. Tennant, T. G. Ward, and W. P. Rowe. 1965. Virus studies with germfree mice. I. Preparation of serologic diagnostic reagents and survey of germfree and monocontaminated mice for indigenous murine viruses. J. Nat. Cancer Inst. 34:371–380.
6. Sever, J. L. 1962. Application of microtechnique to viral serological investigations. J. Immunol. 88:320–329.
7. Lowry, O. H., N. J. Rosebrough, A. L. Farr, and R. J. Randall. 1951. Protein measurement with the folin phenol reagent. J. Biol. Chem. 193:265–275.
8. Wilsnack, R. E., and W. P. Rowe. 1964. Immunofluorescent studies of the histopathogensis of lymphocytic choriomeningitis infections. J. Exp. Med. 120:829–840.
9. Benda, R. V., L. C. Hronovsky, and J. Cinatl. 1965. Demonstration of lymphocytic choriomeningitis virus in cell culture and mouse brain by the fluorescent antibody technique. Acta Virol. 9:347–351.
10. Pedersen, I. R., and M. Volkert. 1966. Multiplication of lymphocytic choriomeningitis virus in suspension cultures of Earle's strain L cells. Acta Pathol. Microbiol. Scand. 67:523–536.

11. Wilsnack, R. E., J. H. Blackwell, and J. C. Parker. 1969. Identification of an agent of epizootic diarrhea of infant mice by immunofluorescent and complement-fixation tests. Am. J. Vet. Res. 30:1195–1204.
12. Kraft, L. M. 1957. Studies on the etiology and transmission of epidemic diarrhea of infant mice. J. Exp. Med. 106:743–749.
13. Dalton, A. J., L. Kilham, and R. E. Aeigel. 1963. A comparison of polyoma, "K", and Kilham rat viruses with the electron microscope. Virology 20:391–398.
14. Fischer, E. R., and L. Kilham. 1953. Pathology of a pneumotropic virus recovered from C_3H mice carrying the Bittner milk agent. Arch. Pathol. 55:14–19.
15. Fraser, K. B., and E. M. Crawford. 1965. Immunofluorescent and electronmicro-scopic studies of polyoma virus in transformation reactions with BHK 21 cells. Exp. Mol. Pathol. 4:51–65.
16. Boss, J. H., and W. A. Jones. 1963. Hepatic localization of infectious agent in murine viral hepatitis. Arch. Pathol. 76:4–8.
17. Horsfall, F. L., and R. G. Hahn. 1940. A latent virus in normal mice capable of pro-ducing pneumonia in its natural host. J. Exp. Med. 71:391–408.
18. Tennant, R. W., and T. G. Ward. 1962. Pneumonia virus of mice (PVM) in cell culture. Proc. Soc. Exp. Biol. Med. III:395–398.
19. Harter, D. H., and W. C. Purnell. 1967. Studies on pneumonia virus of mice PVM in cell culture. I. Replication in baby hamster kidney cells and properties of the virus. J. Exp. Med. 126:251–266.
20. Liu, C., J. Collins, and E. Shaye. 1967. The pathogenesis of Theilers GDVII en-cyphalomyelitis virus infections in mice as studied by immunofluorescent technique and infectivity titrations. J. Immunol. 98:46–55.
21. Smith, C. W., J. D. Marshall, and W. C. Eveland. 1959. Use of contrasting fluorescent dye as counterstain in fixed tissue preparations. Proc. Soc. Exp. Biol. Med. 102: 179–181.

DISCUSSION

DR. FRITZ: You mentioned ecology and hus-bandry. Would you indicate what you consider to be the usual result of nor-mal transmission from mother to offspring for each of the viruses you mentioned?

DR. WILSNACK: I am not sure I am equipped to answer that question. One that I would be able to respond specifically on is LCM, in which intrauterine transmission is the normal and most efficient method of transfer. However, in the other infections, I would hesitate to specify which would be the most important or which the most specific.

DR. HAYDEN: Are the FA reagents for these viruses commercially available, or must you prepare your own?

DR. WILSNACK: The FA reagents reported here are not commercially available.

THE INFLUENCE OF THE CONDITION OF THE LABORATORY ANIMALS EMPLOYED ON THE EXPERIMENTAL RESULTS

*Vaclav Jelinek**

An essential prerequisite conditioning successful work in the field of experimental biological science is a live experimental object. Experience teaches that one of the traits of a live organism is its variability. This is the reason why an evaluation of the results obtained from experimental work is very often confronted with a wide variety of responses conditioned by quantitative deviations. Since variability is a result of the complex nature of living matter, the individual deviations must be compensated for in two ways: by using samples comprising large numbers of experimental objects (as closely related biologically as possible) and by employing mathematico-statistical methods yielding information about the reliability of the experimental results.

When studying the reaction of live organisms to external conditions, we find that differences observed in a reaction are due in part to individual variability of the organisms, but that a variability of a broader character must also be considered. In animals belonging to the same species the latter variability may be conditioned, apart from other factors, also by acquired traits that develop in connection with the adaptability of the organisms to environmental influences. This kind of variability often goes unnoticed because it is encountered mainly by workers whose task it is to carry out serial quantitative tests. That certain experimental investigations cannot be reproduced may be explained by such variations in the traits of experimental animals. The only means counteracting this kind of variability is a standard mode of breeding, keeping,

*Deceased, February 1970.

and feeding of laboratory animals. In any case, when an experiment is carried out on animals whose previous history is not reliably known, caution is indicated, particularly when quantitative tests have to be evaluated.

For more than 20 years we have been performing quantitative tests of activity of sexual hormones. From 1942 on we repeatedly observed striking variations in the reactions of castrated infant female rats to a specific hormone. These variations were particularly conspicuous when estrogenic hormones were tested for activity. Such tests were always performed in a standard way by the weighing method of Bulbring as described by Burn.[1]

Infantile female rats weighing 30–35 g were castrated, and from the third to the seventh day after castration, they were given daily doses of estrogen, to a total of five doses. On the ninth day the rats were killed, the uteri were removed, fixed for 24 hr in Bouin's fluid, then thoroughly pared and dried, and finally weighed. (The relative size of the uterus is expressed in milligrams per 100 g of body weight.) In all tests we used estradiol benzoate as a standard; its activity was checked regularly once a year by comparison with that of an international standard of estradiol benzoate. The results obtained after administration of the international standard of estradiol benzoate exhibit striking variations in the reactivity of the animals (Figure 1).

For all titrations we used albino rats; in the course of time, as the graph shows, rats kept under three kinds of breeding conditions were used in succession. The white columns denote rats of a breed termed "Žalov" (its domicile), i.e., rats of unknown origin bred in Czechoslovakia over many years. The black columns denote rats from the "Konárovice" breed, descendants of an original Dutch strain, fed with the Hammarsten diet. The lined columns denote Wistar rats, descendants of animals imported in the year 1950 and bred in Konárovice ever since, fed with the Larsen diet. The column height expresses the relative size of the uterus after the administration of estradiol benzoate totaling one international unit. Each column represents an arithmetical mean of relative sizes of five uteri prepared as described before.

On first glance, a different reactivity is evident in rats belonging to each of the three breeds. Further, evidently the reactivity of rats belonging to a single breed varies with time. Apparently, however, the response of the Konárovice breed of Wistar rats is the most uniform and exhibits the least variability within the period investigated. This fact, in our opinion, is explained by the fact that since the year 1951 the rats have been fed with a standard diet, known as the Larsen mixture, both at the farm in Konárovice and at our institute. We assume that this diet, the composition of which has been continually chemically controlled, also contributed to leveling the considerable variability among the rats bred in Konárovice (Dutch strain) and previously fed with the Hammarsten diet. The Hammarsten and Larsen diets differ considerably. The Hammarsten diet consists of mixed feed to which green fodder is added daily. The kind of

111

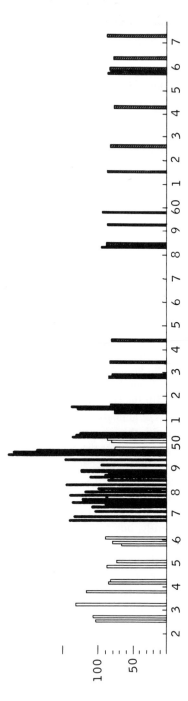

FIGURE 1   Mean relative weights of uteri (mg per 100 g of body weight) of infantile rats after administration of 1 IU of estradiol benzoate (by the Bulbring method). White columns: rats of Žalov breed; black columns: Dutch rats of Konárovice breed fed Hammarsten diet; lined columns: Wistar rats of Konárovice breed fed Larsen diet.

112

fodder added varies according to the season of the year. The Larsen diet is a completely standardized, balanced mixture of concentrated feedstuffs to which no supplements are added. The influence exerted by the diet upon the reactivity of the animals used for hormonal titrations has been described by Rollins and Cole[5] and by Dempsey and Brier.[2]

Figure 1 also illustrates a strikingly elevated reactivity of rats to estradiol benzoate in the year 1949. Analyzing this phenomenon we found that in the spring of that year the rats had suffered from an infectious disease, probably, according to the pathological findings, lymphadenitis, which is described by Klineberger and Steabben.[3] During the course of this disease we observed conspicuous changes of the liver size: first an increase, then a decrease, even in those animals that exhibited no manifest symptoms. An interaction of the infection with the liver function might explain this abnormally elevated reactivity of the rats to the estrogen administered, since it is known that estrogens are inactivated in the liver.

Another phenomenon documenting how external conditions influence the general condition of the laboratory animal has been observed in connection with the transport of animals. When studying the effects of various drugs on the blood count of rats, we saw that the total leukocyte count varied considerably, precluding the production of results that might serve as a satisfactory basis for a quantitative evaluation of the tests performed. In searching for factors that might have caused the considerable variation in the leukocyte count of these rats, we found that the effect of transport can be responsible (Figure 2).

A series of observations, whose results are presented graphically in Figure 2, revealed that the rats transported from the breeding farm in Konárovice to our institute in Prague, a distance of 50 km (= 30 mi), always exhibited higher leukocyte counts per cubic millimeter (black columns) than the rats remaining at the farm (lined columns) and those bred at our institute (white columns), although all these animals belonged to the same Wistar strain. Examinations of the total erythrocyte count proved that rats transported from Konárovice to Prague always exhibited lower values than the nontransported rats did (Figure 3). The tests were always carried out 1 week after the transport. Leukocytosis persisted for 8 weeks and even longer, whereas erythrocytopenia normalized in the course of another fortnight. The connection between the reaction described and the transport has been verified by an experiment on rats previously transported from the breeding farm Dobrá Voda in Slovakia to Prague, a distance of 300 km (= 200 mi). These rats exhibited more marked erythrocytopenia, which persisted for a substantially longer time than it had in rats transported over the shorter distance.

Another effect of transport on these animals was observed when we endeavored to reproduce the method recommended by Sayers and associates[4]

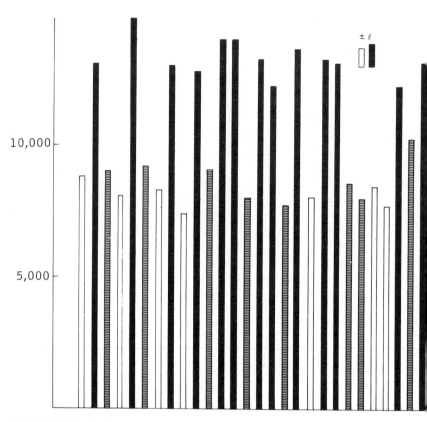

FIGURE 2 Mean leukocyte count per cubic millimeter of blood of Wistar rats (*n* = 10). White columns: rats bred in Prague; black columns: rats bred in Konárovice, 1 week after transport to Prague, a distance of 50 km (= 30 mi); lined columns: rats bred and examined directly at the Konárovice farm.

for a quantitative estimate of ACTH activity. After making several estimates, we found that our results exhibited errors exceeding the limits set by Sayers and co-workers. The error was due to a considerable variance of the ascorbic acid content in the adrenals of the rats studied. Sayers and associates had employed rats belonging to the Long Evans strain. They reported that the adrenals of their rats had total relative weights of 40–50 mg per 100 g of body weight. The adrenals of the male Wistar rats used in our experiments, however, averaged only half those values. A detailed search for the relative weights of adrenals recorded in previous experiments revealed interesting data about the sizes of rat adrenals. These are presented in Figure 4.

The black columns denote mean relative weights (mg per 100 g of body

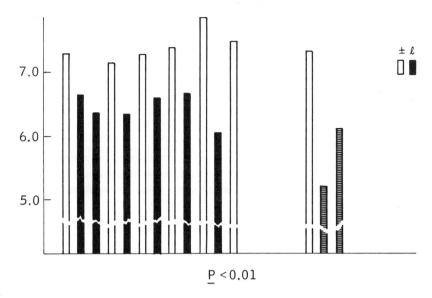

P < 0.01

FIGURE 3 Mean erythrocyte count per cubic millimeter of blood of Wistar rats (*n* = 10). White columns: rats bred in Prague; black columns: rats bred in Konárovice, 1 week after transport to Prague, a distance of 50 km (= 30 mi); lined columns: rats bred in Dobrá Voda, 1 and 2 weeks after transport over a distance of 300 km (= 200 mi).

weight) of both adrenals of male Wistar rats transported from Konárovice. ACTH titrations, involving weighing of the adrenals, were always carried out 14 days after the transport. On first glance a season-dependent variation of the adrenal size is apparent. During the winter, the rats transported from Konárovice have larger adrenals than in summer (always weighed 14 days after transport). The difference between the maximal size values established in summer and in winter is statistically significant. The white columns denote mean relative weights (mg per 100 g of body weight) of both adrenals of male Wistar rats not subjected to transport, i.e., either examined directly at the breeding farm in Konárovice or bred and examined at our institute in Prague. Evidently the season-dependent variation of the adrenal weight is indicated only slightly. In addition, a statistical calculation proved with 90 percent probability an absence of any difference between the respective values established in summer and winter.

Our observations served for improving our subsequent methods. Since those data were obtained, the ACTH activity has been estimated only on nontransported rats. In such cases the estimation error substantially decreased, never exceeding 10 percent, whereas in previous estimates performed on transported rats, the mean error varied between 15 and 20 percent. These experiments

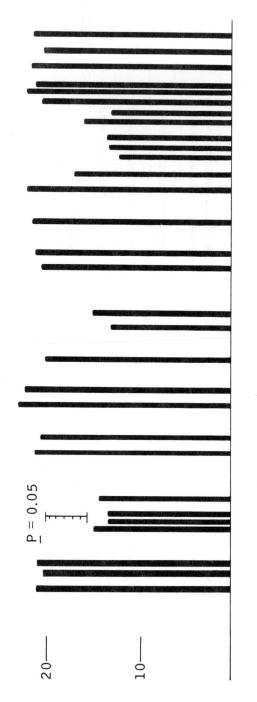

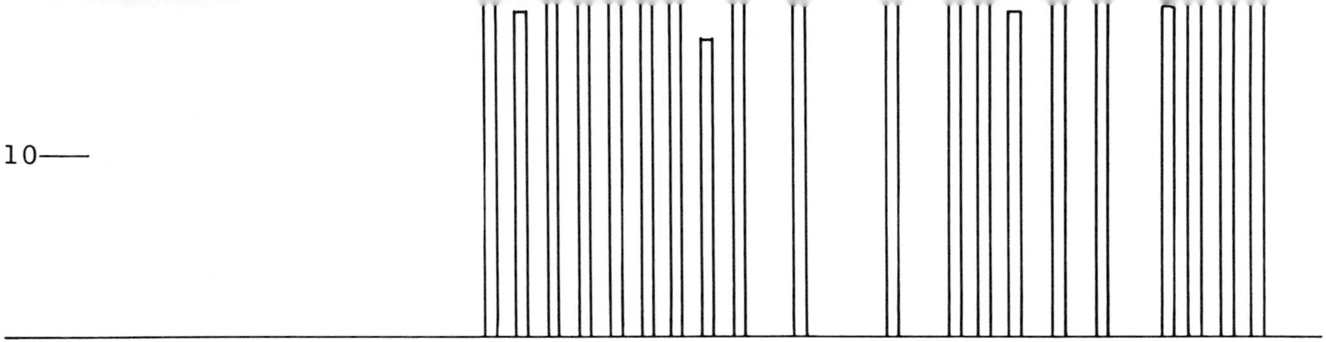

FIGURE 4 Mean relative weights of both adrenals (mg per 100 g of body weight) of male Wistar rats weighing 120–140 g each. Black columns: rats bred in Konárovice and transported 50 km (= 30 mi) to Prague; white columns: rats bred in Prague.

10—

suggest that the size of the rat adrenals might represent a very sensitive indicator of the general condition of this laboratory animal.

Another experiment aimed at establishing the role of current infections affecting the breeds was performed in four groups of male Wistar rats weighing 120–130 g each. Each group comprised 20 animals. One experimental group received intraperitoneally a bouillon culture of *Salmonella typhimurium* at a 10^{-6} dilution, and another experimental group was given a culture of pneumococci at a 10^{-20} dilution. The two control groups received injections of identical amounts of sterile bouillon. The adrenals were removed after 24 hr in one half of the animals of each group and after 60 hr in the other half. (Figure 5).

Both experimental groups exhibited a statistically significant increase in the size of the adrenals. The mouse typhoid infection took its course without manifest symptoms of the disease. The rats with the pneumococcal infection became inert 60 hr after the infection, and three of them died within that time. In the light of these observations, the rat may be said to react, by an increase in the size of the adrenals, to either infection.

The above observations are supplemented by findings from the year 1949 that were obtained from investigations of an epizootic disease that afflicted our breed of rats. Post-mortem findings suggested lymphadenitis, which is described by Klineberger and Steabben (3). In the rats thus affected a considerable enlargement of the adrenals was also observed, even in those animals that did not have manifest symptoms of the disease. Another supplementary contribution are the data compiled on the relative weights of adrenals found in male Wistar rats in whose livers cysticerci of *Taenia crassicollis* were revealed at autopsy. This parasite infected our rat breed in the year 1956 after a short-term breakdown of autoclaves designed for sterilization of the bedding material. In this case also a rather significant increase of the adrenal size was observed.

All these observations suggest that the relative size of the adrenals, regularly checked, represents a useful indicator of the general condition of the rat. The condition of the adrenals may significantly affect the reactivity of the experimental animal.

The condition of the adrenals may, under otherwise favorable experimental conditions, cause a wide variance of the response measured, and therefore, a considerable error of the result. Under less favorable circumstances, the variable quality of the experimental animals could invalidate an entire experiment. To ensure reliable results many aspects of the laboratory animal, its housing conditions, its diet, and other important circumstances must be checked and evaluated carefully in the experimental laboratory.

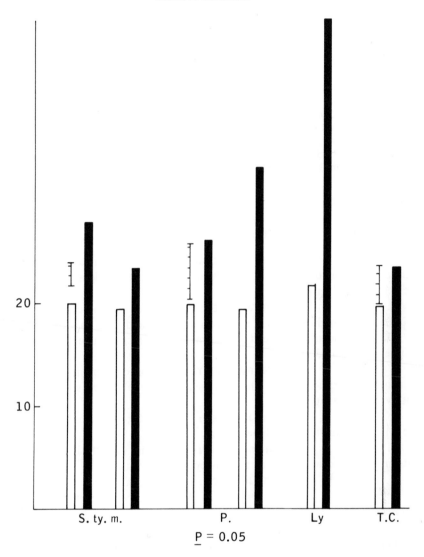

FIGURE 5   Mean relative weights of both adrenals (mg per 100 g of body weight) of male Wistar rats bred in Prague weighing 120–140 g each. White columns: controls; black columns: infected animals. Key to abbreviations: *S.ty.m.*, after intraperitoneal injection of *S.typhimurium* culture at 10^{-6} dilution. *P*, after intraperitoneal injection of pneumococcal culture at 10^{-20} dilution. *Ly*, during epizootic of lymphadenitis. *T.c.*, rats in which autopsy revealed cysticerci of *Taenia crassicollis* in liver.

References

1. Burn, J. H. 1937. Biological standardization. Oxford University Press London, p. 143–146.
2. Dempsey, E., and R. Brier. 1952. Effect of dietary changes on uterine size. J. Clin. Endocrinol. 12:244.
3. Klineberger, E., and D. B. Steabben. J. Hyg. Comb. *37*, 143, 1937. Cited by: The UFAW Handbook on the Care and Management of Laboratory Animals. London 1947.
4. Sayers, M. A., G. Sayers, and L. A. Woodbury. 1948. The assay of adrenocorticotrophic hormone by the adrenal ascorbic acid depletion method. Endocrinology 42:379.

DISCUSSION

DR. GRAFTON: Was there a significant change in altitude between your farm and the institute in Prague that might have increased stress of transport?

DR. JELINEK: In the case of the adrenals, we could not see any differences in the rats transported and those that were not. Our farm and the institute are at approximately the same altitude. There is a difference of only 300 m.

DR. POOLE: You mentioned the hematologic effects of transporting animals. What was the age of the rats? Were such effects observed only in rats?

DR. JELINEK: The age of the rats transported and the rats in our institute was three months.

PROF. SIMIONESCU: What were the histological patterns of the adrenals on their season-dependent variations?

DR. JELINEK: It was not examined.

DR. BECKER: Was there any difference in bedding or insecticides in the transportation?

DR. JELINEK: No, there was no difference in the bedding of the animals transported from our farm to our institute.

COMPARATIVE BIOLOGY IN THE SELECTION OF EXPERIMENTAL SUBJECTS FOR CARDIAC PATHOPHYSIOLOGIC INVESTIGATIONS

Sol M. Michaelson
Bernard F. Schreiner, Jr.

Introduction

More than a century has passed since Claude Bernar‹ (1865) stated,

The constancy of the milieu interieur, (the internal environment) is a necessary condition of free life . . . . all the vital mechanisms, varied as they are, have only one object, that of preserving constant the conditions of life in the internal environment. Experiments made on animals from the physiological, pathological, and therapeutic points of view have results that are applicable to theoretic medicine. Without such comparative study of animals practical medicine can never acquire a scientific character. If animals did not exist, man's nature would be still more incomprehensible.[4]

He goes on to say,

Two things must be considered in the phenomenon of life, first the fundamental properties of vital units which are general, then arrangements and mechanisms and organizations which give each animal species its peculiar anatomical and physiologic form. Among all the animals that may be used some are better suited than others.

Although biologists have presumably been aware of these principles, they have repeatedly failed to take cognizance of them in investigations of animals in which extrapolation to man has been attempted. All too often the selection of a particular animal has been dictated by economy and convenience, or by the personal prejudices of the investigator.

A comparative approach in which mammalian species are compared to and contrasted with man in terms of morphological, physiological, biochemical, and pathological characteristics would appear to afford the likeliest avenue to success from an anthropomorphic point of view. In this review we attempt to present a brief survey of animal models in general and to develop this point of view as it relates to cardiac pathophysiology in terms of functional anatomy, myocardial infarction, and congestive heart failure. In addition to man, comparative data will be reviewed where available in the simian primate, dog, pig, horse, rabbit, rat, and ox.

Animal Models

The comparative approach includes the use of animals with diseases similar to those of man that occur spontaneously or that can be induced experimentally. It also includes the use of those animals that have anatomic, physiological, or biochemical characteristics and integrative or coordinating mechanisms that are most closely comparable to those of man.

In recent years, cybernetics and bionics have provided tools for the investigation of biological phenomena in terms of models or analogs. While these may offer certain mechanistic advantages, functional simulation is not always feasible because of the complexity of the living system which the analog represents. The integrative capacity of the living organism cannot easily be put into a one to one correspondence with elements of a physical system. Furthermore, not all the functions of a living system are capable of performance by an analogous physical system. For example, in living systems some cell activities take place irreversibly and an equilibrated thermodynamic situation may never be reached. Likewise, chemical reactions *in vivo*, unlike those *in vitro*, take place in an open system in which both energy and matter are exchanged with the external environment.[29,47]

Little doubt remains that the use of a living model for study of biologic interactions has the unique advantage of a system endowed with a homeostatic capability. In this system integrated feedback mechanisms are maintained for the control of vital functions extending from intracellular processes to those including the total organism.

To serve its purpose well an animal model must be predictable. An ideal model should be one in which the human disease is reproduced in its entirety within a given time period. Barring the latter, the investigator must attempt to telescope the full process into a shorter period of time or to divide it into component parts amenable to separate segmental study. Since experience so far has failed to discover a single animal species as an "ideal model" and since no experimental method is universally suited for all comparative studies, the investigator must be conversant with the desirable attributes and limitations inherent in the species selected for study.

SOL M. MICHAELSON
AND
BERNARD F. SCHREINER, JR.

The comparative approach can identify a variety of available animal models categorized as experimental or spontaneous.

EXPERIMENTAL MODELS

Models in which attempts are made to reproduce in animals a disease process found in man are termed "experimental." This type is probably the most susceptible to the pitfalls of species selection. A typical example is experimental atherosclerosis in the rabbit, which was subsequently applied to the pigeon, chicken, rat, guinea pig, dog, pig, and simian primate with varying degrees of success.

SPONTANEOUS MODELS

Animal models in which natural diseases that are the counterparts of diseases in man occur are termed "spontaneous." Among these are congenital heart disease in many animals, especially dogs, atherosclerosis in swine and simian primates, congestive heart failure in the dog, high altitude ("Brisket") disease in cattle, and "heaves" in horses as a model for pulmonary emphysema in man.

Cardiac Structure

Detailed descriptions of the coronary circulation in man and other animal species and other aspects of comparative cardiology are available.[8-10,31,34,43,44,60,61,64,75,100,106]

While the gross anatomical features of the cardiac chambers (except for size) are similar within certain limits in most species, the size and distribution of the coronary vasculature differs significantly. This is particularly true of the blood supply to the left ventricle, the interventricular septum, and the specialized pacemaker and conducting tissues. The coronary arterial tree differs in various species in important respects as shown in Table 1.

In the dog the left circumflex coronary artery (LCxCA) gives rise in most cases to the posterior septal and left posterior descending coronary artery (LPDCA). The anterior septal artery is a distinct structure whereas in man 4–6 septal branches arise from the left anterior descending coronary artery (LADCA). The former vessel arises at the division of the anterior descending and the left circumflex arteries where it forms a vessel usually of similar caliber to the other two branches. The right coronary artery (RCA) apparently contributes very little to the normal blood supply of the midportion of the posterior wall of the heart.[60,61] The entire left anterior descending coronary artery and the left circumflex coronary artery do not supply any significant portion of the right ventricle. The posterior descending coronary artery arises from

TABLE 1 Coronary Arterial Tree[a]

Coronary Arteries—Branches	Man	Dog	Pig	Ox	Horse
Left coronary artery					
(LCA)					
Left anterior descending					
(LADCA)	+	+	+	+	+
Anterior septal branches	4–6	–	+		
Left circumflex					
(LCxCA)	+	+	+	+	+
Septal artery	–	+ (1)	±	+ (2)	+ (2)
Left posterior descending					
(LPDCA)	10%	100%		80%	
Right coronary artery					
(RCA)					
Right marginal branch	+	+	+		
Right posterior descending					
(RPDCA)	90%	–	+	+	+
Posterior septal branch	+	–	+		
Right circumflex					
(RCxCA)	+	+	+	+	+

[a](+) = structure present; (–) = structure absent; (%) = structure present in specified number per 100 cases examined.

the left circumflex coronary artery in nearly 100 percent of specimens, whereas in man this artery has a similar origin in approximately 10 percent of specimens.[43]

In the pig the left anterior descending coronary artery gives rise to an anterior septal branch approximately 0.8 cm from the ostium of the left coronary artery (LCA). This originates on the posterior aspect of the vessel and is approximately 30 percent of its size. The right coronary artery provides the main supply to the posterior surface of the heart. It forms the right posterior descending coronary artery (RPDCA) and gives off the posterior septal branch.[61] The right circumflex coronary artery (RCxCA), unlike that of the dog, terminates by descending in the longitudinal sulcus of the diaphragmatic surface of the heart; it also gives off a large descending branch to the central portion of the right ventricle.

In addition to the descending circumflex and septal branches, a large artery leaves the area of the junction between the descending and the circumflex arteries to ramify on the surface of the left ventricle. The termination of the circumflex branch of the left coronary artery in the pig compares favorably with that of the right coronary artery in the dog; it ends in numerous small branches in the region of the coronary sinus.[10]

SOL M. MICHAELSON
AND
BERNARD F. SCHREINER, JR.

In the ox the left anterior and posterior descending coronary arteries come from the left coronary artery. Typically, there are two septal arteries arising one centimeter apart that extend one half to all the way through the septum. There is an equally large artery that arises from the posterior descending coronary artery to supply the atrioventricular node of the interventricular septum.[9]

In the horse the left anterior descending coronary artery arises from the left coronary artery; the right coronary artery gives off the posterior descending coronary artery. Usually there are two septal arteries that extend to the opposite sides of the interventricular septum. Frequently one or two branches from the posterior descending coronary artery are as large as the septal artery.[9,10]

In the dog the atrioventricular node artery is almost always derived from the left coronary artery, whereas in man this origin is found only in 10 percent of specimens. The sinus node artery is derived from the right coronary artery in over 95 percent of the cases, while this source gives rise to the node artery in only 54 percent of cases in man[43] (Table 2).

There are differences in the blood supply to the interventricular septum of animals in comparison to man. These differences are a function of the relative importance of the septal artery, the penetrating branches of the anterior descending coronary artery and the posterior descending coronary artery, and the origin of the posterior descending coronary artery (Table 3).

In the dog, pig, ox, and horse, arterial anastomoses are readily discernible within the septum. The interventricular septum receives most of its blood supply from the left anterior descending coronary artery. The septal artery, a large penetrating artery that arises at or near the junction of the left anterior descending coronary artery and the left circumflex coronary artery, may be absent, single, or paired; it supplies less blood to the interventricular septum than do the left anterior descending coronary and posterior descending coro-

TABLE 2 Arterial Supply to Conduction Tissue

Species	Atrioventricular Node		Sinus Node			
	RCA (%)	LCA (%)	RCA (%)	LCA (%)	RCA+LCxCA (%)	Unknown (%)
Man	90 (discrete)	10	54	42	2	2
Dog	−	100 (several small branches)	95	−	−	−
Pig	>50	−	>50	−	−	−

TABLE 3 Regional Blood Supply to Heart[a]

| Animal Species | Free Ventricular Wall | | | | Right Ventricle | | | Interventricular Septum Arteries |
| | Left Ventricle | | | | | | | |
	Anterior	Posterior	Apex	Free Wall	Anterior	Posterior	Free Wall	
Man	LCA	LCxCA RCA LADCA (terminal)	LADCA (terminal)	LCA 90% RCA 30% (82% of hearts)	RCA (2/3) LADCA (1/3)	RCA	RCA 70% LCA 30% (92% of hearts)	4–6 branches from LADCA LADCA (Anterior 2/3) PDCA (Posterior 1/3) usually from RCA
Dog	LCA	LCA	LADCA	LCA (100% of hearts)	RCA	RCA	RCA	LADCA (4/5) PDCA (1/5) (from LCA) Septal
Pig	LCA	LCxCA RCA	LADCA PDCA		RCA	RCA	RCxCA	LADCA (2/3) PDCA (1/3) (from RCA)
Ox								LADCA (3/4) PDCA (1/4) (from LCA) Septal
Horse								LADCA (2/3) PDCA (1/3) (from RCA) LCA Septal

[a]LCA = left coronary artery; LCxCA = left circumflex coronary artery; RCA = right coronary artery; LADCA = left anterior descending coronary artery; PDCA = posterior descending coronary artery; RCxCA = right circumflex coronary artery.

nary arteries. Approximately 50 to 90 percent of the blood to the interventricular septum comes from the branches of the left anterior descending coronary artery, and the remainder comes from the posterior coronary and adjacent branches of the circumflex and marginal arteries.[9] In man, the interventricular septum is supplied by 4 to 6 anterior septal branches from the left anterior descending coronary artery.[43] In the dog the left anterior descending coronary artery contributes more blood to the interventricular septum than it does in other species.[10]

In the pig, the septal artery may be missing. Anastomoses between anterior and posterior penetrating arteries are common in the midportion of the interventricular septum. Anastomoses are most frequently located between penetrating branches of the anterior and posterior descending coronary arteries in the midportion of the interventricular septum, between the posterior descending arteries and the branches of the right and left marginal arteries. Both the anterior and the posterior descending coronary arteries help supply the interventricular septum.[9]

Normal conduction of cardiac impulses is vitally dependent upon adequate blood supply to the sinus and atrioventricular nodes and to the bundle of His and its branches in the interventricular septum and ventricular myocardium. Occlusion of arteries due to naturally occurring disease or experimentally produced coronary occlusion will interfere with the conducting system and the area of myocardium perfused in different ways in various species. For example, in man and the pig occlusion of the right coronary artery may result in atrioventricular block as well as infarction of the diaphragmatic surface of the left ventricle, whereas in the dog a similar occlusion may affect the sinus node but result in no significant atrioventricular conduction disturbance or infarction of significant portions of the left ventricle.[61]

Cardiac Function

Rather strikingly similar anatomical and functional patterns exist for mammals over a wide range of size from the rat to the horse. Heart weight, components of the heart volume, and heart rate are related to body weight for a large number of species (Tables 4 and 5).

In general, the stroke volume is proportional to heart volume and inversely proportional to heart rate, which is an intrinsic characteristic of cardiac pacemaker tissue. Heart rate is proportional to 0_2 consumption adjusted for body weight. Because of the inertia of blood, adequate ventricular filling may not take place if tachycardia encroaches too severely upon the diastolic interval. Thus, there is a maximal heart rate for all animals beyond which further increase leads to decrease of cardiac output.[96]

TABLE 4 Physiologic Characteristics of Mammals

Species	Common Name	Body Weight (kg)	Surface Area (m²)	Energy Metabolism[a]		Blood Volume (ml/kg of body weight)
				(Cal/kg/day)	(Cal/m²/day)	
Rattus sp	Rat	0.1–0.5	0.03–0.06	120–140 (B)	760–905 (B)	54
Oryctolagus sp	Rabbit	1–4	0.23	47 (B)	810 (B)	56
Macaca sp	Simian primate	2–4	0.31	49 (B)	675 (B)	
Canis *familiaris*	Dog	5–31	0.39–0.78	34–39 (B)	770–800 (B)	87–114
Homo *sapiens*	Man	54–70	1.65–1.83	23–26 (B)	790–910 (B)	65–80
Sus *scrofa*	Pig	100–250	2.9–3.2	14–17 (B)	1,100–1,360 (B)	45–65
Bos *taurus*	Ox	500–800	4.2–8.0	15 (B)	1,635 (B)	70–120
Equus *caballus*	Horse	650–800	5.8–8.0	25 (R)	2,710–2,770 (R)	52–81

[a]B = basal; R = resting.

TABLE 5 Physiologic Characteristics of Mammals

| Species | Cardiac Function | | | | | Arterial Blood Pressure (mm Hg) | |
	Heart Weight (g/100 g)	Heart Rate (beats/min)	Stroke Volume (ml/beat)	Cardiac Output (L/min)	Cardiac Index (L/m²/min)	Systolic	Diastolic
Rattus sp	0.24–0.58	250–400	1.3–2.0	0.015–0.079	1.6	88–184	58–145
Oryctolagus sp	0.19–0.36	123–330	1.3–3.8	0.25–0.75	1.7	95–130	60–90
Macaca sp	0.34–0.39	165–240	8.8	1.06		137–188	112–152
Canis familiaris	0.65–0.96	72–130	14–22	0.65–1.57	2.9	95–136	43–66
Homo sapiens	0.45–0.65	41–108	62.8	5.6	3.3	92–150	53–90
Sus scrofa	0.25–0.40	55–86	39–43	5.4	4.8	144–185	98–120
Bos taurus	0.31–0.53	40–58	244	14.6		121–166	80–120
Equus caballus	0.39–0.94	23–70	852	18.8	4.4	86–104	43–86

The heart size is, over the full range of mammalian size, a constant fraction of body weight, just as the lungs are, although in very small animals there is a tendency for the heart to be larger than predicted. Logarithmic plots show small deviations from the general trend, but, in general, animals capable of severe and prolonged exercise have heart ratios greater than 0.6 (heart ratio = heart weight × 100/body weight), while animals incapable of heavy, steady work have ratios less than 0.6.[6,11]

Cardiac output—the product of stroke volume and heart rate—is directly related to metabolic rate and is proportional to body surface area. Ventilatory minute-volume and cardiac output are both proportional to oxygen consumption at rest. Hence, the ratio of air flow to blood flow is invariant between species.[96] The relationship between the internal radius and the wall thickness of the left ventricle and normal ventricular systolic pressure (130 ± 30 mm Hg) are approximately the same for nearly all mammals.[11,41]

The arterial blood pressure at rest is about the same in all mammals, and the mean pressure probably does not change much with exercise. The product of total peripheral resistance and cardiac output remains the same between species.[96]

For all mammals the relationship of various cardiovascular parameters to body weight (BW) has been described by Kines and associates[49] and is shown in Table 6.

Various mammalian species can be grouped according to the time order of ventricular activation.[39,70,83] In man, monkey, dog, cat, and rat, the ventricles are excited with three general "fronts" of depolarization: (1) initial depolarization (during the first 5 to 10 msec of QRS) of an endocardial shell surrounding the apex of the left ventricle, although simultaneously the interventricular septum is excited from the right ventricular endocardium toward the left;

TABLE 6 Cardiovascular Parameters in Relation to Body Weight[a]

Parameter	Body Weight (BW)
Metabolic rate (MR)	$70 \, BW^{0.75}$
Stroke volume (SV)	$1.1 \, BW^{1.0}$
Heart rate (R)	$203 \, BW^{-0.25}$
Cardiac output (CO)	$223 \, BW^{0.75}$
End-diastolic volume (EDV)	$2.3 \, BW^{1.0}$
End-systolic volume (ESV)	$1.2 \, BW^{1.0}$
Blood volume (BV)	$0.55 \, BW^{0.99}$

[a]Adapted from Kines et al., reference 49.

(2) depolarization of both ventricular free walls (during the next 15 msec of QRS) from subendocardial terminations of the Purkinje fibers toward the epicardium; (3) terminal depolarization of the bases of both ventricles and of the interventricular septum (during the last 5 msec of QRS) in a general apico-basilar direction.

In the horse, cow, pig, sheep, and goat ventricular activation proceeds with only two general fronts of depolarization[21,35]: (1) initial activation, (2) terminal depolarization of the middle and basilar thirds of the interventricular septum (during the final 40 msec of QRS) in a general apicobasilar direction, although during activation of the middle third, two small volumes of myocardium, one comprising the extreme epicardial base and the other near the apex of the left ventricle, are activated simultaneously in a subepicardial to epicardial direction.

It is likely that the two general pathways of ventricular activation responsible for the two categories of body surface potentials result from differences in distribution of Purkinje fibers in the intramural environment of both ventricular free walls.[36]

In the horse, cow, sheep, goat, and dog the ejection phase of the right ventricle is longer than that of the left ventricle. The mitral valve, therefore, closes earlier than the tricuspid. The isometric contraction phase of the left ventricle definitely lasts longer than that of the right ventricle. Thus, the opening of the pulmonic valve ordinarily precedes that of the aortic valve.[89]

There are several cardiopathological disturbances in man for which animals are available with comparable disease states or in which such disturbances can be induced by manipulation. For the sake of brevity, however, only studies of myocardial infarction and congestive heart failure will be discussed to illustrate the concepts and criteria for selection of experimental subjects for biomedical investigation of disease complexes.

Because, historically, the dog, and more recently, the pig have been the animals of choice in cardiovascular studies, emphasis will be placed on these two species in comparison with others.

Myocardial Infarction

Myocardial infarction, which is the most frequent complication of atherosclerosis in man, is now being intensively investigated. Myocardial infarction is a dynamic syndrome characterized as a series of events, the essential feature of which is myocardial necrosis, with or without obvious hemodynamic impairment, which results from either sudden or gradual interference with the blood supply to the area involved. The definition can be applied to all similar pathogenic mechanisms leading to myocardial infarction such as coronary ligation, embolism, thrombosis, atherosclerotic stenosis or occlusion, infectious vasculitis or altered immune response (collagen disease, allergy), as well as obstruction of the coronary sinus outflow tract. There are a limited number of manipulations that, in effect, may lead to obstruction of the intramyocardial coronary circulation, i.e., through the arterioles and capillaries of the myocardial bed. Such obstruction may be equivalent to obstruction of a major coronary artery.[92]

The sequence of pathophysiologic events in myocardial infarction is not well understood. Assessment of the sequence of events requires study of the actions of the infarcted heart under controlled conditions. Since such controlled conditions are extremely difficult to achieve in man, appropriate animal models are required.

The relative incidence of severe coronary-artery disease and myocardial infarction in animals is extremely low. The availability of an animal that develops myocardial infarction comparable to that of man either spontaneously or by manipulation, especially an animal large enough to permit surgical procedures and hemodynamic assessment to be performed with reasonable ease, would be a great asset.

Availability of animal models would permit (1) evaluation and correlation of reproducible pathologic, biochemical, and hemodynamic events of myocardial infarction; (2) analysis of pathologic changes and alterations in tissue metabolism; and (3) study of relationship of changes in pathophysiologic or biochemical parameters to hemodynamic events (e.g., the development of cardiogenic shock or congestive cardiac failure) or to the development of cardiac arrhythmias.

Although there are several reports that are suggestive of myocardial infarction in laboratory animals, pets, or domestic animals, the diagnosis and characterization of the lesion in many of these reports is subject to question.

Interstitial myocardial and focal myocardial necroses are not uncommon in old rats. Interstitial myocarditis is manifested by the presence of edema and infiltrating lymphocytes between the fibers of the myocardium, while focal myocardial necrosis is characterized by the accumulation of lymphocytes in round foci within the areas of necrosis. Spontaneous myocardial infarction apparently has not been reported in the rat.[107]

Systematic production of focal myocardial necrosis, referred to as infarcts and infarctoid necrosis, has been induced in rats by means of a highly saturated-fat diet with added thiouracil and sodium cholate.[97-99] It is noteworthy, however, that in one study, 15 to 20 percent of rats with myocardial lesions had associated thrombosis in the corresponding coronary artery, which showed minimal, or more frequently, no evidence of atherosclerotic change.[99]

SOL M. MICHAELSON
AND
BERNARD F. SCHREINER, JR.

Spontaneous atherosclerosis, myocarditis, and myocardial infarction have been reported in simian primates.[52,73,74,80] Severe atherosclerosis and occasional myocardial infarction after prolonged exposure to high levels of dietary fat has been reported in several species of simian primates.[66,95,108]

Dogs develop a variety of chronic arterial lesions that increase in prevalence and severity with advancing age.[1,12,13,50,57,67] Sclerotic lesions in the intramural and extramural coronary arteries, the aorta, and the pulmonary arteries are described.[53] Lipid accumulations ordinarily are absent or minimal in canine arteriosclerosis.[58]

Although fibrotic valvular changes are observed in hearts from aging humans,[65,71,72] simian primates,[51] and various domestic mammals, it appears that the incidence of severe valvular distortion and atrioventricular insufficiency is much greater in aged dogs than in other species, including man.[5,93,94] Characteristically, the atrioventricular valves, especially the mitral valves, are involved. Mitral stenosis in the dog is rare.[102]

Depending on the degree of valvular incompetence and the duration of the process, cardiac hypertrophy develops in dogs. Intimal thickening associated with hyalinization, which reduces the lumen is observed. The hyalinization, which could be amyloid, may involve the entire wall. Smooth muscle cells are frequently enlarged and increased in number. The basic process is not one of fibrosis. The hyalinization and amyloidosis of small arteries and arterioles with cellular proliferation is identical to that seen in man.[102]

Hyperlipemic dogs exposed to extraordinary stress have developed myocardial infarction.[88] Although there is a relatively large body of information concerning cardiovascular disease in the dog, specific reference to the occurrence of spontaneous myocardial infarction in this species is rare.[17,20,53,76]

In recent years both domestic and miniature swine have been widely used in cardiovascular research. Various investigators have shown that the cardiovascular system and arterial lesions (intimal thickening of coronary arteries and aorta) in swine bear a very close resemblance to those in man.[26,45,59,78,84,85]

Atherosclerosis is described in swine without reference to myocardial lesions.[13,27,28,30,68,110] Luginbühl,[56] however, does report myocardial infarction in swine with spontaneous atherosclerosis.

In horses, myocardial infarction usually in the right atrium is caused by infestation with *Strongylus vulgaris* larvae.[24]

In man, infarcts confined to the upper part of the interventricular septum are seen with considerable frequency at necropsy. Diffuse, generalized, and coronary atherosclerosis may be present, but there may be no evidence of complete occlusion of the arteries supplying the conduction tissue. Destruction of conduction-tissue fibers may not be complete, and occasional surviving cells are seen.[61]

Congestive Heart Failure

Congestive heart failure has been described in dogs, swine, horses, and cattle. Spontaneous heart disease resembling human chronic congestive heart failure affects many dogs; the causes may be dirofilariasis, acquired chronic mitral valvular and myocardial disease, or congenital cardiovascular defects. The cardiac lesion most commonly detected in both clinical and post-mortem examination is chronic valvular disease (fibrosis of the atrioventricular valves), affecting primarily the mitral valve and causing mitral insufficiency, which is associated with various forms of myocardial disease.[48,103]

Heart failure usually results from mitral insufficiency and associated myocardial disease. Hypertrophy and dilation of the left ventricle and atrium occur. Signs and symptoms of left-ventricular failure occur and may be followed by right-ventricular failure. Most cases of combined ventricular failure appear to begin with left-sided decompensation. Another type of left-sided failure is caused by subaortic stenosis. The left-ventricular myocardium in these hearts contains widespread focal areas of necrosis and fibrosis, and advanced intramural arteriosclerosis is present; this eventually leads to generalized heart failure. Right-sided failure unassociated with signs of left-sided failure is most often caused either by congenital lesions, such as pulmonic stenosis, which primarily affect the right heart, or by pulmonary hypertension secondary to dirofilariasis. The remaining cases are caused by cardiac tamponade resulting from pericardial effusion, which may be associated with heart-base tumors, metastatic neoplasms, or an unknown cause.[18,19]

Post-mortem studies indicate that chronic congestive heart failure in aging dogs is associated with progressive fibrosis of the atrioventricular valves, focal areas of myocardial necrosis and fibrosis, and intramural coronary arteriosclerosis. How often the myocardial changes are attributable to ischemia resulting from arterial disease has not been established. The usual result of the valvulaɪ lesions is mitral insufficiency. Atrioventricular valvular stenosis is extremely rare, and clinically significant acquired disease of the semilunar valves is seldom found.[14,15,17]

Some dogs are able to regulate their cardiovascular hemodynamics for years; however, with advancing age or duration of the disease the intensity of myocardial fibrosis and necrosis increases.[19]

Plözlicher Herztod, a disease state in swine is characterized by sudden cardiac failure. Necropsy reveals myocardial degeneration, myocarditis, and pulmonary edema; collapse of the thyroid follicles is also noted. The pathogenesis of the condition is still controversial; there is conflicting opinion as to whether this is a primary disease of the heart, a disease of striated muscle, a manifestation of a generalized vascular disorder, or the result of thyrotoxicosis.[16,62]

SOL M. MICHAELSON
AND
BERNARD F. SCHREINER, JR.

Experimental Procedures

The time period required to induce severe athero-
sclerosis, the unpredictability of atherogenic infarction, and the uncertainty of
its time of occurrence lessen the feasibility of using animals with infarction
caused by atherosclerosis alone. In addition, the species most susceptible to
severe coronary atherosclerosis and infarction (the rabbit) is of such small size
that the application of some investigative techniques is difficult or impossible.[46]

For almost a century, ligation of a dog's coronary artery has been a standard
procedure in coronary research. This technique, however, results in variable
responses, depending upon the size of the artery, the site and speed of occlusion, and the anatomy and condition of other major arteries and of the myocardium. In some studies, where occlusion of the left anterior descending
artery has been the only procedure, mortality has varied from 50 to over 90
percent, largely due to uncontrollable arrhythmias, and the sizes of the infarcts produced have been highly variable.[25,32]

Other techniques have been developed in an attempt to produce myocardial infarction in animals that would be more comparable to that which
develops in man. These include the placement of snares around arteries for
subsequent manipulation[81] and implantation of materials to cause arteritis.
Ameroid, a hygroscopic casein plastic, which swells when in contact with
tissue fluids, can be placed around a major coronary artery, and since the
outside diameter is fixed, the artery is narrowed. By altering the porosity one
can grossly regulate the rate of restriction of the coronary flow.[54]

Gradual occlusion of a coronary artery may result in the development of
an extensive collateral circulation. This situation may be advantageous in some
types of investigations. For example, when one is interested in comparing
subsequent infarction to infarction in man with previously advanced coronary
atherosclerosis, this type of experimental model is particularly useful. Similarly, the evaluation of a controlled exercise program as a therapeutic intervention in ischemic heart disease may be implemented by taking advantage of
knowledge gained in the dog from a similar intervention since rich collateral
anastomoses are a fairly constant feature of human hearts with advanced coronary atherosclerosis.

All these techniques require thoracotomy, which may compromise functional evaluation of the pathogenesis and development of the infarct and which
may lack regional cardiac specificity. Neural and lymphatic pathways are
altered, vigorous postoperative care is essential, and these techniques result
in a fairly high early mortality to useful preparation ratio.

Closed-chest techniques to produce coronary occlusion, ischemia, myocardial infarct, and cardiogenic shock are being evaluated. Multiple small emboli
to the coronary arteries, such as lycopodium spores,[33] microspheres of glass

or plastic,[2,42] or large emboli[37] are introduced into the coronary ostia by selective retrograde arterial catheterization. Wire conductors for electrical or thermal energy that induce thrombus formation have been used.[82] Adenosine diphosphate (ADP) has been infused into major coronary arteries to produce occlusive platelet aggregations and infarction of the myocardium.[79]

These techniques produce obstruction of small vessels, and although attempts have been made to regulate the size and location of the coronary obstruction, desired results have not been uniform. The distribution of the change is patchy, unpredictable, and dissimilar to the lesions usually seen in man; the interruption of flow is acute, and collateral circulation has no time to develop.

It has been known for a long time that exposure to penetrating ionizing radiation results in fibrosis of arteries and veins similar to that which occurs in physiologic aging.[86] Direct exposure of the myocardium of man and animals to x-irradiation results primarily in disturbances of the fine vasculature, with secondary effects on connective tissue and indirect effects on the heart muscle itself, resulting in obliterative vascular changes with aseptic necrosis and hyalin fibrosis, which produces a characteristic picture of cardiac damage as a consequence of tissue ischemia.[38,104,105]

Cobalt-60 or x-irradiation of the heart of dogs results in myocardial necrosis at the irradiation site.[69,91] Macroscopically, the cardiac necrosis is strikingly similar to vascular infarction in the human myocardium. By varying the total radiation dose to produce extensive left-ventricular infarction, severe left-ventricular dysfunction with acute necrosis resembling that of acute ischemic myocardial infarction in man can be produced in 48 hours. The latency period for development of the infarct can be varied to encompass periods comparable to those of abrupt cardiac injury or progressive development of infarction, which would permit study of the sequential stages of progressive heart failure, thus providing time for serial hemodynamic measurements.

Lluch et al.[55] have recently described selective embolization of the circumflex coronary artery with 0.2 ml of metallic mercury to produce infarction of the posterolateral wall of the left ventricle in dogs, with cardiogenic shock and death in 5 to 48 hours.

Critique of Animal Selection

Disease entities are complex. Manifestations that are grouped together as a "disease" are commonly not all present in all cases and apparently "similar" diseases are not the same in all species of animals. Individual species may have some unique characteristic that becomes apparent in the syndrome of a spontaneous or induced disease.

SOL M. MICHAELSON
AND
BERNARD F. SCHREINER, JR.

Some general criteria for choosing or developing an animal model for production and study of myocardial infarction or congestive heart failure follow: (1) The method should be applicable to more than one animal species. (2) The physiologic results must be reproducible. (3) The pathologic process simulates known features of the human disease state. (4) The size and location of the lesion produced are predictable. (5) Thoracotomy is not required. (6) Major metabolic alterations do not develop in the animal.

The type of animal to be selected requires consideration of known biologic characteristics. Some of the larger mammals normally have the same type of intimal structure in their main arteries as is found in man,[109] which suggests that the larger mammal (dog, pig, horse, or cow) might be a more satisfactory model than the smaller one for cardiovascular studies. Animals used in such studies should have biological characteristics similar to those of man, be of adequate size for surgical manipulation, and be docile and easy to handle and house.

Based on the information we now have and the problems we are trying to solve, it appears that the dog and pig are the most likely candidates as models to study myocardial infarction and congestive heart failure. It is essential, therefore, to consider the similarities and differences of these two species in relation to man and the practical problems involved in the utilization of these animals.

When the dog's anterior septal artery is occluded, infarcts involving the upper two thirds of the septum are obtained, with partial destruction of the bundle of His and its branches. Occlusion of the posterior septal artery affects the posterior one fifth of the septal mass and shows evidence of injury to the atrioventricular node and bundle of His. The anterior septal artery is a source of blood supply to the conduction tissue in the dog and is apparently unimportant in the pig.[60,61]

In the pig, the distribution of the coronary bed is anatomically more uniform and more analogous to that commonly found in man than is that of the dog; the artery to both the sinoatrial node and the atrioventricular node arises from the right coronary artery. The posterior septal artery provides most of the functional blood supply to the conduction tissue. Its occlusion is frequently associated with heart block, and the conduction tissue, while not completely destroyed in survivors, is usually damaged to a considerable extent. Anterior septal ligation, on the other hand, does not cause death.[61] The pig, like man but unlike the dog, does not normally have interarterial anastomotic channels smaller than 100μ.

Chronic cardiac failure can be produced in the unanesthetized dog by pulmonary artery constriction, with the development of acute episodes of reduced cardiac output and increased venous pressure.[87]

The dog has a considerable exercise tolerance and resistance to stress, which may make it difficult to produce cardiac failure or to control the magnitude of response when the pulmonary artery is constricted. The pig, on the other hand, has low exercise tolerance, which may tend to exaggerate cardiopathophysiologic responses.

The dog's cardiac configuration, conduction system, and biochemistry are very similar to man's. Because of their size and tractability, virtually all the clinical procedures employed in human medicine may be applied to dogs, and the response to specific therapy will be quite similar.

Information on canine cardiovascular physiology is readily available; for the pig, however, it is quite fragmentary. Nevertheless, what is known should be considered by investigators of cardiac pathophysiology because an important facet of such investigations necessitates hemodynamic assessment of the animal.

Chronic right heart failure may be induced in young swine with quite predictable and consistent results. The rapid growth rate of the piglet effectively increases the relative percentage of pulmonary artery stenosis, and thus the work load of the right ventricle slowly but progressively becomes greater. The physiological alterations are evident early and progress gradually enough to allow for sequential study of the primary and secondary sequelae of the stenosis over a period of 12-15 weeks or more.[63]

In planning and performing cardiovascular experiments with pigs it is important to consider the rapid and excessive excitability of these animals. Swine react to disturbances and fear with rapid and marked increase in blood pressure.[23] The relative heart weight of swine (0.25-0.30 percent) is much smaller than in other animals—horse, 0.7-1.1 percent; dog 0.7-1.0 percent; sheep, 0.5-0.6 percent; cattle, 0.4-0.5 percent; cat 0.47 percent; human, 0.5 percent.[6,7]

The high heart rate of some pigs when electrocardiograms are taken has led to the generalization that the pig has a considerably lower diastolic/systolic quotient (0.7-0.8) than other domestic animals (diastole/systole: horse, 2.12-2.44; cattle, 1.26; dog, 2.1; cat, 1.12; sheep, 1.02; human, 1.37).[90] When this calculation is made for pigs with low heart rates, however, the quotients found are not very different from those of other species.[22,40] The speculation has been advanced that the level of sympathetic tone may have an important influence on the duration of systole and diastole in swine, but this remains to be proven.[23]

The pressure in the right ventricle (51/0 mm Hg) and in the pulmonary artery (51/35 mm Hg) is remarkably high in swine[23] in contrast to the mean pulmonary artery pressure in man (12-25 mm Hg) or the dog (10-20 mm Hg). The high pulmonary artery pressure of swine may be due to the relatively small pulmonary arterial tree and low distensibility of the pulmonary arterial bed.[3,101]

Since the volume of coronary blood flow is much greater during diastole than during systole,[77] the relative duration of diastole becomes of great importance in determining the blood supply to the myocardium. With increasing heart rate, the diminution of the stroke interval is caused mainly by shortening of the duration of diastole. In swine, at heart rates of 210 per minute or more, the diastolic period is shortened so much that atrial contraction starts before ventricular systole is completed.[23]

The relatively small size of the pig's heart in relation to body size would result in excessive strain on the heart under stress. The observation that the myocardial fibers of the pig are "hypertrophied" may be of some significance. Thus, it seems reasonable to conclude that the small heart, small blood volume, and frequently low hemoglobin level are important factors responsible for an unstable circulatory system in swine.[23] The relatively poor thermal regulatory capacity of the pig in comparison to the dog and man may compromise hemodynamic studies, especially those that must be performed while the animal is anaesthetized. There are few superficial veins in swine for administration of anaesthetics and other pharmacologic agents and for blood sampling. The hemiazygos vein of the pig drains into the great cardiac vein, thereby complicating metabolic studies of the heart, unless this vessel is ligated.

Swine increase rapidly in size and soon become difficult to manage. This problem can be partially obviated by using miniature swine when older pigs are desired or when animals are needed for longer periods of study. It should be noted, however, that miniature swine also grow to a fairly large size.

From the foregoing it is apparent that for the resolution of a biomedical problem various animal models may be available. It is also manifest that individual species have general or specific characteristics that make them more or less desirable as investigative subjects. It behooves the biomedical investigator to become aware of the attributes of various animals so that the most meaningful results can be obtained for studying disease in man.

In the specific areas of myocardial infarction and congestive heart failure there is no doubt that both the dog and the pig can serve as useful animal models. Each species has certain advantageous characteristics as well as some deficiencies. Therefore, to obtain the most meaningful information in the investigation of myocardial infarction or congestive heart failure, both the pig and the dog should be studied under the same conditions and in the same manner.

In general, there is little doubt that biomedical study of several species is required to provide the most reliable extrapolation to man. Ideally, one should choose taxonomically unrelated species to bring out generalizations, with the realization, however, that to study multiple species alone will not advance understanding. The main contribution of comparative biology is not to record the same phenomenon in as many different animals as possible, but to select intelligently some that can serve as meaningful comparative models. From a spectrum of species basic information on the comparative reaction of biologic systems can be gained and can be used to elucidate mechanisms of action. For the study of physiologic function, a common parameter such as metabolic rate, body weight, or body surface area could be utilized to provide an index for extrapolation among species. In comparing results of experiments performed in the same or different laboratories standardization of conditions is mandatory. Studies of experimental animal models must be complemented by prospective studies of the naturally occurring disease in animals and in man himself.

A comparative approach is basic to elucidating the nature of vital processes among animals and placing man in his proper biological perspective; it relates the different ways in which various species maintain homeostasis; it characterizes animals particularly suitable for demonstrating specific parameters; and it integrates and coordinates anatomic, physiologic, biochemical, and pathologic similarities of various groups of animals. From a comparative approach we can learn what biologic attributes are unique or common among different animals, we can study interrelations with environmental stresses, and we can find animals that are most suitable for study of important functions and thus provide a basis for biological generalization.

This paper is based on work performed under contract with the U.S. Atomic Energy Commission at the University of Rochester Atomic Energy Project and supported by U.S. Public Health Service Grant 3 R01 HE 08494; it has been assigned Report No. UR-49-*990*.

References

1. Ackerknecht, E., and C. Krause. 1929. Gefässe. p. 1–172. *In* E. Joest [ed.] Handbuch der speziellen pathologischen Anatomie der Haustiere, Vol. V. Paul Parey, Berlin.
2. Agress, C. M., M. J. Rosenberg, H. I. Jacobs, M. J. Binder, A. Schneiderman, and W. G. Clark. 1952. Protracted shock in the closed-chest dog following coronary embolization with graded microspheres. Amer. J. Physiol. 170:536.
3. Attinger, E. O., and J. M. Cahill. 1960. Cardiopulmonary mechanics in anesthetized pigs and dogs. Amer. J. Physiol. 198:346.
4. Bernard, C. 1865. Introduction à l'étude de la médicine expérimentale. Paris, Bailliere. (Trans, Henry Copley Greene, The Macmillan Co., New York, 1927.)
5. Bretschneider, J. 1962. Zur Pathologie und Pathogenese der sogenannten Endocarditis valvularis chronica fibrosa des Hundes. Dissertation. Univ. Giessen, Germany.
6. Brody, S. 1945. Bioenergetics and growth. Reinhold Publishing, Co., New York.
7. Brody, S., and H. H. Kibler. 1941. Growth and development. LII. Relation between organ weight and body weight in growing and mature animals. Univ. Mo. Agr. Exp. Sta. Res. Bull. 328.
8. Christensen, G. C. 1958. Comparative architecture of coronary vessels and cineangiocardiographic studies in domesticated animals. Anat. Rec. 130:285.
9. Christensen, G. C. 1962. The blood supply to the interventricular septum of the heart—A comparative study. Amer. J. Vet. Res. 23:869.
10. Christensen, G. C., and F. L. Campeti. 1959. Anatomic and functional studies of the coronary circulation in the dog and pig. Amer. J. Vet. Res. 20:18.

11. Clark, A. J. 1927. Comparative physiology of the heart. The Macmillan Co., New York.

12. Dahme, E. 1957. Über die Beurteilung der angiopathien bei chronischsklerosierenden Nierenerkrankungen des Hundes. Arch. Exp. Vet. Med. 11:752.

13. Dahme, E. 1962. Blutgefässe. *In* E. Joest [ed.] Handbuch der speziellen Pathologie and pathologischen Anatomie der Haustiere, 3rd ed., Vol. II. Paul Parey, Berlin.

14. Detweiler, D. K. 1962. Wesen und Haufigkeit von Herzkrankheiten bei Hunden. Zbl. Vet. Med. 9:317.

15. Detweiler, D. K. 1963. Cardiovascular disease in the dog. A study in comparative cardiology. Proc. 17th World Vet. Cong., Hanover, Germany, 1:47.

16. Detweiler, D. K. 1966. Swine in comparative cardiovascular research. p. 301-306. *In* Swine in biomedical research. Frayn Printing Co., Seattle, Washington.

17. Detweiler, D. K., K. Hubben, D. F. Patterson, and R. P. Botts. 1960. Survey of cardiovascular disease in dogs, preliminary report on the first 1,000 dogs screened. Amer. J. Vet. Res. 82:329.

18. Detweiler, D. K., and D. F. Patterson. 1965. The prevalence and types of cardiovascular disease in dogs. Comparative Cardiology. Ann. N.Y. Acad. Sci. 127:481.

19. Detweiler, D. K., D. F. Patterson, K. Hubben, and R. P. Botts. 1961. The prevalence of spontaneously occurring cardiovascular disease in dogs. Amer. J. Pub. Health 51:288.

20. DiGuglielmo, L., C. Montemartini, F. Coucourde, A. Schifino, V. Baldrighi, and A. Marchesi. 1962. Spontaneous myocardial infarct in a dog. Mal. Cardiov. 3:283.

21. Durrer, D., and L. H. Van Der Tweel. 1957. Excitation of the left ventricular wall of the dog and goat. Ann. N.Y. Acad. Sci. 65:779.

22. Engelhardt, W. v. 1963. Untersuchungen am Schwein über die Systolen- und Diastolendauer des Herzens und über den Blutdruck in der Ruhe und waehrend der Erholung nach koerperlicher Belastung. Zentr. Veterinaermed. 10:39.

23. Engelhardt, W. v. 1966. Swine cardiovascular physiology–A review. p. 307-330. *In* Swine in biomedical research, Frayn Printing Co., Seattle, Washington.

24. Farrelly, B. T. 1954. The pathogenesis and significance of parasitic endarteritis and thrombosis in the ascending aorta of the horse. Vet. Rec. 66:53.

25. Fautex, M. 1940. Experimental study of the surgical treatment of coronary disease. Surg. Gynecol. Obstet. 71:151.

26. Florey, H. 1963. Atherosclerosis. Endeavor. 22:107.

27. French, J. E., and M. A. Jennings. 1965. The tunica intima of arteries in swine. p. 25-36. *In* J. C. Roberts, Jr., and R. Straus [ed.] Comparative atherosclerosis, Harper and Row, New York.

28. Getty, R. 1965. The gross and microscopic occurrence and distribution of spontaneous atherosclerosis in the arteries of swine. p. 11-20. *In* J. C. Roberts, Jr., and R. Straus [ed.] Comparative atherosclerosis, Harper and Row, New York.

29. Gilstrap, L. O., Jr., J. S. McNeil, L. P. Greenberg, and R. B. Spodak. 1964. A compilation of biological laws, effects, and phenomena, with associated physical analogs. Wright-Patterson Air Force Base, Ohio.

30. Gottlieb, H., and J. Lalich. 1954. The occurrence of arteriosclerosis in the aorta of swine. Amer. J. Pathol. 30:851.

31. Gross, L. 1921. The blood supply to the heart. Paul B. Hoeber, Inc., New York.

32. Gross, F. L., L. Blum, and G. Silverman. 1937. Experimental attempts to increase the blood supply to the dog's heart by means of coronary sinus occlusion. J. Exp. Med. 65:91.

33. Guzman, S. V., E. Swenson, and M. Jones. 1962. Intercoronary reflex: Demonstration by coronary angiography. Circ. Res. 10:739.

34. Halpern, M. H. 1955. Blood supply to the atrioventricular septum in the dog. Anat. Rec. 121:753.

35. Hamlin, R. L., and A. M. Scher. 1961. Ventricular activation process and genesis of QRS complex in the goat. Amer. J. Physiol. 200:223.

36. Hamlin, R. L., and C. R. Smith. 1965. Categorization of common domestic mammals based upon their ventricular activation process. Comparative cardiology, Ann. N.Y. Acad. Sci., 127:195.

37. Hammer, J., and Z. Pisa. 1962. A method of isolated gradual occlusion of a main branch of a coronary artery, in closed-chest dogs. Amer. Heart J. 64:67.

38. Hartman, F. W., A. Bolliger, H. P. Doub, and F. J. Smith. 1927. Heart lesions produced by the deep x-ray. An experimental and clinical study. Bull. Johns Hopkins Hosp. 41:36.

39. Hellerstein, H. K., and R. L. Hamlin. 1960. QRS component of the spatial vectorcardiogram and of the spatial magnitude and velocity electrocardiograms of the normal dog. Amer. J. Cardiol. 6:1049.

40. Hoeller, H. 1959. Vergleichende elektrokardiographische Untersuchungen an normal gefuetterten und eiweissmangelernaehrten Schweinen. Berlin, Muench, Tieraerztl. Wochschr. 72:265.

41. Holt, J. P., H. Kines, and E. A. Rhode. 1965. Pattern of function of left ventricle of mammals. Amer. J. Physiol. 209:22.

42. Jacobey, J. A., W. J. Taylor, G. T. Smith, R. Garlin, and D. E. Harken. 1962. New therapeutic approach to acute coronary occlusion: I. Production of standardized coronary occlusion with microspheres. Amer. J. Cardiol. 9:60.

43. James, T. N. 1961. Anatomy of the coronary arteries. Paul B. Hoeber, Inc., New York.

44. James, T. N., and G. E. Burch. 1958. Blood supply of the human interventricular septum. Circulation 17:391.

45. Jennings, M. A., H. W. Florey, W. E. Stehbens, and J. E. French. 1961. Intimal changes in the arteries of a pig. J. Pathol. Bacteriol. 81:49.

46. Jobe, C. L. 1968. Selection and development of animal models of myocardial infarction. Proc. Symposium, Animal Models for Biomedical Research. July, 1967, Dallas, Texas. Publ. 1954, National Academy of Sciences, Washington, D.C. p. 101–108.

47. Kay, R. H. 1964. Experimental biology: Measurement and analysis. Reinhold, New York.

48. Kentera, D., C. R. Wallace, W. F. Hamilton, and L. T. Ellison. 1964. Hemodynamic response to hypoxia in dogs with chronic pulmonary hypertension. Amer. J. Physiol. 207:650.

49. Kines, H., E. A. Rhode, and J. P. Holt. 1967. Interrelationships between right ventricular volumes, heart rate, stroke volume, cardiac output, heart weight, body weight, body surface and metabolic rate in mammals. The Physiologist, (10)3.

50. Krause, C. 1927. Pathologie der Blutgefässe der Tiere. Ergebn. Allg. Pathol. 22:350.

51. Lapin, B. A., and L. A. Yakovleva. 1963. Comparative pathology in monkeys. Charles C Thomas, Springfield, Ill., p. 166–176.

52. Lindsay, S., and I. L. Chaikoff. 1966. Naturally occurring arteriosclerosis in nonhuman primates. J. Atherosclerosis Res. 6:36.

53. Lindsay, S., I. L. Chaikoff, and J. W. Gilmore. 1952. Arteriosclerosis in the dog; spontaneous lesions in the aorta and coronary arteries. AMA Arch. Pathol. 53:381.

54. Litvak, J., L. E. Siderides, and A. M. Vineberg. 1957. The experimental production of coronary artery insufficiency and occlusion. Amer. Heart J. 53:505.

55. Lluch, S., H. C. Moguilevsky, G. Pietra, A. B. Shaffer, L. J. Hirsch, and A. P. Fishman. 1969. A reproducible model of cardiogenic shock in the dog. Circulation 39:205.

56. Luginbühl, H. 1966. Spontaneous atherosclerosis in swine. p. 347–364. *In* Swine in biomedical research, Frayn Printing Co. Seattle, Washington.

57. Luginbühl, H., and D. K. Detweiler. 1965. Cardiovascular lesions in dogs. Ann. N.Y. Acad. Sci. 127:517.

58. Luginbühl, H. R., J. E. T. Jones, and D. K. Detweiler. 1965. The morphology of spontaneous arteriosclerotic lesions in the dog. p. 3–10. *In* J. C. Roberts, Jr., and R. Straus [ed.] Comparative atherosclerosis, Harper and Row, New York.

59. Lumb, G. D., and L. B. Hardy. 1963. Collateral circulation and survival related to gradual occlusion of the right coronary artery in the pig. Circulation 27:717.

60. Lumb, G., R. S. Shacklett, and W. A. Dawkins. 1959. The cardiac conduction tissue and its blood supply in the dog. Amer. J. Pathol. 35:467.

61. Lumb, G., and H. Singletary. 1962. Blood supply to the atrioventricular node and bundle of His: A comparative study in pig, dog, and man. Amer. J. Pathol. 41:65.

62. Maas, A. 1958. Zur Aetiologie der enzootischen Herztodes der Schweine. Arch. Exp. Veterinaermed. 12:82.

63. Maaske, C. A., N. H. Booth, and T. W. Nielsen. 1966. Experimental right heart failure in swine. p. 377–388. *In* Swine in biomedical research, Frayn Printing Co., Seattle, Washington.

64. McKibben, J. S., and G. C. Christensen. 1964. The venous return from the interventricular septum of the heart: A comparative study. Amer. J. Vet. Res. 25:512.

65. McMillan, J. B., and M. Lev. 1964. The aging heart. II. The valves. J. Gerontol. 19:1.

66. Middleton, C. C., and H. B. Lofland. 1965. Aggravation of atherosclerosis in squirrel monkeys (*Saimiri sciureus*) by diet. Federation Proc. 24:311.

67. Morehead, R. P., and J. M. Little. 1945. Changes in blood vessels of apparently healthy mongrel dogs. Amer. J. Pathol. 21:339.

68. Moreland, A. F. 1965. Experimental atherosclerosis of swine. p. 21–24. *In* J. C. Roberts, Jr., and R. Straus [ed.] Comparative atherosclerosis, Harper and Row, New York.

69. Moss, A. J., D. W. Smith, S. M. Michaelson, and B. F. Schreiner. 1963. Radiation technique for production of localized myocardial necrosis in the intact dog. Proc. Soc. Exp. Biol. Med. 112:903.

70. Penaloza, D., and J. Tranchesi. 1955. The three main vectors of the ventricular activation process in the human heart. Amer. Heart J. 49:51.

71. Pomerance, A. 1966. Pathogenesis of "senile" nodular sclerosis of the atrioventricular valves. Brit. Heart J. 28:815.

72. Pomerance, A. 1967. Aging changes in human heart valves. Brit. Heart J. 29:222.

73. Ratcliffe, H. L. 1963. Phylogenetic considerations in the etiology of myocardial infarction. p. 61–89. *In* T. N. James and J. W. Keys [ed.] The etiology of myocardial infarction. Little Brown, Co., Boston, Massachusetts.

74. Ratcliffe, H. L., T. G. Yerasimides, and G. A. Elliott. 1960. Changes in the character and location of arterial lesions in mammals and birds in the Philadelphia Zoological Garden. Circulation 21:730.

75. Robb, J. S. 1965. Comparative basic cardiology. Grune and Stratton, New York.

76. Roberts, J. C., Jr. 1961. Distribution and severity of spontaneous atherosclerosis in the dog. Circulation 24:1101.

77. Ross, G., A. Kolin, and S. Austin. 1964. Electromagnetic observations on coronary arterial blood flow. Proc. Nat. Acad. Sci. U.S. 52:692.

78. Rowsell, H. C., H. G. Downie, and J. F. Mustard. 1958. The experimental production of atherosclerosis in swine following the feeding of butter and margarine. Can. Med. Ass. J. 79:647.

79. Rowsell, H. C., J. F. Mustard, M. A. Packham, and W. J. Dodds. 1966. The hemostatic mechanism and its role in cardiovascular disease of swine. p. 365–376. *In* Swine in biomedical research. Frayn Printing Co., Seattle, Washington.

80. Ruch, T. C. 1959. Diseases of laboratory primates. W. B. Saunders, Co. Philadelphia.

81. Rushmer, R. F. 1964. Initial ventricular impulse: A potential key to cardiac evaluation. Circulation 29:268.

82. Salazar, A. E. 1961. Experimental myocardial infarction: Induction of coronary thrombosis in the intact closed-chest dog. Circ. Res. 9:1351.

83. Scher, A. M. 1956. Ventricular depolarization in the dog. Circulation Res. 4:461.

84. Seifert, K. 1962. Elektronmikroskopische Untersuchungen der Aorta des Hauschweines. Z. Zellforschung Mikr. Anat. 58:331.

85. Skold, B. H., and R. Getty. 1961. Spontaneous atherosclerosis of swine. J. Amer. Vet. Med. Ass. 139:655.

86. Smith, C., and L. A. Loewenthal. 1950. A study of elastic arteries in irradiated mice of different ages. Proc. Soc. Exp. Biol. Med. 75:859.

87. Smith, D. L., C. A. Maaske, and F. Julian. 1955. Respiratory response to partial occlusion of the main pulmonary artery in the dog. Amer. J. Physiol. 181:341.

88. Sobel, H., E. C. Mondon, and R. Straus. 1962. Spontaneous and stress induced myocardial infarction in aged atherosclerotic dogs. Circ. Res. 11:971.

89. Spoerri, H. 1966. Comparative studies on cardiac dynamics. Cited by Hoernicke, H. Review of the Berlin Symposium on Swine Circulatory System. p. 419–424. *In* Swine in biomedical investigations, Frayn Printing Co., Seattle, Washington.

90. Spoerri, H., and J. P. Siegfried. 1955. Ueber die Systolen- und Diastolendaure bei Haus- und Wildtieren und ihre Beziehungen zur koerperlichen Liestungsfaehigkeit. Helv. Physiol. Acta 13:32.

91. Stone, H. L., V. S. Bishop, and A. C. Guyton. 1964. Progressive changes in cardiovascular function after unilateral heart irradiation. Amer. J. Physiol. 206:289.

92. Straus, R., H. Sobel, S. K. Abul-Haj, and R. J. Kositchek. 1965. Spontaneous myocardial infarcts in treated and nontreated animals. p. 186–195. *In* J. C. Roberts, Jr., and R. Straus [ed.] Comparative atherosclerosis, Harper and Row, New York.

93. Tashjian, R. J., K. M. Das, W. E. Palich, and R. L. Hamlin. 1965. Studies on cardiovascular disease in the cat. Ann. N.Y. Acad. Sci. 127:581.

94. Tashjian, R. J., R. R. Pensinger, K. M. Das, C. F. Reid, and A. A. Crescenzi. 1963. Feline cardiovascular studies. Proc. 100th Ann. Mtg. AVMA 112–113, New York.

95. Taylor, C. B., D. E. Patton, and G. E. Cox. 1963. Atherosclerosis in Rhesus monkeys. VI. Fatal myocardial infarction in a monkey fed fat and cholesterol. AMA Arch. Pathol. 76:404.

96. Tenney, S. M. 1967. Some aspects of the comparative physiology of muscular exercise in mammals. Suppl. I. Circ. Res. 20;21:7.

97. Thomas, W. A., and W. S. Hartroft. 1959. Myocardial infarction in rats fed diets containing high fat, cholesterol, thiouracil and sodium cholate. Circulation 19:65.

98. Thomas, W. A., W. S. Hartroft, and R. M. O. Neal. 1959. Modification of diets responsible for induction of coronary thromboses and myocardial infarcts in rats. J. Nutr. 69:325.

99. Thomas, W. A., K. T. Lee, F. Goodale, R. F. Scott, and A. S. Daoud. 1963. Thrombogenesis, thrombolysis, myocardial necrosis and their relationships to

dietary manipulations. p. 153-168. *In* T. N. James and J. W. Keys [ed.] The etiology of myocardial infarction, Little, Brown, Boston.

100. Truex, R. C., and M. Q. Smythe. 1965. Comparative morphology of the cardiac conduction tissue in animals. Comparative cardiology. Ann. N.Y. Acad. Sci. 127:19.

101. Wachtel, W., L. Lyhs, and E. Lehmann. 1963. Blutdruckmessung beim Schwein. Arch. Exp. Vet. Med. 16:355.

102. Wagner, B. M. 1968. Myocardial disease in man and dog, some properties. Ann. N.Y. Acad. Sci. 147:354.

103. Wallace, C. R., and W. F. Hamilton. 1962. Study of spontaneous congestive heart failure in the dog. Circ. Res. 11:301.

104. Warren, S. 1942. Effects of radiation on the cardiovascular system. Arch. Pathol. 34:1079.

105. Warren, S. 1944. The histopathology of radiation lesions. Physiol. Rev. 24:225.

106. Whipple, H. E. [ed.]. 1965. Comparative cardiology. Proc. Conf. Comparative Cardiology, N.Y. Acad. Sci., New York, 1964, Ann. N.Y. Acad. Sci. 127.

107. Wilgram, G. F., and D. J. Ingle. 1965. Spontaneous cardiovascular lesions in rats. p. 87-91. *In* J. C. Roberts, Jr., and R. Straus [ed.] Comparative atherosclerosis, Harper and Row, New York.

108. Wissler, R. W., L. E. Frazier, R. H. Hughes, and R. A. Rasmussen. 1962. Atherogenesis in the Cebus monkey. I. A comparison of the three food fats under controlled dietary conditions, AMA Arch. Pathol. 74:312.

109. Wolkoff, K. 1924. Über die Altersveranderungen der Arterien bei Tieren. Virchow Arch. Pathol. Anat. 252:208.

110. Zugibe, F. T. 1965. Atherosclerosis in the miniature pig. p. 37-44. *In* J. C. Roberts, Jr., and R. Straus [ed.] Comparative atherosclerosis, Harper and Row, New York.

DISCUSSION

DR. VESSELINOVITCH: Can you comment on some techniques for experimental production of myocardial infarction in miniature pigs and macaca?

DR. MICHAELSON: Yes. There are quite a few studies now in domestic swine and miniature pigs as far as the production of myocardial infarction is concerned. I do not know enough about studies that have been done in simian primates. As far as the studies that have been done in swine, coronary occlusion has been used. Lumb and his associates have studied this fairly well. The techniques seem to work out very well. The results seem to be very comparable to that of man in comparison to the results you get in the dog.

DR. GLICK: Are there any significant differences in the normal cardiopulmonary anatomy between miniature swine and normal domestic swine?

145

DR. MICHAELSON: I am afraid I cannot answer that. I am sure there are many people involved in this conference who are more conversant with the differences between domestic swine and miniature swine than I am. I am sorry. I don't know.

DR. FABRY: Is the difference of relative heart weight in pigs likely to be caused by the excess fat and relative inactivity of the species? Would the wild pig be a better animal for cardiovascular studies?

DR. MICHAELSON: I do not know exactly how much of a difference there is. There is some literature on this, however. In Germany there have been some reports on differences in exercise tolerance between wild swine and domestic swine. There are very definite differences in response. This has been described in the book *Swine in Biomedical Research*, which is the publication of the proceedings of the 1965 conference held at Hanford. These are very well described by people like Luginbühl. I think it is his description that is in that volume.

DR. FOX: The heart is the mirror of emotional reactions. How important is this, and if so, how do you control the species individual differences in emotional reactivity, or in corticovisceral inter-relationship?

DR. MICHAELSON: There is no question that the cardiogenic control is monitored through the higher nervous activity and that emotionality does influence it. We all know this from looking at ourselves. It is extremely important; there is no question about that. Of course, the question of individual differences enters into this. It is quite apparent that there are individual differences, in man as well as in animals. We all have had enough experience with animals to recognize this very easily.

DR. SERRANO: The subhuman primates were not evaluated as a possible model for myocardial infarction and atherosclerosis, although the structure and function of their hearts is probably more similar to humans. Is there some special problem that precludes their use?

DR. MICHAELSON: As far as I know, the simian primate has not really been used very much in myocardial infarction studies. There are problems, as we all know, with their handling, and especially with the smaller simian primates. There is the problem of experimental procedures. Since any good myocardial infarction study requires good hemodynamic assessment, the techniques are not readily available for studying these in simian primates. We have not done it. As far as the coronary circulation is concerned, I do not know that there is that much known about it. There are some detailed descriptions. They have not been studied as well as the dog, man, and the pig. The fact that the monkey is a poor exerciser because of his low cardiac–body weight ratio could make him a little more likely to introduce a deficiency in evaluation. However, I think it should be studied definitely. This has occurred to us many times.

DR. EASTIN: Has the cat been shown to be of any particular value in cardiac research?

DR. MICHAELSON: I have not come across any good studies as far as myocardial infarction is concerned in which the cat has been used. There is no reason why the cat should not be looked at. I personally like cats; they are good animals to work with. Actually, the information I presented was about as much as I could find in a fairly extensive literature search on the problem. I may have missed some papers. In the symposium proceedings, a lot of these things will be detailed much more extensively than I could cover in this presentation.

III

Defining
the Inbred
Laboratory
Animal

GENETICS TODAY

George W. Beadle

I welcome the opportunity to introduce this section on genetics, for in a very real sense, the day for genetics is here. Never has there been so much interest. Never has there been so much progress. Never before could genetics more justifiably claim its crown as queen of the biological sciences. By way of illustration, let me recount a bit of history.

We were all taught that Mendel's classic paper was lost for a third of a century. That is not so. It was not lost; it was unappreciated.

It is well known that Mendel carried on an extensive correspondence with the German botanist Nägeli, and except for ten of Mendel's letters, all of it is lost. Obviously Nägeli was not impressed.

It is known, too, that Mendel received 40 reprints of the famous paper of 1866, which he delivered before a meeting of the Brünn Society. Not only were the members of this scholarly society unimpressed, despite the fact that at a previous meeting a short time before they had listened to a paper on Darwin's theory of organic evolution, but the recipients of the 40 reprints were also unmoved. One wonders if Darwin himself did not get a reprint, for Mendel was aware of Darwin's ideas and that his work provided the basic variability that so puzzled Darwin. If Darwin did receive one, or if he read the paper in the Proceedings of the Brünn Society, which was sent to more than 120 libraries, he too was unimpressed.

One reprint went to Nägeli, another to Kerner, Professor of Botany at Innsbrück. The latter copy was found years later—with the pages still uncut. A third was found in the library of the Dutch botanist Biejerinck, who showed it to deVries, who later said he had "forgotten about it." That is convincing evidence that deVries, too, had not been impressed—at least until he himself had confirming data.

No, Mendel's work was not unknown. Focke's monograph on plant hybridization of 1881 referred to it. So did the American botanist Liberty Hyde Bailey in 1895, although it should be pointed out he did not read the paper, but merely copied the reference from Focke's monograph. Moral: Read the papers you cite.

Mendel's work was known by many who should have appreciated it. The times were just not right. Those were times when living creatures were regarded as much more mysterious than we now believe them to be. Mendel's results said heredity was simple—almost childishly simple. How, then, could it possibly be true?

That, I believe, is why it took so long for biologists in general to appreciate the full significance of Mendel's work.

Even after the rediscovery and confirmation of his findings by Correns, deVries, and von Tschermak, there remained skeptics for many years.

Karl Pearson and his associate Weldon fought bitterly with Bateson on the subject. It has even been said that they were inspired to develop much of modern biometry in an attempt to disprove Mendelism.

Bateson, the articulate and vociferous defender of Mendel, interestingly enough, never, up to his death in 1926, fully accepted the chromosome theory of inheritance.

Prior to his own conversion upon discovering the famous sex-linked white-eye mutation in *Drosophila* in 1910, Thomas Hunt Morgan took a very dim view of genetics. Speaking before the American Breeders Association in 1908 he ridiculed the logic of geneticists by pointing out that they invented factors to explain observed results, then turned right around and alleged the observed results proved the hypothesis.

That white-eye male *Drosophila* turned out to be a pretty persuasive little fellow.

William Morton Wheeler was a famed entomologist of his time, first at the University of Chicago, then at Harvard. As late as the 1920's he described genetics as a tiny insignificant bud on the tree of biology, a bud surely destined to wither away and die.

Shortly after the "rediscovery" of Mendel's work ("appreciation of its significance" is a more apt expression), the English physician Archibald E. Garrod—later, Sir Archibald Garrod—became interested in genetic diseases in man. By 1909 he had pretty well worked out both the genetics and the biochemistry of alcaptonuria—accumulation of alcapton, or homogentisic acid, and its excretion in the urine, causing the urine to blacken on exposure to air. He correctly postulated a defective gene resulting in a defective enzyme and a blocked chemical reaction in the metabolism of phenylalanine and tyrosine. Then, and subsequently, he extended his concepts to cover several other genetic diseases in man, among them albinism, cystinuria, and pentosuria.

He was the father of biochemical genetics—the one-gene-one-enzyme view, though he never used that term.

He published many papers, a classic book in two editions (1909 and 1923) called *Inborn Errors of Metabolism*, and lectured widely on his work, but, like Mendel, he was unappreciated both by geneticists and by biochemists. Why? He, too, was ahead of his time. Life is not simple. Therefore such a simple explanation could not be correct. But it was.

More than a third of a century later, Tatum and I redemonstrated gene-enzyme-chemical reaction relations in *Neurospora.* We did not know of Garrod's work, for none of the genetic or biochemical literature we read referred to it.

Despite what we regarded as very persuasive evidence, there was again widespread skepticism. By 1952, 11 years after our demonstration, I knew of only three geneticists who still really believed in the one-gene–one-enzyme concept, now refined as one-gene–one-polypeptide. I am glad to report I was one of them. Why the unbelievers? That too was a simple idea, therefore it could not be correct.

In 1944 Avery, Macleod, and McCarthy at the Rockefeller Institute published clear evidence that DNA, as pure as they could then make it—which was pretty pure—was capable of transforming *Pneumococcus* bacteria from one serological type to another. Conclusion: DNA is the primary genetic material. The nonbelievers were many. "Too simple", they said. "It can't be true." It was.

In 1953 Watson and Crick reported their now famous paper on the structure of DNA—the most important advance in twentieth century biology in my opinion.

Despite many doubts, they were essentially right.

We know the rest of the story—how DNA carries genetic information in the form of a simple language in which the letters are four, the words are all of three letters, and the total dictionary consists of but 64 words. The definitions of all are *now* known.

We know, too, how a segment of several hundred or more words—a gene—transfers its information to messenger RNA, which moves from the nucleus to the cytoplasm, where in association with ribosomes it serves as a template to order the amino acids in a growing protein chain, each amino acid having been previously labeled with a specific transfer-RNA molecule.

DNA replicates, as Watson and Crick postulated. Two of the enzymes involved, DNA polymerase and DNA ligase, are known. The former has been isolated by Kornberg and associates in more than a half-gram quantity, which was derived from 200 pounds of *Escherichia coli* bacteria grown in about 10 tons of culture medium. It appears to be a single polypeptide chain of about a thousand amino acids. Many details of its properties and action have been worked out in Kornberg's laboratory.

Genetics has come of age. It has indeed earned the title "Queen of Biology." Much of the mystery about the nature of life is now gone, despite the fact that a great deal remains to be learned. The age-old question "What is Life?" has been made relatively meaningless, for we now see that evolution from hydrogen to man is an unbroken sequence. Except by purely arbitrary definitions, we can no longer say what is living and what is not.

What does all this have to do with the immediate interests of laboratory animal researchers in attempting to develop populations of experimental animals in which variability is reduced to a minimum? A great deal, for genetics now goes a long way toward defining what can be done and what cannot.

Experimentation would be far more reliable than it is today if we could produce populations of mice, rats, rabbits, cats, or chickens, in which all animals of one species were exactly alike genetically. Experience has shown that it is not easy.

In organisms that are normally cross bred, as are most of those now being used in experimental work, there are many known or suspected instances in which heterozygosity for a single gene confers a selective advantage as compared with either homozygote. Sickle cell anemia in man is one example in which, in the presence of falciparum malaria, the heterozygote is better able to survive than either homozygote.

The genetic disease cystic fibrosis in man seems to be another example. It has been estimated that about one person in 17 is heterozygous for the recessive gene primarily concerned. There may now be better estimates, but for the argument this is not too important. Thus we might expect one in 289 marriages to consist of two carriers in which one quarter the children will be affected. That would mean a frequency of affected individuals in the population of one in 1,156, assuming no preferential mating or differential fertility. Since few affected individuals survive to reproduce, many genes for cystic fibrosis should be eliminated each generation. To keep the frequency up, we can assume either a high mutation rate or a fairly strong selective advantage of the carrier condition. The latter seems more probable.

If, in the animals used in experimental work, there are many genes for which the heterozygous condition confers a selective advantage, inbreeding would always lead to a decline in vigor. To my knowledge this has been found to be the case in all plants and animals in which the sexes are separated or which are not normally self-fertilized. I believe it is true, for example, for mice, rats, rabbits, guinea pigs, chickens, and corn.

There is good genetic reason to believe that genes conferring selective advantages in the heterozygous condition would arise and persist in such species.

On the other hand, in normally self-fertilized species such as peas, beans, and wheat, there would be far less opportunity for genes of this type to persist. Thus, genomes conferring maximum advantage in the homozygous condition would be preferentially selected.

In corn, in which I have long had a scientific interest, homogeneity in populations is attained by combining relatively homozygous low-yielding inbreds in combinations that give maximum yields. I know that laboratory animal researchers who have a big stake in attaining maximum uniformity in animal populations often take advantage of this vigor and uniformity of F_1 offspring of relatively inbred lines. The problem in this case is, of course, considerably more difficult than that the corn breeder must deal with, and for two reasons. First, the phenomenon of self-fertilization to rapidly produce and test inbred lines is not available; second, the very large numbers of specimens readily available to plant breeders are not possible to maintain in laboratory animal research.

There are still other disadvantages. Unless germfree animals are used, which is expensive in both energy and dollars, the laboratory animal researcher must work with ecological systems that are difficult to keep constant. In addition, animals have nervous systems by which a great deal of information is received, stored, rearranged, and retrieved in ways we are only beginning to understand. But we do know that whatever happens to, and in, the nervous system influences subsequent behavior in ways we seldom understand, but which are clearly important.

Although the behavior of plants does depend on their previous history—mineral nutrition, light intensity and quality, and length of day, for example—they are not frightened by loud noises, made gentle by kind treatment, or inspired to violence by aggressive fellow creatures.

On the other hand, plants cannot contribute to an understanding of man's central nervous system, that fantastically complex and wonderful evolutionary development that enables us to supplement our purely genetic inheritance with the cultural inheritance that so significantly sets us apart from all other organisms on earth, and that enables us to learn about ourselves through experimentation with our fellow creatures.

No other species on earth can do that. The opportunities in laboratory animal research are many and open-ended. I hope they are all explored.

THE USE OF INBRED
STRAINS, F_1 HYBRIDS,
AND NONINBRED
STRAINS IN RESEARCH

Michael F. W. Festing

The research worker now has a bewildering choice
of different types of experimental animals. Thus, not only are there over 200
inbred strains of mice,[16] over 30 inbred strains of rats, and several inbred
strains of other species,[10] but there is also a wide variety of noninbred strains
and breeds of all species. The theoretical number of possible F_1 hybrids between lines is very high, and in practice large numbers of different F_1 hybrids
are used. Finally, all these types may be grouped in any combination in a given
experiment.

In certain types of research the use of inbred strains or F_1 hybrids is essential, but by far the majority of animals used in the United Kingdom come
from noninbred strains.

Although there is now an extensive literature on the relative merits of inbred strains and F_1 hybrids for bioassay,[1-3,5,12] geneticists have given little
guidance on the choice of animals for other types of experiment.

Work on methods of maintaining genetic variability within a population of
experimental animals[13] and the intuitive feeling that a genetically variable
population is likely to be representative of the species as a whole have led to
the widespread acceptance of noninbred strains as the animal of choice for
many types of experiments. Doolittle has stated that, "because of the variability in genotype of animals from random mated populations, generalizations from such populations may be more valid, but only if large samples,
adequately replicated, are employed."[6]

Henderson, on the other hand, concluded that, "... the use of designs including genetic factors is essential in much of animal experimentation," and,

156

". . . the typical procedure of randomly (or conveniently) choosing a particular strain or limited random-bred line of a species for an experiment is not appropriate without specific justification."[9] He also suggests that the use of inbred or random-bred lines be abandoned, except in special cases, and that a species to be studied be represented by the hybrids formed by the complete crossing of several inbred lines.

The purpose of this paper is to examine the relative merits of inbred strains, F_1 hybrids, noninbred strains, and mixed genotypes in various types of research.

Population Characteristics

Some broad generalization about the inbred, F_1 hybrid, noninbred, and mixed populations are given in Table 1. The most important points to note are (1) that a single inbred strain or F_1 hybrid represents only a single genotype. Therefore, if genetic factors are important in the response to the experimental treatment, and if the results are to be generalized to a wider population, more than one strain should be used in the experiment.

TABLE 1 Main Characteristics of Animals Produced by Different Breeding Methods[a]

Breeding Method	Variance		Repeatability	Ease of Production
	Phenotypic	Genetic		
Inbred strain	++++	0	++++	++
F_1 cross	++	0	++++	++
Noninbred strain	+++	variable	+	++++
Mixed genotypes (e.g., from a diallel cross)	++++	++++	++++	+

[a]0 = none; + = very low; ++ = moderate; +++ = high; ++++ = very high.

(2) The genetic variance of noninbred strains is usually unknown, and it is usually difficult to estimate. (3) The productivity of noninbred strains is usually high, and the animals are therefore relatively cheap to produce. (4) The repeatability of samples taken from a noninbred strain segregating genetically may be low.[6]

Types of Experiment

The choice of experimental animal depends largely on the type of experiment that is to be performed. Experiments may be classified into three broad types: (1) bioassay in which interest centers on the substance being assayed; (2) studies in which the population of interest (the target population) is available for experimentation; and (3) studies in which the target population is not available for research.

In bioassay, animals are used as a tool for measuring the concentration (usually) of a substance with a known effect on the animals. Control groups of animals are treated with known concentrations of the substance to give a set of standards for judging the treatment effect. Bioassay is shown diagrammatically in Figure 1.

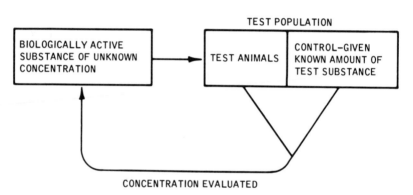

FIGURE 1   Diagram of a biological-assay type of experiment.

The most important point about bioassay is that the animals used are of no direct interest, so the experimenter is completely free (within the bounds of legislation) to choose the animals that give the most economical results. The animals do not have to be representative of a wider population, and in many cases advances in analytical techniques or the use of tissue culture or microorganisms can supercede the use of animals.

TARGET POPULATION AVAILABLE FOR RESEARCH

In the type of experiment illustrated in Figure 2, (a target-population experiment) the research worker is interested in the effect of some experimental treatment on a particular population of animals, the "target population." The experiment is performed on a sample of the target

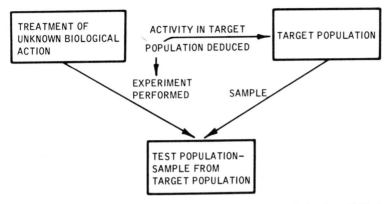

FIGURE 2 Diagram of an experiment in which the target population is available for experimentation.

population, and the results are then used to deduce the characteristics of the whole population. Statistically, the genotype of the target population is often regarded as a fixed effect, though this is not always the case.

This type of experiment is common in animal husbandry, since the main species of economic importance are readily available for research at relatively little cost.

Fundamental biomedical research also falls into this category. Most research of this type is so far removed from direct application to, say, medicine, that the species chosen for the research work becomes a target population in its own right. In this case, the research worker has a considerable amount of leeway in the choice of experimental material, though in many cases practical considerations will largely dictate which species is finally chosen. Once a target population is chosen, there may still be scope for improving the effectiveness of the experiment in the arrangement of the sample of the population to be used for the experiment. This will be considered in more detail later.

TARGET POPULATION NOT AVAILABLE FOR RESEARCH

In the type of experiment illustrated in Figure 3, the target population is not available for research, either for economic or for ethical reasons. Thus, the experiment has to be carried out on samples from one or more "representative populations." The results of the experiment are then interpolated to the representative populations, and these results are in turn extrapolated to the target population. Statistically, the genotype of the representative population will always be regarded as a random variable.

This type of research is occasionally found in agriculture, where it may be more economic to use a "model" experimental animal than to use the species

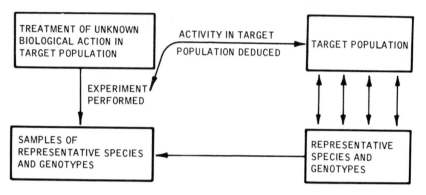

FIGURE 3 Diagram of an experiment in which the target population is not available for research.

of interest. For example, genetic experiments on pigs and cattle can be very expensive, and it may take a long time to get the results, so mice have been used as "model" animals for the larger farm animals.[14] The disadvantages of having to extrapolate from one species to another are in part compensated for by the speed and economy with which results can be obtained.

Screening of new drugs or chemicals that may come in contact with humans also falls in this category, since it is not possible to screen directly in humans.

In this type of research it is frequently necessary to predict the consequences of the experimental treatments in the target population with great accuracy, but the experimenter has a certain amount of freedom to select the representative populations he feels will give a response that closely resembles that of the target population. This freedom extends from various combinations of species down to strains and genotypes within the species.

Sensitivity of an Experiment

When an experiment is performed it is usual to set up the null hypothesis that the treatments have no effect. If the differences between treatments are statistically significant (at some previously determined level of significance) the null hypothesis is rejected and the alternative hypothesis, usually that the treatments have some effect, is accepted. On the other hand, if no significant difference is found between treatments this may be because (a) there is no real difference between treatments, or (b) because there is a real difference between treatments, but the experiment is not sufficiently sensitive to detect it.

The latter type of error (known statistically as a Type 2 error), could have serious consequences in drug screening. Type 2 errors occur when the experiment is too small or the experimental material is too variable. For any given size of experiment, the variability of the experimental material may be reduced by using the genotype as a classification in the design of the experiment. This can be done only when the genotype can be classified, such as when several strains are used or when the experiment is carried out on a within-litter basis in noninbred strains. The increase in precision obtained this way depends on the magnitude of the genetic component of variation for the particular phenotype being measured. Thus, if the genetic component of variation is large, classification of the material by genotype will yield large gains in precision.

Treatment by genotype interactions will also have a strong influence on the sensitivity of the experiment. If these interactions are large, the precision of treatment comparisons will be reduced, even with genotypes included as a classification in the experiment (assuming genotype is a random variable). These effects may be important in certain types of experiments (see reference 11), but they can be detected only when the experimental material is classified by genotype.

Matching the Animal with the Experiment

Having considered the properties of various types of experimental material (inbred strains, F_1 hybrids, and others), the types of experiments that can be performed, and the factors influencing the sensitivity of a given experiment, it is now possible to make some recommendations as to the type of animal that should be used in each case.

BIOASSAY

There is no *a priori* way of deciding which type of animal will be best in bioassay, but certain principles may be followed.

First, the experimental material does not have to be representative of any wider population, so the experimenter has complete freedom of choice.

Second, pilot trials may be used as an objective method of comparing the relative merits of two strains for any given bioassay. The best strain would be the one that would allow the assay to be performed most economically. Since noninbred strains are relatively cheap to produce, they should certainly continue to be tried in this type of experiment; inbred strains and F_1 hybrids should also be used.

Finally, there would generally be no advantage in using a population with a broad genetic base, as precision and cost are the only factors that have to be considered. Thus, a mixed population would have no particular advantages in this type of experiment.

TARGET POPULATION AVAILABLE FOR RESEARCH

In an experiment in which the target population is available for research, the choice of experimental material depends strongly on the nature of the target population. In general, the experimental sample should be a random sample from the target population, but this random sample can be arranged in the experiment according to modern principles of experimental design.

If the target population has a narrow genetic basis, such as an inbred strain, then there is little difficulty in obtaining the experimental sample. Sometimes, however, the target population is much broader, and may even be a whole species. It is impossible to draw a random sample of "mice," for example. This case borders on the next type of experimentation, and in a case like this it may be better to gather some representative genotypes on which to perform the actual experiment. In this case, because of the reduction in precision that is obtained with the use of noninbred strains, representation should be obtained by replication over as many different genotypes as possible, rather than by using a single inbred strain.

TARGET POPULATION NOT AVAILABLE FOR RESEARCH

Lower animals, tissue cultures, and computer simulation are usually cheap and relatively precise, but are not representative of, say, man. Higher animals are expensive but are more representative. The degree of precision obtained with higher animals depends on the type of experimental material available, but in many cases there may be a conflict between precision and representation. If too much attention is given to obtaining a broad representation within a single species, precision may be reduced to such an extent that it will become impossible to demonstrate a significant treatment effect even though the treatment effect is real. This is particularly likely to occur if broad representation is obtained by the use of a noninbred line. It is suggested, therefore, that in this type of experimentation, representation be obtained by replication over species, and that within each species, only a limited number of genotypes, in the form of inbred strains, F_1 hybrids, or partially inbred strains, be used.

An Experimental Illustration

The preceding remarks may be illustrated with reference to a behavior experiment that is outlined in the following paragraphs.

The objectives of the experiment were to examine the effects of previous experience on the "exploratory activity" and "emotionality" of mice and to examine the mode of inheritance of these two characteristics. The target population was "laboratory mice," but because it was not possible to take a random sample from such a population, the experimental group consisted of the progeny of all possible matings between six different inbred strains (excluding the reciprocals), and thus represented 15 different F_1 hybrid genotypes. A total of seven mice of each sex and genotype were examined, giving a total of 210 mice.

"Exploratory activity" was measured by the number of times the experimental subjects reared up on their hind legs when they were placed in the test environment, and "emotionality" was measured by the number of fecal pellets deposited in the test area by the mice during a fixed period.

Treatments involved placing the mice in the novel environment for the first, second, or third time, with gaps of 24 hr between the tests to indicate whether the mice could learn to recognize the test environment, and to adjust their behavior as a result. Treatment means are given in Table 2. The data on "emotionality" were transformed to the square root of $x + 1$ prior to analysis.

TABLE 2   Treatment Means

Treatment	"Emotionality"[a]	"Exploratory Activity"[b]
1	2.53	23.9
2	2.49	20.1
3	2.62	19.9

[a]Heterogeneous at the 5 percent level of significance. Data transformed to $\sqrt{x + 1}$ prior to analysis.
[b]Heterogeneous at the 1 percent level of significance.

Analysis of variance of the data was performed in the usual way.[15] The form of the analysis and the expectations of the mean squares is given in Table 3, and the resulting analyses are given in Table 4.

As the target population was "laboratory mice" it would not have been logical to have used a single genotype in this experiment. The diallel cross gives a mixture of genotypes (though they are all F_1 crosses in this case and, therefore, may not be completely representative of "laboratory mice").

It is possible to use the results of the analysis to work out the relative gain in precision from classifying the animals by genotype as opposed to using the same animals assigned to the treatments at random. This arrangement would be similar to the arrangement of genotypes in a noninbred strain.

Because genotype is considered a random variable, the usual formulas for

TABLE 3 Expectations of Mean Squares Used in the Analysis. Sexes and Treatments Fixed, Strains Random

Source	Degrees of Freedom	Expected Mean Square
Genotypes (G)	g-1	$S^2 + ngS_s^2$
Sexes (X)	x-1	$S^2 + ntS_{sx}^2 + ngt\ T_x^2$
Treatments (T)	t-1	$S^2 + nxS_{st}^2 + ngx\ T_t^2$
G × X	$(g$-1$)\,(x$-1$)$	$S^2 + ntS_{sx}^2$
G × T	$(g$-1$)\,(t$-1$)$	$S^2 + nxS_{st}^2$
X × T	$(x$-1$)\,(t$-1$)$	$S^2 + nS_{sxt}^2 + ngS_{xt}^2$
G × X × T	$(g$-1$)\,(x$-1$)\,(t$-1$)$	$S^2 + nS_{sxt}^2$
Residual	Difference	S^2
Total	sxt-1	

TABLE 4 Analysis of Variance of Exploratory Activity and Emotionality

Source	Degrees of Freedom	Exploratory Activity Mean Square	Emotionality Mean Square
Genotypes (G)	14	313.00[a]	2.4291[a]
Sexes (X)	1	1,841.00[a]	1.1130
Treatments (T)	2	2,203.50[a]	0.9955[b]
G × X	14	165.14[a]	0.6706[b]
G × T	28	46.79	0.1881
X × T	2	26.00	0.8202[b]
G × X × T	28	45.86	0.2367
Residual	540	64.79	0.2086
Total	629	—	—

[a]Heterogeneous at the 1 percent level of significance.
[b]Heterogeneous at the 5 percent level of significance.

comparing the relative efficiency by using a randomized block design instead of a completely randomized design may also be used for estimating the gain in precision by classifying by genotype instead of by assigning the animals to the treatments at random. The appropriate formula follows[15]:

$$\text{Eff.} = \frac{S_{UC}^2}{S_C^2} = \frac{(g\text{-}1)\ M_G + g(t\text{-}1)\ M_{GT}}{(gt\ -\ 1)\ M_{GT}} \cdot \frac{(DF_C + 1)\,(DF_{UC} + 3)}{(DF_C + 3)\,(DF_{UC} + 1)},$$

where: S_{UC}^2 and S_C^2 are the error variances in the two designs, and g and t represent the number of genotypes and treatments, respectively. M_G is the mean square for genotypes, M_{GT} is the mean square for genotype by treatment interactions (the correct error term for the treatment comparison in this case). DF_C is the error degrees of freedom with classification, and DF_{UC} is the error degrees of freedom in the unclassified case.

Using the mean squares for emotionality and exploratory activity given in Table 4, the following estimates for the relative efficiency from classifying by genotype were obtained:

Character	Relative Efficiency
Exploratory Activity	2.64
Emotionality	4.50

Thus, to achieve the same degree of precision in estimating the treatment effects, the experiment would have had to be 2.64 times as large for exploratory activity and 4.50 times as large for emotionality if the animals had not been classified by genotype.

The design of this particular experiment makes it possible to estimate the degree of inheritance of the two characters studied. The method of analysis is that given by Griffing.[8] The following results were obtained:

Character	Heritability (%)	
	Broad Sense	Narrow Sense
Exploratory Activity	85	85
Emotionality	90	63

Thus, both characters were highly inherited, which accounts for the large increase in precision obtained above.

Discussion

The very large increases in precision obtained in the example given through classifying the experimental subjects by genotype would not always be obtained in other types of experiment. Where the character is not genetically controlled, a nonclassified group of experimental animals (such as a noninbred strain) would be as efficient as a classified group. As the magnitude of the genetic control increased, the gains in precision would also increase. Thus, when the character being studied is not genetically controlled, it does not matter which strain is used, but as the genetic control increases, noninbred strains become more and more unsatisfactory experimental material (assuming that they are segregating genetically for the character studied). This is not only because the precision of the experiment will be drastically reduced, but also because information on the genetic control and possible treatment by genotype interactions is confounded with the residual variation. The choice of experimental animal should therefore depend on the degree of inheritance of the character being studied.

The work of Falconer and Bloom[7] is particularly significant in this context. They studied the inheritance of susceptibility to urethane-induced lung tumors in two noninbred strains of mice and found that, ". . . genetic differences among individuals were responsible for 80–90% of the variation in susceptibility." They also found a single major gene controlling the susceptibility of inbred strains of mice to induced lung tumors. These findings suggest that noninbred strains would usually be unsuitable for carcinogenesis studies, and the screening of compounds with possible carcinogenic action.

If noninbred strains are unsatisfactory for screening experiments, and a single inbred strain or F_1 hybrid is also unsatisfactory because it is not representative of a sufficiently wide population, what sort of genetic material should be used? Henderson[9] suggests that a species be represented by a diallel series of F_1 crosses, possibly with pure lines included. Although such a population would be good, it is probably impractical. At this stage it would probably be sufficient to persuade experimenters to use three or four different genotypes in their experiments, some of which could be inbred lines, and some F_1 crosses or even partially inbred strains. This type of experimentation would rapidly build up background information on the relative importance of genetic factors under practical experimental conditions; it would increase the precision of many experiments, and at the same time would ensure that a wide range of gentotypes were tested.

Conclusions

1. Inbred strains and F_1 hybrids should be more widely used in general research to replace noninbred strains.

2. The argument that noninbred strains should be used because they are widely representative of the species is not generally valid because if genetic factors are of minor importance it does not matter which strain is used, while if genetic factors are of major importance, noninbred strains would not only be inefficient, they would not give any indication that the character in question was, in fact, strongly inherited.

3. The best experimental material would be a mixture of different genotypes arranged in a factorial experiment, since this would be representative of several genotypes and would also give high precision and information on the importance of genetic factors and treatment by genetic interactions.

References

1. Becker, W. A. 1962. Choice of animals and sensitivity of experiments. Nature 193:1264-1266.
2. Biggers, J. D., A. McLaren, and D. Mitchie. 1961. Choice of animals for bio-assay. Nature 190:891-892.
3. Biggers, J. D., A. McLaren, and D. Mitchie. 1958. Variance control in the animal house. Nature 182:77-90.
4. Bloom, J. L., and D. S. Falconer. 1962. A gene with major effect on susceptibility to induced lung tumors in mice. J. Nat. Cancer Inst. 33:607-618.
5. Chai, C. K. 1960. Response of inbred and F_1 hybrid mice to hormone. Nature 185:514-518.
6. Doolittle, D. P. 1968. Heterogeneity in a random bred mouse population. (Abstract). Z. Versuchstierk Bd. 10:315.
7. Falconer, D. S., and J. L. Bloom. 1962. A genetic study of induced lung tumors in mice. Brit. J. Cancer 16:665-685.
8. Griffing, B. 1956. Concept of general and specific combining ability in relation to diallel crossing systems. Aust. J. Biol. Sci. 9:463-493.
9. Henderson, N. D. 1967. Prior treatment effects on open field behaviour of mice—A genetic analysis. Anim. Behav. 15:364-376.
10. Jay, G. E. 1963. Genetic stocks and strains. In W. J. Burdette [ed.] Methodology in mammalian genetics. Holden-Day, Inc., San Francisco, Calif.
11. Loosli, R. 1967. Duplicate testing and reproducibility. In R. H. Regamey, W. Hennessen, D. Ikic, and J. Ungar [ed.] International symposium on laboratory animals, S. Karger, New York and Basel.
12. McLaren, A., and D. Michie. 1954. Are inbred strains suitable for bio-assay. Nature 173:686-687.
13. Poiley, S. M. 1960. A systematic method of breeder rotation for non-inbred laboratory animal colonies. Proc. Anim. Care. Panel 10:159-166.
14. Roberts, R. C. 1965. Some contributions of the laboratory mouse to animal breeding research. Anim. Breed. Abstr. 33:339-353.
15. Snedecor, G. W., and W. G. Cochran. 1967. Statistical methods. Iowa State University Press. Ames, Iowa.
16. Staats, J. 1968. Standardized nomenclature for inbred strains of mice. Fourth listing. Cancer Res. 28:391-420.

DISCUSSION

DR. LANE-PETTER: I have several questions here. Do you envisage a role in experimentation for "wild" genotypes?

DR. FESTING: We know that laboratory mice and laboratory animals are, of course, a long way from their wild counterparts, but I do not think it is really necessary to bring in wild genotypes. I think the important thing, when we want, for example, to have an experimental population that is as like humans as possible, is to have a very broad representation of different species. Of course, this should include animals as like humans as possible, but these are generally expensive. But I do not foresee that wild genotypes are likely to be all that important in the future.

DR. LANE-PETTER: Can you compare the F_1 hybrid with the inbred strain in relation to vigor or freedom from deleterious inborn weakness, such as those referred to by Dr. Beadle? What is the advantage to the investigator and his research if he uses the F_1 hybrid?

DR. FESTING: Of course, the F_1 hybrid is much more vigorous than the inbred strain generally. It would probably be freer from inborn weaknesses. This is a function of its vigor. The advantage of the F_1 hybrid, I think, is generally its great phenotypic uniformity compared with the inbred strain. On the other hand, experiments generally depend upon the ratio of sensitivity to the treatment and the phenotypic variation. Inbred strains could easily make up on sensitivity what they lose on phenotypic uniformity.

DR. LANE-PETTER: In your inbred strains, what do you call residual variability? Is it due to residual heterozygosity. Also, do you have an idea of the order of magnitude of this heterozygosity left in the inbred strains that you have studied?

DR. FESTING: No. I think a very, very large proportion of the variation within an inbred strain, the phenotypic variation, is just due to environmental factors. We know, of course, that the moment an animal is born, or even before it is born, it has a unique environment. It has a different environment in the uterus from its brother or sister. It has a different environment the moment it is born. The moment you get social hierarchies, each level exists in a different environment from all others. Then there are all sorts of accidental things that can happen to an animal. Inbred mice are rather sensitive to environmental factors, and this is why they have phenotypic lack of uniformity in many cases.

DR. SERRANO: If one inbred strain is too few to use in an experiment, what is the optimum number of strains to use?

DR. FESTING: I think if you want your experiment to refer to one inbred strain, you have done just right. I would not criticize an experiment for which one inbred strain has been chosen. If, however, you want your experiment to refer to the effects of radiation on humans, for example, I think you could be strongly criticized if you used just one inbred strain of mice, and then used that to say this will be the effect on humans. You would obviously have to use, and I would suggest you should use, not another inbred strain of mice but another species. Using several different inbred strains in a factorial arrangement of experiment would not increase the size or complexity of the experiment very much, but it would provide a lot more information.

BIOCHEMICAL DIFFERENCES
AS A MEANS OF GENETIC
CONTROL FOR INBRED STRAINS
OF LABORATORY MAMMALS

René Moutier

Knowledge of the genotypic formulas of the different inbred strains of laboratory mammals is increasing from year to year; in 1960, for example, the *Standardized Nomenclature for Inbred Strains of Mice*[20] mentioned only color genes and alleles at the *H*-2 locus. The progress achieved in biochemical genetics has revealed an important number of loci, and today most of the main mice strains are defined by about 20 well-known genetic characteristics.[21]

These findings have been obtained by means of a major technique: electrophoresis, particularly starch-gel electrophoresis, of which the great resolving power was fully exploited.

Thus, since 1955, when Ranney[13] described hemoglobin patterns, genetically determined electrophoretic variants were established on starch gel with transferrins,[1,2,18] esterases,[10,11] prealbumin components,[19] and many other proteins and enzymes.

However, this technique is not suitable as a routine because preparing the gel and performing the tests takes too long; simpler media such as acrylamide gel or cellulose acetate appear more suitable for genetic control tests on a large scale.

This paper deals with the possibility of surveying colonies of inbred strains of mice by means of a simplified electrophoresis technique.

Materials and Methods

STRAINS OF MICE AND SAMPLES

The following inbred strains of mice were studied:
A/Gif, AKR/Gif, BALB/c Gif, CBA/Gif, C_3H/He Gif, $C_{57}BL/6$ Gif, $C_{57}BR/cd$ Gif, DBA/2 Gif, SWR/Gif, XLII/Gif, NZB/Gif.

Adult mice from both sexes were examined. Hemoglobin and plasma samples were obtained from blood, and electrophoresis was carried out within 24 hr to avoid denaturation.

Washed kidneys provided good samples of tissues enzymes after being subjected to grinding in 0.5 ml of distilled water, successive freezing–thawing, and storage overnight at 4° C with toluene and centrifugation.

ELECTROPHORESIS

Electrophoresis was performed on standardized "Cellogel" strips (Chemetron-Milano) in Tris-EDTA-boric acid pH 9 buffer.

Hemoglobin or plasma proteins patterns were obtained within 1 hr at room temperature and stained 5 mn with Amido-Schwarz.

Esterase revelation was done according to the procedure outlined by Popp[11] after a run of 55 mn.

NADP enzyme visualization was performed within 1 hr using the Morton Schmukler mixture[8]: 2.5 mg of phenazine methosulfate, 15 mg of NADP, 25 mg of MTT tetrazolium, 50 mg of $MgCl_2 \cdot 6 H_2O$ in 50 ml of Tris-HCl pH 8 buffer. A specific substrate, such as 0.1 M of malic acid for malate dehydrogenase (MDH), 0.005 M of glucose-6-phosphate for glucose-6-phosphate dehydrogenase (G6PDH), or 0.005 M of isocitric acid for isocitrate dehydrogenase (ICDH) was added to this media for incubation at room temperature for a few minutes. Table 1 shows some of the electrophoretic variants found in proteins and enzymes.

Results

BLOOD EXTRACTS

We shall consider here only the *Hbb, Trf,* and *Es*-1 loci. Other loci have been studied by means of starch-gel electrophoresis on this body fluid and have proved to have several allelic forms, such as the prealbumin variants (prelocus) investigated by Shreffler[19] or the erythrocytic esterase *Ee*-2 locus studied by Pelzer[10]; both need sieving media and cannot be tested on "Cellogel" strips.

TABLE 1 Mouse Electrophoretic Variants in Some Proteins and Enzymes

Strain	Hemoglobin	Transferrin	Esterase	ICDH	G6PDH	MDH
A/Gif	Hbb^d	Trf^b	Es-1^b	Id-1^a	$G6pd$-1^b	Mdh-1^a
AKR/Gif	Hbb^d	Trf^b	Es-1^b	Id-1^b	$G6pd$-1^b	Mdh-1^b
BALB/c Gif	Hbb^d	Trf^b	Es-1^b	Id-1^a	$G6pd$-1^b	Mdh-1^a
CBA/Gif	Hbb^d	Trf^a	Es-1^b	Id-1^b	$G6pd$-1^b	Mdh-1^b
C$_3$H/He Gif	Hbb^d	Trf^b	Es-1^b	Id-1^a	$G6pd$-1^b	Mdh-1^a
C$_{57}$BL/6 Gif	Hbb^s	Trf^b	Es-1^a	Id-1^a	$G6pd$-1^a	Mdh-1^b
C$_{57}$BR/cd Gif	Hbb^s	Trf^b	Es-1^a	Id-1^b	$G6pd$-1^a	Mdh-1^b
DBA/2 Gif	Hbb^d	Trf^b	Es-1^b	Id-1^b	$G6pd$-1^b	Mdh-1^a
NZB/Gif	Hbb^d	Trf^b	Es-1^b	Id-1^a	$G6pd$-1^b	Mdh-1^b
SWR/Gif	Hbb^s	Trf^b	Es-1^b	Id-1^a	$G6pd$-1^b	Mdh-1^b
XLII/Gif	Hbb^d	Trf^a	Es-1^b	Id-1^b	$G6pd$-1^b	Mdh-1^b

HEMOGLOBIN: *Hbb* LOCUS, LINKAGE GROUP I

On seeing the pattern (Figure 1) with or without coloration, the strains can easily be subdivided into two classes: The first one, with a single band, expresses the Hbb^s allele in the C$_{57}$BL, C$_{57}$BR, and SWR strains; the second shows two bands, a fast-migrating band, as in the previous case, and a slow-migrating band. Both bands, which are observed in all other

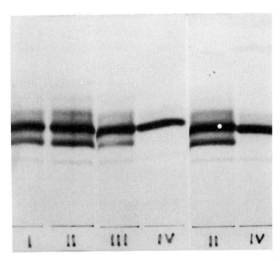

FIGURE 1 Electrophoretic patterns of hemoglobins I and II = XLII strain, III = hybrid (XLII X C$_{57}$BL), IV = C$_{57}$BL strain. The origin is at the bottom (Cathode).

171

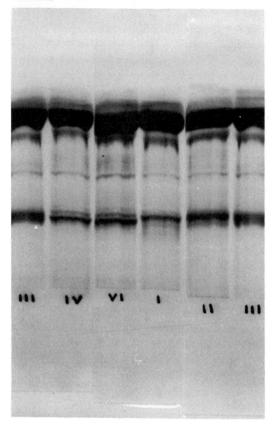

FIGURE 2 Electrophoretic patterns of plasma proteins
I and II = XLII strain, III = mixture C₃H + XLII, IV and
VI = C₃H strain.

strains, particularly the NZB black strain and albino A, AKR, and BALB
strains, are the expression of the Hbb^d allele.

Hbb^s and Hbb^d are two codominant alleles, which were first described by
Ranney[13] and Russell.[17] Hbb^d is more widespread than Hbb^s in laboratory
strains and, it seems, in wild populations as well.[3]

An Hbb^d/Hbb^s hybrid also shows two bands but in different ratios; this
type can be spectrophoretically determined.

TRANSFERRINS: *Trf* LOCUS, LINKAGE GROUP II

Referring to the codominant alleles described,[1,2,18]
the Trf^a allele found in the CBA strain can be distinguished from the Trf^b
allele found in C₃H and all other common inbred strains.

Recently,[9] another strain, the XLII/Gif has been proved, expressing the *Trf^a* allele.

It is difficult to characterize the *Trf^a/Trf^b* hybrid on Cellogel strips, because the *Trf^b/Trf^b* pattern is subject to modifications[6] and sometimes looks like a hybrid pattern.

Figure 2 shows the three types of patterns; the transferrin zone is the first zone from the bottom.

PLASMA ESTERASES: *Es-*1 LOCUS, LINKAGE GROUP XVIII

As described by Popp,[11,12] the *Es-1^a/Es-1^a* pattern of the C_{57}BL and C_{57}BR strains shows a single fast band, while the *Es-1^b/Es-1^b* pattern of all other strains tested provides a double band that is slightly less mobile.

Codominance appears very clearly in the hybrid *Es-1^a/Es-1^b*, which expresses the three bands together (Figure 3).

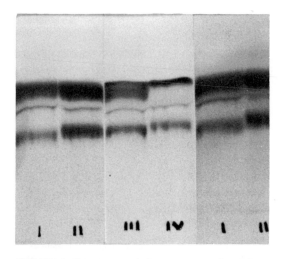

FIGURE 3　Zymograms of plasma esterases. I and II = XLII strain, III = hybrid (XLII X C_{57}BL), IV = C_{57}BL strain.

KIDNEY ENZYMES

Biochemical loci, unlike color loci (e.g., agouti series in mice), rarely show more than two or three allelic forms. In order to characterize each strain, it is necessary to test the largest possible number of loci.

Tissue extracts provide many enzymes that appear under various electro-phoretic forms or "isozymes." A few of these have already been studied, and they prove useful for detection of variants.

ESTERASES
Esterases are nonspecific enzymes that can be re-vealed with naphthyl acetate as a synthetic substrate. Using this technique, they appear in a number of bands on Cellogel strips (Figure 4). Although the sexual dimorphism can be seen very clearly,[15] variations between strains ob-served in the fast-migrating bands cannot be related with certainty to the alleles described at the Es-3 locus.[14]

Better understanding of the metabolic role of these different enzymes will certainly help in selecting a more useful specific substrate.

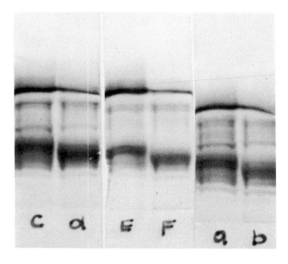

FIGURE 4 Zymograms of kidney esterases
c, d = male and female of C3H strain,
E, F = male and female of C57BL strain,
a, b = male and female of CBA strain.

ICDH: Id-1 LOCUS, LINKAGE GROUP UNKNOWN
Henderson[4] described NADP-IDCH enzyme pat-terns on starch gel.

Crude homogenates from kidney revealed mainly an anodal band with dif-ferent mobilities according to strain, under control of the Id-1^a allele, giving a slow band, or the Id-1^b allele, which produces a faster migrating band.

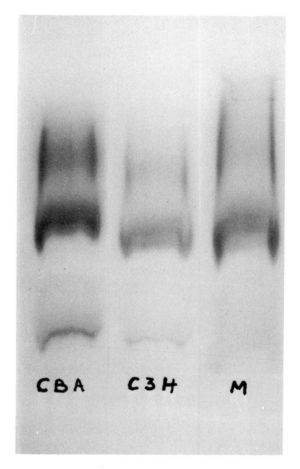

FIGURE 5 Zymograms of isocitrate dehydrogenase (ICDH). M = mixture CBA + C₃H.

The patterns we obtained on Cellogel are resolved in three zones, the second of which showed the highest enzyme activity and variations identical to Henderson's anodal band (Figure 5).

Among the strains not yet tested, strain NZB shows the same mobility as strain $C_{57}BL$ and would have the same allelic form Id-1^a, while XLII strain, like $C_{57}BR$, would carry the Id-1^b allele.

G6PDH: *G6pd*-1 LOCUS, NON-SEX-LINKED

Two alleles $G6pd$-1^a, $G6pd$-1^b [16] control the mobility of a faintly visible band.

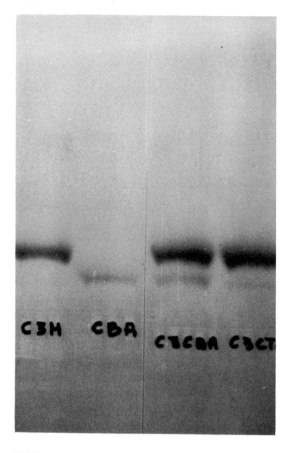

FIGURE 6   Zymograms of malate dehydrogenase (MDH).
C_3CBA and C_3T6 = mixtures C_3H + CBA.

To test this locus with our material, it is better to compare patterns given by the two following mixtures:

1 vol. of sample to be tested + 1 vol. of known solution, $G6pd$-$1^a/G6pd$-1^a
1 vol. of sample to be tested + 1 vol. of known solution, $G6pd$-$1^b/G6pd$-1^b

Differences between homozygote and hybrid patterns appear clearer than between the two homozygote types.

The $G6pd$-1^a pattern was found only in our C_{57}BL and C_{57}BR strains.

MDH: *Mdh*-1 LOCUS, LINKAGE GROUP II

Henderson[5] pointed out the *Mdh*-1^a allele carried by the C$_3$H, CE, and DBA/2 strains and the *Mdh*-1^b of AKR, C$_{57}$BL, and C$_{57}$BR strains.

These findings have been confirmed and extended on "Cellogel": BALB/c and A strains also expressed the *Mdh*-1^a allele, while NZB, CBA, and XLII show the *Mdh*-1^b pattern.

Differences at this locus allow distinction between albino strains as between agouti strains (Figure 6).

Conclusion

The number of these various genetic markers provides a characteristic for biochemical identification of each strain of animals.

A simplified electrophoresis technique like that described above, although inadequate for the detection of new variants, appears useful for routine control, especially with enzymatic markers such as dehydrogenases.

A number of tests may be carried out on a single sample, and several loci can be tested on the same band by adding the different substrates to the media.

The number of lines, sublines, and coisogenic lines is growing every year. In order to preserve the homogeneity of the animal material presented under identical symbols, it is necessary to reach a better understanding of the genotypic formula of the different strains.

Constant progress in enzymology provides an opportunity to extend considerably the probability of detecting individual genes.

References

1. Ashton, G. C. 1961. Serum beta-globulin polymorphism in mice. Aust. J. Biol. Sci. 14:248–253.
2. Cohen, B. L. 1960. Genetics of plasma transferrins in the mouse. Genet. Res. 1:431–438.
3. Heinecke, H. 1964. Hemoglobin types of the wild house mouse. Nature 204:1099–1100.
4. Henderson, N. S. 1965. Isozymes of isocitrate dehydrogenase: Subunit structure and intracellular location. J. Exp. Zool. 158:263–273.
5. Henderson, N. S. 1968. Isozymes and genetic control of NADP-malate dehydrogenase on mice. Arch. Biochem. Biophys. 117:28–33.
6. Klein, P. A. 1966. Starch gel electrophoresis patterns of murine transferrins. Nature 212:1376–1377.
7. Markert, C. L. 1959. Multiple forms of enzymes: Tissue, ontogenetic, and species specific patterns. Proc. Nat. Acad. Sci. 45:753–763.

8. Morton Schmukler, M. D. 1967. Effect of age on dehydrogenase heterogeneity in the rat. J. Gerontol. 22:8–13.
9. Moutier, R. A. 1968. Détermination de quelques caractéristiques génétiques chez des souris de souche XLII/Gif. Exp. Anim. 1(4):261–267.
10. Pelzer, C. F. 1965. Genetic control of erythrocytic esterase forms in *Mus musculus*. Genetics 52:819–828.
11. Popp, R. A. 1962. Inheritance of serum esterases having different electrophoretic patterns among inbred strains of mice. J. Hered. 53:111–114.
12. Popp, R. A. 1965. Loci linkage of serum esterase patterns and oligosyndactylism. J. Hered. 56:107–108.
13. Ranney, H. M. 1955. Filter paper electrophoresis of mouse haemoglobin: Preliminary note. Ann. Hum. Genet. 19:269–272.
14. Ruddle, F. H. 1966. The genetic control of two types of esterases in inbred strains of the mouse. Genetics 54:191–202.
15. Ruddle, F. H. 1967. Tissue specific esterases isozymes of the mouse (*Mus musculus*). J. Exp. Zool. 166:51–64.
16. Ruddle, F. H. 1968. Autosomal control of an electrophoretic variant of glucose-6-phosphate dehydrogenase on the mouse. Genetics 58:599–606.
17. Russell, E. S. 1958. Inherited electrophoretic hemoglobin patterns among 20 inbred strains of mice. Science 128:1569–1570.
18. Shreffler, D. C. 1960. Genetic control of serum transferrin type in mice. Proc. Nat. Acad. Sci. 46:1378–1384.
19. Shreffler, D. C. 1964. Inheritance of a serum pre-albumin variant in the mouse. Genetics 49:629–634.
20. Staats, J. 1960. Standardized nomenclature for inbred strains of mice. Second listing. Cancer Res., 20(2):145–169.
21. Staats, J. 1968. Standardized nomenclature for inbred strains of mice. Fourth listing. Cancer Res. 28:391–420.

DISCUSSION

DR. LANE-PETTER: Are strains A and BALB/c histocompatible?

DR. MOUTIER: I am sorry, but I do not remember the histocompatibility alleles for strains A and BALB/c. I will send you the nomenclature of our laboratory in which these genetic characteristics are indicated.

DR. LANE-PETTER: What is the advantage of this method over skin grafting, which examines simultaneously 15 to 20 loci?

DR. MOUTIER (translated by Dr. Lane-Petter): This is not suggested as an alternative to skin grafting, which examines 15 to 20 loci, but it should be regarded as complementary. It tests other loci, perhaps on other chromosomes, and is an additional test, not an alternative test to skin grafting.

STRAIN AND SEASON DIFFERENCES IN THE REPRODUCTIVE PERFORMANCE OF INBRED STRAINS OF MICE, RATS, AND GUINEA PIGS

C. T. Hansen
W. J. McEleney

Introduction

The factors that affect the reproductive yield from a group of animals can best be determined from long-term studies. However, there has been little information published about the reproductive performance of colonies of inbred strains of animals, particularly for rats and guinea pigs. Experimental data of this kind are often difficult to obtain, especially on a long-term basis. Therefore, data that had been systematically collected from the nucleus colonies of inbred strains of mice, rats, and guinea pigs maintained for the National Institutes of Health research programs were analyzed. The purposes of the study were (1) to determine the differences in the reproductive performance between strains, (2) to evaluate some of the components of reproductive yield, and (3) to study the variability of reproductive performance.

Materials and Methods

The data analyzed were based on the weekly records maintained in the Genetics Unit of the Laboratory Aids Branch inbred nucleus colonies from July 1, 1959, through June 30, 1968, a period of 9 years. The mouse strains studied were A/HeN, AKR/N, AL/N, BALB/cAnN, BRSUNT/N, C_3H/HeN, C_3Hf/HeN, $C_{57}BL/6N$, $C_{57}BL/10ScN$, $C_{57}L/N$, DBA/2N, NBL/N, STR/N, and STR/1N. The rat strains were ACI/N, ALB/N, BUF/N, CAR/N, CAS/N, F344/N, M520/N, OM/N, and W/N. The guinea pig strains were 2/N and 13/N. The strain designations for the three species follow the convention of Jay.[15] These strains were introduced into this colony from previously established strains in 1951 and 1952.

The environment under which these strains were maintained remained essentially the same during the 9 years of the study. The animals were housed separately by species in 9.1 × 9.1-m rooms with temperature averaging 72° F. The mice were housed in 18.4 × 29.2 × 12.7-cm stainless-steel shoe-box cages with perforated lids and the rats in 41.3 × 28.9 × 17.5-cm drawer-type cages with wire mesh bottoms. The guinea pigs were maintained in stainless-steel cages with dimensions of 45.7 × 62.9 × 38.4 cm with drawer-type pans. The cages were cleaned twice weekly. A mixture of soft pine wood chips and cedar shavings was used as bedding for the mice and rats, and soft pine wood chips were used for the guinea pigs.

Purina laboratory feed was offered *ad libitum* to all species. All mice and both pregnant and lactating rats received a twice weekly supplement of fresh washed kale and a porridge consisting of four loaves of stale whole wheat bread and one pound of dried whole milk mixed with tap water. The guinea pigs were given a daily supplement of washed fresh kale and carrots.

Pregnancy was determined by either visual inspection or palpation. Pregnant females from harem matings were isolated in maternity cages that were identical to mating cages. For the rats, a metal insert pan was placed in the maternity cage. Pregnant females in the monogamous matings were not separated from the male; thus, the number of potential pregnancies was greater than in the harem matings. The harem matings made up about 80 percent of the total in all species. Mouse and rat litters were weaned at about 4 weeks of age, and the guinea pigs were weaned at about 3 weeks of age. The isolated females from the harem matings were returned to the mating cage either at the time their litters were weaned or if the pregnancy had failed. A pregnancy failure or loss was defined as a recorded pregnancy failing to wean at least one offspring. In all species the age at which the female breeder was replaced was determined by her reproductive history.

No deliberate selection was made for reproductive performance. Individuals for replacements were selected by pedigree from parents which were directly within the main ancestral line. One person (WJM) supervised the selection of replacements in all strains for the entire study. The females saved for replacement purposes in the colony were included in the succeeding breeder counts irrespective of their age.

The year was divided into 13 equal 4-week periods so that season comparisons could be made for equal time periods for each of the 9 years of the study.

Listed below are the characteristics analyzed:

Percent Pregnancies (%P)   The total number of pregnancies for the 4-week period divided by the female breeder count for that period × 100.

C. T. HANSEN
AND
W. J. McELENEY

Percent Litters Weaned (%LW) The total number of litters weaned for the 4-week period divided by the female breeder count for that period × 100.

Percent Pregnancy Loss (%PL) The total number of pregnancy failures occurring in the 4-week period divided by the female breeder count for that period × 100.

Average Litter Size Weaned (ALSW) The total number of offspring weaned in the 4-week period divided by the total number of litters weaned in that period.

Colony Index (CI) The total number of offspring weaned divided by the female breeder count for the 4-week period.

Replacement Percentage (RP) The total number of female offspring saved for replacement purposes divided by the female breeder count for that 4-week period × 100. The female breeder count included replacement breeders as well.

Results

Since these data were collected under production rather than experimental conditions, a number of problems were encountered in their analysis. In order to make between-strain as well as within-strain comparisons, it was necessary to convert much of the collected data to percentages or ratios. Often data of this kind do not meet the assumptions necessary for the usual test of significances for differences between the means. One of these is that the population variances must be homogeneous. The results of a preliminary analysis between strains within species indicated that this assumption may not hold. Testing these variances using Bartlett's test (reference 24, p. 285) showed heterogeneity of the within-strain variances for each of the characteristics studied in three species. Therefore, the mouse and rat data were analyzed by a technique for testing the hypothesis that the means are the same when the variances are heterogeneous (reference 24, p. 287). This method consists of calculating a pair of weighted mean squares and testing their ratio against the F distribution.

Mice

Tables 1 and 2 present the means and standard deviations, respectively, for the percent pregnancies (%P), percent litters weaned (%LW), percent pregnancy loss (%PL), replacement percentages (RP), average litter size weaned (ALSW), and the colony index (CI) for the 14 mouse strains ranked by their CI over the 9 years of this study. The differences between the strain means for these characteristics were all significant ($P < 0.01$). The within-strain variances were also significantly different ($P < 0.01$) (Table 2).

TABLE 1 Means and Standard Errors for Percent Pregnancies (%P), Percent Litters Weaned (%LW), Percent Pregnancy Loss (%PL), Replacement Percentages (RP), and Time in Days for Colony Turnover, Average Litter Size at Weaning (ALSW), and Colony Index (CI) for the 14 Mouse Strains

Strain	%P $\bar{X}$	SE	%LW $\bar{X}$	SE	%PL $\bar{X}$	SE	RP $\bar{X}$	SE	Days	ALSW $\bar{X}$	SE	CI $\bar{X}$	SE
BALB/cAnN	56.4	± 1.1	51.1	± 1.1	6.4	± 0.6	14.8	± 0.9	(190)	6.49	± 0.11	3.26	± 0.06
C$_{57}$BL/6N	51.1	± 1.2	47.7	± 1.6	13.0	± 1.1	15.9	± 0.9	(176)	5.91	± 0.07	2.64	± 0.08
NBL/N	49.9	± 1.5	49.6	± 1.5	11.4	± 1.1	19.3	± 1.2	(146)	5.31	± 0.06	2.62	± 0.08
AL/N	50.0	± 1.2	40.5	± 1.1	17.9	± 1.2	16.4	± 1.0	(171)	5.78	± 0.08	2.34	± 0.06
C$_3$Hf/HeN	56.0	± 1.3	46.1	± 1.3	20.5	± 1.4	16.8	± 1.2	(168)	4.87	± 0.03	2.25	± 0.07
STR/N	51.3	± 1.5	41.8	± 1.5	22.7	± 1.6	17.8	± 1.2	(157)	5.24	± 0.07	2.22	± 0.09
C$_3$H/HeN	48.8	± 1.1	40.3	± 1.2	19.0	± 1.5	18.2	± 1.2	(154)	5.15	± 0.06	2.10	± 0.07
BRSUNT/N	42.0	± 1.7	38.8	± 1.7	26.0	± 1.9	25.3	± 1.8	(112)	4.88	± 0.07	1.89	± 0.09
A/HeN	53.6	± 1.1	37.8	± 1.1	28.4	± 1.3	15.9	± 1.2	(176)	4.86	± 0.06	1.83	± 0.06
C$_{57}$BL/10ScN	45.9	± 1.1	37.9	± 1.3	17.5	± 1.1	17.0	± 1.2	(165)	4.61	± 0.07	1.74	± 0.06
DBA/2N	44.0	± 1.1	32.7	± 0.9	29.2	± 1.9	16.9	± 1.1	(165)	4.77	± 0.05	1.55	± 0.04
C$_{57}$L/N	41.2	± 1.4	26.2	± 1.2	39.3	± 2.3	20.2	± 1.7	(140)	5.41	± 0.10	1.42	± 0.07
AKR/N	34.2	± 1.4	27.7	± 1.3	29.2	± 2.0	25.8	± 1.9	(109)	5.15	± 0.06	1.41	± 0.06
STR/1N	46.1	± 1.6	34.9	± 1.5	26.4	± 1.8	19.6	± 1.7	(143)	3.79	± 0.07	1.35	± 0.07
Means	47.9	± 1.4	39.5	± 1.5	21.9	± 1.7	18.6	± 1.4	(151)	5.16	± 0.09	2.04	± 0.08

C. T. HANSEN
AND
W. J. McELENEY

TABLE 2   Average Female Breeder Counts (BC), Standard Deviations, and χ^2 Values (Bartlett's Test) for %P, %LW, %PL, ALSW, and CI for the 14 Mouse Strains

Strain	BC	%P	%LW	%PL	ALSW	CI
BALB/cAnN	65	12.0	11.9	6.1	1.18	0.65
$C_{57}BL/6N$	53	12.7	17.2	11.7	0.72	0.83
NBL/N	30	15.9	15.9	12.1	0.65	0.85
AL/N	49	12.9	11.4	13.0	0.86	0.68
C_3Hf/HeN	50	14.6	14.0	15.7	0.34	0.74
STR/N	42	16.1	16.5	17.2	0.75	0.95
C_3H/HeN	55	12.2	13.5	16.0	0.60	0.75
BRSUNT/N	24	18.1	18.0	20.8	0.77	0.92
A/HeN	42	12.0	12.2	14.1	0.67	0.66
$C_{57}BL/10ScN$	38	12.0	14.1	12.3	0.71	0.67
DBA/2N	75	11.6	10.2	15.6	0.54	0.49
$C_{57}L/N$	31	14.6	12.8	24.5	1.05	0.78
AKR/N	46	14.8	13.5	22.0	0.67	0.67
STR/1N	23	17.8	16.2	17.2	0.74	0.95
χ^2 (degrees of freedom = 13)		74.2[a]	83.4[a]	855.2[a]	1,414.2[a]	82.1[a]

[a] $P < 0.01$.

The CI was used as a measure of overall reproductive performance. The average CI for the 14 strains was 2.04, ranging from 3.26 to 1.35 for the BALB/cAnN and STR/N strains, respectively. It was influenced by pregnancy frequency, frequency of litters weaned, frequency of pregnancy loss, average litter size at weaning, replacement percentages, and short- and long-term environmental effects.

Simple correlations of the CI with the %P, %LW, %PL, and ALSW were calculated using the means presented in Table 1. The correlations of 0.74, 0.92, 0.77, and –0.88 of the %P, %LW, ALSW, and %PL, respectively, with the CI all were significant ($P < 0.01$). This indicated that the strains characterized by the highest frequency of pregnancies, weaning the most litters, the largest average litter size at weaning, and the lowest frequency of pregnancy loss were the superior producing strains.

The replacement percentages shown in Table 1 are those that have been found from experience to maintain the maximum CI for a particular strain. These were significantly ($P < 0.01$) different between strains. They ranged from 14.8 percent of the female breeder count for the BALB/cAnN strain to 25.8 percent for the AKR/N strain, with an average of 18.6 percent for all strains. In other words, a complete colony turnover occurred on the average of every 5 months.

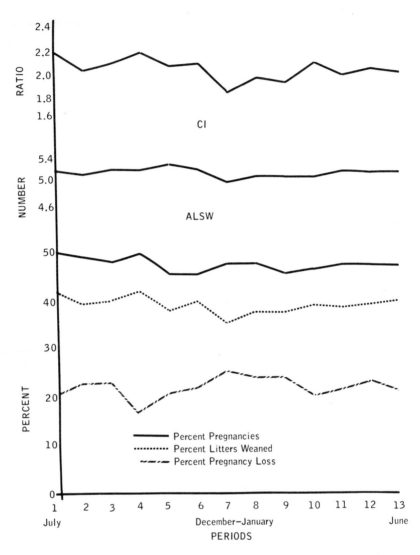

FIGURE 1 The average colony index (CI), average litter size at weaning (ALSW), percent pregnancies, percent litters weaned, and percent pregnancy loss for each of the 13 periods for the 14 mouse strains.

Season differences affected the overall CI for the 14 strains. The average %P, %LW, %PL, ALSW, and CI for all 14 mouse strains in each 13-week period of the year are plotted in Figure 1. The differences between the period means for %LW, %PL, and CI were significant ($P < 0.01$) (Table 3). Since the %P and

TABLE 3 Mean Squares and Test of Significance for Differences between
Period Means for %P, %LW, %PL, ALSW, and CI for the 14 Mouse Strains

		Mean Square				
Source	Degrees of Freedom	%P	%LW	%PL	ALSW	CI
Total	1,637					
Periods	12	244.7	399.7^a	790.9^b	0.7	1.1^a
Error	1,625	236.3	204.2	333.3	1.0	0.6

$^a P < 0.05.$
$^b P < 0.01.$

ALSW did not vary significantly between the periods, the period differences in
%PL were responsible for the changes in the CI. This is confirmed by the significant $(P < 0.01)$ correlations of %LW (0.89) and %PL (-0.84) with the CI
based on the data presented in Figure 1.

On a within-strain basis, the frequency of litters weaned was also the most
important factor affecting the CI (Table 4). The correlations of %LW with the
CI were significant $(P < 0.01)$ for all but one of the 14 mouse strains.

TABLE 4 Between-Period Within-Strain Simple
Correlations of %P (X_1), %LW (X_2), %PL (X_3), and
ALSW (X_4) with the CI (Y) for each of the 14
Mouse Strains

Strain	$r_{X_1 Y}$	$r_{X_2 Y}$	$r_{X_3 Y}$	$r_{X_4 Y}$
BALB/cAnN	-0.400	0.870^b	0.361	0.180
$C_{57}BL/6N$	-0.148	0.342	-0.241	0.025
NBL/N	-0.187	0.914^b	-0.211	0.513
AL/N	-0.011	0.867^b	0.534	0.199
C_3Hf/HeN	0.554^a	0.960^b	-0.547	0.671^a
STR/N	-0.219	0.923^b	-0.368	0.537
C_3H/HeN	-0.588^a	0.964^b	-0.093	0.733^b
BRSUNT/N	0.291	0.946^b	0.059	-0.086
A/HeN	-0.374	0.957^b	-0.128	0.301
$C_{57}BL/10ScN$	-0.347	0.961^b	-0.728^b	0.705^b
DBA/2N	-0.077	0.855^b	-0.091	0.399
$C_{57}L/N$	0.521	0.962^b	-0.298	0.095
AKR/N	-0.365	0.972^b	-0.409	0.394
STR/1N	0.053	0.963^b	-0.076	0.438

$^a P < 0.05.$
$^b P < 0.01.$

TABLE 5 Means and Standard Errors for Percent Pregnancies (%P), Percent Litters Weaned (%LW), Percent Pregnancy Loss (%PL), Replacement Percentages (RP), and Time in Days for Colony Turnover, Average Litter Size at Weaning (ALSW), and Colony Index (CI) for the Nine Rat Strains

Strain	%P		%LW		%PL		RP		Days	ALSW		CI	
	$\bar{X}$	SE	$\bar{X}$	SE	$\bar{X}$	SE	$\bar{X}$	SE		$\bar{X}$	SE	$\bar{X}$	SE
F344/N	44.6	± 1.2	40.4	± 1.1	3.2	± 0.4	15.3	± 1.1	(183)	7.89	± 0.09	3.24	± 0.10
M520/N	46.1	± 1.5	39.6	± 1.1	6.0	± 0.9	16.2	± 1.1	(173)	6.62	± 0.07	2.60	± 0.07
BUF/N	43.1	± 1.4	33.8	± 1.1	20.7	± 1.7	15.6	± 1.0	(180)	6.52	± 0.08	2.19	± 0.07
W/N	31.2	± 1.4	27.1	± 1.3	15.2	± 2.1	15.9	± 1.4	(176)	6.69	± 0.13	1.82	± 0.10
OM/N	33.0	± 1.2	25.2	± 1.0	19.7	± 1.6	18.7	± 1.5	(150)	6.94	± 0.11	1.75	± 0.08
ACI/N	47.8	± 1.4	32.5	± 0.8	28.5	± 1.7	14.8	± 1.1	(189)	5.17	± 0.06	1.70	± 0.06
CAR/N	39.1	± 1.8	25.3	± 1.2	19.1	± 2.0	19.7	± 1.1	(142)	6.05	± 0.12	1.54	± 0.08
ALB/N	31.7	± 1.6	26.1	± 1.3	7.0	± 1.6	17.3	± 1.6	(162)	5.57	± 0.10	1.44	± 0.08
CAS/N	23.8	± 1.6	16.6	± 1.3	25.0	± 3.0	18.7	± 2.1	(150)	4.74	± 0.14	.81	± 0.07
Means	37.8	± 1.6	29.6	± 1.3	16.0	± 1.9	16.9	± 1.4	(165)	6.26	± 0.13	1.90	± 0.10

C. T. HANSEN
AND
W. J. McELENEY

Rats

The means and standard deviations for %P, %LW, %PL, RP, ALSW, and CI are presented in Tables 5 and 6, ranked by the CI for the nine rat strains. The differences between the strain means for these characteristics were all significant ($P < 0.01$). The CI ranged from 3.24 to 0.81 for the F344/N and CAS/N strains, respectively. The average for all nine strains was 1.90. For the rats, the most important characteristics affecting the CI, and in turn the strain ranking, were %P, %LW, and ALSW. The correlations of %P, %LW, and ALSW with the CI, when calculated from the data presented in Table 5, were 0.72, 0.94, and 0.85, respectively, and all of these were significant ($P < 0.01$). The correlation of %PL with the CI (–0.65) was not significant.

TABLE 6 Average Female Breeder Counts (BC), Standard Deviations, and χ^2 Values (Bartlett's Test) for %P, %LW, %PL, ALSW, and CI for the Nine Rat Strains

Strain	BC	%P	%LW	%PL	ALSW	CI
F344/N	64	13.0	11.8	4.5	1.01	1.06
M520/N	28	15.9	11.7	9.8	0.76	0.79
BUF/N	50	15.7	12.1	18.1	0.82	0.79
W/N	30	15.4	14.4	22.8	1.37	1.13
OM/N	44	12.8	11.0	17.8	1.14	0.82
ACI/N	54	15.2	8.4	18.6	0.68	0.67
CAR/N	24	19.0	13.1	22.2	1.26	0.87
ALB/N	24	17.3	14.0	12.0	1.06	0.82
CAS/N	20	17.1	14.6	32.2	1.50	1.13
χ^2 (degrees of freedom = 8)		29.3[a]	50.8[a]	975.6[a]	470.9[a]	1,580.3[a]

[a] $P < 0.01$.

Strain differences were significant ($P < 0.01$) also for replacement percentages (Table 5). The replacement percentages of each 4-week period over the 9-year period of the study averaged 16.9 percent of the female breeder count, and there was a complete colony turnover every 165 days on the average for all strains. The RP ranged from a low of 14.8 percent for the ACI/N strain, with a colony turnover of 190 days, to the high of 19.7 percent for the CAR/N strain, with a colony turnover of 144 days.

Seasonal changes in the CI for the rats were similar to those found in the mice. The means for %P, %LW, %PL, ALSW, and CI for each of the 13 periods

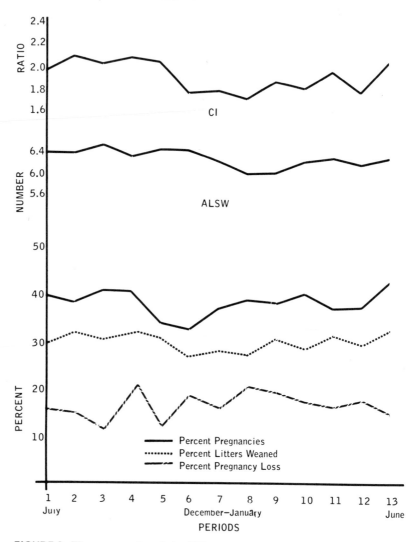

FIGURE 2 The average colony index (CI), average litter size at weaning (ALSW), percent pregnancies, percent litters weaned, and percent pregnancy loss for each of the 13 periods for the nine rat strains.

in the year are plotted in Figure 2. The differences in the period means for all these characteristics were significant ($P < 0.01$) (Table 7). The frequency of litters weaned and the average litter size weaned were more important in affecting the between period CI over all nine rat strains than either the frequency of

TABLE 7 Mean Squares and Tests of Significance of Differences between Period Means for %P, %LW, %PL, ALSW, and CI for the Nine Rat Strains

| Source | Degrees of Freedom | Mean Square | | | | |
		%P	%LW	%PL	ALSW	CI
Total	1,052					
Periods	12	644.1[a]	333.2[b]	631.7[b]	2.3[b]	1.7[a]
Error	1,040	305.0	159.0	360.5	1.2	0.3

[a]$P < 0.01$.
[b]$P < 0.05$.

pregnancies or the frequency of pregnancy losses. The between-period correlations of %LW (0.94) and ALSW (0.58) with the CI were significant at the $P < 0.01$ and $P < 0.05$ levels, respectively, whereas the correlations of %P (0.38) and %PL (–0.42) with the CI were not significant. These correlations were based on the data in Figure 2. On a within-strain basis, the frequency of litters weaned was the most important factor in affecting the within-strain CI (Table 8). Average litter size weaned was also important in some of the strains.

TABLE 8 Between-Period Within-Strain Correlations of %P (X_1), %LW (X_2), %PL (X_3), and ALSW (X_4) with the CI (Y) for Each of the Nine Rat Strains

Strain	$r_{X_1 Y}$	$r_{X_2 Y}$	$r_{X_3 Y}$	$r_{X_4 Y}$
F344/N	0.282	0.881[a]	–0.507	0.558[b]
M520/N	–0.090	0.915[a]	0.046	–0.081
BUF/N	–0.367	0.530[a]	–0.324	0.679[b]
W/N	–0.389	0.743[a]	–0.042	0.330
OM/N	–0.072	–0.112	–0.255	–0.052
ACI/N	–0.077	0.975[a]	–0.435	0.770[a]
CAR/N	0.080	0.952[a]	–0.162	0.473
ALB/N	–0.009	0.942[a]	0.419	0.371
CAS/N	0.573[b]	0.920[a]	–0.018	0.086

[a]$P < 0.01$.
[b]$P < 0.05$.

Guinea Pigs

The means and standard deviations for %P, %LW, ALSW, RP, and CI for the two guinea pig strains are presented in Tables 9 and 10. Data for %PL were not available for the guinea pigs. A modified t-test (reference 25, p. 81) was used to test for the significance of difference between the strain means since there were only two strains and the within-strain variances differed significantly ($P < 0.01$) between the two strains (Table 10). The

TABLE 9 Means and Standard Errors for Percent Pregnancies (%P), Percent Litters Weaned (%LW), Replacement Percentages (RP) and Days for Colony Turnover, Average Size at Weaning (ALSW), and Colony Index (CI) for the Two Guinea Pig Strains

	%P		%LW		RP			ALSW		CI	
	$\overline{X}$	SE	$\overline{X}$	SE	$\overline{X}$	SE	Days	$\overline{X}$	SE	$\overline{X}$	SE
2/N	27.7 ± 1.1		22.0 ± 0.8		18.8 ± 1.0		(260)	2.48 ± 0.03		0.550 ± 0.020	
13/N	19.2 ± 1.1		15.9 ± 1.0		11.5 ± 1.1		(244)	2.69 ± 0.06		0.424 ± 0.027	
Means	23.5 ± 1.1		18.9 ± 0.9		11.1 ± 1.1		(252)	2.58 ± 0.05		0.487 ± 0.025	

TABLE 10 Average Female Breeder Count (BC), Standard Deviations, for %P, %LW, ALSW, and CI for the Two Guinea Pig Strains

Strain	BC	%P	%LW	ALSW	CI
2/N	73	11.6	8.9	0.37	0.22
13/N	62	10.4	10.3	0.64	0.30

differences between the strain means for %P, %LW, ALSW, and CI were significant ($P < 0.01$). The average CI for the 2/N strain was 0.550, and for the 13/N strain, it was 0.424. That is, the 2/N strain weaned an average of 0.6 offspring per female breeder in each 4-week period, compared with 0.4 for the 13/N strain for the 9 years of this study.

Season differences as reflected by the period means (Figure 3) were significant ($P < 0.01$) for the CI. The only important factor affecting the CI was the

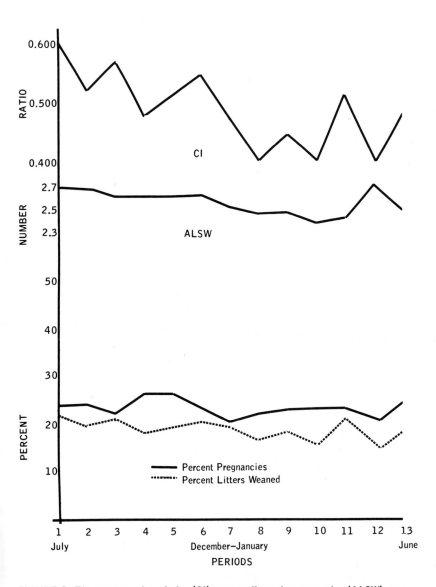

FIGURE 3 The average colony index (CI), average litter size at weaning (ALSW), percent pregnancies, and percent litters weaned for each of the 13 periods for the two guinea pig strains.

frequency of litters weaned. The correlation of the %LW with the CI (0.92) was significant ($P < 0.01$).

Within the two strains, the frequency of litters weaned was again the most important factor (Table 11) affecting the CI. However, for the 2/N strain, average litter size at weaning also contributed significantly ($P < 0.05$).

TABLE 11 Between-Period Within-Strain Correlations of %P (X_1), %LW (X_2), % ALSW (X_3), with the CI (Y) for Each of the Two Guinea Pig Strains

Strain	r_{X_1Y}	r_{X_2Y}	r_{X_3Y}
2/N	0.332	0.857[a]	0.660[b]
13/N	−0.130	0.945[a]	0.213

[a]$P < 0.01$.
[b]$P < 0.05$.

Discussion

The results of this analysis can be considered from two broad aspects: (1) the differences in reproductive yield between strains and some of the factors affecting these yields, and (2) the phenotypic variability within the strains and how this is related to the assumed homozygosity of the strains. The first should be of practical interest to both producers and users of inbred animals. The second poses an interesting problem; if the phenotypic variability observed within the strains reflects a lesser degree of inbreeding in the animals than predicted, this may affect their use in research as well as being of theoretical interest.

FACTORS AFFECTING REPRODUCTIVE PERFORMANCE

The frequency of litters weaned was found to be the most important factor affecting the level of reproductive performance between as well as within the strains of the species included in this study. The frequency of litters weaned was affected by the frequency of pregnancies and pregnancy losses. In the mice and rats, the frequency of pregnancy loss appeared to be more important than pregnancy frequency, whereas in the guinea pigs, the pregnancy frequency was the most important factor affecting reproductive performance.

The role of the frequency of pregnancy loss was most clearly demonstrated

C. T. HANSEN
AND
W. J. McELENEY

by the seasonal effects in the mice. Neither pregnancy frequency nor average litter size at weaning was affected by seasonal differences, whereas the frequency of pregnancy loss appeared to be seasonal, which in turn affected the frequency of litters weaned and the colony index (CI). In the rats, the situation was not as clear, since pregnancy frequency and litter size at weaning as well as frequency of pregnancy loss were affected by the seasonal differences.

In the mice and rats, the cause of the increased frequency of pregnancy loss during the winter months was not clear. Observations of the maternity cages indicated that parturition was normal, but some of the newborn litters were not nursed or cared for. Thus, the greatest proportion of the loss occurred within 24 hr after delivery. The loss of the young in most mouse strains involved complete litters. In the rats, there was an increase in the frequency of differential mortality within the litter, which may have accounted for the greater importance of litter size at weaning than in the mice.

The similarity of seasonal effects on colony yields in the three species suggests a common cause. Perhaps the temperature-control mechanism was not adequate to cope with the extremes of the outside temperature during winter months. Also, seasonal changes in humidity may be involved. Thus, there may have been enough variation in the temperature and humidity within the animal room to introduce a nonspecific stress that inhibited lactation. Light may also have been a factor, since light and dark periods were not standardized.

Average colony age will also affect reproductive yield, although this effect is not season specific. Hinkle and Hansen (unpublished data) studied eight of these mouse strains and found that each strain had a characteristic age of maximum reproduction. When this age had been passed, it was no longer economical to keep the breeder in the colony. However, for the 9 years of data analyzed, replacement of breeding females was on the basis of reproductive failure or death. The replacement percentages shown for each strain were those that have been found from experience to maintain the maximum colony yield.

The average age of the females in the colony has a twofold effect on reproductive yield. When replacement breeders are included in the breeder count, this has the effect of reducing the pregnancy frequency and, in turn, the CI. In a strain characterized by a high replacement rate such as the AKR/N mouse strain, there will be a greater proportion of females in the colony that have not reached reproductive age than in a strain with a lower replacement rate. Moreover, as Hinkle and Hansen found (unpublished data), the frequency of pregnancy loss tends to be bimodal with age, being highest in first parities and in older animals. They also observed, as did Murray,[21] Roberts,[23] and Biggers et al.,[1] that litter size at weaning was smaller in first parities. However, Festing[9] did not find this.

Since these data were based on production colonies rather than experimental colonies, the replacement percentages reflected managerial as well as biological considerations. Colony sizes increased in all strains, with the result that the replacement percentages were greater than would otherwise have been necessary. However, the relative colony sizes remained constant. Therefore, the percentages reported should reflect fairly closely the length of reproductive life of the various strains.

Some of the strains were characterized by obvious pathological conditions, such as the onset of leukemia at an early age in the AKR/N and mammary tumors in the C_3H/HeN mouse strains, that affected their length of reproductive life. The replacement rate for the rat strains that were highly susceptible to chronic respiratory disease, such as the CAR/N, CAS/N, and OM/N, also reflected the death of the males from this condition. The CAS/N rat strain was also characterized by a high frequency of sterile matings, which, according to Hoornbeck,[14] may have been the result of reduced libido in the male. Burack et al.[2] observed from a study of the Albany (ALB/N) strain of rats a high frequency of pregnancy resorptions, which may have contributed to replacing the females because of apparent breeding failure. The usual reason for replacing females in the 13/N strain of guinea pigs was the failure to record further pregnancies after one or two litters had been weaned. Hoar[12] and Hoar et al.[13] have indicated that this strain is deficient in thyroid activity.

Many genetic and environmental factors affect the reproductive yield of a group of animals. However, the traditional measure of reproductive performance, average litter size at weaning, does take all of these into account, since some of them affect the frequency of litters weaned but not the average litter size. The CI provides a reasonably accurate measure of the colony reproductive performance since it reflects the frequency of litters weaned as well as the cost of replacement. It is, however, subject to sampling error, particularly when colony size is rapidly changing. Other approaches of measuring colony yields have been considered by Lane-Petter et al.[17] and Festing.[9] Their approaches have been on an individual female basis and, therefore, are not practical under large-scale production conditions.

The CI is also useful for planning purposes. It can be used for determining space requirements to meet production demands for a given time period. Although the colony indices in this study were calculated for strains that consisted predominately of harem matings, experience has shown these correspond quite closely to those based on monogamous matings. The problem of estimating space requirements for inbred strains of mice has also been considered by Festing and Bleby.[10]

PHENOTYPIC VARIABILITY WITHIN STRAINS

The magnitude of the phenotypic variabilities for reproductive performance between and within the strains in each of the three species is of interest from both a practical and theoretical standpoint. These

C. T. HANSEN
AND
W. J. McELENEY

differences are relevant to the question of whether inbreds are superior to noninbreds for certain kinds of research. Theoretically, the variability found between inbred strains is genetic in origin, and that within strains is environmental in origin. The significant differences found for the between-strain variabilities clearly indicate that there are genetic differences between the strains. However, the question of most interest is the nature of the within-strain variability.

This variability has to be considered within the context of the characteristic concerned. If a characteristic is genetically complex, such as fitness or reproductive performance, as contrasted to one that is simply inherited, it becomes difficult to determine whether the cause of the variability is genetic, environmental, or the result of an interaction between the genotype and environment. This analysis was not designed to assess the relative importance of these factors in contributing to the within-strain variability.

However, some evidence suggests that genetic variability is still present within long-sib-mated strains. Strong[26] compared the reproductive performance of two sublines that had been separated from their ancestral strain after 28 generations of sib mating. He found considerable variation between the two sublines and between the sublines and the ancestral strain in their reproductive patterns.

Loeb et al.[18] found that one of two strains of rats was still heterozygous, as measured by intrastrain skin grafts, after 102 generations of sib mating. In rabbits, Chai[4] found that the percentage of successful intrastrain skin grafts tended to increase as the number of generations of sib mating increased (up to 18 generations), but a wide variation in survival days of the rejected grafts still remained. McLaren and Michie,[20] Deol et al.,[5] and Carpenter et al.[3] have reported subline differences with respect to a number of skeletal characters in inbred strains of mice.

Genetic heterogeneity within long-sib-mated strains can arise from mutation or can be the result of natural selection favoring heterozygotes at some loci. The mutation rate for any one locus is low, but if these effects are taken at all loci over many generations, they could be an important source of within-strain variability. In the studies of Deol et al.[5] and Carpenter et al.,[3] within-strain variability was interpreted as due to mutation. However, Grewal,[7] further analyzing these data by a different statistical technique, concluded that the mutation rate to account for these differences was nearly 1,000 times that estimated from other studies. Also, Wallace (1965) reviewed the factors that might contribute to genetic variability within inbred strains and concluded that mutation was not sufficient explanation to account for the observed variabilities.

Almost all inbreeding studies have shown a reduction in fitness. For example, studies reported by Fekete[8] and Lyon[19] show that the main effect of in-

195

breeding was to reduce the number of young produced because of an increased loss of eggs or embryos before implantation. Duzgunes[6] concluded that losses of this kind occur in a selective manner in that the more homozygous embryos were lost more frequently than the more heterozygous ones. Thus, natural selection was opposing inbreeding, since the more homozygous individuals were less fit. Natural selection can favor heterozygotes at single loci or heterozygotes of segments of a chromosome.

However, the ultimate result of the combined effects of inbreeding and natural selection depends on whether selection is within or between lines. Reeve[22] has shown that natural selection within a line can retard inbreeding progress, but cannot arrest it. Whereas, Hayman and Mather[11] found that when natural selection is operating between lines, a state of balance between opposing forces of inbreeding and natural selection will occur, the balance being determined by the selection pressure.

The techniques used in maintaining inbred nucleus colonies of long-sib-mated strains are designed to minimize the amount of between-line selection. In a high-producing strain, the decision as to the fate of a particular subline is based on its relationship to a common ancestor; in poorer producing strains, reproductive performance is the deciding factor. The technique of sib mating plus elimination of sublines in inbred nucleus colonies has been effective in minimizing the amount of genetic variability within an inbred strain, but it cannot be eliminated entirely. The effectiveness of this technique is probably influenced to a considerable degree by the genetic complexity of the characteristic. The work reported by Strong[26] and the experience of the authors with the strains studied suggest that, with respect to reproductive performance, fixation is not complete. The experimental evidence on this point is limited.

If, however, it can be assumed that strains are homozygous with respect to reproductive performance, then the within-strain variabilities were caused by the environment. Unfortunately, data were not available to make an estimate of the variability due to the genetic and environmental components. Although there is little experimental evidence, it would be reasonable to conclude that traits related to fitness are more sensitive to environmental variations than those that bear little relationship to fitness.

The results of the analysis showed that there were genetic differences between the strains, and the seasonal comparisons revealed that environmental differences were also present. Thus, it is possible that some of the strains were better adapted to the environmental conditions than others, and this would be reflected in within-strain variabilities. That is, there may have been some genotype–environment interactions. Studies of genotype–environment interactions with respect to reproductive fitness for inbred strains are limited. In inbred strains of rats and mice, Hansen (unpublished data) has found a geno-

C. T. HANSEN
AND
W. J. McELENEY

type–environment interaction for reproductive performance and diet. Hansen and McEleney (unpublished data) have found a significant genotype–environment interaction when comparing the reproductive performance of the two inbred strains of guinea pigs fed two commercial diets having different protein levels. This type of interaction is of practical importance in that it will generally be necessary to fit the environment to the strain rather than the reverse.

The results of this analysis present a number of questions that need to be studied in order to ensure that animals used for research are satisfactory. Obviously, the effect of inbreeding on components of reproductive performance is of interest. Of major importance is the source of the within-strain variability: Is it due to genetic factors or to environmental factors, or is it a combination of these? Thus, long-term studies with detailed within-strain analyses designed to separate these components are clearly needed.

Summary

The purpose of this analysis was to obtain some basic information on the factors that might affect the reproductive yield of inbred animals and to measure the population variability with respect to reproductive performance. The analysis was based on 9 years of data collected from sib-mated inbred nucleus colonies of 14 strains of mice, nine strains of rats, and two strains of guinea pigs. The characteristics studied were the frequency of pregnancies, frequency of litters weaned, frequency of pregnancy losses, replacement percentages, average litter size weaned, and the colony index. The colony index (CI) was defined as the average number of offspring weaned per female breeder in a 4-week period. Season effects were studied by dividing the year into 13 equal periods. Significant differences were found between the strains in each species for all the characteristics studied, except replacement percentages in the guinea pigs. The most important factor affecting reproductive performance in all species was the frequency of litters weaned. Significant differences between strains were also found for the within-strain phenotypic variabilities for the various characteristics studied in each of the three species. Some factors that might contribute to the phenotypic variability within long-sib-mated inbred strains were discussed.

Acknowledgment

The authors wish to thank Dr. Janet Hansen for discussions and constructive criticism.

References

1. Biggers, J. D., C. A. Fenn, and A. McLaren. 1962. Long-term reproductive performance of female mice. II. Variation with age of parity. J. Reprod. Fert. 3:313.

2. Burack, E., J. M. Wolfe, and A. W. Wright. 1939. Prolonged vaginal bleeding, fetal resorption, and prolonged gestation in the Albany strain of rat. Anat. Rec. 75:1.

3. Carpenter, J. R., H. Gruneberg, and E. S. Russell. 1957. Genetical differentiation involving morphological character of an inbred strain of mice. II. American branches of the $C_{57}BL$ and $C_{57}BR$ strains. J. Morphol. 100:377.

4. Chai, C. K. 1968. The effect of inbreeding in rabbits. Transplantation 6:689.

5. Deol, M. S., H. Gruneberg, A. G. Searle, and G. M. Truslove. 1957. Genetical differentiation involving morphological characters of an inbred strain of mice. I. A British branch of the $C_{57}BL$. J. Morphol. 100:345.

6. Duzgunes, O. 1950. The effect of inbreeding on the population fitness of S.C.W. leghorns. Poult. Sci. 29:227.

7. Grewal, M. S. 1962. The rate of genetic divergence of sublines in $C_{57}BL$ strain of mice. Genet. Res. 3:226.

8. Fekete, E. 1947. Differences in the effect of the uterine environment upon the development in the DBA and $C_{57}BL$ strains of mice. Anat. Rec. 98:409.

9. Festing, M. 1968. Some aspects of reproductive performance in inbred mice. Lab. Anim. 2:89.

10. Festing, M., and J. Bleby. 1968. A method for calculating the area of breeding and growing accommodations required for a given output of small laboratory animals. Lab. Anim. 2:121.

11. Hayman, B. I., and K. Mather. 1953. The progress of inbreeding when homozygotes are at a disadvantage. Heredity 7:165.

12. Hoar, R. M. 1955. Abortion and stillbirth in thyroxin treated strain 13 guinea pigs. Anat. Rec. 121:311. (abstr.)

13. Hoar, R. M., R. W. Goy, and W. C. Young. 1957. Loci of action of thyroid hormone on reproduction in the female guinea pig. Endocrinol. 60:337.

14. Hoornbeck, F. 1968. Mating success and litter size variation within and between inbred and hybrid generations of rats. J. Anim. Sci. 27:1378.

15. Jay, G. E., Jr. 1963. Genetic strains and stocks. In W. J. Burdette [ed.] Methodology in mammalian genetics. Holden-Day, Inc., San Francisco, Calif.

16. Wallace, M. E. 1965. The relative homozygosity of inbred lines and closed colonies. J. Theoret. Biol. 9:193.

17. Lane-Petter, W., F. M. Brown, M. J. Cook, G. Porter, and A. A. Tuffery. 1959. Measuring productivity in breeding of small animals. Nature 183:339.

18. Loeb, L., H. D. King, and H. T. Blumenthal. 1943. Transplantation and individuality differences in inbred strains of rats. Biol. Bull. 84:1.

19. Lyon, M. F. 1959. Some evidence concerning the "mutational load" in inbred strains of mice. Heredity 13:341.

20. McLaren, A., and D. Michie. 1954. Factors affecting vertebral variation in mice. I. Variation within an inbred strain. J. Embryol. Exp. Morphol. 2:149.

21. Murray, W. S. 1934. The breeding behavior of the dilute brown stock of mice (Little dba.). Amer. J. Cancer 20:573.

22. Reeve, E. C. R. 1955. Inbreeding with homozygotes at a disadvantage. Ann. Human Genet. 19:332.

23. Roberts, R. C. 1961. The lifetime growth and reproduction of selected strains of mice. Heredity 16:369.

24. Snedecor, G. W. 1956. Statistical methods. Iowa State College Press. Ames, Iowa.
25. Steel, R. G. D., and J. H. Torrie. 1960. Principles and procedures of statistics. McGraw-Hill Book Co., Inc. New York.
26. Strong, L. C. 1968. The origin of some inbred mice: Genetic selection of strains for gerontological research. *In* The laboratory animal for gerontological research, NAS Publ. 1591, National Academy of Sciences, Washington, D.C.

DISCUSSION

DR. LANE-PETTER: Do you find any correlation between differential strain productivity and birth weight, weaning weight, or rate of weight gain after weaning?

DR. HANSEN: This is an interesting question. We have some material collected to study this particular point, but we have not gotten around to getting it analyzed yet. There is some reason to believe from other work that there is a negative relationship between rapid growth and lifetime reproductive performance. So there is a question as to whether, rapid growth, postweaning, is desirable for maximum reproductive performance.

We found not in the work reported here, but in some other work, that raising rats under restricted feeding regimes during the period from 3 to 9 weeks of age actually improved reproductive performance. So, maybe maximum growth postweaning is not the most desirable thing. We are going to study it.

DR. LANE-PETTER: Are the diagrams of the seasonal variation based on one or on several years, and are you keeping the climatic conditions constant the year around?

DR. HANSEN: Each point on these diagrams is an average of 9 years for the 14 strains. With respect to maintenance, we do have controlled temperature and humidity mechanisms, although we do not control our light frequency.

DR. LANE-PETTER: In the case of the guinea pigs, there was a marked drop in the colony index January through March. Have you any comment on that?

DR. HANSEN: This is a question that I think might be related to the diet. We have found, and this is rather a sore subject, that it is necessary to supplement the diets with greens, and during the winter months the quality of our greens is generally very poor. Whether this is a factor, I do not know.

DR. LANE-PETTER: While on the subject of season, another question: Can you extract from your data the effect of the season in which the dam was born?

DR. HANSEN: We can, but again this will require an individual within-strain analysis. This, I probably should repeat, is a preliminary analysis to establish a few parameters from which we plan to consider some very detailed analyses, and the age of dam is one point I want to study, although it is confounded by season. It is a rather difficult thing to get out.

DR. LANE-PETTER: What is more important in reproductivity, season or age of female?

DR. HANSEN: I would gather that the age of the female is very important. With respect to pregnancy loss, it tends to be bimodal. In other words, we find a higher frequency of pregnancy losses in first parities, and then as the animals become older, the frequency of pregnancy loss again increases. I should probably point out that the replacement percentages, or the time for colony turnover, were based on what we have found to give us the maximum yield from our colonies. We do not have a selection index that we followed, but we have based this on experience with the strains. Mr. McEleney, who is the coauthor of this paper, has had many, many years of experience, and he seems to be able to pick this very well.

DR. LANE-PETTER: If the environmental factors, temperature, humidity and light, are standardized, how do you explain the seasonal differences?

DR. HANSEN: We have discussed this very much among ourselves, and I really cannot give you an answer. I have some speculations on this point. I think that everybody, whatever his specialty, could probably contribute. I am sure this is in part nutritional. I am sure it is part disease. I think it is probably a nonspecific stress introduced as a consequence of the failure of our control mechanisms, which are not perfect. No doubt there are physiological rhythms involved. It is a very, very complicated question. Maybe we just do not feel as well in the winter and we handle the animals more roughly. I do not know. But it is a very complex question that I think is very fascinating.

DR. LANE-PETTER: Would you have something to say about the season influence in the reproduction of hamsters?

DR. HANSEN: I really do not know very much about hamsters, but I would guess the factors that have affected our mice, rats, and guinea pigs would affect hamsters similarly. Remember, these are inbred strains. The hamsters that I have had experience with are not inbreds, and I am sure that the noninbreds would respond differently from the inbreds.

The other question, of course, that I failed to bring out in the discussion is that we did not have comparable data with noninbred stocks or F_1 hybrids, particularly with respect to the seasonal effects or to the population variability, but I am more concerned with what is going on within the strain than whether inbreds are more or less variable than noninbreds.

DR. LANE-PETTER: How do you determine percent pregnancy, as pregnancy cannot be detected at a very early stage?

DR. HANSEN: This is determined by palpation or gross visual observation. Obviously there are some errors, but we simply look at the animals and after a few months of experience, the people become very, very good in detecting whether the animal is pregnant or not. She simply has a unique appearance that people soon learn to detect.

I should also mention that one person did all of the selection for the replacement breeders during the entire study, and this was Mr. McEleney. He determines the pregnancies and this work is primarily his work. I did the analysis, but he collected the data.

DR. LANE-PETTER: I have two questions that I think you might like to consider together, so I will give them both to you: First, do you expect to analyze infertility to give an added measure of reproductive performance? There are great strain differences in this respect. Second, how much is known about diseases of the reproductive tract, and do you feel this variable may account for differences between strains?

DR. HANSEN: I will answer the second question first. I really do not know very much about this. I am sure there must be some, but not being a clinician, all I can remember is that someone has written about rabbits in which there are known cases of what are euphemistically called social diseases, which affect reproductive performance.

Coming back now to the first question about infertility: Since most of our matings were harem matings, we found very few cases in which we suspected that the male was responsible. In most of the cases, it seemed that the female was the one being affected. What I plan to do, and what I would like to do in each of these strains is a rather extensive analysis. First of all, we need to have some measure of ovulation rates, particularly as to how they are affected by season. We need to know implantation rates or implantation loss and we need to know the loss that occurs between implantation and birth.

Interestingly enough, I feel that there are several critical stages in the reproductive cycle with respect to the young. I feel that one critical stage, probably the first critical stage, is the time of implantation. We know that there is a considerable loss that occurs at implantation. At about 2 weeks, in the case of the mice and rats, we come to another critical phase. Then, obviously, that first 24 hr after birth is very critical. Then there appears to be another critical phase at about 2 weeks of age, and then again somewhere between 4 and 6 weeks. All these characteristics have to be studied, and we would like to get at each one of these in each of the strains and species we are studying.

DR. LANE-PETTER: What change in the data would you expect by introducing a barrier system and by putting them through a barrier system?

DR. HANSEN: I would be particularly interested in this because of our frequency of pregnancy loss during the winter months. I feel that the barrier system might be a much more controlled system with respect to temperature and humidity. Also, particularly in the case of the rats, many strains are susceptible to chronic respiratory diseases, and in the case of the CAS, CAR, and OMN strains, the replacement rates that we found were determined to a considerable extent according to the loss as a consequence of the males being apparently much more susceptible to CRD than the females. I think this would be one big difference for the rat colony.

DR. LANE-PETTER: How closely do you monitor the heat and the humidity in these animal rooms, that is, the actual amounts that are present throughout the seasonal changes?

DR. HANSEN: Unfortunately, we have to keep a close eye on the thermometers in the room. We have not in the past maintained recording thermometers or humidity controls. Just walking into the rooms during the winter months I can feel that the atmosphere is a bit less comfortable than at other times of the year, and I must say this probably depends on how I feel at the time, but I think the animals reflect this apparent difference in temperature and humidity in the form of a nonspecific stress.

IV

Defining
the Genetics
of the
Laboratory
Animal

METABOLIC REGULATION
BY GENES IN THE
LABORATORY MAMMAL

George L. Wolff

Introduction

The polypeptide components of the metabolic system of the laboratory mammal are genetically determined. Dynamic regulation and modulation of metabolic processes result from continuing interaction among these components and between them and macro- and microenvironmental stimuli.

Each gene determines the structure and specificity of a single polypeptide species. This does not, however, imply that each of these polypeptides affects only one biochemical process. Indeed, it is very likely that, with rare exceptions, each polypeptide synthesized under the control of a single gene affects more than one metabolic process. A mutation that alters a particular gene results in an altered polypeptide. Accordingly, a single mutation alters every metabolic process in which the polypeptide under its control is involved. Such effects may be further amplified because of the branched nature of metabolic pathways. Thus, without detailed knowledge of the qualitative and quantitative relationships among all the polypeptides acting in the metabolic network, it is difficult to predict the physiological or biochemical effects of most mutations.

It is also not possible to predict effects on response patterns to any particular experimental treatment without extensive knowledge of the metabolic effects of each mutation. The delineation of metabolic effects induced by a specific mutant must be based on the consideration that the quantitative expression of these effects is always influenced by the strain genotype and phenotype. Thus, the same mutant may induce the same general metabolic

205

effects in two different inbred strains, but the severity of each specific effect may vary independently of the others. The influence of the lethal yellow gene in the mouse on excess fat deposition and spontaneous tumor development in different genomic backgrounds is an example of this principle.

Most mutations that have been studied biochemically alter the structure of a known enzyme or other protein such as hemoglobin, probably by substitution or elimination of a single amino acid. However, the interrelationships of the various metabolic effects produced by such polypeptide alteration have not been studied in the laboratory mammal.

Two general approaches to the elucidation of the relations of diverse metabolic effects induced by a single gene mutation in the mammal suggest themselves. One approach is to screen many different enzymes and other proteins for possible differences in electrophoretic mobility, specific activity, etc., between inbred sibs differing by only the mutant gene. In this manner a specific polypeptide difference that can be related to the presence of the mutation might be detected. If such a difference is found, the model of metabolic interrelations can be developed on this basis.

Another approach, used in development of the model to be described (Figure 1), is to obtain information on each metabolic effect in the mutant and in an allelic phenotype. It then becomes feasible to attempt to develop an internally consistent model of the metabolic interrelations.

The practical value of such a model arises out of the ordering of the various metabolic effects with regard to one another. It aids the detection of relationships between metabolic processes that were not previously obvious. Moreover, by focusing attention on particular relations that appear to be most critical for testing its accuracy, the model aids in the design of experiments to determine the primary gene effect.

Description of the Model

The determination of the metabolic interrelationships of the diversified physiological effects of the lethal yellow (A^y) allele at the agouti locus of the house mouse presents the challenging problem of finding a metabolite that, when altered quantitatively or qualitatively due to the mutation, plays a part in every one of the altered physiological processes, including enhanced tumor growth, coat color pattern, increased body size, excess fat deposition, and insulin resistance. Of course, a regulatory metabolite involved in many metabolic processes would fit this description. However, the problem of determining which particular regulatory metabolite is affected is a major one, unless a clue is furnished by one of the metabolic effects.

In fact, in order to develop a reasonably predictive model it is necessary to

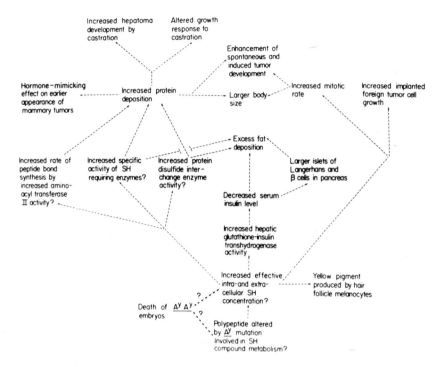

FIGURE 1 Model of proposed interrelations among metabolic effects induced by the A^y mutant in the house mouse. Statements not based on direct evidence indicated by question marks.

exclude the possibility that other metabolites or regulatory entities could be the primary mediators of the gene effect. This can be done if one of the specific metabolic effects can be shown experimentally to depend solely on the specific metabolite or regulatory compound. In the present case, the coat color pattern effect serves this function in suggesting sulfhydryl (−SH) concentration as a specific regulatory entity that may be responsible for the metabolic effects of the "yellow mouse syndrome."

Technical difficulties involved in the determination of *in vivo* intracellular and extracellular ratios of sulfhydryl (−SH) to disulfide (S−S) concentrations have so far prevented a direct test of this basic assumption. However, the available indirect evidence to be presented supports its general accuracy.

Glutathione is a ubiquitous cellular component that constitutes a considerable proportion of the intracellular and extracellular sulfhydryl concentration. Whether the synthesis or degradation of this tripeptide is controlled or influenced by the lethal yellow gene remains to be determined.

HAIR PIGMENT PATTERN

The wild type hair pattern of the mouse and of many other mammals consists of a black or brown hair with a subterminal yellow band and is called "agouti." The melanocytes producing the pigment can synthesize both yellow and black melanin. Silvers[21] demonstrated that the particular pigment produced depends on the hair follicle environment of the melanocyte. The change in pigment from black → yellow → black is apparently related to the concentration of sulfhydryl compounds in the hair follicle[3] or to an effect on the melanocyte's rate of uptake of sulfhydryl compounds by the hair follicle.[4] Yellow pigment synthesis appears to be correlated with a greater rate of uptake and intracellular concentration of sulfhydryl compounds.[4] It thus appears that the most striking visible expression of the agouti locus genes may be due to an indirect effect on the concentration of sulfhydryl compounds in the melanocyte mediated by the hair follicle.

No difference in specific activity of hepatic glutathione reductase between the yellow and nonyellow phenotypes could be detected (Wolff, unpublished data). This suggests that the sulfhydryl level in yellow mice may be elevated due to increased synthesis or decreased degradation of sulfhydryl compounds rather than to increased glutathione reductase activity, which might be induced by increased levels of NADPH.

The structure of the melanosome, the forerunner of the pigment granule, has been studied by electron microscopy.[20] Melanosomes in melanocytes producing black pigment consist of numerous parallel protein fibrils held together in a matrix by regular cross-linking. Black pigment is formed on this matrix and eventually completely covers it, thus forming an eumelanin granule. Melanosomes synthesizing yellow pigment also consist of numerous protein fibrils, but these are randomly arranged and are not cross-linked. Yellow pigment is deposited on the fibrils forming a phaeomelanin granule.

The protein fibrils exhibit tyrosinase activity, which oxidizes tyrosine to dihydroxyphenylalanine and then to dihydroxyphenylalanine quinone, which, after conversion to indole-5,6 quinone, copolymerizes with protein to form melanin.

Three forms of tyrosinase are present in melanocytes during eumelanin synthesis, but only one of these is detectable in melanocytes producing yellow pigment.[15] The two missing tyrosinase forms therefore may represent tyrosinase activity associated with various sections of the regularly arranged and cross-linked protein matrix present in eumelanin-synthesizing cells but absent in phaeomelanin-producing melanocytes.

SERUM INSULIN

Serum-insulin levels were measured in nonfasting yellow (A^ya) and nonyellow (aa) male mice of the inbred YS/ChWf strain.

Serum-insulin levels in nonyellow males were almost twice as high as those in yellow males of the same age and body weight.[28] A possible explanation of this puzzling situation was suggested by the results of Mahler and Szabo[18] who found that a decrease of the plasma glutathione level and the consequent reduction of insulin degradation by cleavage of the A and B chains revealed insulin sensitivity in tissues that had been thought to be insulin resistant, such as kidney. It should be noted here that insulin resistance of lethal yellow mice has been reported.[2]

Since the liver is one of the chief sites of insulin degradation,[18] specific activity of glutathione-insulin transhydrogenase in livers of yellow ($A^y a$) and nonyellow (aa) male YS/ChWf mice was measured. In all cases the specific activity was about 50 percent higher in livers of yellow males than in the livers of their nonyellow sibs (Wolff, unpublished data). The decreased level of serum insulin in the yellow males may thus be related to the increased rate of cleavage of insulin in these mice as compared with the nonyellow males. That insulin synthesis may occur at a higher rate in yellow mice than in their nonyellow sibs is suggested by the increased size of the islets of Langerhans in the pancreata of these mice as well as by the increased size and number of beta cells.[9,11] Such increased insulin synthesis could be a compensatory response to increased insulin degradation.

The increased rate of glutathione-insulin transhydrogenase activity in yellow mice suggested a possible relationship to the difference in structure between yellow and black melanosomes. Glutathione-insulin transhydrogenase is either the same as, or very similar to, the protein disulfide interchange enzyme.[5] The activity of the latter enzyme is also dependent on sulfhydryl concentration. It reduces the disulfide bonds of oxidized enzymes and thus serves to increase the specific activity of many enzymes requiring sulfhydryl for activation. The specific activity of the protein-disulfide interchange enzyme, regulated in part by the sulfhydryl concentration, is thus an important determinant of the metabolic balance of cells and tissues.

In relation to the differential yellow and black melanosome structure it may be that the cross-linking of the protein fibrils in the black melanosome is due to disulfide bonds. If this is true, the absence of cross-links in the case of yellow melanosomes could be due to the increased sulfhydryl concentration within the melanocytes,[4] which would decrease disulfide bond formation by alteration of the redox potential. Increased activity of the protein-disulfide interchange enzyme as a consequence of increased sulfhydryl concentration would also increase the rate of reduction of already formed disulfide bonds.

This portion of the model concerned with coat color pattern and insulin metabolism can be summarized as showing that these metabolic effects can be explained by relating them to the sulfhydryl concentration and secondarily to the activity of an enzyme that regulates the specific activities of many other enzymes and whose activity itself is regulated to a large extent by the level of sulfhydryl compounds in the cells and in the interstitial fluid.

OBESITY

While the lethal yellow mutant induces a tendency to excess fat deposition, the phenotypic expression of this tendency depends on the background genome as well as on the diet.[7] Carpenter and Mayer[2] found that only a high-fat diet produced weight gain differences between adult yellow and nonyellow mice. High-carbohydrate and high-protein diets had no differential weight-gain effects in the two genotypes. Contrary to a previous report,[24] parabiosis of yellow and nonyellow inbred littermates had no effect on excess fat deposition.[25] Growth hormone and prolactin deficiency also had no effect on this aspect of the yellow mouse syndrome.[26] The excess fat deposition often, but not invariably, associated with the presence of the lethal yellow gene is therefore probably a secondary manifestation of the altered regulation of the metabolic network in the yellow mouse.

On a 25 percent greater food intake, fat deposition increased two to three times as much in yellow mice as in their nonyellow sibs.[8] Excess fat deposition and divergence in weight-gain rates do not begin until 8–10 weeks of age,[8,25] which is the end of the maximum true growth period of the animal. In A^y-mice, excess fat deposition resulted in larger fat cells; the number of fat cells was not greater than in nonyellow sibs.[23]

Yellow mice did not become less active than nonyellows until after 80 days of age.[10] Oxygen consumption and weight of 38-day-old yellow and non-yellow mice were not different. At 5 months of age the oxygen consumption per animal was still the same in the two genotypes. However, the resting metabolic rate (cc of $O_2/g^{2/3}$) was lower in yellow than in nonyellow mice because of the greater body weight of the former at this age.[1] These data suggest that decreased activity of older yellow mice is the result, not the cause of obesity. They also suggest that the somewhat greater food intake of yellow mice, which, in the absence of an increased metabolic rate results in adult obesity, may be the result of an alteration of a feedback circuit that normally regulates food intake in accord with the true growth rate. The different serum-insulin levels found in yellow and nonyellow YS/ChWf males at 3 months of age[28] may play a role in this phenomenon.

HOMOZYGOUS LETHALITY

The lethality of homozygous A^yA^y embryos is one of the most striking effects of this mutant. It is also the most difficult to investigate because it occurs about the time of implantation. On the basis of a study of a series of blastocysts from $A^ya \times A^ya$ matings, Eaton and Green[6] concluded that death of the A^yA^y embryos occurred because of "inanition precipitated by incomplete attachment to the endometrium." The cause of this incomplete attachment seemed to be slower differentiation of tropho-blast giant cells in these embryos. Because of this lag, the trophoblast giant cells failed to attach since they were out of phase with the developing endo-metrium. Definition of this gene effect in metabolic terms will have to await the development of suitable techniques for culture of the homozygous em-bryos so that microchemical techniques can be used to differentiate them from the heterozygous yellow and homozygous nonyellow embryos.

REGULATION OF NORMAL AND NEOPLASTIC GROWTH

The other physiological effects of the lethal yellow gene represent, essentially, manifestations of altered regulation of true growth, both normal and neoplastic. Since true growth represents an increase in pro-tein deposition, it is logical to think of these aspects of the "yellow mouse syndrome" as manifestations of altered modulation of protein metabolism. This expresses itself grossly in the form of increased muscle weight,[14] in-creased fat-free dry liver weight,[25] increased body size,[13,14,25] and increased fat-free dry carcass weight.[25]

In protein synthesis, sulfhydryl concentration may influence the rate of peptide bond synthesis by its effect on activation of aminoacyl transferase II. The activation of this enzyme depends on sulfhydryl concentration and is inhibited by oxidized glutathione.[22] Increased intracellular sulfhydryl level would therefore tend to favor protein synthesis by increasing the rate of pep-tide bond synthesis.

SYSTEMIC EFFECT AND MITOTIC RATE

The mitotic rate within the agouti (A-) hair bulb has been reported to increase as yellow pigment was synthesized and to de-crease again as black-melanin synthesis reappeared. In yellow mice the mitotic rate continued to increase during the whole period of yellow-pigment forma-tion.[3] Thus, increased sulfhydryl concentration required for yellow-pigment synthesis may also increase the mitotic rate. The increased number of spon-taneous tumors and their generally earlier age of appearance[13] suggest that the mitotic rate in other tissues may also be increased.

The question of whether the increased mitotic rate was confined solely to the hair bulb or whether there might be a generalized systemic effect favoring cell division was approached by implanting non-strain-specific Sarcoma 37 cells in inbred yellow and nonyellow YS/ChWf males. Known numbers of ascites cells of this tumor were implanted subcutaneously in 3-month-old ani-mals, and the resulting solid tumors were measured. In every case the number of "large" tumors in yellow males was considerably greater than that in nonyellow males. Conversely, the number of "small" tumors was considerably greater in nonyellow males than that in yellow males. Similar results were obtained with the Ehrlich-Lettré hyperdiploid carcinoma.[29] These data indicate that the lethal yellow gene induces a systemic effect that favors an increased mitotic rate of implanted foreign tumor cells.

The presence of such a systemic effect, particularly if it is related to an increased sulfhydryl concentration, as suggested by the pigment work, would have profound effects on the metabolic balance within the whole animal and in all tissues and cells. It would tend to increase the specific activity of many SH-requiring enzymes, in relation to the specific activity of enzymes not affected by thiols, by increasing the rate of spontaneous reduction of disulfide bonds in inactive enzyme molecules as well as the reduction of these bonds catalyzed by the protein-disulfide-interchange enzyme. It would also alter the redox potential within specific cell compartments and thus alter metabolic equilibria and reaction rates of many biochemical processes.[17]

McCaman[19] has described an analogous situation in muscle from mice with hereditary muscular dystrophy ($dydy$). In these muscles most of the NADP-requiring dehydrogenases were more active than in normal $Dy-$ muscle, while the NAD-requiring dehydrogenases had either lower or unchanged activity.

A systemic effect induced by a single gene mutation can thus alter the balance and interrelations of a large number of the biochemical reactions comprising the metabolic network of the animal. Such a mutation produces an essentially "new" metabolic system very differently modulated and regulated than the metabolic system of its wild-type sibling.

SPONTANEOUS TUMOR DEVELOPMENT

The mitosis-promoting microenvironment and increased protein deposition induced by this gene are probably closely related to the enhancement of spontaneous tumor development in yellow mice. Heston and his co-workers have shown that presence of the A^y gene increases the mean number of hepatomas, skin tumors, and lung tumors per animal[12,13,30] on strain backgrounds susceptible to such tumor formation. Apparently the gene induces an increase in the number of tumors that arise in the strain-specific sensitive tissue and does not act exclusively in any one particular tissue. This suggests that here also the gene may induce its effect systemically, i.e., by altering the environment in which neoplastic cells multiply and form tumors.

Mammary tumors develop in virgin agouti (Aa) ($C_3H \times YBR$) F_1 hybrid females at about 15 months of age; in their yellow A^yA virgin sisters mammary tumors develop at about 8 months of age. This is the same age at which breeding females of both genotypes develop mammary tumors.[13] Since the main difference between breeding females and virgins is generally assumed to be the lack of hormonal stimulation of protein synthesis in the mammary glands of the latter, it appears that the metabolic effect of the lethal yellow genotype mimics or supplants the effect of such stimulation in the development of mammary tumors.

EFFECT OF CASTRATION

The lethal yellow gene is also involved in the modulation of protein synthesis by androgen in the male. Data of Heston and Levillain[27] revealed a marked increase in the number of spontaneous hepatomas that developed in castrated yellow $A^y A$ ($C_3H \times YBR$) F_1 hybrid males, while no such effect of castration was observed in their agouti (Aa) sibs. The question arose whether liver growth, carcass growth, or both, in yellow mice might respond differently to castration than in their nonyellow sibs, particularly at very early preneoplastic stages.

To obtain information on this point, yellow and nonyellow male YS/ChWf strain littermates weighing within one gram of each other were castrated or sham-operated at 4 weeks of age. By 16 weeks of age the wet and fat-free dry weights of $A^y a$ livers from castrated males were significantly greater than those in intact males. No difference between fat-free dry weights of livers from castrated and intact nonyellow males was found.[27] These results suggest that castration of yellow males stimulates true liver growth, at least in this inbred strain.

Since androgen stimulates protein synthesis and deposition,[16] castration should decrease protein synthesis. This is indeed the case in castrated nonyellow mice in which the fat-free dry carcass weight was significantly lower than that of the intact nonyellow males. However, there was no effect of castration on the fat-free dry weight of $A^y a$ carcasses. Since the specific biochemical reactions affected by the protein anabolic effect of androgens are unknown, the only conclusion that can be drawn from these observations is that the lethal-yellow gene alters the effect of the withdrawal of androgen on the protein-synthetic mechanism.

These results considered together with the hormone-mimicking effect of the gene in mammary tumor development suggest that the gene induces alterations in the responses of the metabolic network to at least sex hormone stimulation. Such alterations in metabolic response patterns to sex hormones are obviously of major importance in the altered regulation of metabolism in yellow mice. Increased tumor development is undoubtedly also influenced by the indicated systemic effect of the A^y gene.

The suggested influence of the lethal-yellow gene's systemic effect on protein synthesis, per se, may also be involved in the anomalous response of liver and carcass growth to castration as well as in the hormone-mimicking effect of this gene on mammary tumor development. However, since the specific biochemical reactions influenced by any of the sex hormones are not known, it is not possible to suggest specific interrelations between these effects.

Summary

This discussion has described a metabolic model relating the diversified physiological and biochemical effects of a single gene mutation, lethal yellow (A^y), in the mouse to the variation of a single factor, sulfhydryl concentration. Although the model is internally consistent with respect to the available facts, much more evidence must be obtained to establish it on a firm basis.

This work was supported by USPHS grants CA-06927 and FR-05539 from the National Cancer Institute and by an appropriation from the Commonwealth of Pennsylvania.

References

1. Bartke, A., and A. Gorecki. 1968. Oxygen consumption by obese yellow mice and their normal littermates. Amer. J. Physiol. 214:1250–1252.
2. Carpenter, K. J., and J. Mayer. 1958. Physiologic observations on yellow obesity in the mouse. Amer. J. Physiol. 193:499–504.
3. Cleffmann, G. 1963. Die Bedeutung von äusseren Einflüssen auf die Pigmentzelle für die rhythmische Musterbildung im Haar. Arch. Entwicklungsmech. 154:239–271.
4. Cleffmann, G. 1964. Function-specific changes in the metabolism of agouti pigment cells. Exp. Cell Res. 35:590–600.
5. DeLorenzo, F., R. F. Goldberger, E. Steers, D. Givol, and C. B. Anfinsen. 1966. Purification and properties of an enzyme from beef liver which catalyzes sulfhydryl-disulfide interchange in proteins. J. Biol. Chem. 241:1562–1567.
6. Eaton, G. J., and M. M. Green. 1963. Giant cell differentiation and lethality of homozygous yellow mouse embryos. Genetica 34:155–161.
7. Fenton, P. F., and H. B. Chase. 1951. Effect of diet on obesity of yellow mice in inbred lines. Proc. Soc. Exp. Biol. Med. 77:420–422.
8. Gowen, J. W., and J. Stadler. 1955. Influence of genotype on weight control in mice, p. 49–63. *In* Weight Control, Iowa State College Press.
9. Hausberger, F. X., and B. Hausberger. 1966. Castration-induced obesity in mice. Acta Endocrinol. 53:571–583.
10. Hawkins, J. D. 1966. Developmental effects of the A^y substitution on mouse activity level. Psychon. Sci. 4:105–106.
11. Hellerström, C., and B. Hellman. 1963. The islets of Langerhans in yellow obese mice. Metabolism 12:527–536.
12. Heston, W. E., and M. K. Deringer. 1947. Relationship between lethal yellow (A^y) gene of the mouse and susceptibility to spontaneous pulmonary tumors. J. Nat. Cancer Inst. 7:463–465.

13. Heston, W. E., and G. Vlahakis. 1961. Influence of the A^y gene on mammary-gland tumors, hepatomas, and normal growth in mice. J. Nat. Cancer Inst. 26:969-983.
14. Heston, W. E., and G. Vlahakis. 1961. Elimination of the effect of the A^y gene on pulmonary tumors in mice by alteration of its effect on normal growth. J. Nat. Cancer Inst. 27:1189-1196.
15. Holstein, T. J., J. B. Burnett, and W. C. Quevedo. 1967. Genetic regulation of multiple forms of tyrosinase in mice: Action of *a* and *b* loci. Proc. Soc. Exp. Biol. Med. 126:415-418.
16. Kochakian, C. D. 1965. Mechanism of anabolic action of androgens. *In* Mechanisms of hormone action, P. Karlson [ed.], Academic Press.
17. Krebs, H. A. 1967. The redox state of nicotinamide adenine dinucleotide in the cytoplasm and mitochondria of rat liver. Adv. Enzyme Regul. 5:409-437.
18. Mahler, R. J., and O. Szabo. 1967. Insulin action upon insulin-insensitive tissue following impaired degradation of the hormone. Proc. Soc. Exp. Biol. Med. 125:879-882.
19. McCaman, M. W. 1963. Enzyme studies of skeletal muscle in mice with hereditary muscular dystrophy. Amer. J. Physiol. 205:897-901.
20. Moyer, F. H. 1966. Genetic variations in the fine structure and ontogeny of mouse melanin granules. Amer. Zool. 6:43-66.
21. Silvers, W. K. 1958. An experimental approach to action of genes at the agouti locus in the mouse. III. Transplants of newborn A^w-, A-, and a^t- skin to A^y-, A^w-, A-, and aa hosts. J. Exp. Zool. 137:189-196.
22. Sutter, R. P., and K. Moldave. 1966. The interaction of aminoacyl transferase II and ribosomes. J. Biol. Chem. 241:1698-1704.
23. Täljedal, I., and B. Hellman. 1963. Morphological characteristics of the epididymal adipose tissue in different types of hereditary obese mice. Pathol. Microbiol. 26:149-157.
24. Weitze, M. 1940. Hereditary adiposity in mice and the cause of this anomaly. University of Copenhagen. Store Nordiske Videnskabsboghandel, Copenhagen, Denmark, 96 p.
25. Wolff, G. L. Genetic influences on response to castration of liver growth and hepatoma formation. Cancer Res. 30:1726-1730.
26. Wolff, G. L. 1965. Hereditary obesity and hormone deficiencies in yellow dwarf mice. Amer. J. Physiol. 209:632-636.
27. Wolff, G. L. 1970. Genetic influences on response to castration of liver growth and hepatoma formation. Cancer Res. 30:1726-1730.
28. Wolff, G. L., and G. A. Reichard, Jr. 1970. Response of serum insulin concentration to tumor growth in different genetic systems. Hormone and Metabolic Res. 2:68-71.
29. Wolff, G. L. 1970. Stimulation of growth of transplantable tumors by genes which promote spontaneous tumor development. Cancer Res. 30:1731-1735.
30. Vlahakis, G., and W. E. Heston. 1963. Increase of induced skin tumors in the mouse by the lethal yellow gene (A^y). J. Nat. Cancer Inst. 31:189-195.

DISCUSSION

DR. WISSLER: Please go over the diagram with us since it is not readable from here. Also, please discuss the mechanism of increased tumor cell takes.

DR. WOLFF: I am sorry that the slide was not visible. Actually, the point of this diagram (see Figure 1) is merely to illustrate the manifold metabolic effects induced by a single mutation. The individual items for my purpose, at least, and what I am trying to convey, are not really important. I will read these individual items, and then if there are any specific questions, I think it would be better to deal with them than to go over the whole thing in detail.

I will begin reading the diagram from the bottom, starting with the question: Is polypeptide altered by A^y mutation, that is the lethal yellow mutation, and is this involved in sulfhydryl compound metabolism? Somehow this may increase the effect of intra- and extracellular sulfhydryl concentration, and both of these may be related to the lethality of the homozygote.

The increased effect of intra- and extracellular sulfhydryl concentration is postulated as being at the base of this "pyramid." In the first place, increase of sulfhydryl concentration seems to induce yellow pigment formation in hair bulb melanocytes. It also increases the implanted foreign tumor cell growth and increases the mitotic rate. I might say that these are all tentative lines.

In addition, the increased sulfhydryl concentration increases the activity of this enzyme, hepatic glutathionine transhydrogenase, and this may in turn result in a decreased serum insulin level in the yellow males. This then may account for the larger islets of Langerhans and beta cells in the pancreas. The excess fat deposition may be related to the decreased serum insulin level and to the effects in the pancreas.

Here I am postulating that the sulfhydryl concentration increases the specific activity of all sulfhydryl-requiring enzymes, and there are many of these. We know that there is increased protein deposition. This may be related to a possible increase in the rate of peptide-bond synthesis by increased aminoacyl transferase-2 activity. This is a completely speculative suggestion.

The increased protein disulfide interchange enzyme activity may also be involved in increasing the specific activity of these enzymes, in which case it will result in increased protein deposition, and also in excess fat deposition. The increased protein deposition presumably is involved in enhancement of spontaneous and induced tumor development in the larger body size, which is also affected by the increased mitotic rate.

Also, increased protein deposition may be involved in the hormone-mimicking effect on the earlier appearance of mammary tumors, on the increased hepatoma development by castration, and on the altered growth response to castration.

DR. WAGNER: Is there any decrease in latency for carcinogen-induced tumors in the yellow mouse?

DR. WOLFF: I believe there is, but I do not want to make a definite statement on that. Heston has found that more spontaneous and induced tumors develop in the yellow animals, the yellow phenotype, than in the comparable nonyellow phenotype. So it seems to be a matter of increased sensitivity of the tissues of yellow animals to tumor development, so that more tumors of the same type can develop in the yellow mice than in the nonyellow mice.

"MOSAIC POPULATIONS" FOR PHARMACOLOGICAL TESTS AND GENETIC ASSOCIATIONS OF DRUG ACTION

Eszter Cholnoky
János Fischer
Sándor Józsa

Considerations of Genotypes of Laboratory Animals for Pharmacological Tests

Pharmacologists want to obtain a wide range of repeatable information when testing drugs on laboratory animals. Therefore they want to use animals of mixed genotypes, but they wish to reproduce the genotypic composition of the groups in a reliable way.

POPULATIONS OF MIXED GENOTYPE

Usually random-bred populations are used for these purposes. However, random-bred populations might be inbred to an unknown degree, and certain genes affecting drug action might be fixed in them. An example is the case of histamine sensitivity after pertussis vaccine in the non-inbred LAC Grey strain.[1]

Green[2] proposed to "synthesize" a wide range of genotypes by four-way crosses between four inbred strains of mice ($C_{57}BL/6J$, BALB/cJ, DBA/2J, and C_3HeB/FeJ) for irradiation experiments. Cholnoky *et al.*[3] produced a theoretical population of noninbred heterozygous animals carrying all possible alleles of 16 inbred strains in 64 different heterozygous combinations by Robertson and Falconer's method.[4]

In this way the problem of genetic variability can be solved in a satisfactory way. But being faced with the problem of exactness and repeatability of the genetic composition of a group, the geneticist fails to solve the repeatability problem by employing either random-bred or "synthetic" populations, because the members of such groups cannot be characterized individually. Each population is characterized statistically by the gene frequencies and genotype frequencies within that population, and even this characterization is valid only for very large samples.

For example, Doolittle[5] reported a population produced from a cross of C_3HeB, $C_{57}L$, SWR, and 129/R strains that was bred at random for 10 generations with a negligible increase of inbreeding. The frequencies of recessive genotypes a/a, b/b, p/p, and ln/ln were examined, and the frequencies of each generation were compared with those of the total population. Among 40 comparisons, 11 significant deviations from this average were found, although at least 1,400 animals were subject to investigation in each generation.

In our fictitious generation of the strain "synthesized" from 16 different inbred strains, the probability that in a group of 64 animals all 64 possible genotypes of one locus are equally represented is on the order of 10^{-26}, and the probability of having a second group with the same genotypic composition is even more remote, below 10^{-52}.

Both Doolittle's population and the probabilistic considerations refer to the fact that the genotypic composition of groups is far from being repeated under random mating.

GENOTYPICALLY UNIFORM POPULATIONS

Recently there have been many experiments on the sensitivity of different inbred strains to drugs. We refer here only to those of Brown on the reactivity to insulin,[6] on sleeping time responses to pentobarbitone sodium,[6] and on histamine sensitivity,[7] and to those of Fujii et al.[8] on hexabarbital sleeping time in different inbred strains of mice. These experiments have drawn attention to the importance of the genotype in drug action and have aroused interest in the nature of the genetic pattern and the heritability of reaction to drugs. Thus Brown and Wallace[9] found that histamine tolerance is regulated by the genes of one locus.

"Mosaic Populations"

It would be useful to test drugs on a large series of inbred strains because the genotypic composition of a group can be fixed in this way. However, the variability of the reaction of inbred animals is in certain cases greater than that of noninbred ones.[10] On the other hand, one inbred strain or one type of F_1 hybrid represents only one genotype, and the reaction of a strain or an F_1 hybrid is unpredictable. Thus large-scale experiments ought to be performed on a large number of different strains to find the most suitable one. Chai commented on Becker's paper[10] recommending the utilization of small groups of different inbred animals and F_1 hybrids as a preliminary experiment to find the most suitable genotype, although he did not define the number of genotypes and the size of small groups to be employed.

Our aim was to define a population enabling us to form groups with a maximum within-group genetic variability and no between-group genetic variability. Admitting defeat in the case of our reproducing the theoretical population, we came upon the idea of a "mosaic population," which would be produced from all combinative matings between n inbred strains resulting in $n(n-1)$ different F_1 hybrids, reciprocals treated separately. (Treating reciprocals separately seems to be necessary because Brown[6] could find maternal influences in response to insulin by testing reciprocal crosses of $A_2 G$ and CBA mice.) By arbitrarily choosing animals from each cross and adding to each group n different homozygous inbred animals, maximum within-group genetic variability can be achieved and the same composition can be repeated deliberately as many times as the pharmacologist wishes with minimum between-group genetic variability. By labeling the animals, the individual genotypes can be marked directly.

Plan for Experiments on Mosaic Populations

1. *Determine the dose regression of a drug* in three- or four-point analysis. Use for each dose 36 animals produced by 36 combinative matings of 6 inbred strains. Each genotype should be represented once in each group for a dose. This will show the average reaction of the mosaic population and it will give an estimate of LD_{25} and LD_{75}.

2. Form two groups as before, but each genotype should be represented in each group four times. This serves the investigation of genetic factors on drug action. The estimated LD_{25} and LD_{75} are then administered to these two groups.

We recommend LD_{25} and LD_{75} for the following reasons: Each gentotype is represented four times for each dose, the expected death–survival rate within genetically uniform quartets is 1:3 and 3:1, respectively. When the probability of lethality is 0.25, a 4:0 ratio is a significant deviation at the 0.001 probability level, and even a 3:1 ratio can be regarded as nonrandom. The situation is analogous for dose LD_{75} (Table 1). We expect to find the sensitive genotypes by administering LD_{25} and the more tolerant ones by using LD_{75}.

3. Test total groups as to whether they fit the expected death–survival rates (1:3 and 3:1 respectively).

4. Test the genotypically uniform subgroups of four for genotypic differences. If there were no differences in the sensitivity or tolerance among the

ESZTER CHOLNOKY,
JÁNOS FISCHER, AND
SÁNDOR JÓZSA

TABLE 1 The Significance of Death–Survival Rates, Tested by X^2 Test with 1 Degree of Freedom, due to Lethality of Different Doses

Death–Survival Rate		LD_{10}	LD_{20}	LD_{25}	LD_{30}	LD_{40}	LD_{50}
4:0	X^2	36.00	18.50	12.00	9.33	6.0	4.0
	P	< 0.001	< 0.001	< 0.001	< 0.01	< 0.02	< 0.05
3:1	X^2	18.77	7.56	5.33	3.85	1.93	2.00
	P	< 0.001	< 0.01	< 0.05	~ 0.05	> 0.1	> 0.1
2:2	X^2	4.33	2.25	1.33	0.75	0.17	0
	P	< 0.05	> 0.1	> 0.2	> 0.3	> 0.5	1.0
1:3	X^2	1.00	0.06	0	0.04	0.38	—
	P	> 0.3	> 0.8	1.0	> 0.9	> 0.5	
0:4	X^2	0.44	1.00	1.33	1.70	2.25	—
	P	> 0.5	> 0.3	> 0.2	> 0.1	> 0.1	

different genotypes, the expected distribution of genotypically uniform subgroups showing death–survival rates of 4:0, 3:1, 2:2, 1:3, and 0:4 would be a binomial one (Table 2), p being the probability of lethality due to the estimated dose administered and q the probability of survival. The significance of deviations from this expected distribution is tested by a X^2 test with 4 degrees of freedom.

TABLE 2 Expected Distribution of Different Death–Survival Rates in Subgroups of Four

Death–Survival Distribution[a]	4:0 p^4	3:1 $4p^3q$	2:2 $6p^2q^2$	1:3 $4pq^3$	0:4 q^4

[a] p = probability of lethality due to dose; $q = 1-p$ = probability of survival.

Although it is not the single genes, but pairs of genes that act and interact as the genotype, we propose to test each subgroup consisting of $4(2n-1)$ (=44) animals carrying a certain gene to fit the expected death–survival ratio, because significant deviations from this might refer to dominant traits in sensitivity or tolerance. By treating $4(n-1)$ (=20) animals carrying certain genes inherited in common from the mother strain, maternal influences can be analyzed.

Dividing each group of a dose into four subgroups containing each geno-type once, the subgroups are tested to fit the expected 1:3, 3:1 ratios, respectively.

The Realized Experiment

PRELIMINARY CROSSES

In addition to these theoretical considerations, we began to produce a mosaic population from strains A, AKR, BALB/c, CBA, C_3H, DBA/2, NZB, and NZW. At that time the different strains were coded by different colors, and the bicolored labels of different crosses inspired Kállai to call this a "mosaic population." Since then we have found it more manageable to use a numerical code for the genes present in the different crosses (Table 3).

In our preliminary matings, each cross was represented by one bigamous family. Thus the value of separate crosses could not be estimated, having only one male in each, but it supplied us with valuable information about the males and females of different strains in outcrosses at large. We determined the per-centage of the overall production due to males and females belonging to the

TABLE 3 Code Numbers for Eight X Eight Different Genotypes[a]

♀ \ ♂	A (0)	AKR (1)	BALB/c (2)	CBA (3)	C_3H (4)	DBA/2 (5)	NZB (6)	NZW (7)
A (0)	00*	01	02	03	04	05	06	07
AKR (1)	10*	11*	12	13	14	15	16	17
BALB/c (2)	20*	21*	22*	23*	24	25	26	27
CBA (3)	30	31	32	33*	34	35	36	37
C_3H (4)	40	41*	42	43	44*	45	46	47
DBA/2 (5)	50	51	52	53	54	55	56	57
NZB (6)	60*	61*	62*	63*	64*	65	66*	67
NZW (7)	70*	71	72	73*	74*	75	76	77*

[a]* = genotypes realized in experiment.

ESZTER CHOLNOKY,
JÁNOS FISCHER, AND
SÁNDOR JÓZSA

same strain (Table 4). It turned out that the males of the poor breeding strain A are superb and that the production of NZW and NZB females is extremely good in outcrosses, although both A and NZB strains are very difficult to breed.

The offspring obtained from preliminary crosses served to determine the LD_{25} and LD_{75} of Ferrum (Hausmann Laboratories), a parenteral iron complex (Ferropolymaltose, FPM), the dosage of which had been tested on BALB/c and C_3H inbred mice, and on a noninbred population known as Hungarian White Mice.

TABLE 4 Percentage of Overall Production due to Different Inbred Males and Females in Outcrosses

♀	NZW	NZB	C_3H	CBA	AKR	A	BALB/c	DBA/2
	21.9	18.0	13.9	11.3	11.3	10.2	8.5	4.9
♂	DBA/2	A	CBA	C_3H	NZW	AKR	NZB	BALB/c
	21.0	17.4	17.4	13.9	9.6	9.2	8.0	3.5

INCOMPLETE MOSAIC POPULATION

The other purpose of preliminary crosses was to design a plan of even production based on these data. Our aim was to produce at least eight males in each cross. This would have required the 18-fold repetition of certain crosses. Since there was a shortage of breeders and of space, and because *Pharmacopaea Hungarica* prescribes 32 animals for each dose, we elected not to produce all 56 outcrosses and decided to realize only the more advantageous ones, although this meant giving up the possibility of testing maternal influences. Thus we founded an incomplete mosaic population still characterized by different combinations of each gene and by equal gene frequencies. Later we had to exclude all crosses in which DBA/2 mice participated because, for some unknown reason, they ceased to breed both in the inbred strain and in outcrosses. This left us with 21 types of F_1 hybrids and seven inbred genotypes.

A few genotypes had to be rejected from the experiment because they did not produce male offspring in sufficient numbers, others were excluded from evaluation because the dosage was imprecise (the hypodermic needles leaked) and there were no males left for replacement. All in all, 20 different genotypes could be tested, 13 heterozygous and seven homozygous genotypes. For each dose 80 males of similar age (5–7 weeks old) were used.

The purpose of this phase was to determine genotypical dependence of FPM action and to find the most suitable genotypes for further investigations.

The animals were labeled on the day of weaning. Groups of four were

223

TABLE 5 Death–Survival Rates after Different Doses of Ferropolymaltose in a Population Consisting of 20 Genotypes, Each Represented Four Times for Each Dose. Observations of 8 Days. (+ = death, – = survival)

Code of Genotype	600 μg of Fe/g of BW[a]				Death–Survival Ratio	1,000 μg of Fe/g of BW[b]				Death–Survival Ratio
	No. 1	No. 2	No. 3	No. 4		No. 1	No. 2	No. 3	No. 4	
00	+	+	+	+	4:0	+	+	+	+	4:0
21	+	+	–	+	3:1	+	–	+	+	3:1
33	–	–	+	–	1:3	+	+	+	+	4:0
64	–	+	–	+	2:2	+	+	+	+	4:0
41	+	–	–	–	1:3	+	+	+	+	4:0
70	–	–	+	–	1:3	+	+	+	+	4:0
62	–	–	–	–	0:4	+	+	+	+	4:0
63	–	+	–	+	2:2	+	+	+	+	4:0
44	–	+	+	–	2:2	+	+	+	+	4:0
23	–	–	–	–	0:4	+	+	+	+	4:0
11	–	–	–	–	0:4	–	–	–	–	0:4
60	–	–	+	–	1:3	+	+	–	–	2:2
20	–	+	–	–	1:3	–	–	–	–	0:4
77	–	+	+	+	3:1	+	+	+	+	4:0
10	+	–	–	–	1:3	–	–	+	+	2:2
61	–	–	–	–	0:4	+	+	+	+	4:0
73	+	–	+	–	2:2	+	+	+	+	4:0
66	+	–	+	+	3:1	+	–	+	+	3:1
74	–	–	–	–	0:4	+	+	+	+	4:0
22	–	–	–	–	0:4	–	+	+	–	2:2

Total observed					27:53					64:16
Expected					20:60					60:20
x^2					3.26					1.06
Degrees of freedom					1					1
P					> 0.05					> 0.3

Heterogeneity	No. 1 + 2	No. 3 + 4		No. 1 + 2	No. 3 + 4
Observed	13:27	13:27		31:9	33:7
Expected	10:30	10:30		30:10	30:10
x^2	1.2	1.2		0.13	1.2
Degrees of freedom	1	1		1	1
P	> 0.2	> 0.2		> 0.7	> 0.2

[a] Estimated LD_{25}.
[b] Estimated LD_{75}.

ESZTER CHOLNOKY,
JÁNOS FISCHER, AND
SÁNDOR JÓZSA

formed the day before injection. Four animals, two of which represented the same genotype, were housed in each cage. The members of a uniform genotypic quartet were numbered from 1 to 4. Animals 1 and 2 were treated on the same day; animals 3 and 4, on the following day. On the morning of the injection, the animals were weighed and the weights were recorded. One group of 80 animals received an intravenous injection of FPM containing 600 μg of Fe/g of body weight (BW), the other group received 1,000 μg of Fe/g of BW, that is the estimated LD_{25} and LD_{75}, respectively. Observations were made for 8 days.

RESULTS
The results of the experiment are summarized in Table 5. In the group receiving the estimated LD_{25}, 27 animals died, 53 survived; the χ^2 test shows a nonsignificant deviation ($P > 0.05$). In the group injected with LD_{75}, 64 animals died, 16 survived; the deviation is not significant ($P > 0.3$).

We tested the groups divided into genetically uniform subgroups for dependence of drug action on genotypical differences. Results of this analysis are presented in Table 6. In the group receiving 600 μg of Fe/g of BW intravenously, the total shows a significant deviation from the expected distribution ($P = 0.02$) according to the χ^2 test with 4 degrees of freedom. Similarly, in the group that was given 1,000 μg of Fe/g of BW, the deviation from the expected distribution was significant ($P < 0.001$), showing a strong dependence on genotype.

TABLE 6 Deviations of Five Genetically Uniform Subgroups of Four from Expected Distribution after Different Doses of FPM[a]

LD_{25}, $p = 0.25$, $q = 0.75$					
Death–survival rate	4:0	3:1	2:2	1:3	0:4
Observed distribution	1	3	4	7	6
Expected distribution	0.08	0.93	4.22	8.44	6.33

$\chi^2 = 15.454$ with 4 degrees of freedom; $P < 0.01$

LD_{75}, $p = 0.75$, $q = 0.25$					
Death–survival rate	4:0	3:1	2.2	1:3	0:4
Observed distribution	13	2	3	0	2
Expected distribution	6.33	8.44	4.22	0.93	0.08

$\chi^2 = 59.304$ with 4 degrees of freedom; $P < 0.001$

[a]p = probability of lethality due to dose; $q = 1 - p$ = probability of survival.

We analyzed the homozygous and heterozygous subgroups separately for genotypical differences in drug action. The results are summarized in Table 7. The homozygous subgroups receiving LD_{25} show a significant deviation ($P < 0.001$); the distribution of death–survival rates in heterozygous subgroups fits the expected one ($P > 0.95$). Both homozygous and heterozygous subgroups injected with LD_{75} show a significant deviation ($P < 0.001$). The observed distribution has two peaks in both groups. After performing χ^2 tests with 1 degree of freedom on groups carrying one gene of a certain strain in common, no significant deviations could be found after either dose.

TABLE 7   Deviations of Five Genetically Uniform Homozygous and Heterozygous Subgroups of Four from Expected Distribution after Different Doses of Ferropolymaltose (FPM)[a]

		LD_{25} $p = 0.25$, $q = 0.75$					LD_{75} $p = 0.75$, $q = 0.25$				
Homozygotes	Death–survival rate	4:0	3:1	2:2	1:3	0:4	4:0	3:1	2:2	1:3	0:4
	Observed distribution	1	2	1	1	1	4	1	1	0	1
	Expected distribution	0.03	0.33	1.48	2.95	2.21	2.21	2.95	1.48	0.33	0.03
	$\chi^2 = 41.92$ with 4 degrees of freedom; $P < 0.001$						$\chi^2 = 34.59$ with 4 degrees of freedom; $P < 0.001$				
Heterozygotes	Observed distribution	0	1	3	5	3	9	1	2	0	1
	Expected distribution	0.05	0.62	2.74	5.48	4.11	4.11	5.48	2.74	0.62	0.05
	$\chi^2 = 0.64$ with 4 degrees of freedom; $P > 0.95$						$\chi^2 = 27.60$ with 4 degrees of freedom; $P < 0.001$				

[a] p = probability of lethality due to dose; $q = 1 - p$ = probability of survival.

DISCUSSION

Our aim was to find the preferred genotypes(s) for tests of FPM from 13 types of F_1 hybrids and seven inbred types. For comparative toxicity tests the least sensitive genotypes with a mild slope of dose regression seem to be suitable, while, for testing the number of loci controlling

ESZTER CHOLNOKY,
JÁNOS FISCHER, AND
SÁNDOR JÓZSA

the action of a drug, the most sensitive and most tolerant genotypes are taken into consideration (Chai, personal communication to Brown and Wallace).[9]

The tests with quartets of identical genotype receiving LD_{25} are not conclusive because the deviation from the expected distribution is not significant, although it is very close to being so. The distribution of the group shows no significant deviation due to genotype when the expected distribution is calculated for LD_{30} ($p = 0.3$). The homozygous genotypes treated separately show a significant deviation from the expected distribution based upon $p = 0.25$. This might not be due to the genotypical differences but to a "paradoxon of inbreeding" only. The heterozygous groups treated separately gave no significant deviation from the expected distribution calculated with $p = 0.25$.

The group receiving LD_{75} gave more conclusive results. The heterozygous and homozygous genotypes tested separately show similar significant deviations from the expected distributions. The observed distribution shows two peaks, representing (1) genotypes characterized by steeper dose regression and (2) genotypes that are more tolerant to FPM than the average.

Dealing with the different genotypes separately, A, NZB, NZW, and (BALB/cxAKR) F_1 mice are suspected of higher sensitivity to FPM. As mentioned before, the inbred groups have to be evaluated with restriction. It seems to be advisable to double the number of animals representing one inbred strain within the group. For example, NZB males had 3:1 death–survival ratios after both doses, which might have been due to the greater variability of reaction alone. On the previous day, when the animals were grouped, the NZB males had an atrocious fight with their new albino cage mates in all cages. This might have influenced their reaction; they died immediately after the injection. We tested a larger group of NZB males and could not verify the greater sensitivity expected by the reaction to LD_{25} alone. Males of strain A were not available at the time of the experiment, so we have yet to prove the greater sensitivity of this strain. Similarly, a larger group of NZW males was not investigated, and because we lack offspring of both crosses between NZB and NZW mice, we are unable to determine whether this is a sensitive genotype, too. X^2 tests of the subgroup carrying the genes of strain A in different combinations show that even if there is a greater sensitivity in strain A, it must be recessive.

AKR and (BALB/cxA) F_1 males show a greater tolerance to FPM than the average. Tests with LD_{75} on a larger group of AKR males (24 animals) do support the hypothesis that again we are faced with the paradoxon of inbreeding. This leaves us with the (BALB/cxA) F_1 males as tolerant animals for comparative toxicity tests.

A second experiment can be improved by employing LD_{20} and LD_{80} instead of LD_{25} and LD_{75}, on quintets of identical genotype and by employing a third similar group injected with LD_{50}.

Summary

Complete mosaic populations consist of n inbred and $n(n-1)$ types of F_1 hybrids carrying the genes of n inbred strains in equal frequencies but in all possible combinations. In complete mosaic populations maternal influences can be investigated.

Incomplete mosaic populations are the offspring of n inbred strain and $(n/2)$ $(n-1)$ different crosses, the more productive crosses chosen from reciprocal ones.

In mosaic populations each genotype is represented equally.

A plan for experiments on mosaic populations designed to investigate discrete characters like survival and lethality after the administration of a drug is described.

As an example, the determination of the average reaction of a population consisting of 20 different genotypes to FPM is reported. The test with LD_{25} on a group in which each genotype was represented by quartets shows a great variability of homozygous animals and no significant differences of heterozygous ones. LD_{75} tested on a group of the same genotypical composition refers to higher tolerance of $(BALB/cxA)$ F_1 males; the apparent tolerance of AKR males was due to greater variability of the reaction in inbred strains.

Acknowledgments

We are much indebted to the Central Research Institute of the Hungarian Blood Service where we were allowed to carry out our experiments, to the staff of the animal house of that institute, mainly to Mrs. A. E. Fejes for her skilled technical assistance and to the Hausmann Laboratories (Switzerland) for the generous supply of Ferrum to be tested.

References

1. Brown, A. M. 1961. The pattern, sensitivity and precision of the response to insulin in random bred, inbred and hybrid strains of mice. J. Pharm. Pharmacol. 13:670–678.
2. Green, E. L. 1964. Fitness of populations of irradiated mice. Planning of experiments. Genetics 50:417–421.
3. Cholnoky, E., J. Fischer, and S. Józsa. 1969. Aspects of genetically defined populations in toxicity testing. I. A comparative survey of populations obtained by different breeding systems and "mosaic populations." Z. Versuchtierk. 11:298–311.
4. Falconer, D. S., and Robertson. 1968. Stability of strain characteristics. 6. Wissenschaftliche Tagung der Gesellschaft für Versuchstierkunde, Vienna.

228

5. Doolittle, D. P. 1968. Heterogeneity in a randombred mouse population. 6. Wissenschaftliche Tagung der Gesellschaft für Versuchstierkunde, Vienna.
6. Brown, A. M. 1961. Sleeping time responses of mice—random bred, inbred and F_1-hybrids—to pentobarbitone sodium. J. Pharm. Pharmacol. 13:679–687.
7. Brown, A. M. 1962. Strain variation in mice. J. Pharm. Pharmacol. 14:406–410.
8. Fujii, K., H. Jaffe, and S. S. Epstein. 1968. Factors influencing the hexobarbital sleeping time and zoxazolamine paralysis time in mice. Toxicol. Appl. Pharmacol. 13:431–438.
9. Brown, A. M., and M. E. Wallace. 1968. Genetic control of the reaction to histamine in mice after sensitization with pertussis vaccine. Lab. Anim. 2:63–74.
10. Chai, C. K. *In* W. A. Becker. 1962. Choice of animals and sensitivity of experiments. Nature 193:1264–1266.

DISCUSSION

DR. CLARKSON: One of the main aims of experimental design is to reduce within-group variation. This increases precision. Don't mosaic populations reduce precision?

DR. CHOLNOKY: Certainly they do, but the aim of this mosaic population is not to increase precision, but to get a wide range of information, and, based upon the results obtained on mosaic populations, one can pick out the proper genotypical animals.

DR. KADAR: There is quite a bit of similarity between Dr. Festing's work and yours. Was there any exchange of ideas or coordination between your investigations?

DR. CHOLNOKY: I was very much pleased to hear that Dr. Festing accepted at last the idea of the mosaic population, which he calls the mixed genotypes, which is identical. He looked for continuous variation. We tried to find methods to demonstrate the use of mosaic populations with discrete distributions.

INSTABILITY OF CHARACTERISTICS IN INBRED STRAINS OF MICE

O. Mühlbock
W. P. J. R. van Ebbenhorst Tengbergen

One of the more important developments in cancer research was the establishment of inbred strains of mice. Mice are the experimental animals preferred by cancer research workers. The work with inbred strains of mice has not only revealed much about the incidence of cancer and about the genetic factors operating in the development of cancer, but it has also provided investigators with a genetically controlled biological subject. Such material is of irreplaceable value for cancer research.

It is generally agreed according to the recommendation of the Committee on Standardized Nomenclature[1] that "a strain shall be regarded as inbred when it has been mated brother X sister for twenty or more consecutive generations."

At The Netherlands Cancer Institute, the Antoni van Leeuwenhoekhuis, in Amsterdam, inbred strains of mice have been used since 1931. The first two strains given to Dr. R. Korteweg of the institute by Dr. C. C. Little were the DBA strain and the C_{57} BL strain. Many more strains have been added since then. A list of the strains analyzed in this investigation is given in Table 1.

The general principles of keeping these animals are the same from the beginning. The animals were housed in small cages, three or four animals per cage. These cages were kept in small cabins so that the inbred strains could be separated as much as possible from each other. At first, glass cages were used; in recent years cages of the same size are made of plastic. The temperature in the cabins is controlled and kept constant at 22° C. The use of these small cages was based on an investigation of the frequency of spontaneous mammary tumors in the DBA strain. Mammary tumor genesis has been one of the

O. MÜHLBOCK
AND
W. P. J. R. VAN EBBENHORST TENGBERGEN

TABLE 1   Inbred Strains of Mice Investigated

Official Nomenclature (226)	Generation of Inbreeding, April 1967		Origin
	Previous	In Amsterdam[a]	
A/BrA	87	58	G. M. Bonser, Leeds, 1948
Af/A	87	54(53)	O. Mühlbock, Amsterdam
AKR/FuA	?	48	P. Loustalot, Ciba, Basle, 1953
C$_3$H/HeA	? + 35	66	W. E. Heston, N.C.I., Bethesda, 1951
C$_3$Hf/A	? + 35	65(61)	O. Mühlbock, Amsterdam
CBA/BrA	? + 77	70	G. M. Bonser, Leeds, 1948
C$_{57}$BL/LiA	?	98	C. C. Little, Jackson Mem. Lab., Bar Harbor, 1931
DBA/LiA	?	82	C. C. Little, Jackson Mem. Lab., Bar Harbor, 1931
DBAf/A	?	83(45)	O. Mühlbock, Amsterdam
GRS$_1$/A	?	41	Hyg. Inst., Zürich, 1955
LiS/A	?	41	Hyg. Inst., Zürich, 1955
LTS/A	?	44	P. Loustalot, Ciba, Basle, 1954
MaS/A	?	43	Hyg. Inst., Zürich, 1955
020/A	–	146	R. Korteweg, Amsterdam, 1931
TS$_1$/A	?	29	P. Schäfer, Tübingen, 1958
WLL/BrA	? + 29		G. M. Bonser, Leeds, 1950
WLLf/A	? + 29	62(49)	O. Mühlbock, Amsterdam

[a]Numbers in parentheses indicate inbred generations after separation from the original strain.

main topics of research at the institute for the last 30 years. In one experiment, mice of the same strain were housed under the same conditions, 50 to a cage, five to a cage, or one to a cage. The mammary tumor frequency showed very big differences; in the animals kept alone in a cage the mammary tumor frequency was 89 percent, while in the cage in which 50 animals were kept together during their whole life the frequency was only 29 percent.[2]

The composition of the food has, of course, changed over the years. In the beginning the diet consisted of a mixture of different sorts of grains, with milk as fluid; later, commercial pellets were given and tap water was fed *ad libitum.*

During the whole observation period (1931-1966) the mouse colony in The Netherlands Cancer Institute was under the supervision of W. van Ebbenhorst Tengbergen; the care of the animals was the responsibility of P. Kloosterman.

The conditions of our mouse colony have been as constant as possible. It seemed therefore worthwhile to analyze the data of this colony for the inconsistency of the inbred mouse strains, a characteristic that is to be expected of all living material.

Results

MAMMARY TUMOR INCIDENCE IN FEMALE MICE

The study of factors in the genesis of mammary carcinomas in mice was one of the major projects of the last 35 years at The Netherlands Cancer Institute. A review of the tumor incidence in different strains in the various periods is given in Table 2.

The incidence of mammary tumors is higher in breeding females than in virgin females. The appearance of tumors can still be accelerated by a procedure called "force breeding" whereby the animals have pregnancies in rapid succession; the young are removed from the mother immediately after birth.

The most complete data are available from breeding females, so only these are analyzed. The virgin and the forcebred female show the same trend.

The DBA/LiA Strain

The high incidence of mammary tumors (80 percent) in breeding females was observed from 1931 to 1939. The incidence then decreased to 40 percent in the years 1939-1943. In the years 1952-1953 the high incidence of 80 percent was again observed.

It is an established practice to continue breeding an inbred strain by the single line system. For the highest degree of genetic homogeneity it is essential that but one central trunk of the inbreeding tree be maintained.

A comparison of the tumor incidence in generations 16-25 was made in the females of the central trunk line and the females in the side lines.

In the group of breeding females of the central trunk line the incidence was 25 percent (13 out of 52 animals) as compared with the females of the side lines in these inbreeding generations of 54 percent (26 out of 48 animals). The difference is statistically significant. The average tumor age was 12 months in both groups.

From these results one may conclude that a mutation has to be considered responsible for the lower value in the tumor incidence in the DBA strain during 1939-1941. As the incidence rose again in the following period it must be assumed that the mutation had been eliminated.

The A/BrA Strain

The A/BrA strain is characterized by a high mammary tumor incidence in breeding females, whereas virgin females have no mammary tumors. During the years 1948-1951, 82 percent of the breeding

TABLE 2 Mammary Tumor Incidence in Females of Different Inbred Strains of Mice during 1931–1963

| | Virgins | | | | Breeders | | | | Force Bred | | | |
| | | Incidence ma-ca (%) | Average Age at Death (months) | | | Incidence ma-ca (%) | Average Age at Death (months) | | | Incidence ma-ca (%) | Average Age at Death (months) | |
Strain	Number		With Tumors	Without Tumors	Number		With Tumors	Without Tumors	Number		With Tumors	Without Tumors
DBA/LiA												
1931–1935	179	89	13½	11	102	80	12½	12	—	—	—	—
1935–1939	163	62	15	17½	100	81	12	12	—	—	—	—
1939–1943	67	60	17	17	100	40	14	13	—	—	—	—
1950–1951	48	50	15½	14½	93	55	12½	12	—	—	—	—
1952–1953	21	83	17	25	39	81	14	14	—	—	—	—
A/BrA												
1948–1951	46	0	—	13	267	82	12½	18	41	78	12	15½
1952–1955	39	0	—	17	60	52	13	15	30	92	10½	9½
1956–1963	81	0	—	22	181	25	13	17	35	46	11	15½
C₃H/HeA												
1951–1952	—	—	—	—	113	94	8½	6½	—	—	—	—
1955–1956	—	—	—	—	126	97	9	8	—	—	—	—
1959–1963	—	—	—	—	290	78	9	9½	—	—	—	—
C₃Hf/A												
1951–1953	—	—	—	—	80	61	19	20	—	—	—	—
1955–1956	—	—	—	—	74	32	17½	18½	—	—	—	—
1959–1963	—	—	—	—	225	40	18	19½	—	—	—	—

TABLE 2 (Continued)

Strain	Virgins				Breeders				Force Bred			
	Number	Incidence Ma-ca (%)	Average Age at Death (months) With Tumors	Without Tumors	Number	Incidence ma-ca (%)	Average Age at Death (months) With Tumors	Without Tumors	Number	Incidence ma-ca (%)	Average Age at Death (months) With Tumors	Without Tumors
WLL/BrA												
1955–1957	97	48	15	17	46	72	17	17	64	77	16	17
1958	–	–	–	–	27	33	16½	17	–	–	–	–
1959–1963	89	68	17	16½	42	64	17	17	89	85	16	17
LTS/A												
1953–1955	–	–	–	–	33	88	17	16	–	–	–	–
1956–1958	–	–	–	–	25	80	20	15½	–	–	–	–
1959–1963	–	–	–	–	45	55	15	15½	–	–	–	–

females developed mammary tumors. In 1952 a reduction to 52 percent was observed. Tumor incidence remained at this level until 1956. In the years 1956-1963 a further reduction of tumor incidence to 25 percent was noted. This has not changed since that time. In force-bred animals a reduction was seen during the period 1956-1963 when the incidence dropped from 96 to 46 percent.

A further analysis showed that in generations 22 through 37 the descendants of the central line had a tumor incidence of 16 percent (7 out of 43 animals), whereas the descendants of the other pairs had a 36 percent incidence (14 out of 39 animals). Figure 1 is the Pedigree chart of the breeding females.

During the years 1956-1959, in generations 24-32 the low tumor incidence was apparently fixed. The descendants of the central line had a 21 percent incidence of mammary tumors (6 of 29 animals), and the incidence in the other descendants was 38 percent (23 of 60 animals).

The C_3H/HeA and C_3Hf/A Strain

In the C_3H strain a significant reduction of the mammary tumor incidence was noted in the years 1959-1963, when the incidence dropped from 97 percent to 78 percent. In the C_3Hf line, which is free from the mammary tumor agent, a reduction in the mammary tumor rate was noted; the incidence was 61 percent in the years 1951-1953, 32 percent in 1955-1956, and 40 percent in 1959-1963.

The WLL/BrA Strain

The mammary tumor percentage in the WLL/BrA strain was investigated in the years 1955-1963. Three groups were observed: Virgin females had a significant increase in tumor incidence in the second part of the observation period from 48 to 68 percent. The breeding females, however, showed a reduction in tumor frequency in the same part of the observation period. The force-bred females had the same incidence throughout the entire period.

The LTS/A Strain

The breeding females of the LTS/A strain were observed from 1953-1963. During the last 4 years of this period there was a significant decrease in mammary tumor incidence from 90 to 55 percent.

INCIDENCE OF LUNG ADENOMAS IN MALE AND FEMALE MICE

Lung tumor incidence was determined in 11 different lines of mice between 1950 and 1961. In females of the 020/A strain,

235

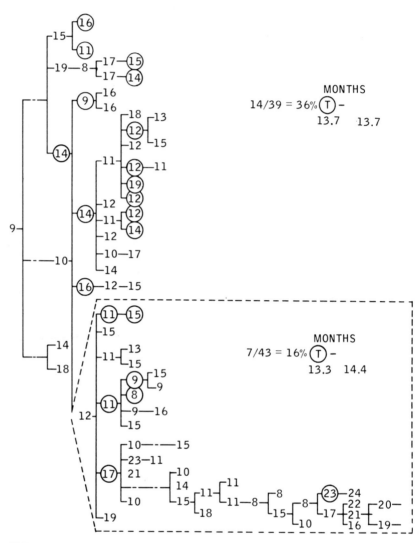

FIGURE 1 Pedigree chart of strain A/BrA.

a lung adenoma incidence of 27 percent was observed in 1937. This incidence increased steadily until it reached 58 percent in 1961. In the males the same significant increase was noted from 41 percent in 1950 to 56 percent in 1959-1961. In the A/BrA strain, the high incidence of lung adenomas showed no change between 1956 and 1961. In the Af/A line only the males showed

O. MÜHLBOCK
AND
W. P. J. R. VAN EBBENHORST TENGBERGEN

an increase in the frequency of lung tumors, from 52 percent in 1956-1958 to 76 percent in 1958-1961.

Especially striking is the increase in the incidence of lung tumors in strains that have only a very low spontaneous tumor rate, i.e., strains $C_{57}BL/LiA$, CBA/BrA, C_3Hf/A, DBAf/A, WLL/BrA, and WLLf/A.

Table 3 gives a summary of the significant changes in the lung tumor rate in these strains. It is apparent that, with a few exceptions, the lung tumor incidence has increased in all strains.

INCIDENCE OF LEUCOSIS IN FEMALE AND MALE MICE OF DIFFERENT INBRED STRAINS, 1931-1963

Most of the 13 lines of mice investigated showed an increase in the incidence of leucosis (Table 4). The higher frequency peaked in the last years of observation. Four lines, DBA/LiA, DBAf/A, WLL/BrA, and Ma/A, showed no significant change in the incidence.

In Table 5 a summary of the data on leucosis is given, and observations on other tumors are reported. The incidence of these tumors fluctuated in the course of the years of observation.

Postnatal Development

Postnatal development follows a fixed pattern: the unfolding of the ears, the beginning of hair growth, the eruption of the lower incisors and then of the upper incisors, and the opening of the eyes. The average age of occurrence of these phases in development was determined in 16 strains in 1942 and in 32 strains and sublines in 1966, using 20 animals per strain. The average age in days for each of these phases in development in all strains together was the same in 1942 as it was in 1966 (Table 6).

The age at which the different steps in the development occur are genetically determined and are not dependent on one another. In the strain WLL/BrA the eruption of the incisors is especially late compared with other strains (lower incisors 9.4 days, upper incisors 11.6 days), whereas all other steps in the development occur at the same average age as they do in other strains.

Changes of the age at which the different phases of postnatal development occurred were observed in some strains; these are reported in Table 7.

Especially striking are changes in the average age at which the phases of development occurred in the $C_{57}BL$ strain when measured in 1938 and in 1966. All phases have an earlier average age in 1966. No age changes were observed in the 020/A strain between 1942 and 1966. In the DBA/LiA strain only the age at which the incisors erupted showed significant changes in 1966 compared with the age in 1938.

TABLE 3 Frequency of Lung Adenomas in Different Strains during 1956–1961

	Females				Males			
			Average Age at Death (months)				Average Age at Death (months)	
Strain	Number	Lung Tumors (%)	With Tumors	Without Tumors	Number	Lung Tumors (%)	With Tumors	Without Tumors
020/A								
1937	277	27	22½	22	–	–	–	–
1950–1951	123	40	20	21	48	41	20½	19½
1957–1958	128	51	15½	13	85	60	16½	11½
1959–1961	133	58	21	21	193	56	19½	19½
A/BrA								
1956	109	56	18½	17	32	69	23	15
1958–1961	55	64	19½	17½	30	76	20	21
Af/A								
1956–1957	75	53	18	17	17	53	18	18
1958–1961	42	56	23½	21½	44	84	21	18
C57BL/LiA								
1931	100	2	23	22	40	0	–	19
1937	158	0	–	21	–	–	–	–

	92	11	23	22	184	22	27	26
1959–1961	92	11	23	22	184	22	27	26
C₃Hf/A								
1956–1958	121	2	19	19½	—	—	—	—
1959–1961	166	10	22	20	—	—	—	—
DBAf/A								
1956–1958	61	1	20	18	—	—	—	—
1959–1961	73	11	20	18	—	—	—	—
AKR/FuA								
1956–1958	222	54	17½	17	32	78	20	20
1959–1961	82	67	20	17½	67	82	21	20
WLL/BrA								
1956–1958	160	2	19½	17	—	—	—	—
1959–1961	158	10	19	17½	—	—	—	—
WLLf/A								
1956–1957	87	1	13	20½	28	0	—	18½
1959–1961	99	8	20	20	36	22	20	19
MaS/A								
1956–1958	23	26	17	18	11	54	22	17
1959–1961	31	52	19	18	19	68	25	21

TABLE 4 Incidence of Leucosis in Different Untreated Inbred Strains of Mice during 1931–1963

Strain	Females				Males			
	Number	Leucosis (%)	Average Age at Death (months)		Number	Leucosis (%)	Average Age at Death (months)	
			With Leucosis	Without Leucosis			With Leucosis	Without Leucosis
C$_{57}$BL/LiA								
1931	100	3	23	22	89	0	–	19
1937	158	5	23	21	–	–	–	–
1949–1951	162	14	19	20½	92	1	24	20½
1957–1958	42	19	19½	20	85	8	21	22
1959–1963	110	25	19½	19	87	14	22	23
DBA/LiA								
1931–1932	92	2	12	15	–	–	–	–
1937–1938	96	2	16	16	–	–	–	–
1950–1951	61	0	–	15½	–	–	–	–
1952–1954	43	0	–	14	–	–	–	–
1955–1959	24	4	11	16	–	–	–	–
DBAf/A								
1948–1949	269	8	17	20	–	–	–	–
1950–1951	59	5	17	17½	–	–	–	–
1952–1954	24	8	17½	19	–	–	–	–
1955–1958	74	8	13½	19	–	–	–	–
1959–1963	134	8	17½	19	–	–	–	–
Af/A								
1956–1957	76	3	17	17	–	–	–	–
1959–1963	172	14	20	22½	–	–	–	–
CBA/BrA								

C₃Hf/A								
1951–1953	192	6	20	20	50	6	23	24
1954–1958	187	14	21	20½	32	21	24	24
1959–1963	260	30	21	20	75	66	26	24½
AKR/FuA								
1954–1958	201	10	19	18	50	4	14	18
1959–1963	89	36	19	18	60	23	21	18
020/A								
1937	279	3	19	22	–	–	–	–
1950–1951	104	2	24	20½	44	0	–	19½
1957–1958	128	1	7	14½	60	0	–	18½
1959–1963	237	10	17	17	213	5	21	20
WLL/BrA								
1956–1958	160	1	18	17	36	5	16½	20
1959–1963	233	2	17	17	47	13	17	20½
WLLf/A								
1956–1957	87	1	19	20½	37	0	–	18½
1959–1963	133	5	21	20½	72	7	19½	19½
LTS/A								
1953–1958	70	0	–	16	–	–	–	–
1959–1963	219	5	17½	16	–	–	–	–
MaS/A								
1955–1958	23	21	16	18	11	9	24	19
1959–1963	225	22	25	22½	33	15	29	23

TABLE 5 Significant Changes in Tumor Frequency in Mice of Different Strains

Leucosis

Strain	Increase	Decrease
A/BrA	ss	
Af/A	ss	
C57BL/LiA	ss	
CBA/BrA	ss	
C3H/HeA	ss	
C3Hf/A	ss	
DBA/LiA	—	—
DBAf/A	—	—
LTS/A	ss	
WLL/BrA	—	—
WLLf/A	ss	
LiS/A	s	
MaS/A	—	—
020/A	ss	—

Lung Tumor

Strain	Increase	Decrease
A/BrA	—	—
Af/A	ss	
C57BL/LiA	ss	
CBA/BrA	ss	
C3Hf/A	s	
DBAf/A	s	
LTS/A	ss	
WLL/BrA	ss	
WLLf/A	s	
MaS/A	s	
020/A	ss	

Liver Tumor

Strain	Increase	Decrease
CBA/BrA	ss	
C3H/HeA	s	
C3Hf/A		ss

Uterus Tumor

Strain	Increase	Decrease
CBA/BrA	ss	
DBAf/A	s	
WLL/BrA		s
WLLf/A		ss

Testis Tumor

Strain	Increase
AKR/FuA	ss

Ovarium Tumor

Strain	Increase
CBA/BrA	ss

Note: $s = P = < 0.05$; $ss = P = < 0.01$.

242

TABLE 6   Postnatal Phases of Development

	Average Age in Days		Minimum Average Age in Days (1966)	Maximum Average Age in Days (1966)
	(1942) (16 strains)	(1966) (32 strains)		
Unfolding of the ears	3.2	3.3	2.5 (DBAf/A)	3.9 (LTS/A)
Epithelial scales (appearance of hairs, dorsum)	5.3	5.4	4.5 (RIII/A)	6.1 (STS/A)
Eruption of lower incisors	6.8	7	5.7 (DBA/HeA)	9.4 (WLL/A)
Eruption of upper incisors	10.6	10.7	8.7 (DBA/HeA)	12 (020/A)
Opening of the eyes	13.8	13.3	11.1 (DBA/HeH)	14.3 (P/A)

TABLE 7   Average Age in Days of Postnatal Development in Inbred Strains in Different Time Periods

	$C_{57}BL/LiA$		DBA/LiA		020/A	
	1938	1966	1938	1966	1942	1966
Unfolding of the ears	4.5 *s*	3.7	2.5	2.6	3.1	3.2
Epithelial scales (appearance of hairs, dorsum)	6 *ss*	4.8	5.4	5	4.8	5.2
Eruption of lower incisors	7.6 *ss*	6.5	6.5	6.4	7.3	6.9
Eruption of upper incisors	11.3	11	9.7 *ss*	11.5	11.8	12.0
Opening of the eyes	15.4 *ss*	13.9	13.2 *s*	12.5	13.7	13

Note: $s = P = < 0.05$; $ss = P = < 0.01$.

Body Weight

The average body weight of mice of different strains is genetically determined. The average body weight of females and males of 28 strains was measured in 20 animals of each group at different periods between 6 months and 24 months of age. The mean body weight from 1,676 females and 1,466 males is given in Figure 2.

In both males and females the body weight increases until the age of 15 months. After that age a decrease of the body weight occurs continuously. In the old age group of 21–24 months the average body weight is less than at the

GRAMS

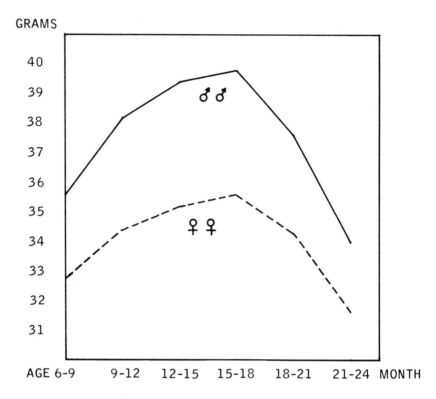

FIGURE 2 Body weights of adult mice.

age of 6 months. In all age groups the difference in body weight between females and males is the same: Males are, on the average, 10 percent heavier than females.

In six different inbred strains the average body weight in the same age groups could be determined in 1950–1953 and 10 years later in 1963. In all strains an increase of the average body weight was noted in 1963: Strain A/BrA = 12 percent, $C_{57}BL/LiA$ = 24 percent, CBA/BrA = 9 percent, C_3Hf/A = 10 percent, DBA/LiA = 4 percent, and 020/A = 17 percent.

The differences in body weight, in grams, between two parallel lines of eight inbred strains of mice are given in Figure 3. In seven of the strains a significant difference in body weight was noted in the two parallel lines. The measurements were done in the same year (1965). Precaution was taken that in both lines animals of the same age group were investigated. It may be assumed that genetic changes, not environmental factors, are responsible for these differences in the two lines.

244

O. MÜHLBOCK
AND
W. P. J. R. VAN EBBENHORST TENGBERGEN

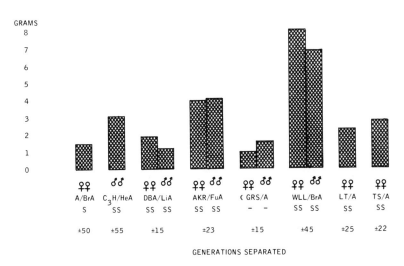

FIGURE 3   Differences in body weights of two parallel lines of eight inbred strains of mice, 1965.

Discussion

For the study of changes in inbred strains, some regulation should be maintained. Preferably, one investigator should have the supervision. The most important requirement is to prevent inadvertent genetic contamination by rigid genetic control and to reduce the environmental variations as much as possible.

Inbred strains of mice are usually considered to be homozygous at more than 99 percent of their loci. In most of the strains observed in this investigation brother X sister matings have been done for more than 20 generations. As far as possible all necessary uniformity was maintained for our investigation of the changes in inbred strains of mice.

General observations of mice within the strains support the belief in uniformity of inbred strains. Tissue and tumor transplantation done at frequent intervals confirmed that mice are as alike as identical twins at the histocompatibility loci. From our observations it was proven that inbred strains of mice do show changes over long periods. The inbred strains, however, in our more than 35 years of observation do not lose their identity to the extent that they are as different as independently derived inbred strains.

Emphasis in our work with inbred strains is on the incidence of so-called "spontaneous tumors," especially on the incidence of mammary tumors in female mice. Our series of observations made it clear that the incidence of

245

these tumors is not constant in a given strain. It is known that although mice of an inbred strain are alike genetically, they do not all develop a given kind of tumor. It appears from the results that this percentage in the inbred strains can vary. Many genes and different nongenetic factors determine whether a tumor will arise.

The question then arises: Is the observed change in the tumor incidence due to a mutation or to inconstant nongenetic factors?

In the case of the mammary tumor incidence, the analysis argues for a mutation. The change in tumor incidence is restricted to a single line, which was kept for the provision of the following generations, whereas in the parallel lines, which were discontinued, this change could not be detected.

The opposite conclusion must be accepted with respect to the changes in the incidence of lung tumors and of leucosis. Almost all strains investigated show an increase in the incidence of both these tumors, and no restriction to a certain line was noted. Therefore, a change in environmental factors seems the more probable cause. There is no direct evidence of which factor must be considered responsible.

A change in the pattern of the postnatal morphological development was also found in some inbred strains. It is likely that this results from a fixation of the mutations that have occurred in the strain. Another possibility would be to assume that in the beginning the strain was still in a heterozygous state. The last assumption seems unlikely: In two of the strains the first observation was made at a time when both strains were already many generations beyond the twentieth inbred generation.

Adult body weight is genetically controlled[3,4] involving probably many genes. In all strains investigated an increase of body weight was noted after 10 years. In parallel lines of seven strains separated for 15–50 generations, a significant difference in body weight was noted. This difference must be considered to be due to a genetic change.

Genetic changes in inbred strains of mice cannot be prevented. Environmental changes may also occur in the course of years. The practical consequence of the instability of characteristics in inbred strains is that control experiments must always be done with the same inbred generation at the same time.

Summary

For more than 35 years inbred strains of mice were kept in the laboratory of The Netherlands Cancer Institute under rigid genetic control and as far as possible under the same environmental conditions. Observations about several characteristics in these inbred strains in different periods were compared. Considerable changes were noted during the course of the 35-year study. Special attention was paid to the changes in tumor incidence in these strains. In some strains a fluctuation of the incidence of mammary tumors was noted. An increase in the frequency of lung adenomas and leucosis in the last three decades could be observed. Other characteristics compared are the phases in the postnatal development of newborns and adult body weight. Out of 17 inbred strains of mice eight are kept in parallel lines for many generations. A study of the changes in adult body weight between these parallel lines kept under the same environmental conditions has been made.

References

1. Staats, J. [For Committee on Standardized Genetic Nomenclature for Mice.] 1968. Standardized nomenclature for inbred strains of mice, Fourth Listing. Cancer Res. 28:391–420.
2. Mühlbock, O. 1951. Influence of environment on the incidence of mammary tumors in mice. Acta Unio Internationalis contra Cancrum, 7:351–353.
3. Grüneberg, H. 1952. The genetics of the mouse, 2nd ed. Martimes Nijhoff, The Hague.
4. Falconer, D. S. 1955. Pattern of response in selection experiments with mice. Cold Spring Harbor Quant. Biol. 20:178–196.

DISCUSSION

DR. MYERS: Do you have any evidence of a possible role of infectious agents on any of the changes you have observed?

DR. MÜHLBOCK: No, we have not. In the case of the mammary tumors, a viral factor is very instrumental. We have observed on many occasions that one can lose this virus factor, which is transmitted by the mother mouse. This could be one explanation for an infectious factor. We have thought that in leukemia a virus infection may occur. We have, of course, in our laboratory, strains with a high leukemia caused by a virus. This suggests the possibility that contamination could occur. It does not seem likely, but the possibility exists. We know that the incidence of leukemia can be elevated by radiation, so this could be another possibility. In the case of lung tumors, one thinks, of course, of air pollution. In fact, we have done some experiments where we could show that air pollution raises the incidence of this lung tumor, but the increase in air pollution required to increase lung tumor incidence is so high that again it does not seem likely. So the answer must be no, we have no actual evidence.

DR. HEINE: Do you have any data about the environmental changes for the time that you described, since 1931—for example, changes in food—and can these data be correlated with those of changes in tumor incidence?

DR. BLEBY: I have a similar question. Could some of the changes over the past 30 years be due to changes in environmental factors, such as diet, disease, and burden?

DR. MÜHLBOCK: As I mentioned, the composition of the food in the last 30 years has certainly changed. In the beginning, the composition of the food was quite different from what we have now. We now use pellets, and we control the composition of the pellets, but there is a variation, and it is definitely a factor. I think the increase in body weight in our animals is due to the better food we give now. We have some data, which I have not mentioned, about newborns and the body weight of the weanlings. They are much bigger than they were 20 and 30 years ago. This is certainly a factor.

The other factor mentioned, "burden," I do not know what this means.

DR. BLEBY: By burden, I mean the absence or presence of disease, the changes in the bacterial flora they carry.

DR. MÜHLBOCK: Oh, I see. No, I could not say this. I would not know. We keep our colony very closed, so there is no infection by salmonella, or whatever we control from time to time, so I would not think so. We once had this problem because we were not so strict. We brought in some mice from the outside and promptly we got an infection of ectomelia[?] so we had to kill all of them. But they were excluded from the registration reported here.

Then, of course, one factor I have not taken up here, is the years of the war in Europe. At that time the conditions in Amsterdam, especially near the end of the war, were very bad, so we had distributed among the co-workers just a very few pairs of the inbred lines just to keep them alive, because it was not possible in the laboratory. But we have excluded this period from our discussion, too.

DR. HANSEN: Have you observed correlated changes in reproductive performance with changes in tumor frequency? How much attention is given to the reproductive performance when selecting replacement breeders?

DR. MÜHLBOCK: Correlated changes in reproductive performance with changes in tumor frequency? We have this in connection with mammary tumors, where there certainly is a correlation between the reproductive performance and the tumor frequency. The reason is that in our experiments we have standardized this, so we always have a fixed number of pregnancies. If we have breeders and we want to determine tumor frequency, we do not go so far as to standardize the number of newborns.

How much attention is given to reproductive performance when selecting replacement breeders? Again in the case of the mammary carcinoma, no, of course we can select only if they breed, but just to keep the mammary tumor incidence high, we select them by early appearance of mammary tumors. In strains that have no mammary tumors, we just select by general appearance, but we make no point of correlating the reproductive performance with the selection.

DR. BOND: Were biochemical changes tested for along with these other changes in growth and tumor frequency?

DR. MÜHLBOCK: No, we have not tested this. I doubt if 25 years ago much was known about this, but we have not done it.

DR. LANE-PETTER: Do you think the variation in mammary tumor incidence could be influenced by the vitamin E content of the diet?

DR. MÜHLBOCK: I cannot answer this definitely. I remember that in some years when we had difficulty keeping certain strains going, we resorted to all these things, like giving them vitamin E, but we have not seen much influence of this because we have not done a study where we really compared the intake of vitamin E with the tumor frequency.

DR. DIETERICH: Please explain further the meaning of the "S" and the "SS," which showed on your slides.

DR. MÜHLBOCK: "S" simply means significant, and "SS" means highly significant.

DR. MOORE: With regard to variation in tumor incidence in breeders of various inbred strains over the years, has the average age of the breeder colony remained stable, or has it changed?

DR. MÜHLBOCK: The age of an animal, or the life-span of an animal is dependent on the strain. I would say that over the years there has not been much change. One strain that has a long life-span (3 years) is the CBA strain. The C-57 black strain has had only 24–26 months, and still has this. So if there are changes, they are certainly not very big differences.

DR. CLARKSON: Were the decreases in the body weight of the old animals related to the heavier animals' having died off first, akin to some observations published by Dr. Berg?

DR. MÜHLBOCK: I do not think so, because in this group the mortality in these inbred strains before 24 months of age is not very great. I do not think this can be influenced by such a factor.

GENE FREQUENCIES IN WORLD CAT POPULATIONS AND THEIR RELEVANCE TO RESEARCH

Neil B. Todd

Introduction

There is a general but erroneous opinion that genetic control in cats cannot be a rigorous requirement in experimental designs utilizing this species. Certainly, there is no inbred cat supply comparable to the isogenic stocks of rats, mice, hamsters, and guinea pigs that are now available to the research community. Today, however, the formal and population genetics of the domestic cat are sufficiently well known that there can be no excuse for ignoring this potential for control. The following discussion will be limited to a consideration of the cats of northwest Europe and their derivatives. It will also be confined to three independent autosomal loci and a sex-linked factor, all of which affect coat colors and patterns. Neither of these limitations is particularly serious, since the majority of cats used for research are drawn from these populations and the four genetic markers have proven sufficient both to equate and to distinguish subunits of these populations.

History of the Domestic Cat

The domestic cat is indigenous only to the Old World. Its introduction into northeastern North America began in the seventeenth century, and it was not established in Oceania until 150 years later. The original populations brought to both continents came principally from England (and to a lesser extent from The Netherlands). While it is beyond the scope of this paper to detail the dynamics of this introductory phase, the evidence available can best be summarized in the conclusion that the genetic character of these populations was fixed and stabilized as a very close approximation of a random sample of the parental populations.

Genetics of the Domestic Cat

For cats, the standard phenotype (wild type), to which comparisons are made, is that of the so-called striped tabby. This phenotype consists of hues of tan, brown, and black. The belly tends to be light tan, often having conspicuous darker spots, while the mid-dorsum is generally dark brown to black. The most distinctive feature is the alternating black (nonagouti) and tan (agouti) stripes arranged vertically on the sides. Each of the four mutants that will be noted below modifies this basic wild type in a particular way.

The first mutant to be considered is sex-linked orange (O). This allele transforms all black pigment to orange. The effect is to produce an orange and buff striped cat, called in the vernacular "ginger" or "marmalade." The two alternative alleles at this locus manifest lack of dominance toward one another so that the heterozygous female is an orange and black mixture known as tortoiseshell. The lack of dominance is apparently related to random deactivation of one X chromosome and has been taken as evidence for the single active X hypothesis.[9] The occasional tortoiseshell male is frequently found to be XXY, corresponding to Kleinfelter's syndrome in the human.[14] Since the three female genotypes produced by this pair of alleles are phenotypically distinguishable, their distribution in the population is used in an analysis of randomness of breeding. The allele frequencies can be computed directly from a tally of individuals in a sample, and the observed and predicted phenotypic distributions based on the Hardy-Weinberg (binomial) distribution can be tested for concordance by a χ^2 test. All populations studied have proven to be random breeding according to these criteria, where the sample size has been adequate.

The second mutant to be considered is nonagouti (a), an autosomal recessive that results in the elimination of all agouti hairs and hence the striping characteristic of the wild type. The most common manifestation of this allele is the solid black cat. This allele is hypostatic to sex-linked orange so that an orange cat homozygous for nonagouti (aa) cannot be distinguished from the heterozygous (a^+a) or homozygous (a^+a^+) alternatives. This is sometimes vividly demonstrated in tortoiseshell females that have solid black patches among striped areas of orange.

A third locus, usually referred to as tabby, affects the dispersal of agouti hairs. In fact, at least three alleles are known that yield, in their order of dominance, Abyssinian "tabby" (t^A), striped tabby (t^+), and blotched tabby (t^b). Abyssinian is the total extension of agouti, striped is the wild type described above, and blotched is a variant of the striped pattern. The latter, in addition to being much darker than the wild type, is easily recognized by the replacement of vertical stripes on the sides by swatches and swirls of black and agouti. All three tabby alternatives are hypostatic to homozygous non-agouti (aa), and a black cat is black regardless of which alleles are present at the tabby locus. Abyssinian is basically irrelevant in the present context, for in northwest Europe and related populations it is rarely found. Except in a few areas where it has been recently introduced through carelessness or indifference on the part of breeders of show cats, it is not encountered. Blotched tabby, in contrast, is a common and almost characteristic mutant of European cats and their colonial relatives.

The final mutant to be considered is so-called blue or Maltese dilution (d), another autosomal recessive. This factor reduces black to gray and orange to cream (tortoiseshell to a mixture called "blue-cream") and is not usually difficult to identify.

A complete description of phenotypes, their genetic bases, and their interactions will be found in a useful synopsis by Robinson.[10] A standard genetic nomenclature and symbolization for mutant factors in the cat has recently been promulgated by Dyte, et al.[6]

Discussion

EUROPEAN POPULATIONS

In Table 1 the results of several gene frequency surveys for European locations are summarized. All except Venice, which is included for comparison, represent samples from the northwest European gene pool. In order to clarify and characterize this population, it will be necessary to consider two determinants of gene frequency distributions in cats. The data from London, Mayenne, and Paris illustrate both phenomena well. Mayenne is a rural district in Brittany some 300 km west of Paris and approximately the same distance south of London. Although it is separated from the latter by the English Channel while at the same time quite accessible to Paris, there can be little doubt that the two urban populations are more similar to each other than either is to Mayenne. This demonstrates the first principal, *viz.*, that geographical barriers may be less important than other factors in influencing gene frequencies in cats. More important is the fact that for the two mutant alleles that result in darker phenotypes, *viz.*, nonagouti (a) and

TABLE 1   Mutant Allele Frequencies in European Cat Populations

| | Phenotype | | | | Approximate | |
	O	a	t^b	d	Sample Size	Reference
London	0.107	0.762	0.814	0.142	700	Searle (11)
Paris	0.066	0.74	0.76	0.33	1,000	Dreux (1)
Mayenne	0.140	0.66	0.58	0.26	300	Dreux (2)
Chamonix	0.12	0.70	0.76	0.31	94	Dreux (3,4)
Marseilles	0.07	0.69	0.70	0.30	272	Dreux (5)
Venice	0.058	0.579	0.483	0.348	210	Searle (12)

blotched tabby (t^b), the rural population of Britanny shows lower frequencies than either Paris or London. Similarly, for the darker wild type alternative of sex-linked orange (O^+) the frequency in London and Paris is higher than it is in Mayenne. Only in the frequency of dilution (d) is there a deviation from the pattern of darker urban and lighter rural phenotypes, with Paris higher than Mayenne. London in its extremely low frequency of dilution differs from all other northwest European and related populations. This may reflect severe conditions of selection in London, or it may be only a coincidental aberration. Unfortunately, no comparable data are yet available for Great Britain. Further clarification and characterization of the northwest European gene pool is provided by comparing the populations of Paris, Chamonix (at the western foot of Mt. Blanc), and Marseilles. Marseilles, which is the most distinct of this trio, and Chamonix are more similar to Paris than the rural population of Mayenne. The fact that both Marseilles and Venice are Mediterranean ports seems unimportant in determining cat gene frequencies, for these two cities differ considerably from each other.

A summary of the northwest European gene pool of the mid-twentieth century is, perhaps, best expressed by giving the outside ranges of mutant allele frequencies and noting that any geographic clines are less striking than internal urban–rural clines. The approximate values are, therefore: $O = 0.066 - 0.140; a = 0.66 - 0.76; t^b = 0.58 - 0.81; d = 0.14 - 0.33$. Finally, before leaving the consideration of this gene pool, it seems pertinent to note a special circumstance surrounding the survey of the population of Mayenne. The 300 cats tabulated in this study had all been transported from a collection point in Mayenne for use in neurophysiological experimentation in Paris. It does not take much imagination to visualize the confusion that might result when an effort is made to duplicate this research in Paris (or anywhere else) employing a sample from a different population.

NORTH AMERICAN POPULATIONS

Turning to the populations of the northeast United States, the data summarized in Table 2 immediately reveal great differences from European populations. On the average, frequencies of lighter mutants are much higher, and darker mutants are lower in this North American group than in Europe. For instance, O is 10 percent higher, and d is 12 percent higher, while a is 9 percent lower and t^b is 33 percent lower. In fact, the only noteworthy single exception to this is the high frequency (0.752) of nonagouti (a) in New York City, which falls high in the European range (0.66 - 0.762). New York is obviously exceptional in this regard since the four other cities, which virtually surround it, vary only between 0.637 and 0.68. On the whole, the five cities thus far studied in the northeast United States, encompassing an area of some one quarter million square miles, are quite homogeneous. Although no rural population has yet been studied adequately, internal differences correlate well with urban intensity, the highest frequencies of dark alleles being found in New York City and the lowest in Columbus, Ohio. The assumption that urban "pressure," as a function of both duration and intensity, is greater for New York than for Columbus is not difficult to support.

As an interesting aside, comparable gene frequency data for San Francisco are included in Table 2. These differ vastly from the northeast United States, but are otherwise inexplicable due to a lack of data for the cats of Mexico (from which this population is directly descended) or of Spain (which was the original source). Still, a cautionary note concerning the differences between the cats of the east and west coasts of the United States is in order. The results of experimentation performed in one place might be difficult to duplicate in another if any significant genetic variable were involved.

TABLE 2 Mutant Allele Frequencies in North American Cat Populations

	Phenotype				Approximate Sample Size	References
	O	a	t^b	d		
New York, N.Y.	0.145	0.752	0.473	0.443	250	Todd (17)
Boston, Mass.	0.193	0.642	0.443	0.426	300	Todd (16)
Rochester, N.Y.	0.196	0.643	0.444	0.421	150	Searle (13)
Philadelphia, Penna.	0.262	0.68	0.48	0.47	115	Tinney and Griesemer (15) Todd, unpublished
Columbus, Ohio	0.289	0.637	0.312	0.501	250	Tinney and Griesemer (15)
San Francisco, Calif.	0.32	0.70	0.50	0.13	120	Searle (13)

OCEANEAN POPULATIONS AND POPULATION CHANGES

Since cats of the northeast United States are derived from cats of northwest Europe, it is apparent that large shifts in the genetic structure of the two populations have occurred. The question is: Which populations have undergone what changes in the 350 years of separation? Remarkably, the answer is provided by the cats of Australia and New Zealand. One hundred fifty years after the arrival of domestic cats in the northeast United States, permanent cat populations were established from European sources in eastern Australia. This was followed half a century later by the introduction of cats (mostly directly from Great Britain) into New Zealand. Although the data from New Zealand are quite fragmentary and represent principally observations on the south island, they are consistent with other findings. In Table 3 the data available for the present populations of these areas have been compiled. It will be noted that frequencies for the alleles in question range between those found today in the northwest European and northeast North American populations: a is close to the former, and d (for Australia) is within the range of the latter, while O and t^b are intermediate.

TABLE 3 Mutant Allele Frequencies in Australian and New Zealand Cat Populations

| | Phenotype | | | | Approximate | |
	O	a	t^b	d	Sample Size	Reference
Brisbane, *etc.*	0.14	0.81	0.68	0.45	400	Moffatt (8)
New Zealand	?	0.81	?	0.33	170	Marples (7)

While cat gene frequencies are affected by the rural–urban cline, they appear otherwise stable for relatively long periods of time and over large areas. On this account it is reasonable to suggest that populations of the northeast United States approximate a sample of northwest European cats of the mid-seventeenth century while the cats of eastern Australia represent another sample made in the early nineteenth century. Thus it follows that the greatest changes in gene frequencies have occurred in Europe. These changes (Figure 1) presumably result from urbanization, which began at a much earlier time in Europe than in the colonial populations. Hence, removal of a sample from Europe emancipated it from the urban selection pressures and arrested the

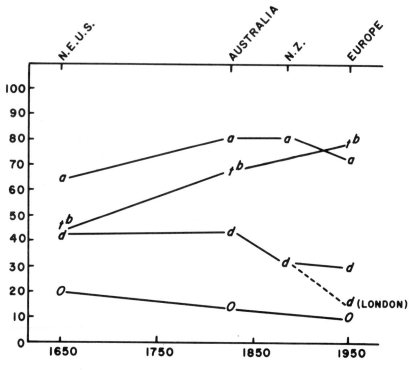

FIGURE 1 Gene frequency shifts in northwest European cat populations during the last three centuries.

change temporarily. Of course the same pressures must be assumed to be occurring now in the United States, as judged by comparative data for its populations, and also in Australia. However, in the absence of rural surveys, the extent of the changes that may have already taken place in these latter areas is undetermined. Therefore, the magnitude of the European changes cannot be fully appreciated, but they may well be greater than is readily apparent. A further complication lies in the nature of the selection pressures that appear to favor darker phenotypes, per se, rather than some pleiotropic effects of alleles that result in darker coats. Since there are several genetic alternatives that produce darker phenotypes, the selection pressures may shift over time from one locus to another, depending on a number of circumstances. Perhaps the best way to visualize the ultimate effect is to derive a "coefficient of darkness" that expresses the frequency, in a population, of the two darkest phenotypes possible for the four principal loci, *viz.*, sex-linked black, orange (O^+, O); agouti, nonagouti (a^+, a); striped, blotched tabby (t^+, t^b); and nondilute, dilute (d^+, d). For males, the predicted frequency of these two phenotypes is given by inserting the appropriate phenotype frequencies in the formula O^+d^+ $(a + a^+ t^b)$. This takes into account the epistatic relationships of the various alleles and gives the actual combined frequency of the two darkest phenotypes. The respective values for the northeastern United States, Australia, and northwestern Europe are 0.34, 0.57, and 0.65. Here, despite local aberrations in gene frequencies, the differences in total effect are clearly seen.

Conclusions

With regard to the utilization of domestic cats for research, at least two sources of error due to genetic variations are possible. The first is, of course, that which is due principally to the rural–urban cline. This can be controlled by taking appropriate measures to define, if not genetically, at least geographically, the animals employed. The second source of error is less obvious, although it is a further manifestation of the rural–urban cline. It is, in fact, the temporal shift caused by urbanization. While this may not be critical at the moment, the acceleration of urbanization portends the time when two samples drawn from successive generations may vary as much as those drawn from different areas at the same time. Although the complications implicit in both the present and future situations may seem discouraging, they are being steadily mitigated through the extension of information on cat populations around the world. If an effort is made to define and standardize samples as a preliminary step to other research, not only will the validity of that research be enhanced, but the task of validating further samples will be eased.

References

1. Dreux, Ph. 1967. Gene frequencies in the cat population of Paris. J. Hered. 58:89–92.
2. Dreux, Ph. 1968a. Gene frequencies in the cat population of a French rural district. J. Hered. 59:37–39.
3. Dreux, Ph. 1968b. Fréquence des gènes chez les chats de Chamonix. Carnivore Genetics Newsletter No. 4:62–63.
4. Dreux, Ph. 1969a. Fréquence des gènes chez les chats de Chamonix (II) (France, Haute Savoie). Carnivore Genetics Newsletter No. 6, in preparation.
5. Dreux, Ph. 1969b. Fréquence des gènes chez les chats de Marseilles. Carnivore Genetics Newsletter No. 6, in preparation.
6. Dyte, C. E., C. E. Keeler, T. Komai, B. W. Moffatt, R. Robinson, A. G. Searle, and N. B. Todd. 1968. Standardized genetic nomenclature for the domestic cat. J. Hered. 59:39–40.

7. Marples, B. J. 1967. Notes on the phenotypes of cats observed in New Zealand and in Thailand. Carnivore Genetics Newsletter No. 3:43–44.
8. Moffatt, B. W. 1968. Cat gene frequencies in two Australian cities. J. Hered. 59: 209–211.
9. Norby, D. E., and H. C. Thuline. 1965. Gene action in the X chromosome of the cat (*Felis catus* L.). Cytogenetics 4:240–244.
10. Robinson, R. 1959. Genetics of the domestic cat. Biblio. Genetica 18:273–362.
11. Searle, A. G. 1949. Gene frequencies in London's cats. J. Genet. 49:214–220.
12. Searle, A. G. 1966. Coat colour gene frequencies in Venetian cats. Carnivore Genetics Newsletter No. 1:6–7.
13. Searle, A. G. 1968. Cat gene geography: The present picture. Carnivore Genetics Newsletter No. 4:66–73.
14. Thuline, H. C., and D. E. Norby. 1961. Spontaneous occurrence of chromosome abnormality in cats. Science 134:554–555.
15. Tinney, L. M., and R. A. Griesemer. 1968. Gene frequencies in the cats of Columbus, Ohio, U.S.A., and a comparison of northeast U.S. populations. Carnivore Genetics Newsletter No. 5:96–99.
16. Todd, N. B. 1964. Gene frequencies in Boston's cats. Heredity 19:47–51.
17. Todd, N. B. 1966. Gene frequencies in the cat population of New York City. J. Hered. 57:185–187.

DISCUSSION

DR. FESTING: Why are darker phenotypes favored in urban areas?

DR. TODD: There is nothing at this point except speculation and some interesting work that has been done by some behavioral scientists in Germany, particularly Paul Leyhausen at the Institut für Tiersikologie. He suggested that cats in their urban environment are subjected to crowding conditions, which disrupt their normal territorial establishment, and they are making all kinds of adjustments—some rather incredible adjustments—to living at these high densities with the innate behavioral patterns that have been evolved for millions of years.

Among these adjustments are moving to nocturnal patrol, because basically cats are diurnal—dusk, actually. Also the temporal dividing of territories must be dealt with. Finally, because they are saturated and overlapping, the old olfactory cues no longer have too much significance, and they come to rely more and more on visual cues for territorial defense. So perhaps in the urban environment it behooves a cat to be less conspicuous. That means to be darker. This is one of the suggestions that has been made thus far.

In this regard, I might say there is no evidence that human selection is playing a role, that is, intentional human preference is playing a role. It tends to be self-defeating. If people like orange cats or black cats or blotched tabby cats, they probably have them castrated or spayed in high numbers and the total biological effect is nil. For instance, in London at the moment something in the vicinity of 70 percent of the tom cats that one encounters in shelters and hospitals are castrated.

DR. WOLFF: Please explain more fully your suggestion that natural selection acts on coat color effect per se rather than on pleiotropic metabolic effects of the specific alleles.

DR. TODD: In the beginning I certainly would have been much happier if I could have said it is just coincidental. As I noted, in a total sampling we look at nine or ten different factors, and at least seven of these are coat-color mutants. There is absolutely no evidence that any of these are linked. That does not prove that they are not, but even if they are, they are at opposite ends of the chromosomes. Basically, the cat has a haploid chromosome set of 19. We have now got 24–25 markers. We have not found linkage yet, so what linkage there is must be very loose. All of these are independent, certainly. They have been pretty well tested, and the coat colors I am talking about are independent of one another. There are epistatic relationships and interactions, but it seems very hard to take all of these as they move, either to wild type or to mutant, but always in the direction of producing a darker phenotype in the urban areas, and say that there is an unknown pleiotropic effect associated with each one that by coincidence is producing a dark coat, when there really are four such effects. In the total sample seven or eight genetically independent factors are responsible for causing a dark coat.

DR. DIETERICH: What about the changes in rarer colors, like Burmese and Havana Browns?

DR. TODD: Burmese is an allele of the so-called albino series. It is also known as dominant Himalayan. It is dominant to Siamese. It is perpetrated and bred by cat fanciers in fair numbers, and thus far it has not been widely liberated into the population at large. These cats are fairly well controlled. Siamese on the other hand have been bred very extensively. For instance, the gene frequency in the cat population at large in Paris is 10 percent. That means phenotype frequency is 1 percent of that seen in stray cats. In Bangkok the gene frequency is only 29 percent, and that is where they came from originally.

Havana Brown is extremely rare, an extremely rare mutant in the cat. It is known basically in the chocolate point Siamese. That is where it came from originally, and it has been placed on a nonagouti–non-Siamese background by breeders. It is still quite rare and not often encountered. In fact, I have never encountered what was unequivocally chocolate brown or Havana Brown with its various names. I have never encountered that in the populations, other than in this chocolate point Siamese or in the lilac, which is simultaneously B and D dilution. But certainly Siamese is getting out, and it is having its effect.

DR. FOX: If cats are relying more on visual cues, why don't they have brighter contrasting colors, and if they are being forced to be nocturnal in urban areas, a bright coat visual signal might also be an advantage, or are there urban predators?

DR. TODD: No, there are no severe urban predators. The cat is his own worst enemy. Basically, to become inconspicuous is to reduce the chances of confrontation. This is the success that is involved. If the cat spends all his time scrapping with his neighbor, which they do when pressed, especially if their territories are liable to overlap, they don't derive too much from displaying and intimidating a neighbor if the neighbor thinks that is his land, too. So I think we could back up a minute and consider a rather curious thing about cats trying to be inconspicuous. There is only a certain group of felids that have the practice of burying feces and urine. These are the true lesser cats of the Old World, the bobcat, lynx, and puma. All other cats use these things to mark their territories. I think perhaps the same thing happened. If the puma wants to overlap the territory of a jaguar he wants to remain inconspicuous, so he buries his excrement and urine and does not use it to mark territories. In this way he remains inconspicuous. Cats fight very severely when crowded in residential neighborhoods, and the object is not the original one of intimidating a neighbor when you are on your own territory because he is easily intimidated when he is on your territory. This breaks down when there is a mutual sharing or an overlap.

DR. BOND: Are you aware of any statistical relationships between the gene data that you have presented and cat leukemias?

DR. TODD: The next step in this whole sequence is to take the pathology reports of Dr. T. C. Jones from Angel Memorial Hospital, on 6,000 cats that have also been tabulated according to genotype. A computer is trying to digest these data now, and there are other problems. I would not want to say anything specific. We did some very preliminary calculations. There were problems in the original data-gathering. We were unable to find any significant association between a genotype or a phenotype and some of the more well known cat diseases. We haven't even considered leukemias yet, although I am sure they are in the pathology material.

Interestingly, we found a very much higher incidence of infection, surgery, chewed ears, and whatnot among males and among the lighter phenotypes; for what it is worth, a higher incidence was detected among red cats, also a higher incidence of disorders associated with the bladder and urethra in red cats. Whether this will prove to be significant, I do not know, but we have this in mind, and hope we can work it up and get some answers. Obviously, the frequencies are quite different. San Francisco has been done, and now there is some very preliminary data in from Dallas. This is another population. This population obviously is not northwest European in origin. It is Mexican in origin, and ultimately traces back to Spain. The gene frequencies are very different for San Francisco and apparently for Dallas. A little bit has come in from Wichita, but not enough to make much sense. If there are, however, relationships between these genotypes and certain kinds of diseases, disorders, and anomalies, veterinarians in one part of the country are going to encounter one kind of problem more frequently than they are in another part.

        DR. SWEENY: What is the expected sex ratio of the orange tiger phenotype?

        DR. TODD: Where are you from?

        DR. CLARKSON: White Eagle Farms.

        DR. TODD: Where is White Eagle Farms?

        DR. SWEENY: Philadelphia.

        DR. TODD: In Philadelphia? Just a minute. In males?

        DR. SWEENY: In orange tigers, how many in a random population do you expect to be males?

        DR. TODD: I don't know whether I can get that slide out. It is about 25 percent for the gene frequency, and that is sex-linked, so 25 percent of the males in this case will be orange. I can't remember what the frequencies were for striped versus blotched alternatives. I think phenotypes 25–20—I can calculate it for you. I have got that information, and I do know that it happens to be one of the things that deviates.

Can I take one more minute for another point?

It did not come up logically, so I will bring it up illogically. One of the people who has done a great deal of this work is M. Phillippe Dreux from the University of Paris. He just came back from an incredible tour to the Kerguelen Islands. The nearest land is the Antarctic. It is 500 miles north of Antarctica in the middle of the Indian Ocean. Here he discovered the first founder effect in cat populations. This, of course, was an uninhabited island, even in the eighteenth century, and it still would be, except the French maintain an air force base there. The cat population there is uniformly black with white spotting. This is the first case of founder effect that has been established, and interestingly, it is in an insular population where one can imagine that the initial introduction was certainly not a nonrandom sample from the metropolitan area.

On his way back from Kerguelen, M. Dreux stopped at Réunion, where the cats look like the cats of Malaysia, a population that was established from Madagascar. We do not know what Madagascar is like yet, but there is a lot of speculation that the early settlement of Madagascar was, in fact, Asian. If it holds up, then the anthropologists can make some use of the fact that even if you can't trace the thing in the blood gene frequencies of the human population, because they have been fouled up subsequently, perhaps the cat gene frequencies will allow it, because they seem to be less susceptible to politics and other things. They will allow these trails to be followed.

Finally, on his odyssey he stopped in Bloemfontein in the Orange Free State, and I guess I should just say that the cats of the Orange Free State look like European cats, and they look like a population that we would project for some time in the early part of the nineteenth century. In fact, it was in the middle of the nineteenth century that Bloemfontein was founded.

THE STANDARDIZED DOG
AS A LABORATORY ANIMAL

Douglas H. McKelvie
A. C. Andersen
L. S. Rosenblatt
Leo K. Bustad

Introduction

Soon after World War II it became obvious that meaningful research into the biological effects of high levels of either internal or external irradiation would be very important. Such research would necessitate the use of a suitable animal model from which resultant data could be analyzed for extrapolation to possible effects in man. Because of its insidious nature, damage from internally deposited nuclides or from low-level exposure to external radiation often requires a relatively long period to develop. Thus an animal with a prolonged life-span was necessary. This requirement eliminated such animals as the mouse, rat, and guinea pig. In addition, the physiological and anatomical features of these animals are not closely related to those of man. The natural choice, some species of nonhuman primate, was ruled out by high cost and difficulties in procurement. The final decision was to use the dog, since it was readily available, easy to handle, adapted well to laboratory environment, and was especially responsive to human care.

The use of nonstandardized, mongrel dogs obtained from pounds and various vendors was excluded. Such animals introduce too many variables, such as undefined age, lack of genetic uniformity, and unknown disease history.[38] The decision then became one of choosing a specific breed suitable for long-term projects. Since the animals were to be used for toxicological studies, large size was not as important a factor as it is for surgical experiments. A short haircoat, even disposition, and adaptability to living in packs were criteria that pointed to the Beagle as a choice for such experimentation.[4]

Whether, in fact, all of these criteria were first considered in the early post-war choice of the Beagle is probably irrelevant. However, the choice has been proved a sound one over almost 20 years of experimentation. As a result, a number of large colonies of Beagles are in use throughout the country today on long-term chronic programs.

Uses of the Beagle

Most of the Beagles in use today are maintained for radiobiological studies under the auspices of the U.S. Atomic Energy Commission (AEC) and the U.S. Public Health Service (PHS). Some of the AEC facilities are located at the University of California at Davis,[3,4] University of Utah,[13] Lovelace Foundation,[22,31] Battelle Northwest, Argonne National Laboratory, and Rochester University. A large colony is maintained under support of the PHS at Colorado State University in Fort Collins.[17] Programs in care and management of such large numbers of Beagles over prolonged periods have been developed to the point of becoming a fine art, and many of the results are forthcoming in a separate book, *The Beagle as an Experimental Dog*,[2] or have already been reported.[25]

This report summarizes some of the information obtained in an 18-year study on a large group of Beagles under controlled conditions. We will include areas that appear applicable for extrapolation to physiological effects and models in man.

The Beagle colony at the University of California at Davis was started in 1951–1952 with a breeding colony of 80 females and seven males purchased from private and commercial breeders. This group of animals and subsequently purchased purebred Beagles have produced two large colonies that are used to ascertain the long-term effects of external (x-rays) and internal (^{90}Sr; ^{226}Ra) irradiation. Although the final results of these experiments are still forthcoming, sufficient information from sham-treated controls and experimental animals is available to warrant discussion of parameters such as growth rate, reproduction, hematology, osteology, behavior, and gerontology.

The x-irradiated colony of 360 experimental females and the original breeding colony were maintained in an outdoor kennel as previously described.[4] In brief, this facility (Figure 1) provided adequate space (200 ft^2) for paired Beagles. Each pen provided shade, two suspended barrels, an automatic waterer, and a crushed-rock surface.

Indoor facilities provide two-dog cages for use while the Beagles are being treated with radionuclides. Pregnant dams are placed in the air-conditioned cages to remain until their pups are weaned, and their progeny are continued in this facility to 19 months of age. After treatment is completed, the Beagles

DOUGLAS H. McKELVIE,
A. C. ANDERSEN, L. S. ROSENBLATT,
AND LEO K. BUSTAD

FIGURE 1 A view of outdoor kennels. Tower visible in the rear is used for observing behavior.

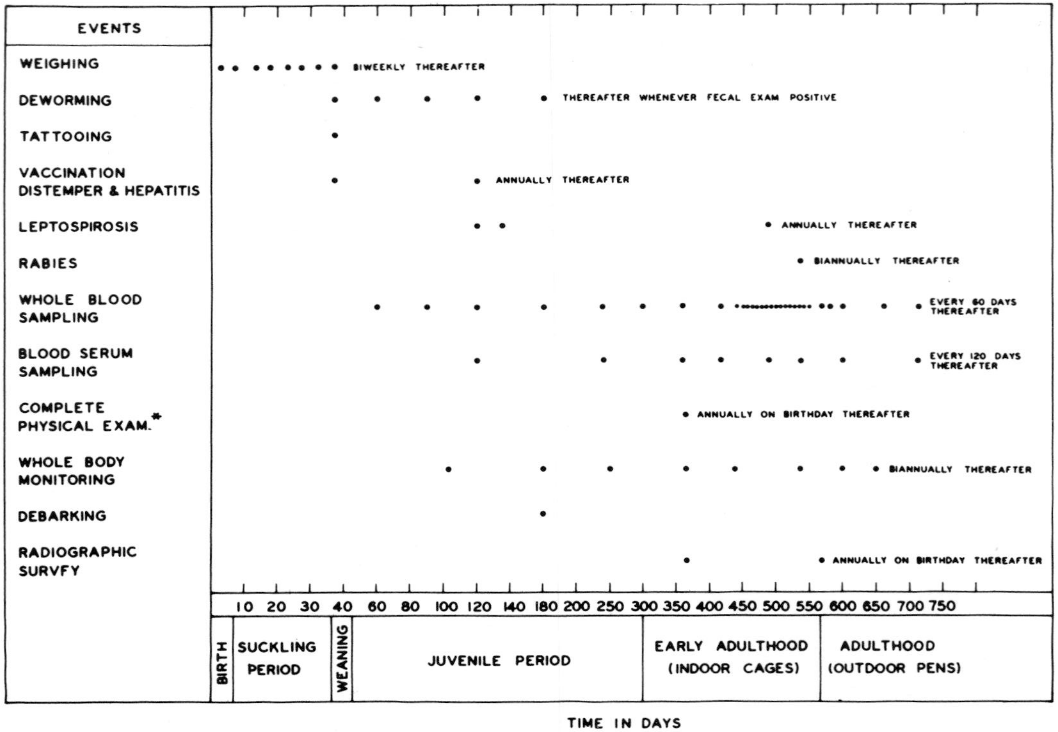

FIGURE 2 Chronological events in the life of a Beagle at Radiobiology Laboratory, University of California, Davis.

DOUGLAS H. McKELVIE,
A. C. ANDERSEN, L. S. ROSENBLATT,
AND LEO K. BUSTAD

are moved to the outdoor pens for life-span studies. These cage facilities have been satisfactorily used for the last 6 or 7 years in the treatment of some 900 Beagles.

Breeding for the radionuclide studies was by controlled random mating to maintain genetic heterogeneity.[29]

A standardized diet was fed to both indoor and outdoor dogs.[37] The quantity was adjusted to the condition of the animal. Oxtails were fed weekly to minimize tartar accumulation on teeth. Records were maintained on forms that would make data readily amenable to computer analysis.[28]

During his lifetime each Beagle is subjected to a number of procedures; these are summarized in Figure 2.[21]

Growth

Body weights were taken throughout the life of each dog (Figure 2). By the use of split litters it was determined that growth rates during the first 14 months of life were essentially the same, whether the dogs were raised indoors or outdoors (Figure 3). Growth of each group

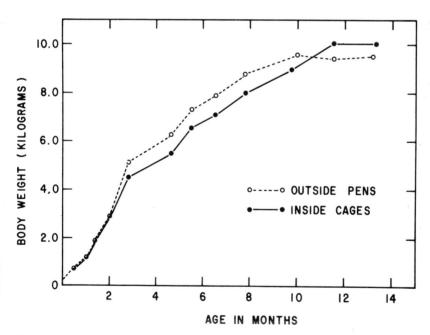

FIGURE 3 Comparison of growth in related Beagles raised in outside pens versus indoor cages.

reached a plateau at 12 months of age. Also, when Beagles raised indoors were placed outside, a 5 percent decrease in weight occurred. Recovery required about 3 months.

Reproduction

Reproduction is not only of great biological significance, it also represents an important set of parameters that are themselves end points in long-term studies.

Estrus occurs in the Beagle at 12 ± 2 months of age. Subsequent cycles average 7.2 months in length until the dog reaches 5 to 6 years of age.[11] Similar estimates have been reported by Smith and Reese.[34] Females are bred on alternate days after proestrus is detected, and again after 2 to 3 days. Conception rates range from 80 to 90 percent. The gestation period is 63 ± 1 day, the dams usually whelping four to six pups (average 5.2 pups). Survival to weaning varies with management, age of dam, and size of litter, and ranges from 65 to 80 percent. Thus, under optimum conditions, dams may be expected to produce about six weaned pups per year.

In a recent analysis of 318 litters whelped in outdoor pens, 5.6 pups were whelped and 3.7 were weaned, a loss of 34 percent. These losses were due, in great part, to the advanced ages of the dams and to adverse environmental conditions encountered outdoors. Most pup losses (66 percent) were neonatal (0 to 48 hr postpartum). Only 20 percent of the dead neonates had gross visible lesions.[24] Some anomalies (e.g., hydrocephalus and cleft palate) were observed, but abdominal hemorrhage was the major finding. Pup losses were higher in first litters and in litters produced after the peak reproductive age of the dam—2 to 4 years; pup production in dams 6.5 years of age was one half that of the maximum attained earlier by the same dams.[1] Neonatal losses were fewest in litters near the average size for the colony.[5]

In this laboratory it has been observed that dams tend to whelp at night in summer and during the day in winter, emphasizing the importance of providing proper shelter during inclement weather conditions. Mortality was reduced in air-conditioned indoor cages by the addition of drapes and adequate mats on the cage floors.

Effective utilization of the Beagle in radiation studies requires an understanding of the basic biology of reproduction. A survey of ovarian specimens indicated that the dog provides an excellent model for the study of developmental changes in the ovary. In most mammalian species the formation of primordial follicles occurs during fetal life. In the dog, however, it is a postnatal process[9] and, as such, facilitates assessment of radiosensitivity of the pup ovary during oogenesis. It has been shown that fractionated x-ray expo-

DOUGLAS H. McKELVIE,
A. C. ANDERSEN, L. S. ROSENBLATT,
AND LEO K. BUSTAD

sures are more effective than single exposures in effecting sterilization of the female, even when single x-ray exposures in the midlethal range are utilized.[7] Preliminary results of recent experiments indicate that low total doses of x-irradiation (210 rads) will sterilize females when given at 10-rad exposures every other day to 42 days of age (Andersen, unpublished data).

An extensive monograph on the Beagle ovary is presently being compiled[10]; this will facilitate interspecies comparison and extrapolation.

Blood

A study of the chemical constituents of Beagle blood has been made over a period of several years. Data illustrating age variations have been published[27]; but, in general, no significant changes with age occur in control Beagles between 1 and 4 years of age. The values for 2-year-old (young adult) Beagles[26] appear comparable to those for man, with the exception of uric acid concentration. Due to additional enzymatic pathways, the uric acid values are significantly lower in the Beagle and most other canine breeds than in man.

Cellular components of Beagle blood have also been studied (Figure 4), and the base-line data have recently been summarized.[2] Mean blood values, of course, vary between laboratories, principally due to environmental factors. Hence, in this report we are showing only the change with age and the effect of varying doses of radiation on the neutrophil component.

The age changes in neutrophils of controls and of dogs fed low levels of ^{90}Sr are indicated in Figure 4A. Increased numbers are noted after the dogs are placed outside. Dogs fed higher levels of ^{90}Sr do not show the aging effects characteristic of controls. Figure 4B indicates the relationship of neutropenia to radiation dose. These data have been analyzed by the probit technique[32] to produce at each age interval estimates of ED_{25} (Figure 4C), the dose required to produce a 25 percent depression in neutrophils relative to controls. These analyses illustrate the utilization of peripheral blood in monitoring and quantifying dose–response relationships. This technique may have some applicability in areas outside radiobiology, such as toxicology.

Marrow

During the ^{90}Sr treatment and post-treatment period of the Beagle, a number of cases of myeloproliferative disorders developed.[18] Extensive studies of these cases have revealed a spectrum of pathologic progression from myeloid metaplasia to myeloid neoplasia. The cases were

269

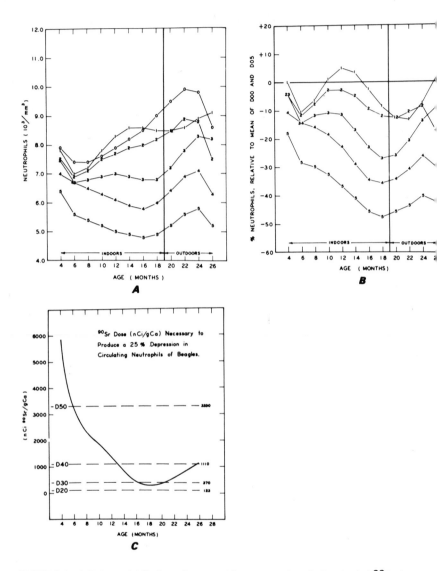

FIGURE 4 *A* Polynomial-fit data of neutrophil concentrations in Beagles fed ^{90}Sr, indicating age effect (o) and neutropenia (3, 4, and 5 levels). *B* Relative neutropenia during
and after 18 months of daily ^{90}Sr feeding. The numbers represent the relative levels of
^{90}Sr. The dose–age relationship may be seen (age at which 25 percent depression is
reached) as well as age to maximum depression. *C* Probit interpolation values for ^{90}Sr
diets necessary to produce 25 percent neutropenia at ages indicated. Horizontal lines indicate feed levels, e.g., the D50 level = 3330 nCi ^{90}Sr/g Ca or ~ 12 μCi/day.

270

DOUGLAS H. McKELVIE,
A. C. ANDERSEN, L. S. ROSENBLATT,
AND LEO K. BUSTAD

similar in some of their general characteristics, but there were minor variations in the extent of organ infiltration and degree of maturity of the aberrant cells. Only occasionally were immature cells noted in peripheral blood smears.

A spectrum of organ infiltration and architectural disruption was observed. The severity of the disease was correlated with changes not only in marrow, but in spleen, liver, lymph nodes, and lungs. Clinically, anemias signal the onset of the disease; the mechanism is undergoing investigation. Considerable interest has been manifest in these dogs as useful models for the study of myeloproliferative disorders of man.[14]

Bone

One of the primary sites of deposition of many radionuclides, including ^{90}Sr and ^{226}Ra, is the bone. As a result, this target organ has been the subject of extensive study. Beagles are an excellent choice for bone studies because of the remarkable similarities of bone development and structure to those of man.

Numerous studies have been undertaken to establish both normal and pathologic descriptions of dog bone. These include microradiography, histology, fluorescent labeling, densitometry, and elaborate chemical analyses.

Microradiography has been of special value in determining turnover of bone and in describing general osteoblastic and osteoclastic activity. Microradiographs were taken of bone specimens 60–100 μm thick, using a specialized camera utilizing about 11-kilovolt peak and 15–19 milliampere x rays.[23] The resultant microradiograph was studied microscopically to determine such parameters as medullary index, hypomineralization, hypermineralization, and resorption cavities. The medullary index is used to give the relative percentage of total cross-sectional area of a bone shaft that is occupied by the medullary canal. Figure 5 illustrates a microradiograph (35X) of a cross section of a femur, with the medullary index determination diameter, and a 250X microscopic examination of part of the same microradiograph from a 3-year-old Beagle, respectively. The latter illustrates four important features of the bone that can be used to demonstrate various phases of bone remodeling and turnover. They are:

1. Hypomineralized osteon, indicating an actively remodeling Haversian system that is nearly complete.

2. Hypomineralized osteon with a large canal, representing a newly developing osteon that has just completed resorption and is now filling in.

3. Resorption cavity or porosity, an enlarged irregular hole that has re-

271

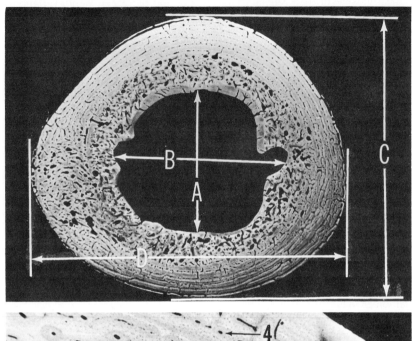

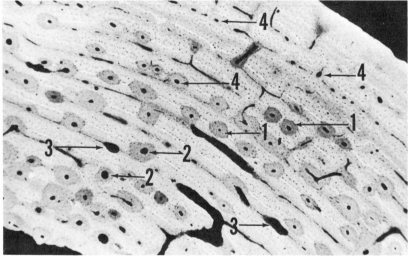

FIGURE 5 (Top) Medullary index determination in the femur of a 3-year-old Beagle, using the formula of Foote[15] (35X). (Bottom) Microradiograph of femur of a 3-year-old normal Beagle (250X) showing 1) hypomineralized osteon; 2) hypomineralized osteon with enlarged canal; 3) resorption cavity or porosity; and 4) nonactive or normally mineralized osteon.

sulted from osteoclastic activity resorbing an area of formerly mineralized bone.

4. Nonactive or normally mineralized osteon, a Haversian system that has completely formed and mineralized.

The relative percentages of each of these types of osteons can be used to compare the turnover rates and the degrees of mineralization in the bone, be it dog or human.

Other studies can also be done directly from microradiographs. For example, density scanning of the radiograph with a minute light beam, preferably monochromatic, can be used to determine bone density per unit volume, provided some type of density wedge is included as a control in the microradiograph.[23]

Linear measurements from radiographs taken at intervals during the rapid growth phase of a given bone, supplemented by material obtained at necropsy, can be used to derive a characteristic growth rate formula. Identical growth curves for left and right tibial lengths of eight Beagles from 50 to 150 days of age are shown in Figure 6. Growth during this period was highly linear, as may be seen from the strong correlation coefficients (r).

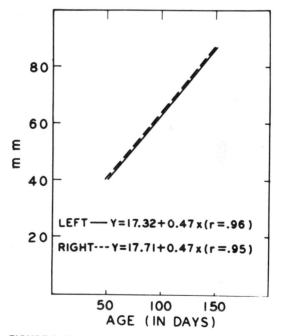

LEFT —— $Y = 17.32 + 0.47x$ ($r = .96$)

RIGHT --- $Y = 17.71 + 0.47x$ ($r = .95$)

FIGURE 6 Growth rates of tibial lengths.

273

Social Behavior

During prolonged maintenance of Beagles in pen environments, there was opportunity to observe development of certain psychological and social behavior patterns. Beagles of the same sex were caged in pairs and observed, primarily for emotional responses and dominance-submission patterns.[35] Among the interesting facts noted was that the achievement of dominance was largely determined by weight advantage. Other observations included the variation in responses toward man. These were classified as friendly, stay-behavior, wary, or aggressive. In paired dogs there was, in general, more friendliness displayed toward man if the pair was neither dominant-submissive nor prone to combative activity, and if both animals shared their environment without friction. Conversely, a combative pair, neither dominant nor submissive, displayed less friendliness and was more prone to "stay" at a distance from the human observer.

Aging and Survival

The relatively long and well-defined life-span of Beagles permits their use in extensive gerontological studies. The stages of the life-span of an animal may be correlated with its reproductive ability. Maximum reproductive fitness in female Beagles, as measured by the number of weaned pups produced per unit time, occurs at about 3 years of age; by 6-7 years of age reproductive fitness has declined markedly due to failures to conceive and to high pup mortality. Females at this time may be considered to be in early senescence. Clinical signs of aging also appear at this time; graying of the hair, skin wrinkling, tooth defects, lethargy, and changes in conformation become noticeable. These signs become increasingly more prominent, and by 10-12 years of age the Beagle may manifest poor vision and lack of vigor, and may have little interest in his surroundings. Although easily recognized, the clinical signs of aging are difficult to quantitate. Yet, some yardstick by which an animal could be "aged" would be a useful research tool. In our laboratory some attempts have been made to semiquantitate changes with advancing age.

Preliminary results indicate that proportions of mucopolysaccharides in costal cartilage change with increasing age.[36] Sclerotic lesions in the kidney increase in severity with advancing age. Progressive intercapillary glomerulosclerosis (IGS) correlates well with age, and the lesion develops independently of vascular changes.[20] The termination of all these processes, i.e., death, is also of great interest to us.

In one study two groups of Beagles were used to ascertain survival curves

274

DOUGLAS H. McKELVIE,
A. C. ANDERSEN, L. S. ROSENBLATT,
AND LEO K. BUSTAD

that might be expected in this breed. The kennel group (57 females) consisted of controls for the x-irradiation program. A field group (215, largely males) was composed of progeny from our laboratory that had been released at 2–3 months of age to private owners. Thus, the two groups afforded a comparison of survival in Beagles kept under different environmental conditions.[6]

Among the field dogs, median survival time from birth was found to be 4.5 years of age, the rather short life expectancy being due to the many accidental deaths observed. When the population of field dogs consisted only of dogs alive at one year (similar to the kennel group) median survival time increased to 7.5 years. However, when accidental deaths were corrected for the median, survival time rose to about 12.5 years, almost identical to that observed for the kennel group. The survival curves were also remarkably similar.

X irradiation given at one year of age shortened life at the rate of 6.7 percent per 100 rads.[8] The survival curves of the dogs are shown in Figure 7. The cumulative survival rate of the control dogs to 8 years of age (7 years postirradiation) was 80 percent. Recent analyses[33] in the radionuclide study in-

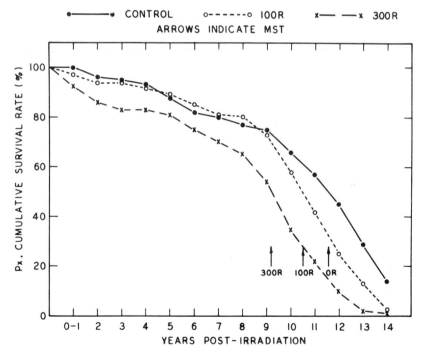

FIGURE 7 Cumulative survival rates for controls and x-irradiated Beagles. Arrows indicate median survival time for each group.

dicated 98 percent survival to 8 years of age. Included were controls and dogs receiving low levels of ^{90}Sr and ^{226}Ra. The reasons for the large difference between survival rates have not been defined, but they undoubtedly include factors associated with better colony management, the result of experience gained over the years. Also, dogs in the radionuclide experiment were maintained in indoor cages for the first 19 months of life, while in the x-ray study, the dogs were housed in outdoor quarters.

Analyses of mortality rates[8] revealed that x irradiation had produced an acceleration of aging. Given levels of mortality were observed first among the dogs exposed to 300 rads and then 1 and 2 years later among those exposed to 100 rads and the control dogs, respectively. Qualitatively, however, pathologic findings were similar in all groups. These facts may be utilized in construction of models of aging, provided adequate parameters can be found. Our laboratory and many others are devoting considerable effort to finding adequate parameters for the Beagle.

Discussion

No attempt has been made here to present either broad base-line data on the Beagle or a review of the literature. Rather, emphasis has been given to presentation of areas in which the dog, in particular the Beagle, may serve as a very useful model. While it is evident that the special areas of interest of the authors have been stressed, an endeavor has been made to present some of the more important work of many people over an 18-year span.

The Beagle has been shown to have wide utility in long-term experiments. It is possible to study major systems over an extended period, making gerontological studies feasible from an economic as well as from a scientific point of view. In areas where little or no data exist relative to man, and all that is known has come from smaller, short-lived species (e.g., rodents), use of the Beagle represents a meaningful compromise.

Although we, and others, have studied reproduction in the dog, much remains to be done. For instance, the dog ovary may prove quite useful since the stages of oogenesis do not overlap, but are discrete. Also, oogenesis occurs postpartum rather than during fetal life, as in many other species.

Many of the chemical constituents in Beagle and human plasma are similar. The metabolism of pharmacological compounds and toxicological agents by Beagles may provide clues as to possible effects in man.

The finding of myeloproliferative disorders in ^{90}Sr Beagles offers oncologists a new model since these syndromes are relatively rare in domestic animals, although they are less rare in man. Scintillation camera scanning has re-

DOUGLAS H. McKELVIE,
A. C. ANDERSEN, L. S. ROSENBLATT,
AND LEO K. BUSTAD

vealed patterns of hematopoietic impairment that could not be seen in any other way (Goldman and VanDyke, personal communication, 1969). This technique has also, in the case of ^{226}Ra-damaged bone, revealed evidences of neoplasia before the lesions were apparent radiographically,[19] a finding of obvious importance to human medicine.

The relationship between the bones of dog and man is one of the most valuable areas for extrapolation. The development of bone in both animals is similar and, adjusting for rapidity of maturation, the extrapolation of radiation effects (e.g., osteolytic lesions, osteogenic sarcoma) from one to the other could be made with a high degree of validity. The effects on man of stress from such products as internally deposited radionuclides can be approximated by studies of hypomineralization, turnover rates, and histological specimens of bones of the dog. For example, an average absorption cavity in the dog takes roughly three weeks to form, this being the period required for tunneling or excavation. The building of the new osteon, including partial mineralization of the organic matrix, requires 6 to 12 weeks.[30] Frost et al.,[16] using tetracycline as a marker, found the mean osteon formation time in a 57-year-old man to be 5 weeks. Comparative alterations of osteon formation time in the dog can be thus extrapolated to man.

Summary and Conclusions

The necessity to do relevant and meaningful research on the effects of toxicological agents that produce prolonged or chronic conditions has resulted in an attempt to choose a biological model from which data can best be extrapolated to man. The demands on animal models include: relatively long life-span, relative genetic uniformity, physiological similarity to that of man, economy of procurement and maintenance, availability, and ease of husbandry in a laboratory environment, known age, and health history. Although some species of nonhuman primate might better fulfill some of these needs, their limited availability, difficulty in handling, and unknown history have ruled them out. A reasonable choice is the dog and, by virtue of certain considerations, the Beagle has been chosen as the most desirable breed. Over 20 years of long-term experiments have proven the Beagle to be a valuable asset in research involving radiobiological studies. The most important considerations have been developed around studies of aging, blood, bone marrow, and bone pathology. The similarity of the responses of dog (Beagle) and man to various radionuclides is probably best exemplified in the production of myeloproliferative disorders, and in bone pathology and the processes of aging.

Although it may be said that the dog does not replace man in any sense of

277

the word, there is little doubt that, until the nonhuman primate can be commercially produced and standardized in sufficient quantity, the standardized dog will continue to be a good choice as the experimental animal for chronic toxicological studies. Even miniature swine,[12] which are rapidly becoming a biological model of merit, must be further miniaturized and defined in order to compete with the standardized dog.

This report was supported by the U.S. Atomic Energy Commission and filed with that agency as document UCD 472-227.

References

1. Andersen, A. C. 1965. Reproductive ability of female Beagles in relation to advancing age. Exp. Gerontol. 1:189–192.
2. Andersen, A. C. [ed.] 1969. The Beagle as an experimental dog. Iowa State University Press.
3. Andersen, A. C., and M. Goldman. 1960. Outdoor kennel for dogs. J. Amer. Vet. Med. Ass. 137(2):129–135.
4. Andersen, A. C., and G. H. Hart. 1955. Kennel construction and management in relation to longevity studies in the dog. J. Amer. Vet. Med. Ass. 126(938):366–373.
5. Andersen, A. C., D. H. McKelvie, and R. Phemister. 1962. Reproductive fitness of the female Beagle. J. Amer. Vet. Med. Ass. 141(12):1451–1454.
6. Andersen, A. C., and L. S. Rosenblatt. 1965. Survival of Beagles under natural and laboratory conditions. Exp. Gerontol. 1:193–199.
7. Andersen, A. C., and L. S. Rosenblatt. 1968. Effects of fractionated whole-body X-ray exposure on reproductive ability and median survival of female dogs (Beagle). p. 11.1–11.14. In Proc. Symposium on Dose Rate in Mammalian Radiation Biology, D. G. Brown, R. G. Cragle, and T. R. Newman, [ed.] CONF-680410.
8. Andersen, A. C., and L. S. Rosenblatt. 1969. Effect of whole-body X-irradiation on the median life-span of female dogs (Beagle). Rad. Res. (In press).
9. Andersen, A. C., and M. E. Simpson. 1960. Development of the ovary in the Beagle. Radiobiology Laboratory Annual Report, UCD 472-103.
10. Andersen, A. C., and M. E. Simpson. In press. Development of the ovary in the dog (Beagle).
11. Andersen, A. C., and E. Wooten. 1959. The estrous cycle in the dog. p. 359–397. In H. H. Cole and P. T. Cupps [ed.] Reproduction in domestic animals. Vol. 1. Academic Press, New York.
12. Bustad, Leo K., and Roger O. McClellan [ed.]. 1966. Swine in biomedical research. Frayn Printing Company, Seattle.
13. Dougherty, T. F., B. J. Stover, J. H. Dougherty, W. S. S. Jee, C. W. Mays, C. E. Rehfeld, W. R. Christensen, and H. C. Goldthorpe. 1962. Studies of the biological effects of Ra^{226}, Pu^{239}, Ra^{228}, (MsTh), Th^{228} (RdTh) and Sr^{90} in Beagles. Rad. Res. 17:625–681.
14. Dungworth, D. L., M. Goldman, and D. H. McKelvie. 1968. Development of a form of myelogenous leukemia in Beagles continuously exposed to Sr-90. Symposium on Myeloproliferative Disorders of Animals and Man (Richland, Wash.).

15. Foote, J. S. 1916. A contribution to the comparative histology of the femur. Smithsonian Institution, Washington, D.C.

16. Frost, H. M., A. R. Villanueva, and H. Roth. 1960. Measurement of the bone formation in a 57-year-old man by means of tetracyclines. Henry Ford Hospital M. Bull. 8:239–254.

17. Garner, R. J. 1967. Collaborative Radiological Health Laboratory. 4th Annual Progress Report. College of Veterinary Medicine, Colorado State University, Fort Collins, Colorado.

18. Goldman, M., D. L. Dungworth, J. F. Wright, J. E. West, J. W. Switzer, and H. Tesluk. 1968. Myeloproliferative Disorders in Sr-90 Burdened Beagles. p. 72–74. Radiobiology Laboratory Annual Report, UCD 472-115.

19. Goldman, M., and D. C. VanDyke. 1968. Bone-seeking radionuclide effects as demonstrated by scintillation camera scanning. p. 66–70. Radiobiology Laboratory Annual Report, UCD 472-115.

20. Guttman, P. H., and A. C. Andersen. 1968. Progressive intercapillary glomerulosclerosis in aging and irradiated Beagles. Rad. Res. 35(1):45–60.

21. Hosein, A., A. C. Andersen, D. H. McKelvie, and L. K. Bustad. 1967. p. 22–25. Biography of a Beagle at the Radiobiology Laboratory. Radiobiology Laboratory Annual Report, UCD 472-114.

22. McClellan, R. O. 1966. Selective summary of studies on the fission product inhalation program from July 1965 through June 1966. p. i–ii. In R. G. Thomas [ed.] AEC research and development report. LF 33 (Lovelace Foundation, Albuquerque, N.M.).

23. McKelvie, D. H. 1968. The sensitivity of immature and adult dog bone to acute and chronic irradiation. PhD Thesis, University of California, Davis.

24. McKelvie, D. H., and A. C. Andersen. 1963. Neonatal deaths in relation to the total production of experimental Beagles to the weaning age. Lab. Anim. Care 13(5): 725–730.

25. McKelvie, D. H., and A. C. Andersen. 1966. Production and care of laboratory Beagles. J. Inst. Anim. Technol. 17(1):25–33.

26. McKelvie, D. H., S. Bentley, and S. Munn. 1967. Serum chemistry analysis in Beagles given Sr-90 and Ra-226. p. 36–39. Radiobiology Laboratory Annual Report, UCD 472-114.

27. McKelvie, D. H., S. Munn, and S. Bentley. 1968. Serum chemistry values in Beagles treated with Sr-90 and Ra-226. p. 25–30. Radiobiology Laboratory Annual Report, UCD 472-115.

28. McKelvie, D. H., and F. T. Shultz. 1964. Methods of observing and recording data in long-term studies on Beagles. Lab. Anim. Care 14(2):118–124.

29. McKelvie, D. H., F. T. Shultz, J. W. Parcher, and L. S. Rosenblatt. 1966. Random selection of Beagles to maintain heterogeneity and minimize bias in a lifespan experiment. Lab. Anim. Care 16:(4):337–344.

30. McLean, F. C., and M. R. Urist. 1961. Bone, an introduction to the physiology of skeletal tissue. University of Chicago Press.

31. Redman, H. C. 1965. Experimental animal housing for the fission product inhalation program. p. 1–24. In AEC research and development report. LF 23 (Lovelace Foundation, Albuquerque, N.M.).

32. Rosenblatt, L. S., and M. Goldman. 1967. The use of probit analysis to estimate dose effects on postirradiation leukocyte depressions (a preliminary report). Health Phys. 13:795–798.

33. Rosenblatt, L. S., and S. W. Bielfelt. 1969. Survival of Beagles in the internal emitter program. Radiobiology Laboratory Annual Report. UCD 472-116.
34. Smith, W. C., and W. C. Reese, Jr. 1968. Characteristics of a Beagle colony. I. Estrous Cycle. Lab. Anim. Care 18(6):602–606.
35. Solarz, A. K. 1965. Classification of emotional responses and their relation to dominance-submission in adult Beagle dyads. Psychol. Rep. 16:1253–1258.
36. Tsai, Huan-Chang C., R. J. Della Rosa, and N. Nix. 1968. Biochemical studies on bone and other connective tissues. 1. Biochemical survey of the organic matrix of bone and cartilage. p. 45–47. Radiobiology Laboratory Annual Report. UCD 472-115.
37. Wolf, H. G., R. J. Della Rosa, and A. C. Andersen. 1966. Nutritional management of a large experimental Beagle colony. Lab. Anim. Care 16(4):309–315.
38. Zinn, R. D. 1968. The research dog. J. Amer. Vet. Med. Ass. 153(12):1883–1886.

DISCUSSION

DR. KADAR: Was there any genetic consideration given to your breeding program, and, another question, besides hydrocephalus and cleft palate, have you found any "normally occurring" eye abnormalities such as cataract?

DR. MC KELVIE: The genetic consideration was based primarily on our outbreeding program; in order to avoid getting involved in too many genetic problems with a dog colony that obviously was not inbred enough to be of concern, we did maintain an outbreeding program. That is the only consideration.

The hydrocephalus and cleft palate are being studied now in some of the laboratories that I mentioned. Other than these, there were some congenital cataracts found in the original x-irradiation colony that Andersen reported on. We have not seen any evidence of those in the colony at the strontium and radium group.

DR. FABRY: What drainage do you have in the outdoor facilities shown, and how do you remove excreta?

DR. MC KELVIE: There is pretty good drainage there. Actually, the crushed rock is on a base, about 8 inches of crushed rock on a dirt base, and it soaks down. Leaching is very good in the Sacramento Valley. Admittedly, these are things you have to think about when you develop a new facility. The crushed rock is replenished as needed, but the caretakers go in and remove the excreta every day, carry it away, and have it buried.

As far as the drainage is concerned, we have had no real problem except in heavy rains, once in a while it will puddle up. But the material usually leaches right down into the ground.

DR. CLARK: What chemical determinations other than uric acid did you include in the 9-year period?

DR. MC KELVIE: There is a report out on that in one of the issues of the *American Journal of Veterinary Research* (September 1966). This included a considerable number of chemicals. I cannot name them all, but I think we had around 14, including protein by electrophoretic separation into the globulins, glucose, bilirubin, total protein, and BUN. There were quite a few, and most of them compare quite favorably with human data.

DR. SMITH: Would you comment on the laboratory dog being developed at Oregon.

DR. MC KELVIE: Yes, I think it is a great thing. The laboratory dog being developed at Oregon, as you probably know, is primarily from the Labrador strain. I think we need two types of dogs. I have emphasized here the Beagle, again emphasizing primarily toxicological studies and pharmacological studies. I think the one that is being developed at Oregon has been geared primarily for the studies of surgery, and having been associated with the Oregon Medical School in this program, I think it is going very well. I think they are developing a terrific animal. Dr. Fletcher reported at last year's ALAAS meeting in regard to percentage of survivals of heart surgery and implantations of various equipment. The percentage of survival in Labradors up there was considerably more than among pound dogs. This in itself, with good statistical proof, is something we have needed for a long time.

V

**Environment
and
Behavior**

ANIMAL EXPERIMENTATION VERSUS HUMAN EXPERIMENTATION

John C. Eccles

As one can see from the title of this symposium, we are concerned very much with the medical applications of animal experimentation. It has been traditional that quite a lot of the experimentation related to medicine is done with human subjects. As we become more and more expert in applying experimental procedures to animals, in finding the right animals, and in carrying out these procedures under the appropriate scientific conditions, there will be progressively less necessity for human experimentation. We must try to diminish human experimentation as far as possible by substituting animal experimentation.

I think that no serious and hazardous experimental procedures should be done on humans unless it is quite impossible to carry them out on animals. Such a statement severely limits human experimentation. Furthermore, experimenting on humans in any way that involves risk, should be preceded by experiments on animals by which the technical procedures can be developed.

We all agree that man is an animal, yet he is a very special kind of animal. He is separated by an immense gulf from the animal stocks from which he has evolved. That is not so obvious to many people today, yet it is absolutely true. The amazing feature of the evolutionary story is that man has become so enormously separated from his nearest relatives.

As an illustration, we have to look not only at the moral code, but also at the legal code. Legally there are the severest penalties, death sentences, for example, for treating humans in a way that constitutes humane treatment of an animal. There is a world conscience about such things, and the moral and

285

legal codes derive from the generally accepted belief that there is an immense gulf between man and animal.

On the other hand, human experimentation must be carried out because it is essential in many respects to advance therapeutic procedures in man. I think it is also necessary to carry out human experimentation in order to further our understanding of the very special experiences that men share.

Nevertheless, we must plan to minimize human experimentation and maximize animal experimentation, and we must define quite rigorously the conditions under which human experimentation can be carried out. This is something that the societies concerned with animals, animal care, and animal experimentation should understand. They should recognize, moreover, how important animals are and how significant they are in minimizing human experimentation. The more effective animal research centers can become and the more facilities they can provide, the more they will be able to eliminate the really dangerous and damaging forms of human experimentation.

Let us also consider allowable forms of human experimentation. Many such experiments are quite without risk. For example, during the final three terms of physiology, which I taught from 1944 to 1951 at the medical school of the University of Otago, New Zealand, the students performed 85 percent of the total experiments on themselves. I had three three-hour courses on the systematic study of pain—muscle pain, skin pain, all kinds of inflammatory pains, and periosteal pain. Of course, a great many procedures like this can be performed on human subjects without any danger and with little discomfort. I designed these courses originally because we were so short of animals, and I discovered that it was far better to use the students as subjects. They investigated their own muscle contractions, stimulated their nerves, and tested their reflexes. This experimentation can be effectively and appropriately done by medical students on themselves in order to give them some feeling for their eventual work in neurology. Similarly, excellent experiments on respiration and urinary secretion can be carried out on human subjects.

I believe, then, very much in this kind of innocuous human experimentation, and medical courses should be developed along these lines. This may relieve the problem of providing large numbers of rabbits or other animals for medical students to use in experimentation.

There are also advanced levels of experimentation that require human subjects. I refer, for example, to the experiments on different kinds of perception, for example, color vision, color matching, problems relating visual perception to retinal pigments and their bleaching, and after-images. These experiments can be done only on humans experimenting on themselves. The necessity of human experimentation derives from the fact that no other animal can report to you what was seen. At best, an animal's behavioral performance is a very inadequate measure of its perceptual experience. This is true not only for vision, but for all senses, e.g., for investigations on touch and pain. Often this kind of work is done on volunteers. A friend of mine subjected some volunteer medical students to an aseptic operation in which small cutaneous nerves in the forearm were dissected down until there was only a single fiber left. He was then able to test their sensory perceptions aroused by stimulating these single fibers in cutaneous nerves. I think this is about the limit to which one can ask students to participate!

I have known people who have done the most incredible investigations on themselves—people who study sensation produced by brain stimulation by putting stimulating electrodes on different parts of their heads, and then passing powerful electrical currents through those electrodes. I don't like those kinds of experiments, but of course the experimenters have been their own subjects. They are courageous people. I have also known scientists to take all kinds of drugs, for example, curare, strychnine, and chloralose, often to highly dangerous doses, just to get some further understanding of drug intoxication.

To give you another example there are the many schools of respiratory physiology. For example there was the school of Haldane, Douglas, and Priestley, which I was associated with when I was a student at Oxford in the 1920's. Respiratory work of this kind is now carried out in a much more sophisticated manner and has enabled men to live under such extreme conditions as at high altitudes or in the depths of the ocean. But such research is also of great importance in the understanding of respiratory diseases. Much of this work has to be done on humans because cooperation of the subject is necessary to the success of the experiment.

There are a few amusing stories of what happens when investigators try to do things on themselves. You may have heard of John Scott Haldane, J. B. S. Haldane's father, who investigated respiration using alveolar air sampling. He developed a special gas-analysis apparatus for analyzing the oxygen and carbon dioxide concentrations of respiratory gases. He was an expert in the use of this equipment and was very critical of anybody else in their style and skill in using his apparatus. On one occasion he was in a low-pressure chamber and insisted on performing the examinations then and there on his own expired air. To their horror the investigators looking through the window saw the old man skillfully and with great enjoyment pushing the pyrogallic acid into the mercury and then mixing that with the potash. They realized it was time to pull him out. He was suffering from severe anoxia and was really quite ill. Things like this can happen when experimenters start to work on themselves, so they have to be watched!

A contemporary researcher has also performed respiration experiments using himself as subject. Dr. Tom Sears had recording electrodes implanted in his expiratory and inspiratory intercostal muscles, and he then served as his own "prima donna." He showed that when holding a long note, having first taken a breath, the first respiratory stage is a progressive relaxation of the inspiratory muscles, and then as they relax fully, the expiratory muscles begin to work, becoming progressively stronger. So, instead of getting a prima donna to sing for him, he put the electrodes into his own expiratory and inspiratory muscles, sang the note himself, and produced for the first time a picture of just how a prima donna does keep a long note going. Of course, he could have interpreted these results only on the basis of the animal experiments in which he is such a master. On the basis of these animal experiments he was able to give us an account of the human mechanisms that act during phonation.

There are other kinds of experimentation that cannot be done satisfactorily on animals, and that cannot be done by the experimenters on themselves. For such experiments volunteer subjects are required.

An example of this experimentation is the work on brain stimulation performed by Penfield, Jasper, and Libet, among others. Penfield had to map out the various parts of the cerebral cortex that were associated with somesthetic sensations, vision, hearing, or speech. This work was dependent upon the work started during the last century using lower primates at first, and then using the highest primates, anthropoid apes. On the basis of this work, mapping of the cerebral cortex can be performed quickly and effectively. This mapping is required for therapeutic procedures, for example, for the localization of epileptic foci, and then for their treatment.

Penfield was able to accomplish his excellent work because it was done originally on anthropoid apes late in the nineteenth century. I remember the story about the first gorilla that was worked on by Sherrington in Liverpool, with Harvey Cushing assisting. They had to work with the animal under light anesthesia, and Harvey Cushing was taking no risks, so he brought along a big gun, which he put on the shelf over the fireplace. What shocked him was that the gorilla, as it was coming out of the anesthesia, had a good look at the gun, carefully examining it in all its length. There were a few frightening moments for Cushing who wondered whether the gorilla would get the gun first!

In this rather homely way the earliest experimental work was done on the primate brain in the latter years of the last century. It is this work that has made possible the fine experiments of Penfield and others on the mapping of the human brain.

In California today, Libet is doing most ethical and careful work on the sensations evoked in human subjects by varying the parameters of gentle electrical stimulation of the somesthetic area of the cerebral cortex. This work relates to some most interesting philosophical problems about how man actually derives conscious sensations from the complex patterned impulse transmission in the neural machinery of his brain. It is much more complicated than anybody had imagined. This type of investigation has to be done on human subjects who can report what they feel. But this is done only on volunteers whose brains are exposed for the purpose of a therapeutic operation, for example in a patient suffering from Parkinsonism. It is explained to the patient that the experiment has nothing whatsoever to do with the therapeutic procedures. Nevertheless, almost all volunteer under those conditions to give half an hour of their time on the operating table unanesthetized.

These examples illustrate what can be done with humans, but the basis for this human work lies in the animal experimentation that has already been carried out.

Now I come to animal experimentation. As long as we have a wider range of understanding of what animals can be used for, then humans are required less and less to undergo any of the dangerous situations involved in some types of experiments. The experiments that I have mentioned already carry no danger at all because every safeguard is exercised.

There is general agreement that with animal experimentation the only safeguard we would require is that the animal not be subjected to pain or a situation that might give rise to pain. Of course, animals can have all kinds of violent reactions to a situation, but whether they actually feel pain we can never know. We have to give the animal, shall we say, the benefit of the doubt. If it is reacting to a given stimulus that would give rise to pain in humans, then it must be assumed that the animal is suffering. We must always be careful about this, but with reference to the use of anesthetics in animals, how far down the animal scale do animals have sensations in any way resembling those of humans? I would readily agree to include all mammals, and also birds because they have very highly developed nervous systems. After that we come to the lower vertebrates, reptiles, amphibia, and fish. But some vertebrates or cordates have such simple nervous systems that it seems pointless to wonder if they suffer pain. Also, it is generally agreed that, if an animal is decerebrated, taking away the cerebral hemispheres, no matter how it reacts to noxious stimuli, it is not feeling pain. This is based upon human knowledge that, when this part of the brain is destroyed, the patient enters a perpetual coma, and never complains of pain or reacts as if he were suffering. We can assume, therefore, that pain and the nociceptive reactions associated therewith occur only at the highest levels of the nervous system. I am excluding of course the nociceptive reflexes that occur particularly well at the spinal level.

I want to just raise another interesting point. It is indeed fortunate, as I mentioned, that there is no missing link alive today—nothing between man and the highest anthropoid apes. Furthermore, we know he will never be discovered. A hundred years ago this was not so. I had to address UNESCO and write something for their journal *Impact* a year or two ago. I told them that, although there are difficult problems in the world today, we should contemplate what problems would be created if *Homo erectus* or *Homo habilis* or some other species of primitive man with a brain half the size of that of *Homo sapiens* also existed on our planet. These primitive men would be tool users, with intelligence enough to develop tools of a primitive kind, but clearly they would be of a lower order of intelligence than any of the races of men existing today.

I cannot explain how it was arranged that these very interesting and marvelous creatures ceased to exist. I daresay our ancestors exterminated them. But they do not exist, so we do not have this problem. It is quite easy to say that all existing animals except man have no rights to exist in the sense that we can kill them humanely for experimentation, or food, or for whatever we wish.

In contemplating the existence of a "missing link," however, I have often thought that somebody should write a novel on this theme. I discovered the other day that it has been done in a novel by Vercors, called *You Shall Know Them*, which purports to describe a tribe of primitive men, Pithanthropus Erectus, which had been discovered in the highlands of New Guinea by some Australian explorers. These primitive people, who were in the evolutionary stage of *Homo erectus* were called Tropis. They gave rise to no employment problems because they were docile and clever enough to learn and do simple things. The Australian manufacturers planned to import them and breed them and use them to run their industrial plants as unskilled labor.

In the novel a journalist named Templemore plans to discover if Tropis should be treated as human or not. He sets about doing this by the following rather drastic procedure. Some of these Tropis had been brought to Sydney, eventually to be taken to London to the Zoological Society for investigation. Templemore decides that he will do something quite fantastic. He has Derry, a young female Tropi, artificially inseminated with his own semen. She brings forth a child, or tropi, or whatever you will, which he calls his child. He has this child duly named and entered in the birth registry. Then he takes it to London with him. The other Tropis and Derry, the mother, are housed in the London Zoo. He takes this little creature into his own apartment, where he kills it humanely. He then calls the police and tells them what he has done. The police have no idea of what to do in this case. They say it was not a child. He says, "It is. Here is its name and birth registery, duly signed by a doctor." The story is designed to explore the question concerning the moral and legal relationships of man and animal and to emphasize the differences we almost subconsciously believe to exist between man and animal. Just imagine how difficult the situation legally, morally, and theologically could be if such hominids did exist. Fortunately we do not have to deal with such problems!

There are, however, several problems relative to unethical human experimentation. For the most part such experiments are carried out by people who are loathe to use primates or by people who do not have the cost structure for primate investigations and are unwilling to ask for it. One such experiment involves chronically implanted electrodes in the human brain. In animals this is a marvelous technique. It is being employed at the National Institutes of Health with monkeys in which electrodes have been implanted in the various parts of the cerebral cortex and the cerebellum. With microelectrode recording the investigators study the responses of single nerve cells in the cerebrum and cerebellum while the animals carry out trained procedures of various kinds. It is beautiful work and can lead to new levels of understanding of the mode of action of the brain.

It does, however, require an immense cooperative effort, because the people involved must ensure that highly trained animals will be kept under the best conditions and be well nurtured. Under ideal conditions the animals can go on for many weeks or months learning and being experimented on. Animals can learn only when they are comfortable and happy, so such investigations demand highly skilled experimenters and technicians.

The value of this kind of work in understanding the nervous system is becoming more and more apparent. The more perfect animal experimentation can become, the more obvious it will be that no one should subject human beings to chronically implanted electrodes. Surprisingly, it is cheaper to do the experiment on human subjects. Patients come along with some kind of psychosis or neurosis, and the doctors say "Yes, now we have to study your brain more. We will have to drill some holes and put some things in." A cap is placed over the assemblage of implanted electrodes, and these people go about their work at home and so on, and come in every now and then for recordings from these electrodes.

I regard this work as quite unethical. Because of these buried implanted electrodes, the brain will never be the same and many people will suffer from epileptic seizures. Because this procedure is destructive, and not at all therapeutic, it is unethical. I do not countenance any destruction or damage to the brain under conditions that are not related to the therapeutic treatment of the patient. This kind of human experimentation should be stopped completely, and to remove the need for it, we must establish primate centers. The centers we have in this country are very good ones, but we need more facilities in primate centers, so no one will be tempted to do these unethical human experiments.

Experiments on humans are required in the field of therapeutic innovation when new drugs are being tried, but in all cases these drugs should be given only after extensive investigations on animals. The animals will have to be chosen carefully, which will require certain knowledge of which kinds of animals are appropriate for experimenting with various kinds of drugs. What is really needed is a battery of different kinds of animals.

We are familiar with what happens when therapeutic innovation goes wrong, as it did with thalidomide. In this case the actual therapeutic trials on animals had failed as a safeguard, but now we are more alerted to these dangers.

I would recommend that wherever possible the investigator first try the drug on himself. For example, with LSD, the discoverer inadvertently experimented on himself; when he repeated the performance to verify the effects, he had a very severe reaction, because the extreme potency of this substance was not yet appreciated.

When it was suggested that I take part in this effort to define the laboratory animal, I was immediately attracted. I believe that the goal of all laboratory animal research is to find those animals in which we can best understand the diseases of man and then to discover the appropriate treatments. It is only because of the most advanced experimentation on animals that basic biological and medical sciences have progressed to where they are now, and we can look forward to the time when this will become one of the goals of our society. So the spinoff includes primarily basic understanding, from which will be derived clinical understanding and therapeutic treatment.

So, the task facing the laboratory animal researcher is a tremendous one. It is one that will become greater in the future. It will become necessary to provide a much wider variety of animals for specific purposes. Certain problems in living organisms are best investigated with invertebrates, like the nerve impulse and synapse in squid or the ganglia of aplysia. So a wide variety of animals, an almost zoological collection, must be made available for research. This will provide a tremendous service not only in advancing biological and medical science, but also in saving humans from suffering the risks and the travail that come when unethical experiments are done upon them.

To conclude, I will quote from the Helsinki Declaration on Human Experimentation, a document prepared by the World Medical Association. This document defines the conditions under which human experiments can be done. Almost all the conditions laid down do not apply to animals experimentation. On animals all we have to do is to take safeguards about pain, and then we can do what we wish on the animal, but on humans this is not so.

First:

Clinical research must conform to the moral and scientific principles that justify medical research and should be based on laboratory and animal experiments or other scientifically established facts.

That is to say, in all of this work, animal experimentation must go as far as possible on the appropriate animals before humans are considered at all.

Other principles relate to clinical research and are not really relevant to my present theme.

Clinical research cannot legitimately be carried out unless the importance of the objective is in proportion to the inherent risk to the subject.

No one makes these statements about animal experimentation.

Every clinical research project should be preceded by a careful assessment of inherent risks in comparison with the foreseeable benefits to the subjects and others.

There are also special precautions about the clinical research and relationship to the personality of the subject and what is going to happen after the experiment. None of this applies to animals.

In conclusion, the work of the laboratory animal scientist is going to become more and more important and more and more appreciated by the people of the world. It is generally agreed that animal investigations lead to the understanding and treatment of human diseases. In these days of political and social stress we can be grateful that there is agreement on some general principles that transcends all political differences.

EFFECTS OF REARING CONDITIONS ON THE BEHAVIOR OF LABORATORY ANIMALS

M. W. Fox

Introduction

The purpose of this paper is to focus attention on a number of experiments that demonstrate how environmental factors can affect the brain and behavior of laboratory animals. Such environmental factors may constitute highly significant experimental variables, and if they are not recognized and controlled, the significance of the results may be questioned. These environmental variables—all part of laboratory animal husbandry—include cage size, population density per cage, provision for exercise or varied stimulation in an "enriched" or "impoverished" environment, methods of rearing, handling and socialization, and long-term effects of selective breeding and of environmental influences over successive generations, which may modify a particular strain or gene pool. We must also consider some general questions like what is the most humane way to keep laboratory animals, and how representative are many laboratory species of their wild relatives or, indeed, of a natural population of animals.

GENETIC INFLUENCES AND SELECTIVE BREEDING

In working with an individual strain, will we not be collecting data from a very narrow spectrum of subjects relative to the broader range of variability that exists within natural heterogenous populations? Mayr[17] emphasizes that through developmental flexibility the genotype is buffered in such a way that genetically very different individuals subjected to the same environmental influences are phenotypically similar. This buffering effect (together with recessiveness and hybrid vigor) essentially protects the gene pool from environmental pressures that could otherwise result in a depletion of the gene pool through selection. In some cases, such selection occurs and results in overspecialization. In inbred strains, the buffering effect may be minimal, so the gene pool will be more exposed to environmental selection pressures. Variability (genetic) is a natural biological phenomenon, and without it, evolution would be arrested or protracted. A good deal of individual variability is attributable to meiosis, a process by which the parental chromosome number is reduced by one half; following fertilization, the resultant offspring, although bearing species and strain characteristics, have, in addition, unique gene combinations as a result of meiosis. In this way, variability and individuality within a given gene pool (of a normal rather than an inbred population) is maintained.[12] Research with selected strains in which there is a high probability that the offspring either carry or manifest a particular trait is, of course, invaluable, provided we are concerned with such a trait experimentally or are controlling for it in our experimental design. And yet, how often do we see a drug toxicity study, for example, conducted on a single inbred strain. Why select such a strain? Not to evaluate genetic effects, surely, but merely to reduce the variability of the results (a pseudosophistication), and to reduce the number of subjects needed. (Also, only males may be used because females have the added variable of estrus, and it is a point to consider that most of what the American psychologist knows is the psychology of the male rat.) In reducing variability we are reducing the biological significance of our results if they are to be applied as biologically relevant to a natural population. The results are only of value within a very narrow sample of a normal population (see Figure 1).

Where appropriate, therefore, it may be advantageous to use an inbred strain of known genetic background and as an additional control, either a "wild" strain or a heterogeneous random-bred line. In this way, the direction of the effects of experimental treatment that may be influenced by a particular genotype would be recognized and appropriate controls instituted.

ENVIRONMENT–GENOTYPE INFLUENCES

It has been shown by a number of investigators that environmental influences can be transmitted over successive generations following exposure of only the mother to such influences.[5] Handling or gentling makes rodents more tractable, and if instigated prenatally to the mother, it will affect the offspring. Conversely, traumatic stimulation (electroshock or a conditioned emotional reaction) can result in offspring that are more emotional.[13] Comparable stressful environmental or husbandry influences could clearly influence an experiment. For example, if in one laboratory pregnant rats and their offspring are regularly handled and in another

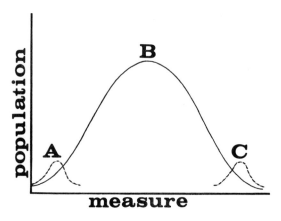

FIGURE 1 Theoretical schema of distribution of a given trait or measure in a normal or random population (B) and in two inbred populations (A and C), which may bias results or misrepresent the true biological significance of a particular phenomenon in a heterogeneous population.

laboratory the entire animal facility is semiautomated and the subjects are handled only when they are caught and restrained (with a gloved hand) for experimentation, the results from these two laboratories might be quite different.

The genotype principally determines the direction of effect of such environmental influences on the eventual outcome or phenotype. These environmental influences have been well documented.[21] It is clear that if we are going to go to great lengths to establish uniform strains of laboratory animals, we are wasting time and money until we also standardize the type of environmental experiences these strains receive during pregnancy, early in life, and throughout their growth until they are used in experiments. It is extremely difficult to work with rats that have had little or no prior handling, and yet they are used in experiments and we publish results on "genetically uniform" individuals whose emotional-autonomic reactivity, because of lack of prior handling, was explosive at the time of testing. This factor must affect the validity of those results. Another laboratory practicing regular handling of all subjects (once per day at irregular times each cage is opened and the attendant touches the rat on the head long enough to say "Hi, *Rattus rattus Norvegicus* and then closes the cage) may come up with very different results. I say *may*, because few experiments have been conducted to evaluate the effects of differential handling. Susceptibility to radiation, gastric ulcers, tumors, and stress (of restraint) can be influenced by handling early in life.[21] Emotional reactivity in the adult can influence many physiological measures—blood pressure,

296

heart rate and blood eosinophil, and corticosteroid levels, for example. More chronic emotional stress may reduce immune responses and body defense mechanisms against bacteria and toxins, and prolong wound healing, principally as a result of elevated ACTH release and a reduction in phagocytic activity.[14] Selye's syndrome may result, especially where a certain experimental treatment is of some duration and the subject is of a particularly susceptible genotype or has been reared so that the genotype–environment interaction during development produces a more susceptible phenotype (see Figure 2A and B). Handled rats can withstand terminal starvation or swimming stress far longer than nonhandled controls, for example, but their survival time may be shorter if they have been excessively handled (stressed) earlier in life.

HANDLING, GENTLING, AND SOCIALIZATION

Methods of rearing and handling early in life clearly influence emotional reactivity of animals when they are handled for experimentation in later life.

Prior handling greatly facilitates later handling: In rats, this is due to a re-

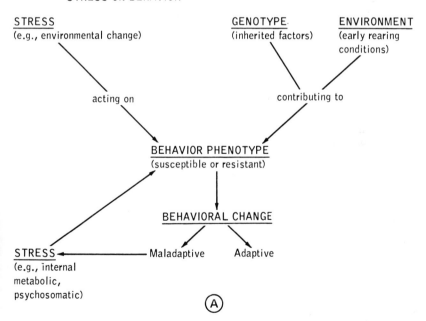

FIGURE 2A Effects of genetic background, environment, and stress on behavior.

PRENATAL AND POSTNATAL INFLUENCES ON BEHAVIOR:
"FAMILY CIRCLE EFFECTS"

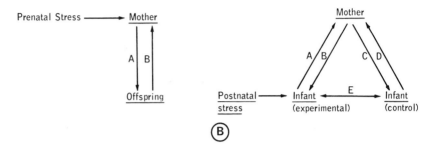

FIGURE 2B   Prenatal and postnatal influences on behavior: "Family circle effects."
A, B, C, D, and E represent interaction feedback between mother and offspring and
differentially treated offspring.

duction in emotional reactivity via the adrenal pituitary axis. If instigated early
in neonatal life, the neurohypophysis may be permanently affected so that
ACTH is released in a "graded" fashion to increasing intensities of stress
rather than in an "all or none" fashion as occurs in nonhandled controls.[15,16]

Handling or more correctly, "gentling,"[21] of older animals results in increased docility, partly as a result of habituation of the emotional fright,
flight, or fight response when first handled. Again, in some strains early handling has a greater effect than in others, indicating that the direction of effect
is influenced by the genotype–environment interaction.[13] In higher mammals,
such as the dog, gentling early in life results in socialization,[25] and experiments have shown that dogs raised without human contact until 12 weeks of
age are extremely shy and difficult to handle: They are literally as wild as
wolves. No experiment could be contemplated on such subjects except to evaluate the effects of not being handled earlier in life. The emotional and physiological reactions of such animals would influence the direction of results in
either an acute or a chronic experiment. Yet we persist (in some laboratories)
in fighting to restrain a rat or dog to obtain the "meaningful" blood sample,
or we are disappointed if the subject dies suddenly after induction of anesthesia.

Two more points for consideration come up here, namely the variables of
the handler, or animal technician, and those of animal selection for the laboratory, in terms of animal docility and fertility.

HANDLING PROCEDURES

The animal may be fighting restraint, not because
it has not been socialized, but because the handler is ignorant by virtue of in-

adequate instruction. On the other hand, the animal may have been exposed only to one handler and become socialized or imprinted to him only and is intolerant of strangers. Animal handlers should, therefore, be carefully instructed and rotated among various units to avoid any over-attachment or even favoritism for one animal, which has on many occasions disrupted the experiment—nutritional studies, for example, where subject and handler share their lunch.

Animal technicians should be impressed with the fact that their charges are randomly assigned "experimental," "control," and "breeding colony" numbers, that they should receive equal treatment (if you accidentally drop one, you should drop all of them), and that they are social animals—be they rats, cats, or dogs—that respond to handling and to voice. Handling procedures should become standardized as part of laboratory animal care[14] to the same extent as we now standardize room temperature and humidity, light–dark cycles, cage design, nutrition, and genotype.

SELECTION FOR DOCILITY AND FERTILITY

The second point is that of selection of laboratory animals: Other than selection for particular traits (hemophilia, maze brightness, muscular dystrophy, diabetes, or spontaneous leukemia), there is selection for docility and fertility. Selection for docility or ease of handling may involve selection for hypoemotional–adrenal reactivity: Richter[23] has emphasized the point that the laboratory rat may be hypoadrenal and hypergonadal as a result of selection. Hypoadrenalism may also increase the tolerance of crowding so that more animals can be housed together; conversely, this tolerance of some individuals under crowded conditions may be correlated with these animals' having small, unhypertrophied adrenal glands. Social stress of crowding may be reduced through selective breeding in laboratory rats, and also in intensively housed farm animals, notably pigs (which still react to transit stress, however). Animals are less aggressive and are therefore more tolerant of one another when housed together in large numbers; crowding stress is consequently reduced. Selection plus early handling, habituation, or exposure learning may also serve to reduce mass fright–flight reactions to sudden noises or to strangers. (An additional sound blanket of music, sound-insulating tiling, and location of work areas at some distance from the animal colony are also of importance.)[6]

Hypergonadism with high fertility, plus selection for good maternal behavior to reduce postnatal mortality may result in other endocrine changes that may influence the significance of laboratory animals in terms of normality relative to a natural population. Bronson and Marsden[2] observed that the Bruce effect (pregnancy block) was difficult to demonstrate in highly inbred strains of mice; selection for fertility in crowded colony conditions may therefore have profound and ramifying effects on the neuroendocrine system.

LONG-TERM EFFECTS OF ENVIRONMENTAL INFLUENCES

This leads on to another issue, namely, what are the effects over successive generations of the laboratory animal facility environment, which is essentially a particularly intensive and specialized form of domestication?

Denenberg and Rosenberg[5] have confirmed earlier studies that have shown how environmental influences may be "transmitted" over several successive generations. Handling or stressing a pregnant rat has a demonstrable effect on her offspring and without further handling, the effects of such treatment can be shown 4 or 5 successive generations later. Darwin[3] observed that the domesticated rabbit has a much smaller brain than its wild counterpart, and he attributed this to the effects of domestication and the fact that a captive rabbit would not have the environment in which all its instincts would be stimulated. Bennett et al.[1] and others have demonstrated marked structural and biochemical changes in the brain and behavior (learning) deficits in rats raised in complete social isolation. Rats raised under standard laboratory conditions showed relatively similar deficits compared with rats raised in an "enriched" environment in which they were in social groups and had numerous "play" objects. In other words, it has been shown that the genetic capacity or potential for brain and behavior development can be influenced by the environment. We may now ask what do those uniform rats (and other species[7]) that have been propagated for generations in the impoverished, sterile environment of the "standard" cage represent and what are they biased towards.

THE ISOLATION-EMERGENCE SYNDROME

Emergence from isolation in higher animals such as the dog, is characterized by two types of reaction, depending partly on the age of the subject and on its breed (it must be added that more timid breeds are more severely affected than others).[11]

One reaction is a mass fear response, in which the animals freeze and adopt bizarre postures, (Figure 3) defecate, urinate, and may even show defensive aggressive "fear-biting"; they are also touch-shy, sound-shy, and sight-shy in the layman's terminology. The other type of reaction is one of hyperexploratory behavior, in which the dog runs around hyperactively (rather like a much younger naive pup) and approaches objects that normally would be avoided. Melzack and Scott[18] reported impaired pain perception in such isolation-raised dogs; they would repeatedly stick their noses into a candle flame for example. It has been proposed that in this type of reaction, excessive arousal actually increases the pain threshold. EEG and evoked potential studies of pups emerging from isolation into a more complex environment[8] reveal that the

300

FIGURE 3 Whirling in aroused, isolation-raised Beagle.

brain is in a state of arousal, which may be a pathophysiological state, for spindle-like bursts of activity were recorded (Figure 4). These bursts disappeared rapidly as the subjects recovered from the effects of isolation. They had only been isolated for 7 days, from 4 until 5 weeks of age, and recovery was evident in most subjects after a further 4–7 days in the normal animal room environment. Isolation for longer periods up to 16 weeks of age produces almost permanent behavioral change; adaptation following emergence from isolation is protracted, and symptoms of the isolation syndrome may persist permanently in innately susceptible individuals or breeds. Fuller found little protection in raising dogs in pairs in isolation, as did Bennett *et al.*[1] with rats.

This isolation syndrome has been emphasized because many dogs bred and raised under relatively impoverished conditions may be removed to a more complex environment, for example, the testing room or the laboratory animal facility, where they react to the novelty and complexity of the new surroundings in one of two ways described above. In addition to showing symptoms of the isolation emergency syndrome, they may also have received little handling and socialization early in life, which in no way enhances their usefulness as research subjects, especially in long-term experiments.[8]

EEG RECORDINGS OF ISOLATE & CONTROL DOGS
5 WEEKS 4 DAYS OLD

CONTROL - ALERT   ISOLATE - ALERT

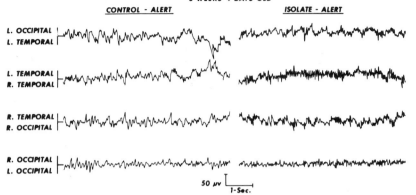

FIGURE 4   Fast-frequency spindle-like activity in isolation-raised dog, on emergence from isolation and persisting for several days thereafter.

If pups are raised with plenty of human contact to ensure that they are socialized, but are not removed from their pens until 12 weeks of age, they show some of the symptoms of the isolation syndrome.[10] They actually prefer to remain in their pens even when the door is left open, and when first placed in an observation arena with various play objects, they withdraw and do not explore. Pups placed in the arena for a mere 15 minutes at 5 and 8 weeks of age are "protected" in that they actively explore the arena at 12 weeks of age (Figures 5 and 6).

Sackett[24] has found comparable effects in monkeys raised under varying degrees of deprivation. The practical implications of these experiments are brought home by the familiar statement that random-source (pound) dogs are usually much easier to manage in the laboratory than those dogs born and raised specifically for research. It is not only socialization that influences these different reactions, but also the fact that random-source dogs have been exposed to a more enriched and varied environment; they have prior associations, which facilitates their habituation and adaptation to novel visual and auditory (and possibly olfactory) stimuli in the laboratory. Of course, random-source dogs with wide genetic variability and no known prior life history are of limited value for many research projects. What we should consider is the type of programmed life history we should provide for specifically raised animals that have been selectively bred for research purposes. Some varied stimulation to prevent possible brain changes and behavioral degeneration may be wise, not purely for humane reasons, but also for the validity and justification of using such animals for biological research.

ACTIVITY & VOCALIZATION
(LONGITUDINAL GROUP)

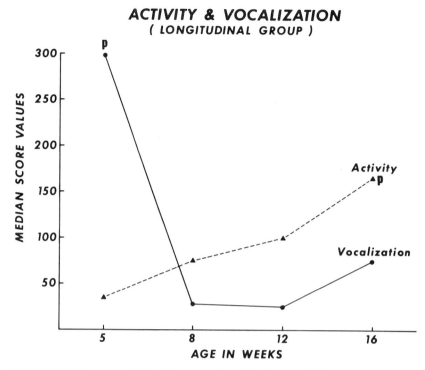

FIGURE 5   Distress vocalization decreases and activity increases in laboratory-raised pups exposed to novel environment at successive ages (cf. Figure 6).

EXERCISE

Many people claim that exercise is important, but animals in nature that have fed, are warm, and are not afraid of predation or are not sexually frustrated, do not exercise; exercise *per se* is an unbiological activity at variance with the law of conservation of energy. Wild animals either play (either with each other, by themselves, or with appropriate inanimate objects) or they sleep, if all other basic drives are satisfied.

The answer may be not to provide confined animals with exercise alone, such as a treadmill (the value to many species of such stereotyped "neurotic" activity is questionable), but with varied stimulation, such as social interaction (even visually in opposite cages may help) and objects to manipulate. Dogs might be walked around the animal facility, leash trained, and exposed to trucks and dollys, to strangers, and to a variety of audiovisual stimuli of varying intensity and complexity. This would be especially advantageous and appropriate where long-term experiments were to be conducted in which the subject would have to adapt to a new set of complex and novel stimuli.

303

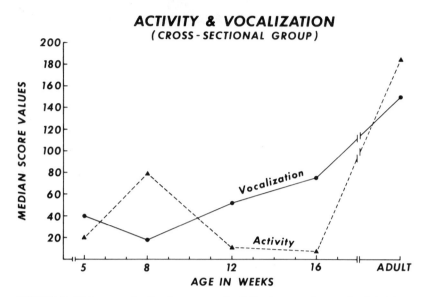

ACTIVITY & VOCALIZATION
(CROSS-SECTIONAL GROUP)

FIGURE 6   Distress vocalization increases and activity decreases (except in adults) in laboratory-raised pups after 8 weeks.

Essentially, this issue of prior experience in a particular environment involves the animal's ability to adapt to a new environment. The narrower the range of prior experience, the more limited would be the animal's ability to adapt to more varied or totally different conditions. Each environmental change, from the rearing kennels to the laboratory animal facility to the animal research unit, entails some degree of adaptation (to psychophysiological stress); adaptation would be more protracted and psychophysiological stress more pronounced where there is a greater dissonance between consecutive environmental conditions or where the animal has had a restricted, experientially impoverished prior history and is suddenly plunged into a more complex environment. Consecutive environmental conditions should be of comparable qualitative and quantitative complexity to reduce dissonance and contingent adverse effects. Some time should be allowed to enable the subject to adapt to the new conditions.

THE PSYCHOPATHOLOGY OF ISOLATION AND DEPRIVATION

Consider a wild animal suddenly being removed from all that it has learned and from all that it has evolved to interact with; translocated from one socioenvironmental milieu to another, which in no way simulates a "natural" environment. Here I personally question the humaneness

304

of such treatment, especially in the higher primates. In so many zoos and laboratories we see bizarre behavior patterns emerging as maladaptive attempts to adapt to confinement; fear is overcome through habituation, gentle (or predictable) handling, and food reward, but in the frustration of captivity, thwarted attempts to escape may become stereotyped fixations, where the animal paces or weaves backwards and forwards. Other abnormalities may develop—automutilation, hyperphagia, coprophagia, hyperaggression, hypersexuality, and disruption of reproductive and maternal behavior.[19,20] Stereotyped movements may also develop as a means of providing varied stimulation; even children hospitalized for extended periods will develop such patterns in order to "enrich" their environment by varying their sensory input and motor output. Monkeys and dogs raised under restricted conditions frequently develop such stereotypes. We may postulate a mechanism of perceptual-motor homeostasis, where the organism seeks stimulation (or provides its own) to maintain this homeostasis. The quality and quantity of stimulation may be lower in an animal raised in relative isolation than in one raised under more natural conditions, the "set point" being higher in a feral animal. A wild animal suddenly placed in an environment in which perceptual input is lower, both qualitatively and quantitatively, and in which motor activity is restricted, will experience a perceptual-motor homeostatic imbalance. Psychophysiological depression, anorexia nervosa, increased susceptibility to infectious diseases, and psychosomatic disturbances may occur during initial stages of adaptation. Adaptation to this situation essentially involves "institutionalization" in which the set point is gradually lowered and the animal adapts to a lower level of stimulation. (Indeed, this is a problem in many mental hospitals; by the time the "neurosis" is cured, the patient has adapted to the relatively monotonous and impoverished environment typical of so many over-crowded and under-staffed institutions, and the clinicians now have to deinstitutionalize and rehabilitate the patient.)

The adaptation period may be prolonged in animals derived from random sources or wild conditions. Recently captured monkeys may take up to 12 months to develop some degree of physiological stability. There is a clear advantage in using animals "raised for the laboratory" under standardized relatively impoverished cage conditions, for their adaptation period will be considerably shorter.

Apropos of this separation-depression and psychophysiological stress, several workers[9,22] have shown that if group-raised rats are isolated from each other in separate cages, this isolation can be stressful to such an extent that they are more susceptible to drugs or toxins. Some "controls" even die, and on *post mortem* they show lesions diagnostic of Selye's syndrome.

We should be mindful, therefore, of how we alter the socioenvironmental structure of our subjects before experimentation and during the experiment

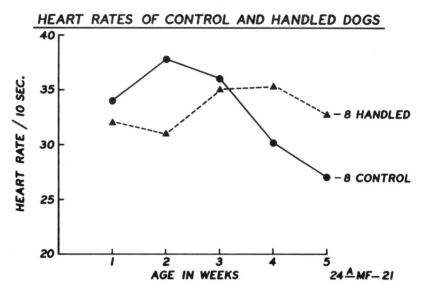

FIGURE 7 Psychophysiological stress following handling of pups early in life: marked tachycardia (increased sympathetic tone), with reduced vagotonia in handled pups.

itself. Adaptation may be facilitated by selective breeding, socialization and handling (Figure 7), and rearing under standardized socioenvironmental conditions that match conditions under which the animals will be kept prior to and during experimentation. If the experiment involves confinement (and even more variables, such as those experienced in a space capsule in which the subject must execute certain procedures while experiencing weightlessness), it is imperative that psychophysiological adaptation and stability to these environmental conditions be established prior to count down. Simulate take-off before take-off, otherwise there will be compounding interacting and uncontrolled variables that cannot be separated from the dependent variables of the experiment.

Conclusions

We have considered a number of issues concerning the husbandry of laboratory animals—important issues that are supported by a few experiments but that require a good deal more investigation before we can instigate some international standards to protect the quality of biological research and advance our understanding of biological phenomena and of the interaction of complex interrelated psychophysiological processes. We must not forget that all mammals have a "psyche" as well as a physiology, and ignorance is no excuse for inhumane treatment. Cruelty is based on ignorance and, more rarely, indifference, and the insidious growth of ignorance and indifference of the experimenter toward his subjects is a common syndrome. How rarely does a scientific investigator visit the animal facility: Perhaps he is better kept out of there anyway. But there must be close liaison between the personnel in charge of the animals and the investigator who is going to use them in a particular experiment. Such communication is too often lacking. Indeed, the many generalizations in this paper are intended not as dogmatic statements but as issues that should be seriously considered and evaluated on the basis of available experimental evidence. A final global generalization: Accepting the established fact that environmental influences can be transmitted across successive generations[5] and that the environment can result in a gradual selection (genetic drift) within the gene pool (in spite of any buffering influences on the genotype as proposed by Mayr,[17] we see the inevitable result. Namely, foundation stock of the same genetic background is used to provide foundation animals in laboratories that are geographically separated and that employ different methods of animal husbandry within different animal facility environments. Within only a few generations, the different laboratories will have their own distinctive lines, a fact of which they are aware only indirectly, in terms of disease susceptibility or infertility problems. The infertility problem will soon be resolved by comparing husbandry methods and identifying the intervening variable. More subtle unrecognized changes influencing the gene pool and phenotype may affect the experimental results so that one laboratory cannot verify the findings of another with the same strain of animals. Intervening variables may not be identified, and colleagues or collaborators become alienated by conflicting data that actually reflect conflicting methods of animal husbandry.

In conclusion, we must re-evaluate the quality of our research in terms of our quality of animal care. Standardization of experimental procedure is as important as standardization of all aspects of animal care and of the genotype-environment–phenotype interaction. The quality of the animal must match the assumed quality and sophistication of the experiment. The responsibility for achieving this balance lies with the investigator. This responsibility should be met soon and it should be met on an international scale with standardization after rigorous discussion and appropriate experimental evaluation.

Summary

A multiplicity of interacting variables can influence dependent variables. The intention of this paper has been to demonstrate how some variables that often are ignored are related to the selection, rearing, and handling of laboratory animals and can have profound and long-lasting effects influencing the significance of dependent experimental measures. This has been a one-sided (or strongly weighted) argument in support of a total re-evaluation of the significance of animal research in general and in developmental and psychobiological research in particular. A more rigorous investigation of these variables is urged as a necessary prerequisite for subsequent international standardization of laboratory animal science.

References

1. Bennett, E. L., M. C. Diamond, M. R. Rosenzweig, and D. Krech. 1964. Chemical and anatomical plasticity of the brain. Science 146:610–619.
2. Bronson, F. H., and H. M. Marsden. 1964–1965. Endocrine responses to the social environment. p. 98. *In* Jackson Laboratory 36th annual report, Bar Harbor, Maine.
3. Darwin, C. 1815. The descent of man. Rand McNally, Chicago.
4. Denenberg, V. H., and A. E. Whimbey. 1963. Infantile stimulation and animal husbandry: A methodological study. J. Comp. Physiol. Psychol. 56:877–878.
5. Denenberg, V. H., and K. M. Rosenberg. 1967. Nongenetic transmission of information. Nature 216:549.
6. Fox, M. W. 1965. Environmental factors influencing stereotyped and allelomimetic behavior in animals. Lab. Anim. Care 15:363–370.
7. Fox, M. W. 1966. Natural environment: Theoretical and practical aspects for breeding and rearing laboratory animals. Lab. Anim. Care 16:316–321.
8. Fox, M. W. 1967. The effects of short-term social and sensory isolation upon behavior, EEG and averaged evoked potentials in puppies. Physiol. Behav. 2:145–151.
9. Fox, M. W. 1968. Socialization, environmental factors and abnormal behavioral development in animals. Ch. 19 *In* M. W. Fox [ed.] Abnormal behavior in animals. W. B. Saunders, Philadelphia.
10. Fox, M. W., and J. Spencer. 1969. Exploratory behavior in the dog: experiential or age dependent? Develop. Psychobiol. 2:68–74.
11. Fuller, J. L., and L. D. Clark. 1968. Genotype and behavioral vulnerability to isolation in dogs. J. Comp. Physiol. Psychol. 66:151–156.
12. Hirsch, J. 1963. Behavior genetics and individuality understood. Science 142:1436–1442.
13. Joffe, J. M. 1965. Genotype and prenatal and premating stress interact to affect adult behavior in rats. Science 150:1844.
14. Kurtsin, I. T. 1968. Physiological mechanisms of behavior disturbances and cortico-visceral interrelations in animals. Ch. 7 *In* M. W. Fox [ed.] Abnormal behavior in animals. W. B. Saunders, Philadelphia.
15. Levine, S. 1967. Maternal and environmental influences on the adrenocortical response to stress in weanling rats. Science 156:258–260.
16. Levine, S., and R. F. Mullins. 1966. Hormonal influences on brain organization in infant rats. Science 152:1585–1592.
17. Mayr, E. 1963. Animal species and evolution. Harvard University Press, Cambridge, Mass.
18. Melzack, R., and T. H. Scott. 1957. The effects of early experience on the response to pain. J. Comp. Physiol. Psychol. 50:155–161.

19. Meyer Holzapfel, M. 1968. Abnormal behavior in zoo animals. Ch. 25 *In* M. W. Fox [ed.] Abnormal behavior in animals, W. B. Saunders, Philadelphia.
20. Morris, D. 1966. Abnormal rituals in stress situations. The rigidification of behaviour. Phil. Trans. Roy. Soc., Ser. B. 251:327–330.
21. Morton, J. R. 1968. Effects of early experience, "handling and gentling" in laboratory animals. Ch. 17 *In* M. W. Fox [ed.] Abnormal behavior in animals. W. B. Saunders, Philadelphia.
22. Nutrition Reviews. 1966. One or many animals in a cage? Nutr. Rev. 24:116–119.
23. Richter, C. P. 1954. The effects of domestication and selection on the behavior of the Norway rat. J. Nat. Cancer Inst. 15:727–738.
24. Sackett, G. P. 1965. Effects of rearing conditions upon the behavior of Rhesus monkeys (*Macaca mulatta*). Child Develop. 36:855–868.
25. Scott, J. P., and J. L. Fuller. 1965. Genetics and social behavior of the dog. University of Chicago Press, Chicago.

DISCUSSION

DR. DIETERICH: For the ultimate adaptation of adult laboratory animals, should they be "institutionalized" or "socialized"?

DR. FOX: That is a very good question. I am working with wild canines now, and they are in an impoverished laboratory environment. I am happy with them because they are well socialized. They were raised under these relatively impoverished conditions. I have had adults brought in from the wild and they never really adapt to captivity or the impoverished environment. I think this early plasticity in the developing system is an advantage to us in getting the animals early in life and adapting them to the laboratory. This is probably especially so in primate research. I think in such cases we will probably speak more of habituation rather than socialization.

DR. HOWLAND: Have you any experience with slow environmental transition versus abrupt transition from the forest to individual animal cages that would recommend one method of abrupt transition?

DR. FOX: I really do not know anything along this line.

DR. HENEGHAN: Do premature animals or man require socializing types of stimuli during the early neonatal period, or are the basic auditory and other stimuli that they have received *in utero* sufficient?

DR. FOX: No, apparently the intrauterine stimulation does not make a difference, but this has not been demonstrated in mammals higher than the rodents, in fact, no kind of prenatal stimulation has been. In general, the increments of experience are first based on tactile and olfactory stimuli in animals that are very immature at birth. This forms a stepping stone for subsequent imprinting or socialization based on visual and auditory cues. So there are increments of experience in different sensory modalities that establish and consolidate this bond.

DR. GARLICK: Can social behavior be altered significantly, and is it possible to retrain social behavior following a period of critical anoxia in growing animals—that is, anoxia in the absence of anesthesia?

DR. FOX: Social behavior can be altered significantly. Heddiger describes a two-ton moose that was hand-raised. It reached sexual maturity and tried to copulate with its keeper. There are many such problems in nature based upon imprinting.

Can social behavior be altered significantly? Yes. We go back to Freud. Is social behavior retraining possible? It is extremely difficult. Animals do tend to regress. Some socialized dogs gotten between 6 to 8 weeks of age, if subsequently isolated, say, from 4 months until 6 months of age in large breeding kennels, will regress and refuse to come out. They remain shy. They are the ones who probably innately are rather timid to begin with, and they need greater increments of experience to compensate for their deficits.

Can critical anoxia in a growing animal influence this? Gershin and Berkson at the Pediatrics Institute in Chicago were looking at short-term anoxia nitrogen induced in kittens, and found that early in life this brief anoxia had a stimulatory effect that caused the cats to mature somewhat faster. There were interacting variables of handling in the experimental procedure that did make them more social. This is the only reference that I know of.

DR. TAYLOR: How many times should rats be handled in the first 5 days of life to influence temperament?

DR. FOX: Three minutes a day. They are taken out of the nest separately and are put into a little metal pot with a few shavings in the bottom for three minutes at laboratory temperature—room temperature. It has a remarkable effect. It is a very critical period. There are similar changes with castration and thyroid administration at slightly later time sequences, but, again, these are very critical periods.

DR. LANE-PETTER: How do you explain the observation that a colony of rats or mice, difficult to handle on first introduction, can become very docile after many weeks or months, too soon for genetic change, too late to be attributed to individual adaptation?

DR. FOX: Thank you, Dr. Petter. We do have the human element here, and it is important to consider the effects of the handler. Handlers do improve with experience. I have had psychology graduates try to restrain dogs for me, and the dogs defecated and hit the ceiling. Within a matter of weeks, however, the animal trains the handler in how to handle the animal, often through negative reinforcement. The handler even gets bitten, but eventually, there is almost a rapport. Read any of the French work on animal hypnosis. Some people simply cannot put a rabbit out, others can. No matter how hard some try, they can never do it. There is a certain ability to communicate—to read the animal's signals. Some of my friends always get bitten by rats. There are particular odors that are released when a rodent is frightened. There is a fear pattern which was demonstrated last year in rats, and when they are unduly disturbed, other rats in the colony get this warning signal. It is a definite pheromone, and it probably operates in pigs, too. So if you disturb one, it is just like a whole stack of cards falling.

DR. SABOURDY: In some dog-breeding colonies, the animals are kept and bred in cages with no possibility of exercising. Do you think this might have an adverse affect on their psychological development? Do you think that such animals are better adjusted to the conditions they will meet in the laboratory when they are used for experiments?

DR. FOX: If they are raised under those conditions I think they are well suited to go into a comparable situation because this will reduce the dissonance and shorten the time necessary for adaptation. This would provide a metabolic and behaviorally stable system because there is not an extreme environmental change. The humane aspects, however, may be questioned. I have mentioned Bennett's work and that of his co-workers who have shown that there are even structural changes in the rodent's brain when it is raised under these impoverished conditions. Heaven help us if we have something like the Beagle, say, in 200 years time, which is no longer a dog at all, but is a strange little thing that resembles a dog, and yet has a very shriveled brain, like Darwin's rabbits. I wonder.

DR. GRAFTON: How do you handle newborn rodents as recommended without a high rate of cannibalism by the mother?

DR. FOX: Apparently the strains used do not have the problems of cannibalism at all. There is the odd rodent that will cannibalize its offspring; in certain strains of mice they also do this, but it is very rare. Usually there is selection against this, especially in Dr. Lane-Petter's situation. If he had cannibalism, he would not be in business, so he selects against it. These animals have an incredible tolerance for one another's offspring. We see this in cattle and pigs too. This, too, is through selection. Wild strains, on the other hand, will very often kill each other's offspring.

DR. SEDA: Did you investigate the type of gonadotrophins secreted by animals being exposed to prenatal stress conditions?

DR. FOX: No. They looked simply at the onset of vaginal opening, histological development of the testes, and so on, no exact measures on the gonadotrophins. The ACTH measures have been well documented during critical times. This is very solid, indeed.

DR. MALEK: Would you recommend any extension in common weaning dates for cats and dogs so that they may develop better social relationships with their mothers and litter mates?

DR. FOX: It is very often important to take the puppies away from the bitch, especially if she is a fearful timid type who reacts fearfully to you. There is a good deal of postnatal transfer of behavior. It is important to keep puppies together as a litter, but you have to watch out for the strain. The Beagle is very tolerant and you can keep large groups of litters together. But in some strains of wire-haired fox terriers, for example, if you have five together, the fifth will be killed. In kittens it is a different thing; weaning can be very protracted. They are still nursing from the mother at 6 months of age. Our usual system was to keep the cats in a large cat colony, and put the kittens into a nursery around about 4 months of age and wean them quite late. Puppies you can wean as early as five weeks with no problems.

DR. ROTH: Do you know of significant changes in primate reaction between small-unit animals such as chimpanzees and large troop animals such as baboons, when they are held in impoverished environments?

DR. FOX: I don't quite understand the question. "Do you know of significant changes in primate reaction between small-unit animals?" I don't quite understand what you mean by reaction.

DR. ROTH: Reaction to the environment.

DR. FOX: Yes, such studies are very well documented by Harlow, especially in terms of social behavior when they are raised in the impoverished conditions. They are excessively aroused when they meet each other. Sexual behavior is impaired. They tend to be hyperaggressive. They will often withdraw into an almost schizoid or autistic reaction. Similar behavior has been noted in some strains of dogs, too, following such deprivation. Does that answer the question?

DR. ROTH: Not quite. I was trying to find out if there was a difference in large-troop animals as compared with small family units in reaction to their natural environment.

DR. FOX: We don't know. Sackett is going to look into this, especially in terms of the ability of these isolation-raised animals to adapt to more complex environments, and he is looking at the issues of social organization in various groups, too. This could be a variable when you have a large group of animals and you are getting away from uniformity because some are low in the pecking order, and perhaps are already in a pretty bad state of vigilance and half beaten up. So because of their social interactions there are some advantages in separating them from physical contact with each other. But most of these social animals we believe, but it has to be tested, do have a drive for visual contact, at least, apropos of the Butler box test for the dog where the animals do show operant performance to see their conspecifics.

EFFECT OF SOCIAL ENVIRONMENT ON THE DEVELOPMENT OF MOUSE ASCITIC TUMORS

Roger-Paul Dechambre

Introduction

Several studies have been performed on the influence of environment on the development of spontaneous tumors in mice.

Andervont[1] found that the incidence of spontaneous mammary tumors in mice was related to the number of animals in a cage. Isolated mice developed cancer earlier and at a higher incidence than those kept eight in a cage.

Muhlbock[5] compared the incidence of spontaneous mammary tumors when the mice were housed in zinc cages containing fifty or five animals or in glass jars containing five or one animals. The respective tumor incidences were 29, 56, 67, and 84 percent. The same author demonstrated that the use of an exercising wheel reduces the incidence of mammary cancer in cages of five female mice.

In the field of endocrinology, Christian[2] reported that grouping of mice results in an enhancement of adrenal gland weight which reflects a stimulation of the hypophysoadrenal system.

Our study was initiated to determine the effect of social environment upon transplanted ascites tumors in mice and to understand the mechanism of the observed phenomenon.

Material and Methods

We use Ehrlich or Krebs II ascites tumors. We graft them upon C_3H, $C_{57}Bl$ or Swiss strains of mice. The mice receive an intra-

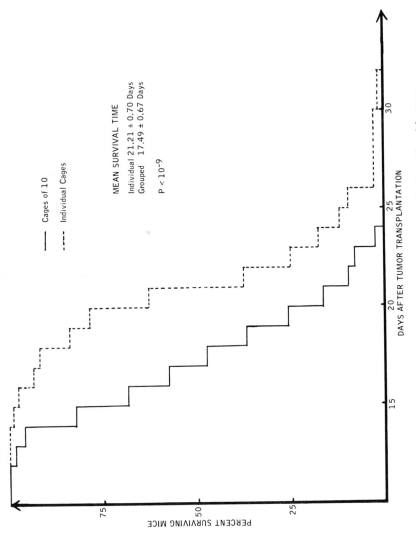

FIGURE 1 Influence of grouping on the mortality of mice grafted with Krebs II ascitic tumor.

314

peritoneal injection of 0.5 ml of cell suspension containing $5-10^6$ to 10^7 cells. The animals, 2-3 months old, are distributed according to a table of random numbers into two groups. Those of the first group are kept in individual cages; those of the second one are housed 10 in a cage. Each experiment groups at least 60 animals. All cages measure $28 \times 18 \times 14$ cm. All animals are kept in the same room at $21° \pm 1°$C. Water and food (UAR pellets) are available *ad libitum*. All mice are weighed daily.

About 2 weeks after tumor transplantation, the mice are killed and weighed. The ascitic liquid is removed, and the body is weighed again. The weight of the ascitic liquid is obtained by difference. Adrenal glands are removed and weighed immediately after the sacrifice of each mouse.

The statistical interpretation is carried out using the average comparison test.

Influence of Social Environment Upon the Mortality of Mice Grafted with Krebs II Ascitic Tumor

For this experiment we use virgin female mice of the C_{57}Bl/6 strain. Each experimental group (grouped versus isolated) includes 80 animals. The dead animals are registered daily at a fixed time. Figure 1 shows the evolution of mortality in the two groups. The average survival time for segregated mice (21.2 ± 0.7 days) is significantly higher ($P < 10^{-9}$) than for grouped ones (17.5 ± 0.7 days).

Influence of Social Environment Upon the Development of Ascites Tumors in Mice

EVIDENCE OF THE PHENOMENON

The total weight of isolated mice becomes higher than that of mice grouped in cages of 10. Figure 2 shows the evolution of Krebs II ascitic tumor in segregated versus grouped Sw female mice. Identical results are obtained with Ehrlich or Krebs II ascites tumor, using mice of C_3H or C_{57}Bl/6 strains, male or female. The weight of control ungrafted mice does not differ for single or grouped ones (Figure 3). The body weight, after removing the ascitic liquid, is the same for the two groups (Table 1 and Figure 3). Thus, the difference observed between the weights of isolated and grouped mice is mainly related to the volume of ascitic liquid, which is significantly larger (about 4 ml) for isolated animals.

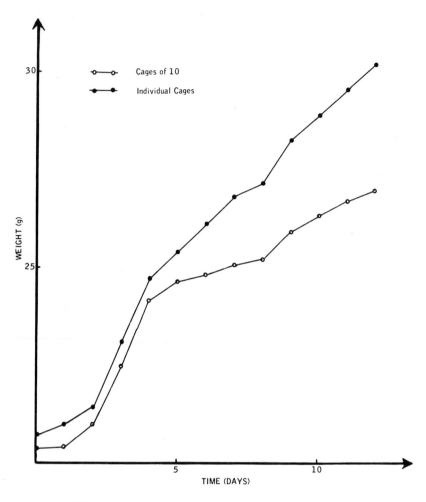

FIGURE 2 Effect of grouping on Krebs II ascitic tumor evolution.

RELATION BETWEEN THE EVOLUTION OF
TUMORS AND MORTALITY

We observe a significantly positive correlation between the survival time of each mouse and its weight at a fixed time of the tumor evolution (for instance 12 days after tumor transplantation). Figures 4 and 5 illustrate this observation, which is found both in grouped mice and in isolated ones. The larger is the volume of ascitic liquid, the longer is the survival time.

316

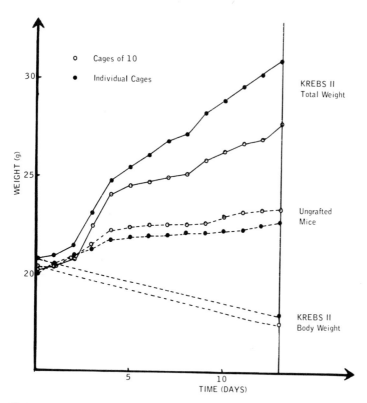

FIGURE 3 Evolution of the weight of mice sacrificed 13 days after tumor transplantation (each punch is the average of 30 determinations).

TABLE 1 Body and Ascite Weight of Grouped versus Isolated Mice[a]

Weight (g)	Isolated Mice	Grouped Mice (10 per cage)	Difference	ϵ	P Values[b]
Total	38.73	33.88	4.85	5.32	$< 10^{-7}$
Body	23.90	23.46	0.44	0.80	NS
Ascite	14.83	10.35	4.48	6.30	$< 10^{-9}$

[a] Mice sacrificed 12 days after Krebs II ascites tumor transplantation.
[b] NS = $P > 0.05$, not significant.

317

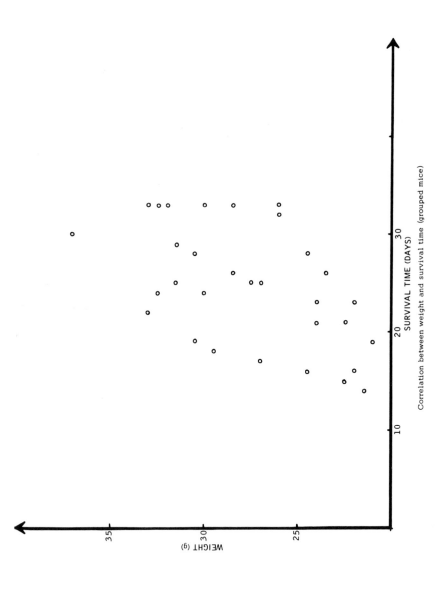

SURVIVAL TIME (DAYS)

WEIGHT (g)

Correlation between weight and survival time (grouped mice)

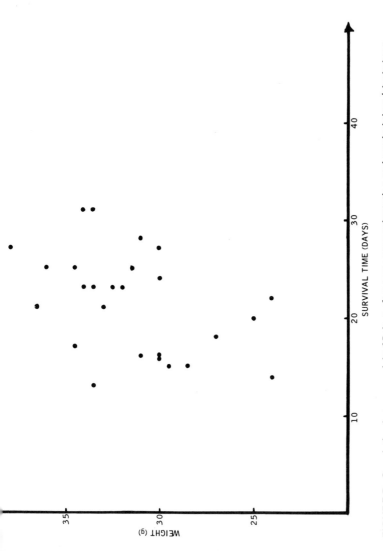

FIGURE 5 Correlation between weight 12 days after tumor transplantation and survival time (singly kept mice).

319

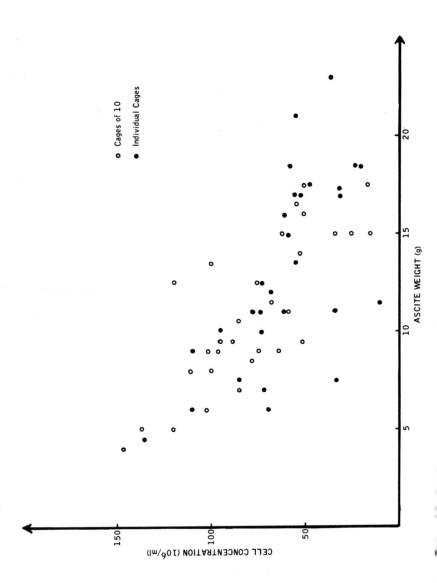

ANALYSIS OF THE PHENOMENON

Influence of Population Density Upon the Number of Tumor Cells

A negative correlation is observed between cell concentration and ascitic liquid volume. At different periods after tumor transplantation (4, 7, 11, and 14 days) we kill the mice and we calculate the cell concentration with a Malassez slide. The larger the volume of ascitic fluid, the lower the cell concentration (Figure 6).

The cell concentration decreases during the tumor evolution, whereas the liquid volume increases. In other words, we observe a progressive dilution of the tumoral cells (Figure 7).

The population density does not appear to affect the total number of tumoral cells. In isolated mice, the liquid volume is larger, but the cell concentration is less than in grouped ones. The product of both (total number of cells) does not differ (Figure 8, Table 2).

Influence of Nutritional Factors

A difference in food intake would be a likely hypothesis to explain the difference of ascitic tumor weight. To test this hypothesis, we weigh daily the average amount of food (pellets) consumed by each mouse. The results are summarized in Table 3. It can be seen that there is no significant difference in the food intake of isolated and grouped mice. To confirm this result we underfeed mice bearing ascitic tumor, giving them pellets for only 2 hours a day. Thus all the animals must eat the same amount of food. We still observe that the isolated animals are heavier than the grouped ones (Table 4).

Therefore, we conclude that the nutritional factors have no influence upon the observed phenomenon.

Influence of Temperature

We demonstrated that the temperature has no influence using the following approaches.

1. Animals were housed in rooms at different temperatures ($17°$-$24°$ C) and did not display any difference in their tumor evolutions.
2. When we put a piece of cotton in the cages, allowing the mice to build a warm nest, we did not observe any variation in the tumor development.

Influence of an Infectious Disease

An infectious factor inhibiting the growth of the ascitic tumor might be postulated. This factor would be present, for instance,

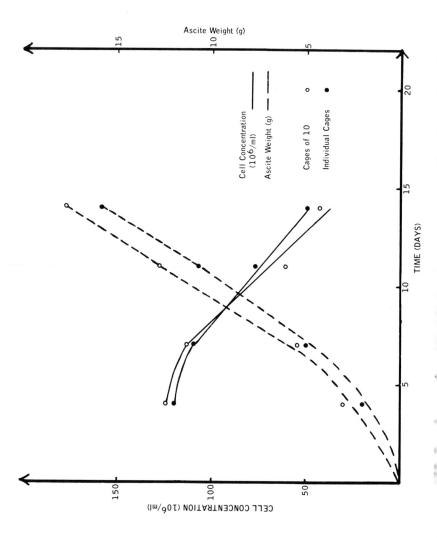

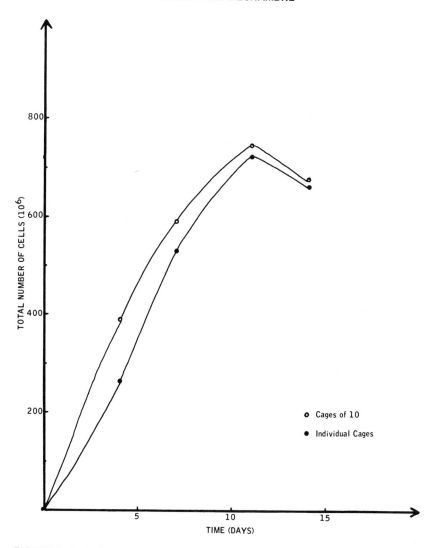

FIGURE 8 Evolution of total number of cells in Krebs II ascitic tumor.

TABLE 2 Total Cell Number 10 Days after Tumor Transplantation

	Ascite Volume (ml)	Cell Concentration (10^6/ml)	Total Number of Cells (10^6)
Isolated mice	10.9	59	640
Grouped mice	7.7	88	675

323

TABLE 3 Daily Consumption of Food of Isolated versus Grouped Mice
(Krebs II Ascitic Tumor)

	18 Days before Grafting	Day of Grafting	10 Days after Grafting	11 Days after Grafting
Weight of isolated mice (g)	26.8	28.2	39.4	42.6
Weight of grouped mice (g)	25.9	27.0	35.8	36.7
P values[a]	NS	NS	$P < 0.03$	$P < 10^{-4}$
Weight of food (g) isolated mice	3.79	5.39	6.53	6.06
Weight of food (g) grouped mice	3.53	4.32	5.76	5.93
P values[a]	NS	NS	NS	NS

[a]NS = $P > 0.05$, not significant.

TABLE 4 Development of Krebs II Ascitic Tumor on Underfed Isolated
versus Grouped Mice

	2 Days before Grafting	Day of Grafting	10 Days after Grafting	11 Days after Grafting
Weight of isolated mice (g)	22.4	22.7	28.2	28.8
Weight of grouped mice (g)	22.9	22.6	25.4	25.1
P values[a]	NS	NS	$P < 0.04$	$P < 0.01$
Weight of food (g) isolated mice	2.93	2.94	2.64	2.71
Weight of food (g) grouped mice	2.90	2.84	2.99	2.60
P values[a]	NS	NS	NS	NS

[a]NS = $P > 0.05$, not significant.

only in 10 percent of the mice. In grouped mice this factor would contaminate
all animals in the cage, inhibiting the growth of their ascitic tumors, whereas
in segregated mice only one animal out of ten would display such a decrease
in its ascitic liquid volume.

We grafted Krebs ascitic tumor in grouped mice. During the following 5
days, the mice remain together, i.e., subjected to the hypothetic contamination,
After 5 days, half the mice are isolated. Figure 9 shows that there is no difference between mice that are first grouped and then segregated ($10 \rightarrow 1$) and
mice that are always segregated ($1 \rightarrow 1$). This experiment rules out the influence of an infectious factor.

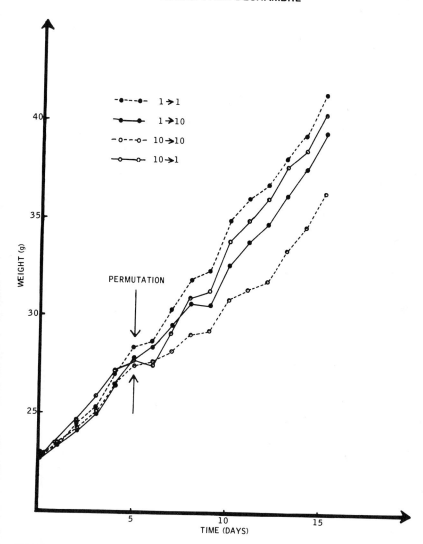

FIGURE 9 Effect of permutation carried out 5 days after tumor transplantation.

Influence of Jostling and Mutual Compression

Nine days before tumor transplantation, 80 mice are grouped and 80 mice are segregated. At the time of grafting, 40 grouped mice are segregated and 40 isolated ones are grouped. We observe no difference (Figure 10) between mice that are first isolated and then, at the time of grafting grouped (1 → 10) and mice that are first grouped and then isolated

325

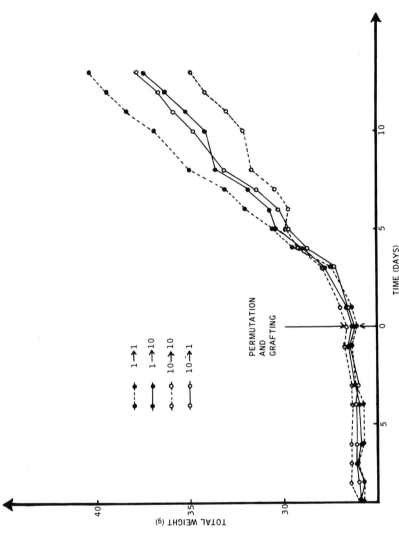

FIGURE 10. Effect of permutations carried out the day of tumor transplantation

$(10 \rightarrow 1)$. This points out that jostling does not interfere with the development of ascitic tumor. Furthermore, we see that mice isolated 9 days before tumor grafting $(1 \rightarrow 1)$ develop a more voluminous ascitic liquid than those segregated the day of the transplantation $(10 \rightarrow 1)$.

The last two experiments establish that the increase in ascitic volume is related to the isolation, which is dominant with respect to the grouping.

INTERPRETATION OF THE PHENOMENON

We have seen that the external factors cannot be retained for explaining the increase in ascitic liquid volume induced by isolation. It is plausible to assume that internal factors, especially endocrine factors, play a major role. Two experiments are performed to test this assumption.

Study of the Adrenal Gland Weight

The weight of adrenal glands is known to reflect the activity of hypophysoadrenal system. In 18 experimental series grouping more than 1,000 mice we notice the following facts:

● Isolated mice always have adrenal glands of higher weight than grouped mice. This difference is, most of the time, statistically significant (Table 5). An adrenal weight increase is also observed for control isolated mice, without ascitic tumor as has been reported by Weltman[7] and Hatch.[4]

● In several series, we observed a positive correlation between the adrenal weight and the ascitic tumor volume of every mouse, whatever its social environment (Figures 11 and 12). We never found a negative correlation.

Stimulation of Adrenal Activity

We inject mice with ACTH, which is known to be a specific stimulating factor of adrenal activity.

Table 6 shows that ACTH-treated mice have heavier adrenal glands and a

TABLE 5 Ascite and Adrenal Weights of Grouped versus Isolated Mice[a]

Weight (g)	Isolated Mice	Grouped Mice	Values of P
Ascite	24.56	18.61	$< 10^{-4}$
Adrenal	6.46	5.82	< 0.01

[a]Mice sacrificed 14 days after Krebs II ascites tumor transplantation.

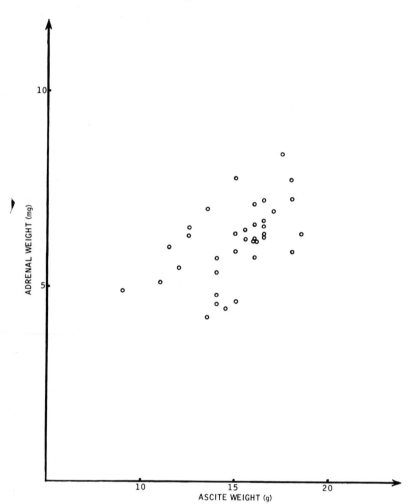

FIGURE 11 Correlation between adrenal weight and ascite weight (grouped mice, 10 per cage) 16 days after tumor transplantation.

more voluminous ascitic fluid than untreated ones. That finding is in accord with the works of several other authors. Takeuchi[6] has shown a parallelism between decrease of adrenal weight and inhibition of Ehrlich tumor following hydrocortisone treatment. Fujiwara[3] found a relation between hypersecretion of ACTH and ascitic tumor development.

328

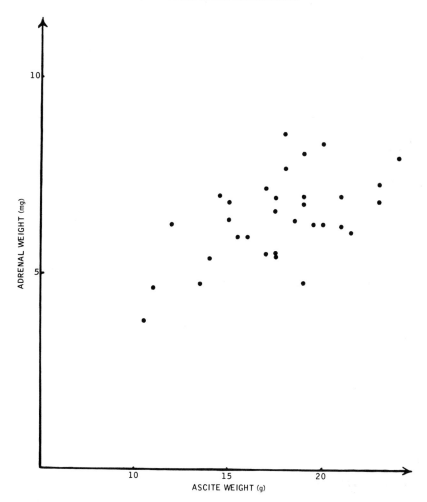

FIGURE 12 Correlation between adrenal weight and ascite weight (singly kept mice).

TABLE 6 Influence of ACTH on Adrenal and Ascite Weights[a]

		Adrenal Weight (mg)	Ascite Weight (g)
Isolated mice	Treated with ACTH	6.99	18.53
	Untreated	6.33	17.41
Grouped mice	Treated with ACTH	6.80	15.53
	Untreated	5.86	14.34

[a]Mice sacrificed 14 days after Krebs II ascites tumor transplantation.

329

Conclusion

We have pointed out that the volume of ascitic liquid is larger (about 4 ml) for mice isolated in individual cages than for mice grouped in cages of 10. This phenomenon cannot be explained by external factors such as feeding, temperature, infectious disease, and jostling. The influence of endocrine factors was established.

The increase of ascitic tumor volume in isolated mice may be explained as follows: The isolation induces a stress that stimulates the hypophysoadrenal system. From this enhanced activity results an increase in the ascitic tumor liquid.

The importance of these facts is obvious:

● It is noteworthy that the evolution of a malignant disease may be altered by psychological agents

● In the experiments of experimental cancerology, using laboratory mice or in therapeutic tests, it is important to place in each cage the same number of animals. Otherwise, the differences observed in tumor evolution may result from differences in population density.

This work illustrates the importance of psychological environment upon the physiopathology of the laboratory mouse.

References

1. Andervont, H. B. 1940. Influence of environment on mammary cancer in mice. J. Nat. Cancer Inst. 4:579.
2. Christian, J. J. 1955. Effect of population size on the adrenal glands and reproductive organs of male mice in population of fixed size. Amer. J. Physiol. 182:292.
3. Fujiwara, S. 1965. Modifications of the adrenal function of mice bearing Ehrlich ascitic tumor (in Japanese with English summary). Sapporo Med. J. 28:273.
4. Hatch, A., G. S. Viberg, T. Balazs, and H. C. Grice. 1963. Long-term isolation stress in rats. Science 142:507.
5. Muhlbock, O. 1951. Influence of environment on the incidence of mammary tumors in mice. Acta Unio Intern. c. C. 7:351.
6. Takeuchi, J., S. Kano, and H. Tauchi. 1965. Influences of age and sex on the effect of hydrocortisone upon the Ehrlich tumour and the adrenal gland. Brit. J. Cancer 19:353.
7. Weltman, A. S., A. M. Sackler, B. B. Sparber, and S. Opert. 1962. Endocrine aspects of isolation stress on female mice. Federation Proc. 21:184.

DISCUSSION

DR. LANE-PETTER: Are the observed differences between 1 and 10 mice per cage due entirely to social causes, or could they be due partly to thermal or density factors, both of which can cause stress?

DR. DECHAMBRE (Dr. Sabourdy interpreting): The density factors do not seem to be so important. He has tested with various numbers of animals, with mice groups, and has not observed much difference.

Mice have been placed in large cages so as to have the same area as they would if they were isolated, and no difference has been observed. In other words, the group mice had at their disposal the same area as isolated mice, and no difference was observed.

Concerning the thermal factor, this has already been partly answered in the paper. Some mice have been put at different temperatures, 17 degrees and 24 degrees, and no differences were observed. In other experiments, mice have been given some nesting materials so as to be able to increase their thermal environment, and no differences were observed in that case.

MISS HUGHES: Have you studied the survival time for isolated mice, groups of 4, 6, and 10? Is there a correlation?

DR. DECHAMBRE (Dr. Sabourdy interpreting): No studies have been made on the survival time in relation to the number of animal groups. Only the difference in weight has been studied, and no differences have been observed, except, of course, differences compared with the isolated mice.

DR. HESTEKIN: What do you use to initially stimulate ascites fluid production?

DR. DECHAMBRE (Dr. Sabourdy interpreting): The ascites is initially stimulated by an injection of tumor cells in suspension— 5 to 10 million in each mouse.

EFFECT OF NOISE IN THE ANIMAL HOUSE ON EXPERIMENTAL SEIZURES AND GROWTH OF WEANLING MICE

W. B. Iturrian

Mice have an acute sense of hearing, and the deleterious effects of explosive noises such as hammering, bell ringing, and banging of metal cages have been recognized in rodent breeding rooms.[25] Furthermore, certain strains of mice are so audiosensitive that convulsions are provoked by intense noises. Susceptibility to audiogenic seizures, as these convulsions are called, is believed to be controlled genetically, and only certain strains of rodents are susceptible. The intensity, frequency, and duration of the sound appear to be important determinants in provoking audiogenic convulsions and interfering with breeding performance. Our experiments indicate an additional parameter.

This discussion concerns the potential of certain housing conditions to produce a marked susceptibility to sound-induced convulsions, to prolong audiosensitivity, and to alter electroshock and chemoshock seizure patterns and the rate of growth in stocks of mice [CAW:CF-1 (SW)] * known to be resistant to audiogenic seizures.[22]

Nearly all mice (over 90 percent) of genetically susceptible inbred strains (e.g., the DBA/2 strain) exhibit seizures when first subjected to sound, provided they are over 15 and less than 40 days of age. Death usually follows the occurrence of a maximal seizure.[24] Animals are considered genetically susceptible if they exhibit a clonic–tonic seizure upon initial exposure to sound.[6,24] The practice of specifying the first trial is not always observed, and realization of this may clarify a portion of the controversial literature concerning sound-induced convulsions.

*Offspring of mice received from Carworth Inc., New City, New York.

Pretest exposure to sound enhances or reduces susceptibility, depending upon the temporal parameters of the treatment. "The classical priming method of physiology has shown the number of animals convulsing varies with the duration of the conditioning stimulus and the condition–test stimulation interval," according to Bevan.[3] Generally, both the duration of the conditioning stimulus and the condition–test interval have been short, being only a few seconds.[3,7] We have demonstrated a dramatic increase in seizure susceptibility in CF#1 mice by using a condition–test interval of days.[12] Seizure susceptibility was markedly influenced by age, prior auditory conditioning, the duration of sound, and especially the interval in days between the initial stimulus and subsequent exposure to sound. Recently, these observations have been confirmed by use of similar procedures in other strains of mice.[5,10,14] This response of sensitized mice has been designated the "Audioconditioned Convulsive Response" (ACCR) to distinguish it from the genetically controlled audiogenic seizure and from nonauditory sensitizing procedures.[13]

Effects of Age and Condition-Test Interval

Sound-resistant strains of mice such as the CF#1 showed a convulsive incidence of about 5 percent upon the first exposure to sound; this incidence was found to be independent of age. However, upon the second exposure to sound, 3 days after the initial exposure, the incidence of convulsive behavior was greatly altered and was found to be profoundly influenced by age (Table 1). Previous auditory stimulation was found to be essential

TABLE 1 Effect of Age and Prior Auditory Stimulation upon Convulsions in CF#1 Mice

Age (days) When Conditioned	Incidence of Convulsions (%)		Number of Animals
	First Exposure to Sound	Second Exposure 3 Days later[a]	
12	0	0	53
20	4	90 ± 4 (tonus 65 ± 7)	63
30	8	31 ± 3 (tonus 3)	61
45	5	3	92
60	2	2	97

[a]± standard error.

333

TABLE 2 Reproducibility of Audioconditioned Convulsions in Six Independent Sample Groups of Animals

Number in Group	Age When Conditioned (days)	Condition–Test Interval (days)	Total Incidence (% of total)	Clonus (% of total)	Maximal Seizure (% of total)	Death (% of total)
A = 50	18	2	86	26	60	20
B = 60	18	2	91	29	55	15
C = 36	18	4	45	8	37	16
D = 38	18	4	42	3	33	19
E = 23	22	2	48	13	22	9
F = 33	22	2	46	18	24	9

for the genesis of convulsions. Twenty-day-old mice audiosensitized 3 days previously exhibited 90 percent seizure activity; a high percentage of the animals tested had maximal seizures. At 30 days of age, incidence had decreased greatly, and the seizures were predominately clonic.

Seizure activity begins, after a latent period of about 6.6 sec, with an explosive burst of wild running that continues for 3–5 sec. This episode ceases abruptly with the mouse leaping into air (and occasionally emitting an audible cry), after which convulsive spasm, clonic convulsion, or clonic–tonic convulsion occurs. The tonic convulsion is characterized by hind-leg extension; the animal may die after such a tonic, or maximal, seizure. The primary objective of our research has been the assessment of these maximal convulsions in relation to experiments in behavioral physiology and pharmacology.

To ascertain the relative reproducibility of the seizures, animals of the selected age were randomly assigned to various groups. They were audiosensitized and exposed to the test stimulus 2 or 4 days later. The observed frequencies of seizure components are recorded in Table 2. The incidence, pattern of seizure, and uniformity of response between groups of a particular age and condition–test interval appear to be remarkably constant.

Sound-induced seizures have not been standardized because so many physiological and psychological factors alter incidence and severity that it is difficult to duplicate experiments quantitatively.[8] The primary difficulty may be that the stimulus has lacked sufficiently precise definition. Little concern has been directed toward specifying the physical properties of the sound stimulus or clarifying the significance of nonauditory characteristics. Fuller and Wimer[8] suggest that the physical dimensions of the stimulus are not critical determinants, provided it is sufficiently intense and of proper frequency range.

We have found, however, that the characteristics of the sound inducing the audiosensitivity influence the incidence and severity of subsequent seizures. Therefore, mere measurement of sound intensity may not be sufficient for standardization from bell to bell, as other characteristics apparently are important in determining the seizure pattern. Furthermore, after extensive use, a bell apparently changes frequency without affecting intensity, as shown by decreased incidence of maximal seizures (Figure 1). The latency of the different seizure components was also altered as the bell changed tone. An increasing latency is correlated with decreasing seizure susceptibility and severity.

Despite the variety of factors that affect seizure susceptibility and severity, it is possible to standardize the procedure for sound-induced convulsions if proper precautions are taken. Therefore, each bell was tested and was checked periodically to provide a reference for responses throughout the investigation. The criteria for standardization were seizure incidence of approximately 90 percent with 60 percent maximal seizures among 20-day-old CF#1 mice tested 48 hours after audiosensitization. An electric doorbell (6 cm in diame-

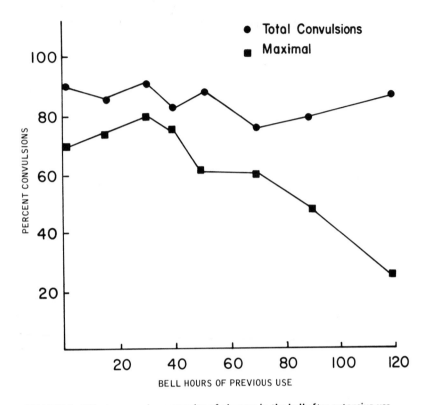

FIGURE 1 Effect upon seizure severity of changes in the bell after extensive use.

ter) driven by a 3-V battery will produce approximately 95 db (relative to $2 \times 10^{-4} dyn/cm^2$) in the glass testing chamber (25 cm in diameter and 15 cm deep). Bells were discarded when they failed to operate as specified. The severity of the response is affected by the characteristics of both the audioconditioning stimuli and the test stimulus.

Environmental conditions must also be rigidly controlled if results are to be reproducible. Mice of the selected age are obtained from dated pregnancies and raised in an isolated location with controlled lighting, humidity, and temperature.[13] Extreme caution is necessary to protect the animals from environmental sound.

The audioconditioned convulsion was characterized in detail at the period of greatest susceptibility to maximal seizures. A total of 1,182 animals were divided into subgroups of at least 10 mice each and subjected to an initial 60 sec of sound stimulus (audioconditioning). The mice were subsequently exposed individually to a second sound stimulus 1, 2, 3, 4, or 5 days later (condition–test interval). Seizure incidence was markedly influenced by the age of the animal when first exposed to sound (Figure 2). Mice audioconditioned at 12 days of age or less did not exhibit seizure upon the second exposure at any condition–test interval. The incidence of convulsive behavior increased in animals conditioned at 13 days of age, peaking at 18 or 20 days of age and declining thereafter. A 2- or 3-day condition–test interval was also found to contribute to high incidence. No difference in seizure susceptibility between sexes was noted, nor does season appear to affect the incidence of convulsive response.

In the non-drug-treated animals, death occurred only after a maximal seizure. The highest death rate occurred in subjects tested at 20 days of age: 20 percent of these succumbed. The mechanisms of susceptibility to seizure and capacity for recovery appear to be independent (Figure 3). The death risk after seizure was higher with a 1- or 5-day condition–test interval than with an interval of 2 or 3 days.

Effect of Animal-House Noises

CF#1 mice audiosensitized at age 18 days and tested at age 20 days displayed 90 percent ACCR susceptibility. A third exposure at age 30 days produced 42 percent convulsions; however, if the mice are not exposed to the test at age 20 days, only 5 percent convulse when tested on day 30. Once an overt convulsion has been elicited by sound, the susceptibility to sound-induced convulsions persists for a prolonged period. Susceptibility can persist for several weeks if the animals are repeatedly exposed to the bell at 2-day intervals (Figure 4). The fact that seizure susceptibility persists once an animal experiences a convulsion indicates that audiosensitivity and seizure may reflect separate mechanisms. Thus, to assess a causal relationship between convulsive activity and changes in neurophysiological or neurochemical processes accurately, all secondary changes due to a previous convulsion must be partitioned.

Unless mice are subjected to a second bell sound they show a transitory audiosensitivity that lasts about 4 days (Table 3). This period of audiosensitivity (as measured by susceptibility to maximal seizures) can be profoundly influenced by environmental situations that may occur during housing of the mice. If mice audioconditioned at age 18 days and tested 5 days later are exposed to environmental noise at age 20 days, markedly different results are obtained. In particular, impact noises, such as hammering on metal, or various other noises, such as dogs barking, are most effective in prolonging seizure susceptibility, even though they may be brief in duration and not sufficiently intense to produce seizure activity in the 20-day-old mice. An electric drill

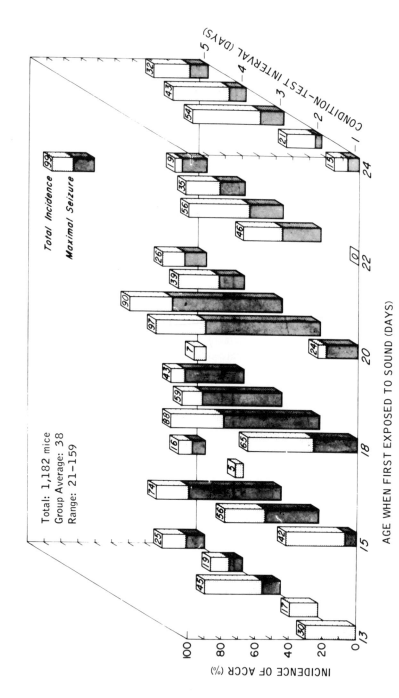

FIGURE 2 Profile for seizure susceptibility among CF #1 mice on the second exposure to sound. Seizure incidence on the first exposure to sound was less than 5 percent, and those subjects were not tested further.

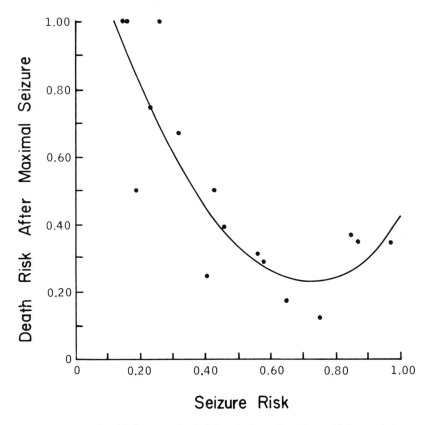

FIGURE 3 Relationship between death risk and seizure risk. The multiple correlation coefficient is 0.8748, and the quadratic component is significant at $P \leqslant 0.01$.

employed in the animal room or a garbage can lid carelessly banged results in over 90 percent seizure incidence instead of the 7 percent expected at the fifth day condition–test interval (Table 4).

Extraneous noise in the animal room can also serve as an audioconditioning stimulus in a manner similar to the initial exposure to the bell. The intensity of the extraneous conditioning stimulus may influence the incidence and severity of seizures produced by the test bell (Table 5). A galton whistle (95 db) adjusted to two different frequencies demonstrated different effectiveness and differences in latency of the convulsions.

Finally, it should be noted that a single 60-sec audioconditioning exposure to the bell has a marked effect upon the growth of weanling mice even though no overt seizures occur (Figure 6). The effect is most apparent 4 days after audiosensitization. The gross appearance changes and the coat becomes un-

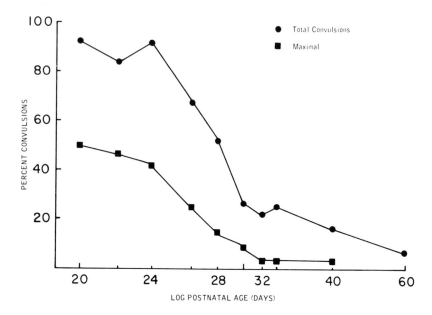

FIGURE 4   Effect of intermittent stimulation on seizure pattern. The stimulus was repeated every 2 days for eight trials at 40 and 60 days of age.

TABLE 3   Profile for Seizure Susceptibility among CF#1 Mice Subjected to an Initial Bell Sound at 18 Days of Age and Tested 1 to 6 Days Later

Condition–Test Interval (days)	Number of Animals Exposed	Incidence of Convulsions (%)[a]	
		Maximal Response	Total
0	416	1	3 ± 1
1	23	26 ± 1	65 ± 10
2	208	57 ± 6	88 ± 4
3	36	47 ± 12	58 ± 9
4	74	35 ± 9	43 ± 9
5	54	0	7 ± 4
6	21	0	5 ± 5

[a] ± standard error.

340

TABLE 4 Modification of Seizure Susceptibility among Mice Initially Exposed to Various Sounds at 18 Days of Age and Tested for Seizures 5 Days Later

Intertrial Stress at Age 20 Days	Number of Animals Exposed	Incidence of Convulsions (%) (2nd Exposure to Bell)	
		Maximal Response	Total
Controls	54	0	7
Concrete drill noise	14	21	93
Galton whistle	14	57	86
Banging can lid	10	50	90
Barking dogs	12	75	75
Barking dog	11	45	64
Bell	43	51	92

TABLE 5 Effect of Noises in the Animal Room in Eliciting Seizure Susceptibility

Source of Initial Sound[a]	Number of Animals Exposed	Incidence of Convulsions (%) (Exposure to Bell 2 Days Later)	
		Maximal Response	Total
None	21	0	5
Galton whistle			
setting 8	11	9	27
setting 16	11	37	37
Fire Alarm	18	67	77
Alarm clock	8	13	25

[a]Duration of initial sound approximately 1 min.

kempt. The mice recover and mate and reproduce in an apparently normal manner. Noxious stimuli are known to produce changes in appetite[20]; endocrine secretions are probably also affected.

Lane-Petter[16] has suggested that music could mask the deleterious effects of cage-changing and other disturbing noises upon reproduction. It is not known if music affects the period of audiosensitivity or the rate of growth.

However, if the duration of the initial exposure to the bell was increased from 60 sec to 48 hr and the CF#1 mice were tested after a 48-hr condition–test period, none of the animals convulsed (Table 6). Nor did testing these subjects a third time result in seizures. However, a genetically audiogenic seizure-susceptible strain (DBA/2J)* subjected to a continuous bell sound from age 10 days to age 18 days responded with a maximal seizure, provided the bell had been turned off for one hour prior to testing. Mice less than 12 days of age, including seizure-susceptible strains, do not exhibit sound-induced seizures. No seizures were observed in any of the mice during the period of continuous sound, a fact that is attributed to masking.[1] Apparently, an important but un-recognized relationship exists between studies of nonseizure audiogenic stress and those of audiogenic crisis.

TABLE 6   Effect of Duration of the Audioconditioning Stimuli upon Seizure Incidence in DBA/J and CF#1 Mice

Age (days) When Conditioned	Duration of Conditioning Stimuli	Incidence of Convulsions (%)	Number of Animals
CF#1			
18	10 sec	84[a]	26
18	60 sec	88[a]	181
18	48 hr	0[a]	23
		0[b]	23
DBA/J			
10	200 hr	100	18

[a] Second exposure 2 days after initial stimuli.
[b] Third exposure 2 additional days later.

Effect of Audioconditioning on Electroshock and Chemoshock Seizures

We were curious about the possible effect of audioconditioning on other experimental seizures commonly used in the pharmacology laboratory. Twenty-day-old mice that had previously been audioconditioned (60 sec at 95 db) were subjected to a battery of seizure-evoking procedures, and their responses were interpreted in terms of neuro-physiological mechanisms. The effect of maturation on the development of sound-induced, chemoshock, and electroshock convulsion threshold and pat-tern was also studied. The pattern of maximal seizures induced by electro-

*Offspring of mice received from The Jackson Laboratory, Bar Harbor, Maine.

shock,[23] pentylentetrazol, and sound are remarkably similar at a given age. With maturation, the threshold for low-frequency electroshock[4] and for minimal pentylenetetrazol[2] decreased with increased age and weight (Figure 5). There was no significant difference in maximal electroshock seizure threshold with aging, but the ratio of extension–flexion duration in the seizure pattern decreased markedly with age. Audioconditioning produced profound changes in the ontogeny of low-frequency electroshock and chemoshock thresholds, but maximal seizure threshold and pattern were unaffected (Figure 6). Ontogenic pattern of maximal electroshock seizure was not altered by electroshock alone. However, in sound-sensitized mice, seizure pattern in the second and third maximal electroshocks was altered in that the extensor component was lengthened.

The study of behavioral development has several important as well as fascinating parameters. It is clear from the present investigation that the pre-

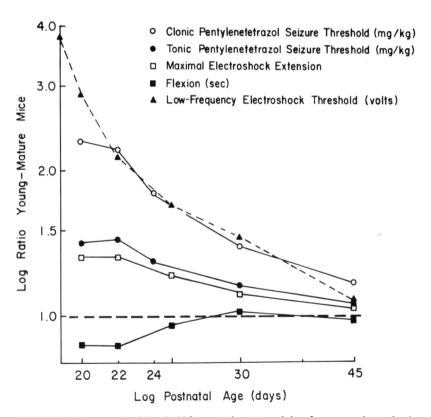

FIGURE 5 Ontogenesis of threshold for pentylenetetrazol, low-frequency electroshock seizures and the pattern of maximal electroshock seizures.

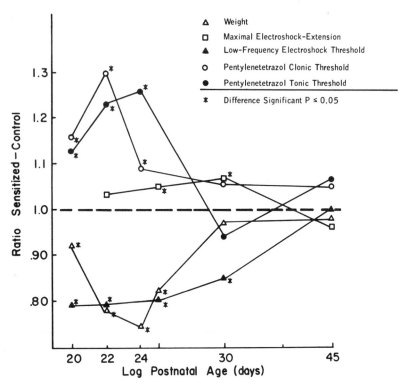

FIGURE 6 Effect of audioconditioning on threshold for pentylenetetrazol, low-frequency electroshock seizures, weight, and maximal electroshock seizure pattern.

juvenile period in mice (age 15–26 days) is characterized by a multiplicity of complex interacting maturational changes. The period may be represented as one of overgeneralized responsiveness, a time when neural organizations of the central nervous system are integrated to permit adaptation to the environment. It is well known that supranormal stimulation may cause morphological, biochemical, and behavioral changes[11] and consequently may alter the pattern of neurological maturation. The development of audiosensitivity after initial exposure to the bell, therefore, may be a disruption in the time-dependent integration of neuronal organization.

Effects of Nonauditory Sensitizing Stimuli

Postural stimulation[17] and several other forms of stress, including anoxia and drugs, were used to help define the relative speci-

344

ficity of sensitization.[13] The most effective stimulus that induced audiosensitivity (as measured by maximal seizure susceptibility 2 days later) involved audible sound. Anoxia, however, also had some effect. The incidence of death upon the second exposure to the bell (4 days after audiosensitization) was greatly increased when anoxia was the conditioning stimulus; over 50 percent of these animals died. The relationship between anoxia and sound-induced convulsions warrants further investigation. This view is strengthened by the observation that respiratory arrest and death are not results of increased seizure susceptibility.

Mice subjected to postural disequilibrium exhibited slightly increased incidence of clonic seizure activity not demonstrated in drug-preconditioned individuals. Placing mice on a 1-cm film of mercury for 60 sec produced long-term effects on body weight and gross appearance of the mice, but none of these animals convulsed during the initial exposure to the bell. This suggests separate neuronal mechanisms affecting growth and seizure susceptibility.

Effect of Drugs on Audioconditioning

Theoretically, pharmacological protection from the effects of audioconditioning should be afforded by several classes of drugs, the pharmacodynamic mechanisms of which might elucidate the mechanism(s) of the facilitated audiosensitivity. Such protection might result from several contributing factors: blockade of the response to the initial bell; inhibition of the unknown neuronal, endocrine, and metabolic processes responsible for the slow development of subsequent audiosensitivity; impairment of hearing and possibly other sensory organs; and modification of any one of the numerous factors involved in seizure production itself.

Prototypes of several drug classes were evaluated. After each drug was administered, the mice were audioconditioned during its peak activity. Two days later, the mice were exposed to the second bell; if the drug had protected the mice from the effects of the initial sound, no convulsions would occur when the second stimulus was presented. None of the drugs tested effectively protected the mice from the effects of audioconditioning. Several drugs, mainly tranquilizers, prolonged seizure latencies and reduced the incidence of maximal responses. It is difficult to determine whether the apparent decrease in seizure severity was due to impairment of the audioconditioning process or to residual anticonvulsant or tranquilizer impairment of the seizure response. Because of the reduced metabolic and excretory potential of young mice, the latter remains a possibility.

When the interval between medication and testing was 3 days, none of the drugs effectively altered seizure incidence. Indeed chlorpromazine and chlor-

345

diazepoxide increased the occurrence of fatal convulsions. Phenobarbital given shortly (90 min) before an experimentally produced seizure is a potent anti-convulsant.[22] Rümke,[19] however, has reported a "proconvulsant" effect 2 days after phenobarbital administration. Thus the observed effects of a drug may depend on when observations are made, in terms not only of hours but of days.

Effect of Drugs on Seizure Response

A multitude of pharmacological agents promote or inhibit the convulsion induced by the second exposure to the bell. In general, convulsants promote seizures, while anticonvulsants, sedatives, and tranquilizers inhibit the onset and severity of seizures. The duration of the latent period prior to the onset of running or tonic extension and the duration of tonic flexion seem to offer the best measures of drug modification of seizure severity. The ACCR of young mice is apparently modified by drugs in the same manner as the audiogenic seizure in genetically susceptible strains.

ACCR in Other Age Groups, Strains, and Species

In animals in which ACCR has been demonstrated, the optimum condition–test interval (i.e., the interval most conducive to maximal seizures) varies greatly in different age groups, strains, and species. We have found that the condition–test interval changes from strain to strain in young mice (Table 7). Our current investigations suggest that ACCR occurs also in mature mice and that susceptibility may be cyclic. The Sprague-Dawley rat also exhibits ACCR, but the optimum condition–test interval is difficult to characterize since it varies between litters within the strain. All litters from a given breeding pair possess the same effective condition–test interval, however. Testing rats on alternate nights with combinations of sound and flashing light apparently enhances convulsive behavior.[9] The adolescent baboon can be convulsed by flashing light, and this seizure response also appears to be a function of the interval (in days) between test sessions and possibly of some general cycle as well.[15]

Conclusions

It is surprising that a time–space parameter in animal responsiveness, such as the ACCR, has only recently been recognized. The

TABLE 7   Variation among Strains in the Sensitization–Test Interval Effective in Inducing Maximal Seizures

Strain	Incidence of Maximal Seizures (%)	Sensitization–Test Interval (days)[a]
CBA/J[b]	30	3
	80	4
	75	5
C$_{57}$Bl/6[c]	25	6
	35	7
	21	8
CF-1	68	2
	65	3
	15	4
Swiss-Webster[c]	22	2
	15	3

[a] Mice audiosensitized at 20 days of age.
[b] Offspring of mice received from the Jackson Laboratory, Bar Harbor, Maine.
[c] Offspring of mice received from The Small Animal Science Department, Oregon State University, Corvallis, Oregon.

occurrence of this phenomenon, however, suggests that research domains such as "temporal pharmacology" and "temporal toxicology" may become as important in elucidating the so-called individual differences in laboratory animals as present-day research into biological rhythms[21] and the psychophysiology of early experience.[18]

In the meantime, on a more immediate and applied level, our experiments on audiosensitivity suggest that animal room noise is a variable that is generally uncontrolled in most experiments in comparative medicine. It is a variable that must be controlled if uniform results are to be achieved, especially in the behavioral sciences.

Acknowledgments

The author wishes to thank Dr. H. D. Johnson for helpful criticism in the preparation of this paper. The advice of Dr. G. B. Fink is greatly appreciated.

The work reported in this paper was supported in part by NIH Biomedical Grant and USPHS Grant ES00040.

References

1. Arnold, M. B. 1944. Experimental factors in experimental neurosis. J. Exp. Psychol. 34:257–281.
2. Bastian, J. W., W. E. Krause, S. A. Ridlon, and N. Ercoli. 1959. CNS drug specificity as determined by the mouse intravenous pentylenetetratrazol technique. J. Pharmacol. Exp. Therap. 127:75–80.
3. Bevan, W. 1955. Sound precipitated convulsions: 1947–1954. Psychol. Bull. 52:473–504.
4. Brown, W. C., D. O. Schiffman, E. A. Swinyard, and L. S. Goodman. 1953. Comparative assay of antiepileptic drugs by psychomotor seizure and minimal electroshock threshold test. J. Pharmacol. Exp. Therap. 107:273–283.
5. Collins, R. L., and J. L. Fuller. 1968. Audiogenic seizure prone (asp): A gene affecting behavior in linkage group VIII of the mouse. Science 162:1137–1139.
6. Fuller, J. L., C. Easler, and M. E. Smith. 1950. Inheritance of audiogenic seizure susceptibility in the mouse. Genetics 35:622–632.
7. Fuller, J. L., and M. E. Smith. 1953. Kinetics of sound-induced convulsions in some inbred strains of mice. Amer. J. Physiol. 172:661–670.
8. Fuller, J. L., and R. E. Wimer. 1966. Neural, sensory and motor functions, p. 609–629. In E. L. Green [ed.] Biology of the laboratory mouse. McGraw-Hill, New York.
9. Goldberg, D. 1952. Audiogenic seizures and related behavior in the albino rat. M.A. thesis, Emory University, Atlanta, Georgia.
10. Henry, K. R. 1967. Audiogenic seizure susceptibility induced in $C_{57}Bl/6J$ mice by prior auditory exposure. Science 158:938–940.
11. Himwich, W. A. 1962. Biochemical and neurophysiological development of the brain in the neonatal period. Internat. Rev. Neurobiol. 4:117–158.
12. Iturrian, W. B., and G. B. Fink. 1967. Conditioned convulsive reaction. Federation Proc. 26:736.
13. Iturrian, W. B., and G. B. Fink. 1968. Comparison of bedding material: Habitat preference of pregnant mice and reproductive performance. Lab. Anim. Care 18:154–158.
14. Iturrian, W. B., and G. B. Fink. 1968. Effect of age and condition-test interval (days) on an audio-conditioned convulsive response in CF#1 mice. Develop. Psychobiol. 1:230–235.
15. Killiam, K. F., E. K. Killiam, and R. Naquet. 1967. An animal model of light sensitive epilepsy. Electroenceph. Clin. Neurophysiol. 22:497–513.
16. Lane-Petter, W. 1963. Animals for research, Academic Press, New York, p. 1–20.
17. Naruse, H., M. Kato, M. Kurokawa, R. Haba, and T. Yabe. 1960. Metabolic defects in a convulsive strain of mouse. J. Neurochem. 5:359–369.
18. Newton, G., and S. Levine. 1968. Early experience and behavior, Charles C. Thomas, Springfield, Ill. 785 p.
19. Rümke, C. L. 1967. Some remarks on the pharmacology of anaesthia. p. 557–567. In M. L. Conalty [ed.] Husbandry of laboratory animals. Academic Press, New York.
20. Sackler, A. M., and A. S. Weltman. 1963. Endocrine and behavioral aspect of intense auditory stress. Colloq. Int. Cent. Nat. Recherche Sci. (Paris), 112:255–288.
21. Sollberger, A. 1965. Biological rhythm research, American Elsevier, New York. 461 p.
22. Swinyard, E. A., A. W. Castellion, G. B. Fink, and L. S. Goodman. 1963. Some

neurophysiological and neuropharmacological characteristics of audiogenic-seizure susceptible mice. J. Pharmacol. Exp. Therap. 140:375–384.

23. Toman, J. E. P., G. M. Everett, and A. H. Smith. 1955. Use of electroshock seizure latency for pharmacological testing in mice. Federation Proc. 14:391.

24. Vicari, E. M. 1951. Fatal convulsive seizures in the DBA mouse strain. J. Psychol. 32:79–97.

25. Zondek, B., and I. Tamari. 1967. Effects of auditory stimuli on reproduction. p. 4–19. *In* G. Wolstenholme and M. O'Connor [ed.] Ciba Foundation Study Group No. 26, Little, Brown and Co., Boston, Mass.

DISCUSSION

DR. STOWE: Was hypervitaminosis A attempted to reduce the incidence of convulsions?

DR. ITURRIAN: I have not tried hypervitaminosis A. I have tried B_{12} and most of the other vitamins. To induce a vitamin deficiency requires time, and it takes a young animal a long time to recover. Theoretically, this tool should be very nice for studying this type of thing. However, there is a paradox; there is no correlation between the pharmacological response of weanling mice and that of adult mice. A drug that acts as a tranquilizer for an adult mouse is quite often a convulsant for a weanling mouse. The anticonvulsant doses required on these animals for even electroshock is 20- or 30-fold what is required in the adult. The convulsant doses of drugs, however, are a tenth of what you would expect from any logical body-size surface or metabolic rate calculation. So, the real problem here is that because of the immaturity of the system, you have to do time-spaced studies. This is the reason I have been looking for this type of responsiveness in adult animals, rather than trying to work with weanlings.

DR. POPE: After an accumulated 120 hours of 95 decibels, is there damage to the auditory apparatus? Can the mouse still hear, or can it hear as well?

DR. ITURRIAN: This would be a very difficult question to prove conclusively. I did not show the data in my paper. I know you are interested in the DBA strain, but I do not have that information. In the CF_1 mouse that is going to be sensitized, however, if the bell is rung for a long time, the animals do not convulse. I can give further demonstration of the fact that they are not deaf in that if I make the condition-test interval less than 12 hours, this interferes with convulsions that would be produced at the third exposure to the bell.

DR. LANE-PETTER: Have you found any variation in audiosensitivity at different times of day, and have you found it to be influenced by social factors?

DR. ITURRIAN: Yes, more in terms of anticipation than pilot study, we do all of our investigations in the early evening in the twilight hours of the day to have maximum responsiveness. At least this is when the best response occurs in the sensitive strains.

DR. LANE-PETTER: Have you also investigated "beat rhythm" noise?

DR. ITURRIAN: Beat rhythm noise has already been investigated by others; back in the 1940's there was a group interested in what we call reflex epilepsy. Unfortunately, back in 1955 there was a meeting of the psychologists who were working in this field at the time. Because there were a lot of contradictory statements in the literature, the psychologists came up with a set of terminology that cannot be evaluated today as they decided, by committee, that previous exposure to the bell made no difference. So the data are available, and I think you could extrapolate back to it if you were interested.

DR. RICH: What is the frequency output of the bell used?

DR. ITURRIAN: The output of the bell is a mixed frequency. It is a very complex mixture of oscillating sounds. It is very difficult to analyze. It appears from initial studies that 12 to 15 kilohertz are the critical frequencies, but these have to be oscillating back and forth. White noise is not effective at all. Perhaps something that would be directly related to this is that the pain threshold for mice, as these experiments would suggest, is considerably lower than we had previously thought. For example, if you plot the desynchronization of the EEG against intensity of the sound source, you go along with a very flat curve until you get to about 80 decibels, and then the curve climbs logarithmically to about 100 decibels. The animals are being sensitized by things as low as 80 decibels. I chose 90 decibels to get away from the threshold effect.

DR. HARRINGTON: Have you done any work using very long condition-test intervals, that is, weeks or months?

DR. ITURRIAN: Not only do you have to adjust the condition-test intervals, but keep in mind that the maturation system is also changing. Dr. Fox points out in his paper one critical period in the development of mice—that of about five days. I am trying to show another critical period—that of 18 to 22 days. I suggest there is another one at about 50 days. But the interval that is most effective changes with age. The pattern of convulsion also changes.

DR. KELLEY: In a breeding colony would you suggest controlling noise level by avoidance of noise or a constant level of continuous noise?

DR. ITURRIAN: I have tried using background music. I find that my mice still convulse. The incidence, however, is somewhat changed. I did not mention that I have tested several hundred different types of things to induce audiosensitivity. Sound is the only one that is effective, highly effective. Anoxia, 30 seconds in 95-5/CO_2-O_2 does induce some sensitivity. I have not investigated the condition-test interval with anoxia. A very interesting sidelight of this is that the weight-loss experience seems to be carried by a different sensory modality than the audiosensitivity. Dysequilibrium, that is, putting mice on a film of mercury for 60 seconds, produces weight losses almost identical to this exposure to the bell. In fact, if you suspend the mouse in the chamber during audiosensitization, you quite often do not see the weight change. Even though these methods are quite closely related, they are separate neuronal mechanisms.

DR. GALLOWAY: What is the effect of transportation on the production of convulsions in mice used for toxicological studies?

DR. ITURRIAN: I really wish I knew. I have found, however, in buying animals from Jackson Laboratories—animals that I know when I raise them in my own colony will convulse at a specific interval—that the few hours they spend in the airplane seems to be doing the same thing that my long exposure to the bell does. However, if I have them shipped prior to 12 days of age, they convulse in a normal manner. But the effect on drugs is not known. I did mention in my paper that the toxicity of compounds is entirely different in audiosensitized mice than in normal mice. Phenobarbital at very low doses does not kill any of the young mice. If they are sensitized, they will all die.

There is another problem. If phenobarbital, a very potent anticonvulsant, is administered, 90 minutes later it is a good anticonvulsant, but within 3 days after administration it is a convulsant. So in these young mice there are several superimposing problems.

DR. WEISSENBORN: What is the normal level of noise in decibels in the facility in which this work was done? Also, can you relate the noise level in decibels to the distance between the test animal and the source of noise; and then, please indicate what pathological lesions were found in animals that died.

DR. ITURRIAN: The normal level in my quarters is about 70 decibels. In a normal laboratory, using plastic cages with stainless steel tops, it gets up to 85 decibels, 90 decibels. I always keep my audiosensitized mice not in the animal quarters, but rather in my office in an isolated location.

Relate the noise level to the distance. The animal during this running pattern is changing. The figure of 90 decibels that I use here is an average. It will change as much as 5 decibels at different places as the mouse moves during the running phase of the convulsion. If you stop the bell immediately after the convulsion has proceeded into the wild uncoordinated phase, the animals will run and go into a typical tonic–clonic convulsion. In fact, this is a very good method if you want none of the animals to die. Neuropathological lesions are not found.

DR. BARTH: Do you consider the cage as a sound barrier?

DR. ITURRIAN: I have done some work with this, and Maryland Plastics, I believe, has a filter that has been tested. The animals that are held in this seem to show a decreased incidence of convulsions. It is decreased more to the clonic type of behavior. However, I must point out that the severity of the convulsion is affected by so many, many things. Prenatal administration of drugs increases almost uniformly the incidence of maximal convulsions. Handling the mice prior to 18 days of age will increase the incidence of convulsions. If you put a nonsensitized mouse in with a DBA mouse, he will convulse. In fact, I have some data with rats that would suggest that perhaps this is a defense mechanism, a warning system. I have shown, by holding a rat that has convulsed in a separate chamber while it is going through this cyclic pattern, during which it emits a cry at about 3 per second, other rats will go into convulsions. This leads me to believe that this held rat is emitting a distress signal.

THE SIGNIFICANCE OF
THE PHYSICAL ENVIRONMENT
FOR THE HEALTH AND
STATE OF ADAPTATION
OF LABORATORY ANIMALS

Wolf H. Weihe

Introduction

In the Constitution of the World Health Organization health is defined as "a state of complete physical, mental and social well-being and not merely the absence of disease or infirmity."[44] This definition is applicable to laboratory animals, at least mammals, as well as to man. Because their chief function is to serve as models for man in laboratory investigations, we also have to know about the health of laboratory animals if we want to know about the health of man.

In recent terminology health or well-being means fitness for dealing with any stress. An individual in good health can withstand a great amount of stress from conditions and changes in the environment or can drive itself to high performances of many kinds without lasting traces of strain in the body. Fitness is a state of high adaptability. Hence, health represents the ability of an individual to adapt to the environment with ease. To adapt with ease means that there is no strain after stress. Health then is absence of strain, not only because there is no disease but also because there is fitness to deal with the environmental conditions.

In physiological terms, adaptability means that under the influence of stress, homeostasis can be maintained without disarrangement by means of the multiplicity of integrated functions at various levels of organization of the body.[32,34] The continuation of a stress can lead to a new steady state through the process of adaptation. It depends on the quantity of the stress, on the one hand, and on the adaptability of the individual, on the other, whether the stress can be tolerated and a new steady state can be reached. A stress that is mild for an individual with good adaptability may constitute a severe stress that cannot be tolerated for any length of time by an animal with poor adaptability. In the case of good adaptability, leading to adaptation to a given stress, the new steady state is characterized by a return to normalization of physiological functions and by an increased tolerance to this or similar stresses.

Tolerance is based on limits. The limits are those stress conditions at or beyond which adaptability is exhausted and one or more physiological functions fail, so that homeostasis cannot be maintained; the response becomes strain leading to failure, damage, or death.[32,34,37] Applying this concept of limits of tolerance[37] to the definition of health, it can be stated that in healthy animals there are wide limits of tolerance at any state of adaptation, while in diseased animals these limits are narrow.

If there are limits of tolerance to environmental stress, there must be environmental conditions of least strain or comfort. While the limits of tolerance change according to the adaptability of an individual, the conditions of least strain should be constant for a given species, independent of the limits of tolerance of the individual. In practice it is difficult to determine the limits of the comfort zones because they are characterized by a minimum level of response. In order to detect the zones of least strain it is useful to expose a variety of individuals of one species, which will differ in their adaptability, to a stress of increasing quantity. A variety of responses related to the physiological functions involved will be found, ranging from severe damage or failure, at the one extreme, to hardly any response, at the other. The most sensitive indicators in such a set-up would be the individuals with little adaptability because of their narrow limits of tolerance. Conversely, animals with high adaptability would be the least sensitive indicators because of their wide limits of tolerance, or, to use a more common term, their high resistance.

Among laboratory animals, high adaptability is found in physically healthy individuals at the prime of life and low adaptability in very young or old animals, inbred animals, and sick animals with wanted and unwanted diseases. The incidence of unwanted disease is declining in modern animal experimentation because of improved housing conditions and the numerous measures employed to prevent contamination and spread of infections in animal colonies. Wanted diseases are an essential feature of the animal model. They can be inherent functional failures, such as diabetes in mice and Chinese hamsters or jaundice in rats of the Gunn strain, or acquired failures, such as those resulting from contamination with micro-organisms.

Adaptability implies the presence of environmental factors. Some environmental factors are always present, so a continuous interaction exists between them and the organism, while other factors occur occasionally or only under special conditions. Permanent factors are those of the biosphere on which the life of the animal depends, e.g., oxygen and food for energy metabolism. Occasional factors are those of an experiment such as the introduction of a particular micro-organism or a chemical substance.

The Physical Environment

The physical environment comprises the climatic conditions of the biosphere. In the particular case of laboratory animals living in captivity mainly under indoor conditions it consists of the climatic conditions within their ecosystem, that is to say, the room climate of the animal room and the cryptoclimate produced by the cage system. Climatic conditions can be conveniently described by means and variations of meteorological elements or factors.[11,26,35] The meteorological factors are usually divided into the classical or nontrivial factors and the accessory or trivial factors. Of the classical factors, those that deserve the greatest attention are the thermal factors of temperature, humidity, air motion, and radiation because they affect the heat budget of the body. The accessory factors are, for example, air ions, air electricity, electromagnetic fields, and fluctuating phenomena. These may have a significant biological effect, but under normal indoor conditions their strength is such that this is comparatively small or absent.[23,33]

In this paper only the thermal factors of temperature, humidity, air motion, and radiation will be considered. They are the dominant factors of the indoor climate. These are the factors that determine the metabolic activity of the animals and all other physiological functions related to thermoregulation and are dealt with by the air-conditioning engineers.

Climatic Adaptation

Climatic adaptation is both genetic and physiological. All animal species are adapted genetically to a particular climate through evolution.[13] Genetic adaptation has made them fit to deal with the common changes in the native climate during their lifetimes. This means that, though the animal must continuously adapt physiologically to climatic changes, its limits of tolerance are such that this can be done without leaving it unfit to cope with other stresses of life. In the case of laboratory animals living in captivity, genetic adaptation takes place inadvertently while breeding under local climate conditions and leads to deviation of the specific genotype. For any species taken into captivity at different places from a natural habitat, in the course of time many Mendelian populations will be formed that are genetically adapted to their given environmental conditions.

The climatic conditions within breeding colonies and laboratories through-out the world vary over a wide range from hot to cold. Suppose that geneti-cally adapted individuals of one species were distributed throughout labora-tories in different climate zones; the responses of the individuals would dem-onstrate a steep gradient of physiological adaptation from cold to heat. It might be argued that differences in genetic and physiological climatic adapta-tion no longer exist because animals are bred under constant temperature con-ditions. This might apply to one country, particularly if its financial means are sufficient to provide standard breeding rooms and laboratories, but it does not apply to the increasing international exchange of animals, as there are different national ideas about the room climate that should be provided.[42] If these na-tional recommendations are strictly followed, climatic tolerance will become particularly pronounced in inbreeding. This may cause much trouble when other climate conditions are provided because the tolerance of inbred animals to climate changes is less than that of outbred animals.[14]

Physiological adaptation is a necessity for life because homeotherms are open thermodynamic systems with input, throughput, and output of energy.[25,40] Much of this energy occurs in the form of heat from oxidative processes during metabolism. As the ambient thermal conditions must be lower than the body core temperature, a continuous flow of heat is carried from the core of the body to the environment. The basic principle of physio-logical temperature adaptation is to reach a thermoneutral steady state be-tween the body and the environment in which heat production and heat loss are balanced while the core temperature remains unchanged.

Homeotherms maintain their heat balance with the environment by the following means of heat regulation:

1. Physical or insulative heat regulation, i.e., regulation of heat transport to the body surface by changes in the peripheral circulation, insulation by depo-sition of a subcutaneous fat layer, and length and thickness of the fur

2. Chemical or metabolic heat regulation, i.e., increased or decreased heat production by the amount and quality of the food intake, muscular work, and shivering and nonshivering thermogenesis

3. Behavioral heat regulation, which is strictly physical heat regulation, i.e., huddling together or isolation of animals, crouching or stretching out on the floor, licking or wetting of the fur by rolling in excreta, etc., to change the effective body surface and heat dissipation.

In small laboratory animals physical heat regulation plays a less important role than in large animals. Heat loss controlled by the extent of the peripheral cir-culation occurs at the relatively small surfaces that are not covered by fur, such as the tail in mice and rats and the ears in rabbits. Well known is the structural adaptation of these bare parts in growing mice and rats. These animals develop short tails in the cold and long tails in the heat.[14] All such species have one trait in common—they are not able to sweat like large animals and man.[6,22]

Heat production of homeotherms depends on the effective body surface, the temperature gradient between the body and the environment, and the surface insulation according to the equation

$$H = A \frac{T_b - T_a}{I},$$

where H is heat production, A = body surface, T_b = body temperature, T_a = ambient temperature, and I = insulation.[17]

Heat production is lowest when at rest and fasting. This heat production is just high enough to provide the required heat for the maintenance of constant body temperature at the given ambient temperature, which is called neutral temperature. The temperature zone in which the minimum metabolic rate or minimum heat production is found is called the thermoneutral zone (Figure 1). Minimum metabolic rate or minimum heat production mean the same thing but refer to different methods of investigation. Metabolic rate is determined

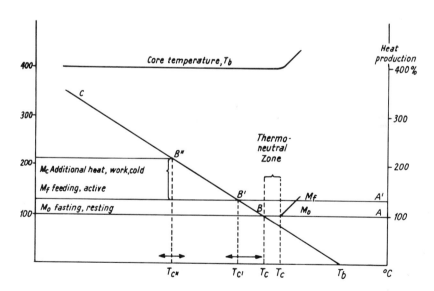

FIGURE 1 Demonstration of the thermoneutral zone $T_c - T_c$ for fasting animals at rest and left shift of the lower critical temperature at T_c', T_c'' with increase in metabolic rate, M_f, M_c.

357

by oxygen consumption, and heat production is calculated from it (indirect calorimetry) or is determined directly (direct calorimetry). The minimum metabolic rate for the fasting animal at rest is expressed by the symbol M_o. It is arbitrarily set as 100 percent in Figure 1 to provide a base line for comparisons between different animal species.[36] The proper definition of heat production would be kcal/(hr·m²·C°) or, indirectly, as mO_2/hr·kg^n.

The thermoneutral zone of M_o is limited by the lower and upper critical temperatures, T_c - T_c. These are the temperatures at which the means of physical functions to maintain homeothermia are exhausted and insulation is at its maximum. Therefore, at lower ambient temperatures there is a linear relationship between temperature on the abscissa and heat production on the ordinate. At higher temperatures heat production increases according to the Van't Hoff-Arrhenius law as a result of a rise in body temperature.

The metabolic rate increases above the M_o level in a regular pattern during the activity and feeding period of the animal to a level M_f from B to B' in Figure 1. The increased metabolic rate at B' causes a shift of the lower critical temperature T_c to $T_{c'}$ on the abscissa of Figure 1. This is easily conceivable because since there is more heat produced by the animal it will tolerate lower ambient temperatures. As a further step, metabolic rate can increase by additional work or under the influence of cold to a level M_c from B' to B''. Consequently, the lower critical temperature will be shifted further to the left from $T_{c'}$ to $T_{c''}$.

The shift to the left of the lower critical temperature with the physiological increase of metabolic rate by the animal's activities such as feeding and exercise is opposed by the enforced increase of metabolic rate from exposure to cold. If the diagram of Figure 1 is read on the abscissa from right to left, the required heat to maintain thermal balance at a certain low temperature can be found on the ordinate at the point of intersection of this temperature with the heat production versus ambient temperature gradient, e.g., B'' in Figure 1.

The lower critical temperature can be extended to a lower value by adaptation to cold. Adaptation can take place either by an increase of metabolic rate as in Figure 1 from B' to B'', which is called metabolic adaptation, or by a reduction of heat loss through better insulation, which is called insulative adaptation. The two kinds of adaptation are demonstrated in Figure 2. With improved insulation the minimum metabolic rate can be maintained at T_c' instead of T_c. The difference T_b - T_c' is greater, and the thermal gradient $T_b B' D$ becomes flatter. Since the increase in insulation reduces the heat flow from the animal, it also has the effect of shifting the upper critical temperature to the left, i.e., towards lower environmental temperatures.

When the metabolic adaptation has been accomplished, the animal lives at a permanently increased metabolic rate above the M_o level. If such animals are brought back to thermoneutrality temperatures, their metabolic rates will be

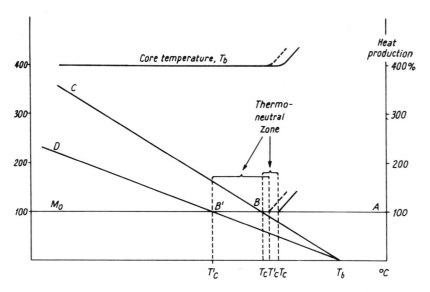

FIGURE 2 Demonstration of metabolic adaptation *BC* and insulative adaptation *B'D*, causing left shift of T_c to T_c' at lower ambient temperature.

higher than before, and it will take 2 to 3 weeks for their metabolic rates to come down to the previous levels. While with cold exposure the tolerance to cold increases in animals by metabolic adaptation, their heat tolerance decreases. In general, the increase in tolerance to one stress decreases the tolerance to the opposite stress.

It is of importance to know that insulative adaptation is not observed in laboratory animals of less than 2 kg of weight under conditions of caging in breeding colonies and laboratories.[15] The common response of animals up to the size of young rabbits is metabolic adaptation.[15,16,43] Metabolic adaptation can commence within minutes, starting with shivering thermogenesis and muscular exercise.[17,19] Shivering thermogenesis, which is easily reversible, is slowly converted into nonshivering thermogenesis, which is slowly reversible and represents a well-established adaptation process.

It was mentioned by Lee[28] and repeated by Gelineo[12] that rabbits held at 17°C showed a 27 percent higher metabolic rate at 28°C than controls that had been living at that temperature for long periods. This increased metabolic rate persisted for several weeks. If one assumes that the rabbits at 17°C were at thermoneutrality, one must infer that these animals adapted to heat on exposure to 28°C. It is usually requested that animals be kept for several weeks at such high temperatures as 28–30°C before the M_o is determined.[4] At present it is not possible to say whether the experimentally determined upper temperature of the thermoneutral zone for each animal species does represent physiological conditions of no heat adaptation or whether it is the highest temperature at which permanent heat adaptation is still possible. This is the upper limiting or tolerance temperature because a further increase of ambient temperature will increase core temperature and lead to death.

Below the lower critical temperature, T_c, core temperature can be maintained at a constant value down to the lower limiting or tolerance temperature beyond which core temperature will decline. So, while there is a lower critical temperature, T_c, and a lower limiting temperature, with a wide temperature range in between, there should be an upper critical temperature and an upper limiting temperature. According to present knowledge, the upper critical temperature seems to coincide with the upper limiting temperature.[8]

Limiting temperatures, where core temperature either falls or increases, are provided only under experimental conditions in the laboratory.

The thermoneutral zone also comprises the zone of thermal comfort. For the provision of comfortable conditions for the animals in breeding colonies and laboratories, the lower critical temperature should be known.[8,17,25] In Figure 3 the lower critical temperatures for six species of laboratory animals are plotted on the M_o line, and the heat production versus temperature gradients are drawn for an average core temperature of 38° to 38.5°C.[4] The steepness of the different gradients reflects the influence of insulation on heat

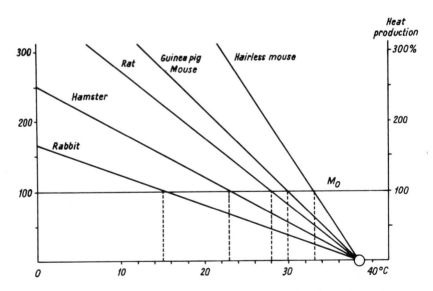

FIGURE 3   Lower critical temperature and thermal gradient for six species of laboratory animals.

loss. The heat loss is largest in small animals because of the little heat capacity of the small body mass and the high body surface–body mass ratio. Among these animals the hairless mouse has the steepest gradient because of the lack of fur for insulation. The opposite applies to the rabbit, which has thick fur and a smaller body surface–body mass ratio. The diagram is based on the investigations of Scholander $et\ al.$[36] (1950) and Hart.[15,17]

If the heat production gradients are extended they will pass the marks where the metabolic rate is doubled or tripled above M_o, and eventually will reach a point of maximum metabolic rate, M_{max}. The maximum metabolic rate can be reached by either physical exercise or extreme cold exposure, or both.[17] Cold is the most stimulating force. M_{max} can be maintained for only short periods of up to 15–30 min, after which the animal is exhausted or core temperature declines. The common laboratory animals cannot exceed an M_{max} of 4–7 times their M_o.[24,43] It may be mentioned here that the figures given for the mean metabolic rate of laboratory animals by Brewer,[5] using factors of 2–3 times M_o values, are definitely wrong.

Cage Systems

These basic physiological mechanisms in animals must now be considered with respect to the actual living conditions of laboratory animals. The indoor living conditions of laboratory animals are determined by the room and the cage system. Consequently, there are two climates, the room climate and the cage climate or cryptoclimate, between which equilibria are formed. The difference between these climates and the speed of reaching equilibrium depends on the cage system. Two cage systems are found, the cage system with free air exchange and the cage system with limited air exchange.

The cage system with free air exchange is made up of cages with all sides of wire mesh or bars through which air can flow from every direction. Cage air can easily be mixed with or exchanged for room air. The cage floors are made of wire mesh, with no bedding, and the only resistance to the air flow is the animal itself. Excreta are removed from below the cage and will never remain in the room for any length of time.

The cage system with limited air exchange exists in many variations. It has a compact floor with bedding material spread out on it and one or more compact walls. The remaining sides, e.g., the walls and top of the cage, consist of wire mesh. Free air exchange is hindered by the compact walls and floor of the cage. The bedding material is used by the animal for insulation in building either a nest or a cover in cold ambient conditions. An animal in such a cage system is partially protected against the room climate and can utilize the available bedding material for the formation of a microclimate inside the cage.

This is effected by the mixing of the excreta with the bedding material result-ing in an increase of water vapor from evaporation and heat production from bacterial decomposition of the excreta.

Thermal Equilibria Between Animal and Room Climates

In discussing the thermal balance between the heat production of the animal and the thermal conditions of the room, three ex-amples can be considered. For simplicity the thermal conditions are expressed as temperature only. The examples are featured in Figure 1.

THERMAL CONDITIONS FOR MAINTENANCE OF THERMAL BALANCE IN THE FASTING AND RESTING ANIMAL = M_o CONDITIONS

At the ambient temperature of the thermoneutral zone T_c - T_c in Figure 1 an accumulation of heat in the animal will occur or an enforced effort to get rid of heat will be made during the activity and feed-ing period. Heat accumulation is associated with an increase in body tempera-ture. To some extent this is a physiological phenomenon during the activity phase.[2] However, in this temperature zone, feeding and activity may become a strain for the animal, and the animal will respond with an avoidance of physical exercise and reduction of food consumption to the minimum neces-sary level. This is a form of heat adaptation. For such an animal there is nei-ther the stimulation of feeding nor that of activity. It is obliged to keep as quiet as possible to avoid overheating by its own vital activities.

THERMAL CONDITIONS FOR MAINTENANCE OF THERMAL BALANCE IN THE FEEDING AND ACTIVE STATE = M_f CONDITIONS

Ambient temperatures suitable for the feeding and active animal are between T_c and $T_{c'}$ in Figure 1. They are cooler than for the resting animal. The extent to which this animal will feed and run about to reach thermal balance depends on the given temperature. It will be easy for the animal to get rid of any additional heat (M_f - M_o) produced from feeding and activity. This animal, however, will have to adapt during the resting phase to the relatively cooler conditions by both behavioristic and metabolic heat regulation. In behavioristic heat regulation the animal will keep the tail cov-ered by its body, will roll up and hide the head to prevent heat loss by de-creasing its effective body surface. If this should not suffice, resting metab-olism will be increased to balance the heat loss.

THERMAL CONDITIONS THAT NECESSITATE METABOLIC ADAPTATION IN THE FEEDING AND ACTIVE ANIMAL = M_c CONDITIONS

Ambient temperatures that will provoke metabolic adaptation with permanent stimulation of activity and increased food intake are between $T_{c'}$ and $T_{c''}$ in Figure 1. The temperature of the environment is permanently lower than that necessary to balance the heat production of the animal during the feeding period, so the animal has to increase its metabolic rate to make up for the heat loss. As the means of behavioristic heat regulation are limited under these conditions, chemical thermogenesis will be activated. More energy is required, and consequently the animal will increase its daily food consumption, and it will be stimulated to be physically more active than animals under the M_o and M_f conditions to gain heat from muscle work. As a result, this animal should be physically fit. On return to M_o conditions its metabolic rate will be above normal for the period of reacclimatization.

Temperature can fulfill the function of a "Zeitgeber"[2] in the particular case of combining M_o and M_f conditions. If the temperature of the environment were to be adjusted to correspond to the activity and resting cycles of the animal, it should be lowered at the beginning of the activity and feeding phase and increased when there is less heat production during the resting phase within the range $T_c - T_{c'}$ of Figure 1, as indicated by the horizontal arrow. Such an animal would find the thermal environment almost perfectly suited to its life pattern and would not be hindered in active life as under M_o conditions; nor would it have to make much use of behavioristic heat regulation as under M_f conditions. These conditions would be of least stress. From the practical point of view, of course, it would be extremely expensive to provide such conditions in an animal room.

The thermal conditions of the three examples would be similar for animals in cage systems with limited air exchange, but the dimensions would differ. Animals would suffer more under M_o conditions if there were a cryptoclimate inside the cage with warmer thermal conditions than the reference temperature in the room. If there is a heat accumulation inside the cage it would be better for the animals to be kept under M_f room temperature conditions. As a result of the formation of a cryptoclimate, which is always warmer than the room climate because of the heat produced by the animal, there may still be M_o conditions in the cage when there are M_f conditions in the room. In any case, the animal has a much greater chance to make full use of behavioristic heat regulation, such as covering itself with insulating bedding material during the resting period, and would not be forced to increase its metabolic rate during rest like the animal in the cage system with free air exchange. Therefore, animals in cage systems with limited air exchange and bedding provided are more protected against heat loss in cooler conditions but more stressed under warmer conditions. Depending on the difference between the room climate and the cryptoclimate, there will be a metabolic difference between animals held in such cage systems and those in cages with free air exchange.

Recommended Comfort Temperature Range for Minimum Metabolic Adaptation

From the previous considerations it must be concluded that the temperatures around the lower critical temperature of the feeding and active animal (M_f) should be the most satisfactory ones for maintenance.[30]

The question arises as to where the lower critical temperatures for active animals of the different species lie. Two averages must be calculated, the length of the activity phase and the increase of metabolic rate during this time. If the daily length of the activity period is taken as 12 hours, the mean increase of metabolic rate will be about 30 percent above the M_o values. This increase will be as high as about 50 percent in the case of very active and fast growing animals, or below 30 percent in the case of inactive and old animals. An increase of less than 30 percent also occurs in animals that suffer from diseases associated with reduced appetite, have had operations or extirpation of endocrine glands involved in metabolism, or are forced to inactivity.

The critical temperature range between M_o and M_f, using the +30 percent increase as a mean, is indicated in Figure 4 by the hatched fields. Depending on the steepness of the heat production gradient for each animal species, the fields are either round, as for hairless mice, or flat, as for rabbits. They are more extended on the ordinate for mice and rats because these species can be very active, at least for some periods of their life. Though mice and guinea pigs show the same lower critical temperature, the increase of metabolic rate with feeding and activity seems to be less in guinea pigs than in mice, so the field for this species alone should actually be smaller. Though these temperatures seem rather high for guinea pigs, they are based on the information given in the literature.[17,20]

For the guinea pig the heat production gradient will actually be flatter than for the mouse in spite of similar T_c's for both species.[19] This is due to the difference in core temperature T_b which is about 39°C for the guinea pig and 38°C for the mouse. Therefore the hatched field is placed more to the left and becomes identical with that of the rat.

For rabbits, the zone reaches as low as 12°C. Such low temperatures apply particularly to older animals with much fat deposition and large body mass. In general, the lower critical temperature decreases with increase in body mass. This shift is more pronounced in large animals than in small ones such as the

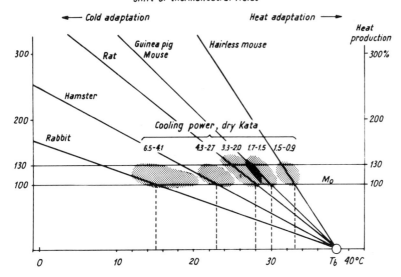

FIGURE 4 Recommended comfort temperature range for minimum metabolic adaptation of animals in cages with free air exchange.

common laboratory animals. An exception even for the larger species is the early period of life before heat regulation is fully developed and the lower critical temperature for the naked young is near body temperature. The shift of the lower critical temperature with increasing weight has been clearly shown for the pig by Heitman et al.[18]

The fields given in Table 1 and Figure 4 are recommended for cage systems

TABLE 1 Recommended Comfort Temperature for Minimum Metabolic Adaptation of the Single Animal in Cages with Free Air Exchange

Species	Comfort Range (°C)	Cooling Power Dry Kata[a] ($mcal\ cm^{-2}\ sec^{-1}$)
Hairless mouse	31–34	1.5–0.9
White mouse	26–31	2.7–1.5
Guinea pig	24–29	3.3–2.0
Rat	24–29	3.3–2.0
Hamster	21–26	4.1–2.7
Rabbit	12–21	6.5–4.1

[a]These values are slightly higher than those resulting from calculated values using the equation by Hill,[3] which is independent of the instrument.

with free air exchange. They will be more extended to the left if cage systems with limited air exchange and large groups of animals are used. In general, a left shift of 3° to 4° C is recommended. If this is applied one arrives exactly at the ambient temperatures that are most frequently suggested for maintenance.[42] It may also be necessary to extend the fields to the right to provide warmer conditions in the case of infected animals with a disturbance of thermoregulation, possibly due to endotoxins. Either their heat production is below the 100 percent line or their heat loss has increased. In this case thermoneutral conditions can be re-established by an increase of the ambient temperature.

It is obvious that the usual recommendations for maintenance temperature for mice, rats, and guinea pigs are too low when the animals are kept singly in cages with free air exchange. The recommendations of 20–22° C in Europe provide conditions that are too cool and will force the animal to cold adaptation.

Strict temperature control can be recommended only for single animals in cage systems with free air exchange. In all other cases a moderate fluctuation of thermal conditions following outdoor weather changes seems permissible, as long as the variations do not last so long that a new steady state in metabolic adaptation can develop. Moderate, short lasting fluctuations act as a climatic stimulus that might be particularly recommended for growing animals to improve fitness.

Animals respond to changes of floor temperature faster than to changes in air temperature.[20] Therefore, if heated floors are provided, the air temperature can be reduced. This is practiced in cat breeding and in experimentation. The floor preference temperature is identical with the thermoneutral zone temperature. Recently, Morrison and Warman[29] have described an improved temperature-gradient chamber based on the original temperature-gradient chamber by Herter.[20] They found two maxima of preference temperature within a 24-hr period, one at a higher and one at a lower temperature for a small Arctic mammal. The concept demonstrated in Figure 4 is supported by their data.

Measurement of Thermal Conditions

For reasons of simplicity, thermal conditions have been referred to only by temperature thus far in this discussion. However, they are dependent on four factors, namely, temperature, humidity, air motion, and radiation, because these factors affect the four channels of heat loss from the animal: radiation, convection, conduction, and evaporation. The combined effect of temperature and air motion is called the cooling power of the air.

Cooling power is measured as heat flow in mcal cm^{-2} sec^{-1}. A very useful instrument, suited for measurements inside small cages, is the katathermometer introduced by Hill.[3,21] The katathermometer is essentially an alcohol thermometer with a very large bulb 1.5 cm in diameter and 5 cm high and with only two scale divisions etched on its stem. The bulb is heated in hot water until the alcohol has risen to the upper reservoir. It is then quickly dried, and the thermometer is exposed at the point where the cooling power is to be measured. Care must be taken that the body of the experimenter is not protecting the bulb from air streams in the vicinity. The time for the alcohol to fall from the upper 38°C to the lower 35°C division is measured with a stopwatch. The cooling power H = mcal cm^{-2} sec^{-1} is calculated by dividing the instrument factor (etched on the stem of the thermometer) by the measured cooling time in seconds. From the resulting kata value and the dry temperature, air speed can be read from nomograms. The katathermometer loses heat by radiation and convection when dry, and by radiation, convection, and evaporation when the bulb is equipped with a wetted cloth covering (wet-bulb katathermometer).

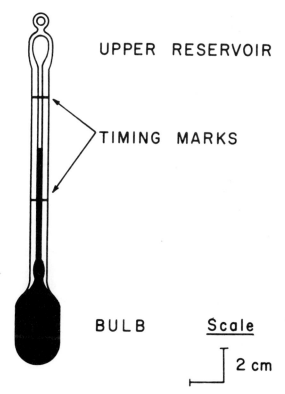

UPPER RESERVOIR

TIMING MARKS

BULB

Scale

2 cm

FIGURE 5 Katathermometer for measuring cooling power of the air. Vertical scale is twice horizontal scale. From B. A. Hertig, *Adaptation of Domestic Animals*. E. S. E. Hafez (ed.), Lea & Febiger, Philadelphia, p. 330, 1968.

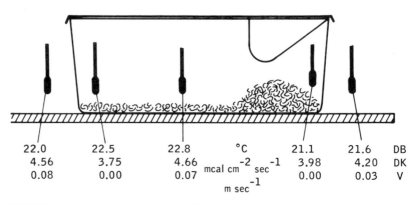

22.0	22.5	22.8	°C	21.1	21.6	DB
4.56	3.75	4.66	mcal cm^{-2} sec^{-1}	3.98	4.20	DK
0.08	0.00	0.07	m sec^{-1}	0.00	0.03	V

FIGURE 6 Dry kata measurements inside and outside a cage with compact walls and bedding material. There is slow air motion around the cage and in the center, but still air in the front and rear of the cage.

In Figure 6 dry kata values from measurements in a common animal room inside and outside a cage with compact walls are presented. It will be noted that there is some air motion around and in the center of the cage, while the air is still in the angles between the floor and the front and back walls. In cages with free air exchange such pockets of still air do not exist.

In animal rooms there is always some air motion in conjunction with the ventilation system. The air motion is either a turbulence or a directed air stream with air velocities up to 0.5 m sec^{-1} as provided by the modern laminar flow system.

That the energy requirement of an animal is affected significantly by weather-dependent irregular changes of air humidity under constant temperature conditions can be demonstrated in a few examples. Two groups, 25 male and 25 female growing rats, 5 animals per cage, on wood shavings, were held in a conventional room with constant temperature at 21° C. Each animal was weighed in the morning between 08:00 and 09:00 hr. The deviations of morning body weight from the expected value for each individual within 10-day periods were counted (sign-Test) and plotted on the graph in Figure 7. The numbers of daily weight deviations run parallel in the two sexes ($P < 0.05$ if $n > 18$). In correlating these deviations with the weather-dependent changes of humidity in the room, it was found that with decreasing humidity to 35 percent RH, the rats consumed about 5 percent more food than with increasing humidity to 75 percent RH.[41] In another experiment, the deviation was evaluated quantitatively for two groups of 21 male rats each. One group was housed in a room with an open north-facing window, and one in a room with

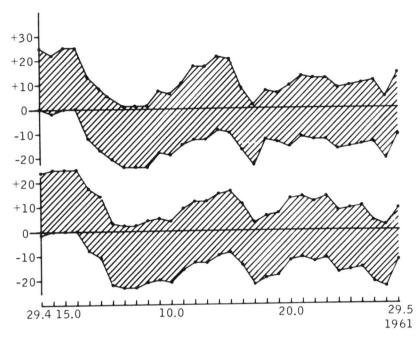

FIGURE 7 Mean daily variation of morning body weight from the expected value of male (top) and female (bottom) rats in a conventional animal room without air conditioning (*n* = 25 in both groups). Ordinate indicates positive or negative variation.

the temperature constant at 21° C, but with uncontrolled humidity. The animals were weighed twice daily, first between 08:00 and 09:00 hr and second between 17:00 and 18:00 hr. The individual weights were plotted for 10-day periods, and regression lines and standard deviations were calculated (Table 2). The variance of weight was always greater in the morning than in the after-

TABLE 2 Deviation in Grams of Body Weight from Mean of Individual Expected Values during a 10-Day Period for Two Groups of Growing Male Rats (*n* = 25)

Hour		Periodic and Aperiodic Changes	
		Temperature + Humidity	Humidity
Night	18:00–08:00	3.15	2.20
Day	08:00–18:00	1.62	1.60

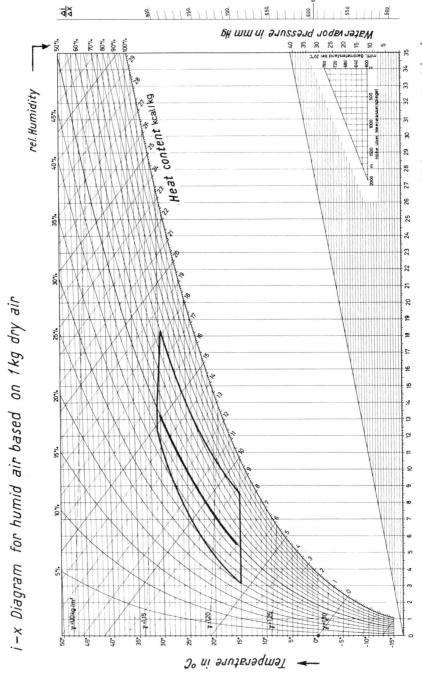

i-x Diagram for humid air based on 1kg dry air

rel. Humidity

Temperature in °C

Heat content kcal/kg

Water vapor pressure in mm Hg

Water vapor content in g/kg

370

noon. This was because of the previous night feeding. It was greater with un-controlled temperature and humidity and less when only the humidity was uncontrolled.

The results are interpreted as follows: With lower air humidity more water is evaporated, and therefore more heat is lost with the expired air. Therefore, animals consume more food and are more active. The effect of air humidity on activity was demonstrated in a study in which the activity of mice kept at constant temperature but uncontrolled air humidity was measured during a 30-min observation period. Measurements over 4½ years were correlated with weather-dependent humidity changes, which showed that with increased humidity the animals were less active and that with decreased humidity they became more active.[38]

Variations of relative humidity are biologically less significant at low temperatures than at high ones. With increasing temperature, the relative humidity range should be smaller. On the i-x diagram in Figure 8 an area is marked out from 30–80 percent RH at 15°C to 45–70 percent RH at 31–29°C, which is physiologically the most satisfactory temperature–humidity range. The decline of the upper temperature line with increasing relative humidity corresponds with the sensation of sultriness in man, i.e., before sweating occurs.[3] The field corresponds with the earlier description of climatograms for mice and rats based on physiological parameters such as growth.[42]

Humidity affects the survival rate of airborne micro-organisms, which seems to be lowest at 50 percent RH.[1] With a relative humidity of about 50 percent, the risk of infection from viable airborne micro-organisms is minimized, and unwanted effects in the animals, such as ring-tail disease in young rats under dry air conditions or decreased resistance to bacterial infections occurring in humid air, are avoided. To draw attention to this optimal relative humidity independent of temperature, the 50 percent RH line was also marked inside the recommended temperature–humidity range of Figure 8.

In small animals most heat is lost by radiation, convection, and conduction and less by evaporation, depending on the cage system. Therefore, in practice the dry kata values of cooling power of the air are more important than the wet kata values. As there are no wet outer surfaces in small animals, wet kata values do not directly apply, but they can be used to deduce air humidity as long as air motion and temperature do not change simultaneously.

As the critical temperatures for animals are determined in still air, it is possible to define the recommended temperature range in Figure 4 in dry kata values. The temperature range for the different species and the corresponding kata values are given in Table 1 and are inserted on top of the hatched fields in Figure 4. Cooling increases with air velocity. Therefore, with increasing air velocity, kata values will also increase. To maintain the cooling power of the air, temperature must increase with increasing air velocity. In Table 3 kata

TABLE 3   Cooling Power (Dry Kata[a]) at Still Air and Low Air Velocity

Temperature		Air Velocity (V = m sec^{-1})						
°C	°F	0.00	0.05	0.10	0.15	0.20	0.25	0.30
34	93.2	0.9	1.0	1.1	1.2	1.3	1.3	1.4
32	89.6	1.2	1.3	1.4	1.5	1.6	1.7	1.8
30	86.0	1.7	1.9	2.2	2.4	2.5	2.6	2.7
28	82.4	2.2	2.5	2.8	3.0	3.2	3.4	3.6
26	78.8	2.7	3.1	3.5	3.7	4.0	4.2	4.4
24	75.2	3.3	3.7	4.2	4.5	4.8	5.0	5.2
22	71.6	3.7	4.3	4.8	5.2	5.5	5.8	6.1
20	68.0	4.3	4.8	5.4	5.8	6.3	6.6	6.9
18	64.4	4.8	5.4	6.0	6.6	7.0	7.4	7.7
16	60.8	5.4	5.9	6.7	7.3	7.7	8.1	8.5
14	57.2	5.9	6.5	7.3	7.9	8.4	8.9	9.3
12	53.6	6.5	7.1	7.9	8.6	9.2	9.7	10.2

[a]These values are slightly higher than those resulting from calculated values using the equation by Hill,[3] which is independent of the instrument.

values for air velocities up to 0.3 m sec^{-1} in the temperature range from 34 to 12° C are presented.[3] It can be seen that within this range of air velocity, kata values change little at high temperatures but much at low temperatures.

The cooling power measured with the dry- and wet-bulb katathermometer can be used to give exact information about the climatic conditions inside and outside an animal cage during the experiment. The method has been applied by Sundstroem[39] in investigations on the effect of tropical climate on rats.

Since the measurement of humidity is not possible with the katathermometer, wet-bulb temperature should also be recorded in experiments where the state of metabolic adaptation of an animal should be known. Wet-bulb temperature is measured outside the cage because this method depends on having an air stream of constant velocity blown along the bulb.

Practical Significance of Metabolic Adaptation in Laboratory Animals

The very recent thermal experience of an animal is decisive for the response during an experiment. If reacclimatization is required, this will last for periods of up to several weeks in fully metabolically adapted animals.[43] Inbred animals from different strains genetically adapted to different thermal conditions will develop different levels of metabolic adaptation when brought into a standard climate of any laboratory. When used in a comparative study, such as testing a standard drug, strain differences may be due to different metabolic activity but not to specific strain susceptibility. In such studies in mice and rats, metabolic adaptation should be avoided by providing thermal conditions not lower than the temperatures recommended here (Table 1, Figure 4). Much information is now available about the effect of temperature on the response of animals to chemical compounds such as drugs,[7,9,10,27] where the state of metabolic adaptation of the animal is not clearly stated. Fuhrman and Fuhrman[10] have pointed out that for most drugs, when they were tested in animals over a wide temperature range, a bell-shaped sensitivity curve is observed, with the least effectiveness and lowest toxicity at a mean temperature between $17°$ and $30°$ C and the highest toxicity at low and high temperatures. This may be misleading if the animals used were not fully adapted before the test to the different temperatures. Their metabolic rate will increase at low temperatures and also at high temperatures if they had previously been moderately cold adapted, as is commonly the case. Therefore, at both temperature extremes, because of accelerated cell metabolism, the distribution and metabolism of the drugs is faster than in animals from the normal maintenance room temperature of $20°$ to $22°$ C.[43]

Conclusion

To summarize, it can be stated that for the optimum state of health with minimum adaptation to either cold or heat the best thermal conditions would be those of thermoneutrality for the feeding and active animal. To improve the fitness of an animal, cooler maintenance conditions are recommended.

References

1. Anderson, J. D., and C. S. Cox. 1967. Microbial survival. p. 203–226. *In* P. H. Gregory and J. L. Montheith [ed.] Airborne microbes. Seventeenth Symposium of the Society for General Microbiology, Cambridge University Press, Cambridge, England.
2. Aschoff, J. 1963. Comparative physiology: Diurnal rhythms. Ann. Rev. Physiol. 25:581–600.
3. ASHRAE. 1967. Handbook of fundamentals. American Society of Heating, Refrigerating and Air-Conditioning Engineers. New York.
4. Bartholomew, G. A. 1968. Body temperature and energy metabolism. p. 209–354. *In* M. S. Gordon [ed.] Animal function: Principles and adaptations. MacMillan, New York.
5. Brewer, N. R. 1964. Estimating heat produced by laboratory animals. Heat. Pip. Air Condit. 36(10):139–141.

6. Ederstrom, H. E. 1966. Temperature characteristics: Homoiothermic animals. p. 1. *In* P. L. Altman and D. S. Dittmer [ed.] Environmental biology. Fed. Amer. Soc. Exp. Biol., Bethesda, Maryland.

7. Ellis, T. M. 1967. Environmental influences on drug responses in laboratory animals. p. 569-588. *In* M. L. Conalty [ed.] Husbandry of laboratory animals. Academic Press, London and New York.

8. Folk, G. E., Jr. 1966. Introduction to environmental physiology. Environmental extremes and mammalian survival. Lea & Febiger, Philadelphia, 308 p.

9. Fuhrman, F. A. 1963. Modification of the action of drugs by heat. Arid Zone Research (UNESCO) 22:223-238.

10. Fuhrman, G. H., and F. A. Fuhrman. 1961. Effects of temperature on the action of drugs. Ann. Rev. Pharmacol. 1:65-78.

11. Gates, D. M. 1965. Energy exchange in the biosphere. Harper & Row and John Seatherhill, New York and Tokyo, 151 p.

12. Gelineo, S. 1956. Contribution à la connaissance de la calorification du lapin. Bull. Acad. Serb. Sci., Cl. Math. Nat. (4):1-16.

13. Gelineo, S. 1964. Organ systems in adaptation: The temperature regulating system. p. 259-282. *In* D. B. Dill [ed.] Adaptation to the environment. Handbook of physiology, Section 4, Williams & Wilkins, Baltimore, Maryland.

14. Harrison, G. A., R. J. Morton, and J. S. Weiner. 1960. The growth in weight and tail length in inbred and hybrid mice reared at two different temperatures. Phil. Trans. B., 242:479-516.

15. Hart, J. S. 1957. Climatic and temperature induced changes in the energetics of homeotherms. Rev. Canad. Biol. 16:133-174.

16. Hart, J. S. 1961. Physiological effects of continued cold on animals and man. Brit. Med. Bull. 17:19-24.

17. Hart, J. S. 1963. Physiological responses to cold in nonhibernating homeotherms. p. 373-406. *In* C. M. Herzfeld [ed.] Temperature: Its measurement and control in science and industry. Vol. 3, Part 3, Reinhold Publishing Corp., New York.

18. Heitman, H., Jr., C. F. Kelly, and T. E. Bond. 1958. Ambient air temperature and weight gain in swine. J. Anim. Sci., 17:62-67.

19. Herrington, L. P. 1954. Biophysical adaptations of man under climatic stress. Meteor. Monogr. 2(8):30-42.

20. Herter, K. 1936. Das thermotaktische Optimum bei Nagetieren, ein mendelndes Art- und Rassenmerkmal. Z. Vergl. Physiol. 23:605-650.

21. Hertig, B. A. 1968. Measurement of the physical environment. p. 324-337. *In* E. S. E. Hafez [ed.] Adaptation of domestic animals. Lea & Febiger, Philadelphia.

22. Hill, J. R. 1961. Reaction of the new-born animal to environmental temperature. Brit. Med. Bull. 17:164-167.

23. Israel, H. 1950. Zur biologischen Wirkungsmöglichkeit luftelektrischer Faktoren. Deutsch. Med. Wschr. 75:202-205.

24. Jansky, L. 1966. Body organ thermogenesis of the rat during exposure to cold and at maximal metabolic rate. Federation Proc. 25:1297-1302.

25. Kleiber, M. 1961. The fire of life. An introduction to animal energetics. John Wiley & Sons, New York, London, 454 p.

26. Landsberg, H. 1954. Bioclimatology of housing. Meteor. Monogr. 2(8):81-89.

27. Laroche, M. J. 1965. Influence of environment on drug activity in laboratory animals. Fed. Cosmet. Toxicol. 3:177-192.

28. Lee, R. C. 1942. Heat production of the rabbit at 28°C as affected by previous adaptation to temperatures between 10° and 31°C. J. Nutr. 23:83-90.

374

29. Morrison, P., and N. Warman. 1967. A thermal-gradient chamber for small animals with digital output. Med. Biol. Engr. 5:41–45.

30. Mount, L. E. 1968. The climatic physiology of the pig. Arnold Publishers, London, 271 p.

31. Nomura T., C. Yamauchi, and H. Takahashi. 1967. Influence of environmental temperature on physiological functions of the laboratory mouse. p. 459–470. *In* M. L. Conalty [ed.] Husbandry of laboratory animals. Academic Press, London, 459–470.

32. Odum, E. P. 1959. Fundamentals of ecology. Saunders Company, Philadelphia and London, 546 p.

33. Pavlik, I. 1964. Significance of air ionization. p. 317–342. *In* S. Licht [ed.] Medical climatology. Elizabeth Licht, Publisher, New Haven, Connecticut.

34. Prosser, L. 1964. Perspectives of adaptation: Theoretical aspects. p. 11–26. *In* D. B. Dill [ed.] Adaptation to the environment. Handbook of physiology, Section 4, Williams & Wilkins, Baltimore, Maryland.

35. Sargent II, F., and S. W. Tromp. [ed.] 1964. A survey of human biometeorology. Tech. Note No. 65, World Meteorological Organization, Geneva, 113 p.

36. Scholander, P. F., R. Hock, V. Walters, and L. Irving. 1950. Heat regulation in some Arctic and tropical mammals and birds. Biol. Bull. 99:237–258.

37. Shelford, V. E. 1962. Paired factors of the physical environment operating on the sensitive periods in the life history of organisms. Int. J. Biometeorol. 5:44–58.

38. Stille, G., H. Brezowsky, and W. H. Weihe. 1968. The influence of the weather on the locomotor activity of mice. Arzneim.-Forsch. 18:892–893.

39. Sundstroem, E. S. 1930. Contributions to tropical biochemistry and physiology. II. Supplementary experiments on rats adapted to graded levels of reduced cooling power. Univ. Calif. Publ. (Berkeley) 7:103–195.

40. Von Bertalanffy, L. 1951. Theoretische Biologie. II. Band. Stoffwechsel, Wachstum. 2. Aufl. Franke-Verlag, Bern, 418 p.

41. Weihe, W. H., H. Brezowsky, and F. R. Schwarzenbach. 1961. Der Nachweis einer Wirkung des Klimas und Wetters auf das Wachstum der Ratte. Pflügers Arch. Ges. Physiol. 273:514–527.

42. Weihe, W. H. 1965. Climatograms for mice and rats. Lab. Anim. Care 15:18–28.

43. Weihe, W. H. 1971. Die Bedeutung von Temperatur Feuchtigkeit und Stromungs-geschwindigkeit der Luft bei der Haltung von Ratten und Mäusen.

44. World Health Organization. 1967. Basic documents. Eighteenth Ed., World Health Organization, Geneva.

DISCUSSION

DR. FOX: What about cage position in relation to others in the 5 or 6 tier rack situation?

DR. WEIHE: Yes, this is a common question, because it is difficult to determine how to arrange the cages to avoid a thermal gradient, which always exists in the animal room if you have a vertical arrangement of the cages. The statisticians recommend the Latin square arrangement, which is a good thing. Perhaps the best solution would be to use a room where you know your thermal conditions and where you can be sure of equal air velocity along the cages. However, I have done extensive measurements using the katathermometer within animal rooms, and I have found that the range is actually very narrow in a decently ventilated animal room even when the cages are arranged vertically. But the thermal conditions depend on the locality, and in my opinion measurements should be made. Otherwise, the Latin square arrangement should be used.

DR. FOX: Thank you. I have another question. Animals do employ behavioral responses to control body temperature, such as huddling together and building a nest. Will such considerations make your single animal in cage data too limited for group-raised animals or nest-dwellers?

DR. WEIHE: I hope I can answer this question properly because I did not quite get it. The thermal regulation of the animal is definitely of the greatest significance, and all of this pertains to singly kept animals compared to the open cage systems. If animals are grouped together of course they decrease the radiant surface, particularly if they huddle together. An animal that is stretched out on the floor, of course, can lose much more heat by conduction than an animal that is rolled up. A rat that keeps its tail underneath its body will prevent heat loss from the tail surface, and all animals do so. So, if we provide a cage with limited air exchange and bedding and have low ambient temperature conditions or high cooling conditions, the animal can still easily regulate its thermal condition.

DR. LANE-PETTER: If animals are raised and kept strictly within their thermoneutral zone, will they subsequently be less able to adapt to temperature changes, for example, while under experiment?

DR. WEIHE: I do not think so. There is no evidence that an animal that has been kept under these more limited conditions is less resistant or has less tolerance to extremes. I do feel, however, and there is proof for this, it is better not to live exactly under these conditions but to go a little bit off. In other words, the English view, with which I sympathize very much, is to have slightly lower ambient conditions. For us coming over here to the United States, it is a great strain to adapt to the higher average ambient temperature, for example, of this room, because we are all cool-adapted coming over from Europe. The same applies to an animal.

It is, of course, better to introduce a short-lasting variance of 2 or 3 days of cooler conditions, which is sort of a challenge to the homeostatic system, rather than to provide a canned climate without any variance. But it is certainly important not to forget the seasonal changes. Humidity is the one factor to introduce seasonal changes between summer, winter, spring, and fall because the humidity changes in the air will make the animal adapt on a long-range scale. If people say that they have no seasonal changes in their environment, the humidity conditions within the animal room should be investigated.

DR. TAYLOR: With all other factors being the same, what difference in temperature would be required if the number of air changes per hour were raised from 10 to 30?

DR. WEIHE: The other factor that should be considered is mainly the velocity, because air velocity increases with the increase in air exchanges within the room. We have done extensive measurements of counter-cooling power within the clean-bench system. The clean-bench system is a very useful one for maintenance of animals protected against contamination because it employs the highest air velocity of any animal-maintenance system, which is up to 0.5 meters per second. Using the right cage, still air can be provided within the cage, under even such high air velocity. But the moment the air exchange within the room is increased, air velocity is the prime factor that has to be considered. In other words, wire cages should not be used.

DR. RICHTER: How long after higher heat levels occurred was the high incidence of mucoid enteritis in rabbits seen? Is this observation reproducible experimentally?

DR. WEIHE: Yes. The important thing is this did not occur, of course, right away. There was an incubation time, of 8 days, and this was reproducible, because we were lucky enough to have several heat waves during the last year, and each time we could observe the same thing. Our rabbits are kept at 15 to 17° C. Under these fairly constant cooling conditions, mucoid enteritis occurred exactly 8 days after the beginning of the increase of ambient temperature in every case.

DR. MITCHEL: Who manufactures the kata-thermometer?

DR. WEIHE: I don't know of an American manufacturer, but I know this is produced in England and in many other countries on the Continent. I would recommend that you ask your local meteorologist or anyone dealing with air conditioning. They will know how to get one.

DR. FESTING: What are the implications with regard to the cage design?

DR. WEIHE: I think there are many implications, and one must really choose a cage according to the climatic conditions within the room, according to the facilities, and according to the research one is going to do, not according to its price. I would recommend for the rabbit, because usually the conditions are too warm, a wire cage, because heat loss is much easier. There is direct air exchange. But for smaller animals I would definitely recommend a cage with a compact floor to put in bedding, so that the animal can make full use of its behavioral heat regulation.

DR. GOLDERSON: If given their own choice, will animals seek out the thermal conditions that you have described as optimal, or will they select some nonoptimal condition?

DR. WEIHE: I think there is general agreement among physiologists who have done this work that animals will look for the optimal conditions, and actually that is how we learn what the optimal conditions are. But the animals, of course, live in conditions that are not just optimal, as humans do, but they do it for only a short time, so they won't have to adapt.

DR. GREENWOOD: Has the period of physiological stabilization of laboratory animals been specifically evaluated against time and degree of environmental change under controlled management?

DR. WEIHE: Yes. In reviewing the literature, I have read about 800 papers on this. I have found that most people do not consider the period of physiological stabilization for their experimental designs. Most of the papers published on the effect of temperature on drug testing do not consider adaptation, as we do nowadays, to be the period of time to which the animal was exposed before the drug was actually given. In many experiments, the animals were exposed to the experimental temperature for only half an hour before the drug was applied. This of course means that the animal is just in the state of adaptation. Very few experimental animals have been adapted for at least 3- or 4-week periods, so that one could be sure that they had been fully adapted. I hope this is the answer to your question.

VI

Endocrinology

DIFFERENT CHARACTERISTICS
OF RAT STRAINS:
LIPID METABOLISM AND
RESPONSE TO DIET

Mary W. Marshall
Anna M. Allen Durand
Mildred Adams

In several published reports, the response of our laboratory strain of rats (BHE) was shown to be uniquely different, with respect to lipid metabolism, from that of a strain of Wistar rats fed the same diets. The strain responses varied with type of carbohydrate or fat or level of dietary cholesterol. Studies were carried out to learn to what extent the strain responses were genetic or nutritional. Because extensive work with rats is done in many laboratories to explain human food needs, knowledge and study of specific genetic characteristics is essential for proper interpretation of results and for establishing a base line upon which the effect of diet on individual response can be measured. Because some of the responses of the BHE rat were associated with spontaneous occurrence of nephrosis, genetically nephrotic, genetically lean rats, or lines with high or low levels of serum cholesterol, were inbred from the BHE strain and studied. Characteristics such as body size, food intake, body measurements (such as lengths, girths, and skinfolds), organ weights and liver composition, voluntary activity, blood and body composition, and incidence of nephrosis (gross and microscopic) varied with strain, age, and sex of the rats. Parallel studies with three additional strains commonly used, as well as with BHE and Wistar rats, have provided evidence of marked differences in their responses to dietary carbohydrate.

One of the more urgent needs in nutrition research is to determine why individual variations in nutritional requirements exist, and to what extent, and to learn whether the various requirements are the result of genetic determinants or of adaptations due to differences in metabolic patterns of the individual resulting from long-standing dietary practices. For many years, Williams and co-workers[1,2] have been a "voice in the wilderness" calling for recognition of the need to abandon the "average" requirement for specific nutrients. The increasing identification of individuals with metabolic defects such as galactosemia, phenylketonuria, diabetes, and the various hyperlipidemias[3] provides increasing evidence that our population consists not only of a few individuals with isolated instances of metabolic disorders but of a large number of metabolically aberrant individuals, many of whom do not show signs of abnormal response until middle life. (Indeed, some do not show evidence of clinically abnormal response at all but have adaptive mechanisms that permit the "normal" function of their vital processes.) Recently, it was suggested that studies are revealing an increasing number of individuals with a plethora of abnormalities in lipid and glucose metabolism. These individuals are said to be "carbohydrate sensitive" and to have a "high risk" potential for atherosclerosis.[4] The advances of medical science have assured that the number of such individuals will grow, and their specific dietary requirements will undoubtedly play an increasingly important role in our evaluations of food needs.

The rat has been used extensively in nutrition studies because of its similarity to the human being in the processes related to food utilization; requirements for many nutrients have been established, and their metabolic processes studied in detail. Rats of many strains have been used, and it cannot be determined whether the disparity in results reported from tests of the effectiveness of similar diets are due to the variety of conditions used, such as age, sex, length or time of the study, or variations in ingredients, or to the strain of rat. Relatively little has been written about physiological–pathological correlations in the rat, but there is a voluminous literature on such correlations in the mouse. Although it would seem logical that both long- and short-term studies are desirable to determine function and requirements of nutrients, there is a more recent recognition that results of short-term studies may not be compatible with those from animals fed the same diets for long periods or for the duration of their life-span. It is well known that expression of certain genes, e.g., diabetes, multiple sclerosis, and muscular dystrophy, in humans may not be evident in some individuals until after maturity. Thus, it would not be surprising to find certain characteristics in older animals that are not evident in younger ones.

In early long-term studies with our laboratory strain of rat, BHE, in the Human Nutrition Research Division,[5] degenerative changes in the kidney were associated with certain types of diets, particularly one containing 25 percent cooked, dried egg. It was not until these studies were repeated with another rat strain that a genetic–diet interaction was apparent.[6–8] A Wistar strain fed the same diets usually did not show the extensive degenerative kidney changes, and when these changes did occur, it was in animals much older than the BHE rats. In most studies, lipid components of blood and tissues were found, on

MARY W. MARSHALL,
ANNA M. ALLEN DURAND,
AND MILDRED ADAMS

the average, to be higher in BHE than in Wistar rats. The variations in lipid metabolism frequently seen in BHE rats did not occur invariably. The spontaneous development of nephrosis in individual rats was responsible for many of the variations. It became necessary, therefore, to identify characteristics of both strains to determine whether nephrosis was an inherited characteristic and to what extent it contributed to dietary response. Such studies were necessary to provide a base line on which response to diet could be measured. In addition, the production of substrains with specific characteristics would provide models of unique individuals for study of individual dietary requirements. Pairs of rats differing in their history of kidney damage were selected from BHE stock, and by continuous matings of full sibs in successive generations, two distinct inbred lines were produced[9]: One line was prone to nephrosis; the other was prone to kidney defects of another nature.

After feeding diets differing in their nutrient content to the BHE and Wistar rats, to the two inbred lines of rats, and to other rat strains,[10–20] it became evident that various rat models exist, or can be readily produced, with differences in metabolic patterns or in diet responses, that can be used to study the factors that may cause variations in nutritional response. Because of the similarities of mammals and the universality of the genetic code, our studies of the various rat strains may provide useful information in interpreting differences in nutritional requirements of human beings.

Rat Strains: General Characteristics

The BHE strain previously described[6] has been bred in the laboratory for approximately 43 generations. During the past 10 years, certain procedures were adopted to assure production of rats of fairly predictable response. These rats obviously were closely related, but full- or half-sib matings, as well as all albino matings, were consciously avoided. Also, restricted numbers of animals, about six rats per litter, from the same families were kept for on-going colony. Second-litter lineage was followed with young reduced at birth, at random, to 10 per litter. One-to-one matings were made, and individual records were kept of all rats. Wistar rats were obtained from the Dairy Cattle Research Branch[6] and have been maintained under conditions identical to those of the BHE rats during the past 10 years. Holtzman and Sprague-Dawley rats were purchased from commercial suppliers and were kept only for a limited time. Data reported on Columbia, Charles River, and Long-Evans rats were obtained by contract with another laboratory.[20]

One of the rat strains from which the BHE rat was derived, the Yale strain, has been described as having a diabetic trait[22,23] and is said to be subject to genetic obesity.[24] The responses of the BHE rat to diets containing sucrose or elevated levels of fat, particularly unsaturated fat or cholesterol, indicate the potential for similar defects in this strain. The manifestation of nephrosis in individual stock-diet-fed BHE rats with accompanying increases in liver, kidney, adrenal, and thyroid size and increase in blood lipids indicates the genetic character of the defect.

Some of the features of the BHE rat and of rats inbred from the BHE stock are illustrated in Figures 1–6. All rats described were killed at 300 days of age unless otherwise specified. Figure 1 shows a fat BHE male rat fed a stock diet. A fat BHE male fed a purified diet containing 52 percent sucrose, 30 percent lactalbumin, 15 percent corn oil[11] is shown in Figure 2, while Figure 3 shows a BHE male rat with nephrosis fed the same diet. Figure 4 shows a BHE male rat with nephrosis fed a stock diet; note absence of fat. The gross appearance of normal and nephrotic kidneys of stock-diet-fed BHE rats is shown in Figure 5; a typical liver and cut sections of BHE kidneys are also shown (Figure 6).

At necropsy, while the rats were under amytal anesthesia, blood was removed by heart puncture. Gross examination of organs was made, and imme-

FIGURE 1 A 300-day-old
fat BHE male fed Purina
Lab. Chow.

MARY W. MARSHALL,
ANNA M. ALLEN DURAND,
AND MILDRED ADAMS

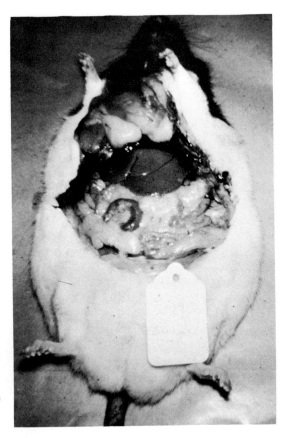

FIGURE 2 A 400-day-old fat BHE male fed a purified diet containing 52 percent sucrose, 30 percent lactalbumin, and 15 percent corn oil. (From Marshall *et al.*, reference 11.)

diately thereafter organs were removed and weighed, and portions were placed in neutral buffered formalin. Not more than 5 min was required to complete this procedure in each rat. Histological examination was made of selected portions of the liver, right kidney, left adrenal and thyroid lobe, and heart after sectioning, paraffin imbedding, and staining with hematoxylin and eosin. Periodic-acid-Schiff and Feulgen stains were done on selected sections. Tissues of approximately 1,000 BHE rats, divided evenly by sex, were examined. Of these, about 20 percent of the 300-day-old rats had nephrosis, most of it occurring in the males. Only rats that had not lost more than 30 g from their maximum weight were included. When kidneys were nephrotic, both were equally affected. The surface of these kidneys was mottled and had a brownish color that became lighter as the disease progressed. Gross nephrosis was graded ± to 4+, depending upon color, extent of mottling, and increase in size.

Upon histological examination, kidneys were graded for nephrosis from

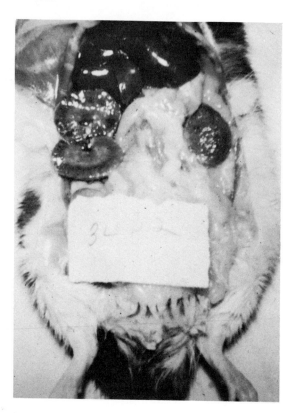

FIGURE 3   A BHE male
rat with nephrosis. Same
age and diet as in Figure 2.

0 to 4+, as shown in Figure 7. Grade 0 (Figure 7a) indicates no significant findings, though occasionally a single small hyalin cast was seen; this was not considered unusual for the rat. Grade 1 (not pictured) has scattered hyalin casts in slightly dilated tubules with flattened epithelium. Grade 2 (Figure 7b) is similar to Grade 1 but with more extensive involvement and patches of lymphocytic infiltration. In Grade 3 (Figure 7c), tubular dilation is increased, in some areas appearing cystic, with moderate proliferation of tubular epithelium, fibrosis, and some hyalin casts. Grade 4 (Figure 7d) shows extensive cystic dilation of most tubules, distention of Bowman's space, islets of hyperplastic tubular epithelium, rare hyalin casts, diffuse interstitial fibrosis, and many normal glomeruli, but some shrunken and vacuolated ones. Characteristically, a stock-diet-fed BHE rat exhibited nephrosis without the extensive enlargement or hyperplasia that occurred when purified diets containing sucrose, fructose, cholesterol, or highly unsaturated oils were fed. Nephrosis and hydronephrosis were rarely seen in 300-day-old Wistar rats in our laboratory, but

MARY W. MARSHALL,
ANNA M. ALLEN DURAND,
AND MILDRED ADAMS

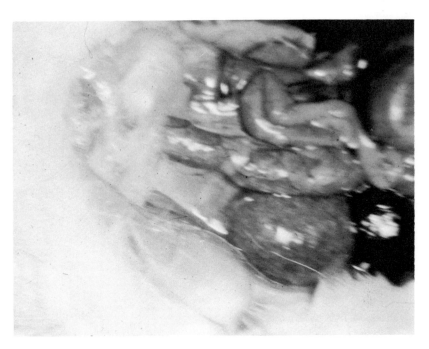

FIGURE 4   A 300-day-old BHE male rat with nephrosis fed Purina Lab. Chow. Note absence of fat.

FIGURE 5   Gross appearance of normal and nephrotic kidneys of a stock-diet-fed 300-day-old BHE male rat.

387

FIGURE 6   A typical liver and cut surface of normal kidneys.

chronic kidney disease could be produced in old rats or under specific dietary conditions.[18]

A common finding in both BHE and inbred rats was hydronephrosis of varying degrees. Thirty percent of this group of BHE rats and 37 percent of inbred rats exhibited hydronephrosis that was usually bilateral, but in some instances, only one kidney was affected. These kidneys were graded mild, moderate, or severe, depending upon the degree of pelvic dilation and destruction of kidney substance. Some kidneys were cystic, and some contained well-formed stones or fluid containing granules of calcium. Usually, hydronephrosis was unassociated with nephrosis, and only occasionally were hyalin casts found. As seen in Figure 8, there was calcium in the pelvis (Figure 8a), or in the collecting tubules (Figure 8b). Heavy granular deposits, some in laminated spherules, were frequent in the dilated calyces (Figure 8c). In a number of these kidneys, the pelvic mucosa over large calcium plaques was markedly hyperplastic (Figure 8d). Pyelonephritis was not found in these kidneys. Gross examination revealed a definite line of demarcation at the cortico-medullary junction in a number of BHE kidneys. This finding was more common in the inbred line-1 rats than in BHE rats. After 24 generations of inbreeding in this

MARY W. MARSHALL,
ANNA M. ALLEN DURAND,
AND MILDRED ADAMS

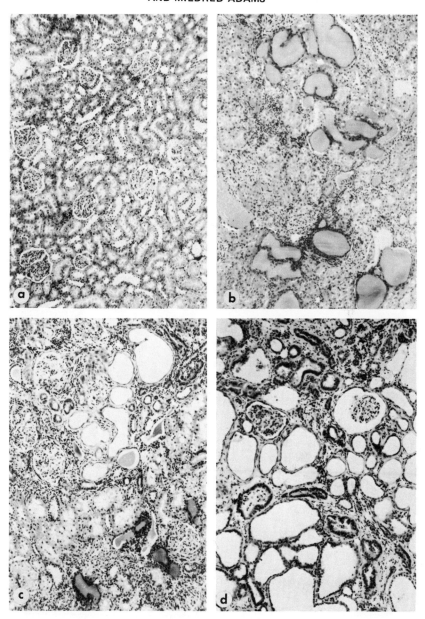

FIGURE 7   Normal and nephrotic kidney; hematoxylin and eosin X 125. (a) Normal kidney. (b) Nephrosis ++; hyalin casts in dilated tubules, flattened epithelium, lymphocytic infiltration. (c) Nephrosis +++; moderate hyalin casts, cystic dilatation of tubules, epithelial proliferation in some tubules, lymphocytic infiltration and fibrosis. (d) Nephrosis ++++; extensive cystic tubules.

389

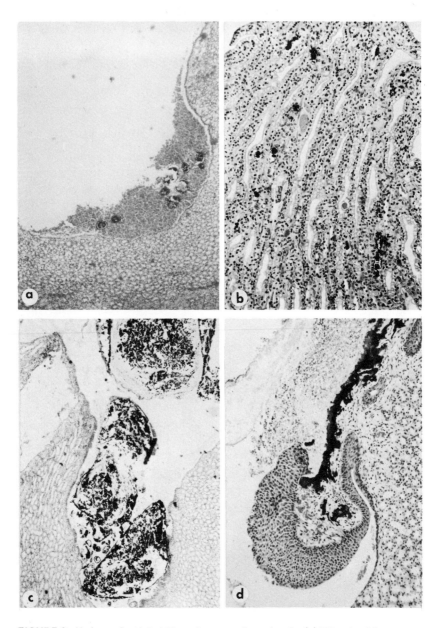

FIGURE 8   Hydronephrosis in kidney, hematoxylin and eosin. (a) Dilated pelvis containing blood and small laminated concretions, X 52. (b) Calcium in tubules of papilla, X 125. (c) Heavy deposit of calcium granules in calyx, X 52. (d) Hyperplasia of pelvic mucosa over calcium plaque, X 125.

MARY W. MARSHALL,
ANNA M. ALLEN DURAND,
AND MILDRED ADAMS

line, each animal exhibited the characteristic. This has tentatively been identified as ceroid. Line-2 inbred rats were more prone to nephrosis, and after eight generations of full-sib inbreeding, some rats died at 200 days of age with 4+ nephrosis.

The enlargement of the liver was essentially a hypertrophy, though occasionally, small scattered patches of fat were seen. Some fatty hemorrhagic necrosis was seen also in the small lobes. Necrosis and hemorrhage were frequently seen in the adrenal gland in the line-1 inbred rats. Thyroid size was variable; it increased in size, as did the adrenals, with severity of nephrosis. Corpora amylacea, small laminated hyalin bodies, were seen in the thyroids of approximately 10 percent of BHE rats, most centrally located and variable in number and size, with slightly more in the female than in the male. These were presumably degenerative in nature. Because only one such body was seen in 225 Wistar rats, the finding was interpreted to indicate a more sluggish thyroid in BHE than in Wistar rats. Thyroids of Wistar rats, usually smaller than those of BHE rats, were very cellular, as were the thyroids of some BHE rats. Parathyroids were not usually enlarged unless nephrosis was severe (4+), and this was not always the case. Various stages of atrophy were seen in the pancreas, but because the pancreas was not routinely examined, this finding will be evaluated later.

Stock-diet-fed BHE and Wistar rats usually had similar average body weights up to 300 days of age (Table 1), but male BHE rats at the same age exhibited

TABLE 1   Live Body Weight, in Grams, of Three Strains of Rats at Three Ages: Stock Diet[a]

Age (days)	BHE	Wistar	Inbred Line 1
MALES			
50	232 ± 6	218 ± 5	197 ± 6
	(29)	(34)	(29)
100	364 ± 5	367 ± 5	396 ± 6
	(37)	(38)	(28)
300	499 ± 5	485 ± 4	536 ± 5
FEMALES			
50	169 ± 6	168 ± 5	150 ± 6
	(27)	(33)	(29)
100	240 ± 5	229 ± 5	242 ± 6
	(38)	(42)	(31)
300	319 ± 5	299 ± 4	306 ± 4
	(45)	(52)	(62)

[a]Data from Marshall et al.[14]; numbers of rats in parentheses; stock diet: Purina Laboratory Chow.

more body fat for their weight than Wistar rats (Figure 9). Although male line-1 inbred rats were largest after 50 days of age, they were the leanest of the three strains ($P < 0.01$). The increase in body fat in BHE rats, even in those fed the stock diet, was influenced to a great extent by the occurrence of nephrosis. After severe damage to the kidneys, the BHE rat lost weight, and both blood lipids and body lipids decreased.

Simple body measurements were made on live rats of three strains at three ages[14] to identify characteristics associated with fat or lean rats. The measurements differed among strains at the different ages. Multiple regression equations produced for predicting body fat from body measurements such as lengths, girths, and skinfolds were different for the sexes as well as for the strains at different ages. This finding suggested differences in fat patterning or distribution resulting in different phenotypes for the different strains. Because the leanest strain, line-1 inbred rats, was bred from the BHE strain, it is suspected that a potential for the gene for leanness also existed in BHE rats.

Mean organ weights of BHE rats are characteristically larger and more variable than those of Wistar rats when fed a stock diet[6,7] and particularly when fed purified diets.[6-8,18] Organ size of BHE rats is influenced also by type of stock diet (Table 2). Litters of rats divided at weaning and fed one of two stock diets showed changes not only in liver and kidney size but in serum cholesterol as well. The relationship between liver and kidney size, degree of nephrosis, and serum cholesterol level was generally seen in the two lines of inbred rats as well (Table 3). More will be said later about serum cholesterol.

On the average, hearts of Wistar rats are larger in relation to their body size than hearts of BHE rats.[25] Inbred line-2 rats prone to nephrosis had smaller hearts than line-1 rats not prone to nephrosis (unpublished data). Because differences in oxidative capacity could influence heart size as well as body composition, a study was done to quantitate the voluntary activity of the BHE and Wistar strains. Table 4 shows not only strain differences in voluntary activity, but sex and age differences, as well as interactions of these factors. The peak of activity, as well as total amount of activity, differed with sex. Females were much more active than males.

In another study, it was found that excretion of large amounts of urine protein, previously thought to be uniquely characteristic of BHE rats, was also characteristic of Holtzman rats fed certain diets (Table 5). Although the urine protein of BHE and Holtzman rats was elevated at an early age, it could be modified by changing the nature of the diet. Elevated levels of urine protein were found in the inbred lines of rats and were shown in some cases to be familial, and to be influenced by diet as well.[9] Urine protein was considerably elevated in nephrosis-prone line-2 rats of the seventh generation as early as 200 days of age and in male line-1 rats that were not as prone to nephrosis but that showed other kidney defects (Table 6).

MARY W. MARSHALL,
ANNA M. ALLEN DURAND,
AND MILDRED ADAMS

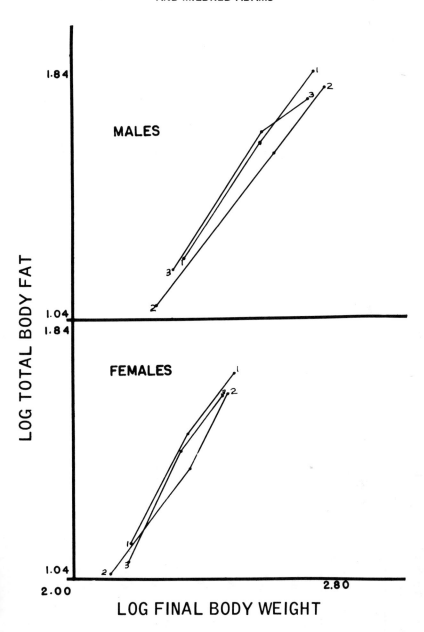

FIGURE 9   Association of body fat with body weight of three strains of rats. (1) BHE rats. (2) Inbred Line-1 rats. (3) Wistar rats. All were fed Purina Lab. Chow until 300 days of age.

TABLE 2 Effect of Stock Diet on Body and Organ Weights, Nephrosis, and Serum Cholesterol of Male and Female BHE Rats at 300 Days of Age

Number of Rats	Diet of Parents[a]	Diet of F$_1$[a]	Body Weight (g)	Liver–Body Weight (%)	Kidney–Body Weight (%)	Number with Nephrosis[b]	Nephrosis Score	Weight of Left Adrenal (mg)	Weight of Left Thyroid (mg)	Weight of Heart (g)	Serum Cholesterol (mg/100 ml)
BHE MALES											
14	AF	AF	448 ±12	3.9 ±0.12	0.81 ±0.03	8/14	0.7	20.4 ±0.53	12.7 ±0.59	1.25 ±0.04	138 ±10
11	AF	PC	445 ±13	3.9 ±0.20	0.70 ±0.03	3/11	0.5	18.0 ±0.70	10.1 ±1.12	1.23 ±0.03	134 ±18
19	PC	AF	513 ±10	3.3 ±0.05	0.67 ±0.01	8/19	0.4	18.2 ±0.60	11.4 ±0.58	1.42 ±0.02	100 ±3
23	PC	PC	507 ±11	3.1 ±0.07	0.64 ±0.01	5/21	0.2	19.5 ±0.84	8.0 ±0.39	1.38 ±0.02	89 ±3
BHE FEMALES											
9	AF	AF	279 ±7	3.7 ±0.09	0.75 ±0.02	0/9	0.0	32.3 ±1.58	9.9 ±0.47	0.92 ±0.02	144 ±29
8	AF	PC	278 ±10	3.9 ±0.11	0.70 ±0.02	0/8	0.0	32.8 ±1.84	7.7 ±0.49	0.94 ±0.02	194 ±38
18	PC	AF	299 ±10	3.6 ±0.08	0.72 ±0.01	4/18	0.2	32.4 ±1.10	9.4 ±0.34	0.94 ±0.02	101 ±6
16	PC	PC	316 ±7	3.4 ±0.07	0.66 ±0.01	0/14	0.0	31.6 ±1.26	8.1 ±0.41	1.00 ±0.02	100 ±4

[a]AF = Animal Foundation diet; PC = Purina Laboratory Chow.
[b]Histological evaluation of nephrosis.

TABLE 3 Change in Size of Kidneys, Degree of Nephrosis (Gross), and Serum Cholesterol Level with Increase in Liver Size of Healthy 300-Day-Old Inbred Rats: Stock Diet[a]

Line and Liver Weight Range (g)	Number of Rats	Liver–Body Weight (%)	Kidney–Body Weight (%)	Gross Nephrosis, Score	Cholesterol (mg/10 ml)
MALES[b]					
Line 1					
15–18.9	15	3.5	0.71	0.4	122
19–20.9	12	3.9	0.79	1.2	137
21–22.9	10	4.0	0.81	1.6	161
23–24.9	10	4.2	0.88	2.0	183
25 and over	10	4.9	0.99	2.9	216
Line 2					
15–18.9	13	4.1	0.88	1.6	231
19–20.9	23	4.5	0.98	1.7	290
21–22.9	17	4.8	1.02	2.3	350
23 and over	8	5.1	0.99	2.5	314
FEMALES					
Line 1					
8–11.9	21	3.9	0.80	0	100
12–13.9	26	4.2	0.80	0.4	108
14–15.9	19	4.4	0.82	0.2	134
16 and over	6	5.2	0.93	0.5	206
Line 2					
8–11.9	10	4.0	0.86	0	138
12–13.9	16	4.6	0.88	0.4	194
14–15.9	12	5.4	0.97	1.0	240
16 and over	9	6.4	1.12	2.0[c]	359

[a]Stock diet, Animal Foundation diet.
[b]Line-2 males with livers in excess of 27 g died early or lost too much weight to be included.
[c]Six rats were included in this group.

TABLE 4 Voluntary Activity of Male and Female Rats of Two Strains of Rats at Three Ages: Stock Diet[a]

| Age (days) | Miles Traveled (Males) | | Miles Traveled (Females) | |
	BHE	Wistar	BHE	Wistar
21	4.3 ± 1.2	9.2 ± 1.2	7.5 ± 3.1	15.5 ± 2.9
	(18)	(18)	(17)	(18)
100	4.3 ± 1.0	6.0 ± 1.2	31.4 ± 6.2	50.5 ± 6.3
	(18)	(18)	(17)	(18)
300	2.8 ± 0.5	2.2 ± 0.4	48.6 ± 12.1	40.5 ± 7.9
	(18)	(18)	(17)	(18)

[a]Stock diet, Purina Laboratory Chow. Rats were placed in individual cages, with access to wheels, for two periods of 3 weeks each at weaning and at 100 days of age; the final period was for 4 weeks. At all other times, the rats were housed in individual cages.

TABLE 5 Urine Protein, mg per Day, of Male Rats of Three Strains, at Two Ages, Fed Different Diets[a]

Diet	BHE	Hotzman	Wistar
155 DAYS OLD			
Stock-AF	119 ± 15	217 ± 44	38 ± 5
SP	63 ± 6	58 ± 7	23 ± 2
SPE	164 ± 28	161 ± 20	35 ± 6
Stock-PC		118 ± 21	
325 DAYS OLD			
Stock-AF	308 ± 44	406 ± 79	84 ± 19
SP	93 ± 17	108 ± 23	25 ± 4
SPE	292 ± 72	373 ± 36	35 ± 10
Test difference ($P < 0.05$) = 58			

[a]Data from Marshall and Hildebrand.[6] Stock diet, AF, Animal Foundation diet; PC, Purina Laboratory Chow. SP, semipurified basal diet; SPE, SP diet with 25 percent cooked dried egg.

Differences in Blood and Liver Components

When the animals were fed either a stock diet or a semipurified diet containing 25 percent egg,[19] livers of 90-day-old BHE and Wistar rats were found to have different levels of glucose-6-phosphate dehydrogenase, beta glucuronidase, and alkaline phosphatase. Some strain differ-

MARY W. MARSHALL,
ANNA M. ALLEN DURAND,
AND MILDRED ADAMS

TABLE 6   Liver and Kidney Weights in Relation to Body Weight and Urine Protein of Three Generations of 300-Day-Old Rats of Two Inbred Lines: Stock Diet[a]

Line	Generation[b]	Number of Rats	Liver–Body Weight (%)	Kidney–Body Weight (%)	Urine Protein (mg/day)
MALES					
Line 1	6	9	3.9 ± 0.14	0.75 ± 0.03	–
	7	11	3.9 ± 0.10	0.72 ± 0.02	541 ± 59
	8	21	4.1 ± 0.11	0.82 ± 0.04	595 ± 61 (13)
Line 2	6	4	4.8 ± 0.30	0.98 ± 0.02	–
	7	8	4.7 ± 0.20	1.00 ± 0.07	1074 ± 65
	8	20	4.4 ± 0.09	0.99 ± 0.05	930 ± 56
FEMALES					
Line 1	6	12	4.1 ± 0.06	0.84 ± 0.02	–
	7	12	4.4 ± 0.07	0.83 ± 0.02	89 ± 14
	8	18	4.5 ± 0.15	0.84 ± 0.03	334 ± 78 (13)
Line 2	6	9	4.4 ± 0.27	0.84 ± 0.03	–
	7	9	4.4 ± 0.30	0.83 ± 0.03	330 ± 79
	8	15	4.8 ± 0.22	0.95 ± 0.04	645 ± 99 (10)

[a]Stock diet, Animal Foundation diet.
[b]Eight-generation data from Marshall et al.[9]; rats in sixth and seventh generations were healthy survivors at 300 days; urine protein of seventh generation measured at 200 days.

ences were seen also in liver enzyme levels (Figure 10). The significance of these differences in the production of changes in body or liver composition cannot be assessed at present. It is well known that adaptations occur that may either influence metabolic patterns or result from changes in metabolism.

Serum protein components of 330-day-old BHE, Holtzman, and Wistar rats were found to vary with strain and type of diet (Table 7). A component migrating ahead of albumin (moving-boundary electrophoresis) and designated as prealbumin[12,13,17] was commonly seen in BHE rats, but it was found also in other strains. Lowered levels of total protein and serum albumin and elevated levels of nonprotein nitrogen and beta globulin were seen frequently in nephrotic rats, some of which were not losing weight.

Mean values of some other constituents of serum of BHE and Wistar rats are shown in Tables 8 and 9. Except for triglycerides and free-fatty-acid levels, values for the BHE male rats were usually more variable than those for Wistar male rats; values for females of the two strains did not show as many differences as did values for males. One reason for the similarity between the two

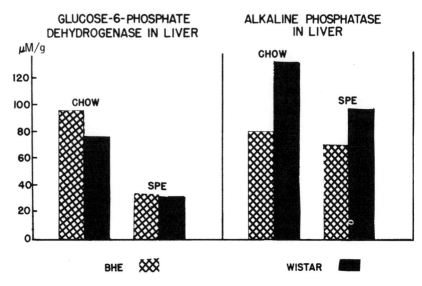

FIGURE 10   Effect of diet and strain on enzymes in liver of 90-day-old BHE rats. Diet, Purina Lab. Chow; SPE, semipurified diet with 25 percent whole egg. (From Chang *et al.*[19])

sets of female values is that the incidence of nephrosis in BHE females was much lower than it was in BHE males, and nephrosis was practically nonexistent in the Wistar females. The largest effect of nephrosis was seen on the serum cholesterol, although total lipid and triglycerides also were elevated in nephrotic rats prior to loss of weight.

Table 10 shows the influence of nephrosis on some mean values of serum cholesterol of rats fed a standard stock diet. A large variation in some of the values was produced because of the various degrees of severity of nephrosis. Table 11 shows the results of a study of 10 generations of BHE rats. Serum cholesterol level was significantly influenced by sex (males higher than females), type of stock diet (Animal Foundation diet higher than Purina Laboratory Chow), and nephrosis (nephrotic rats higher than non-nephrotic ones). There were no changes in serum cholesterol levels of BHE rats that could be attributed to hydronephrosis. An analysis of data for 698 rats of five generations of the two inbred lines (both lines fed Animal Foundation diet) showed an influence of line and sex on serum cholesterol due to presence of nephrosis, but no change due to hydronephrosis. Another analysis of 742 representative rats from 17 generations of line-1 inbred rats fed the two different stock diets showed effects on serum cholesterol level due to type of stock diet and occurrence of nephrosis, but the results of this analysis, unlike the others, showed a

TABLE 7 Serum Protein Components of Three Strains of Male Rats Fed Two Different Diets[a]

	Total		Protein Components (% of total)[c]			
Strain[b]	Protein (g/100 ml)	Nonprotein nitrogen (mg/100 ml)	Prealbumin	Albumin and Alpha-1 Globulin	Beta Globulin	Gamma Globulin
STOCK DIET[d]						
BHE	5.9	42	6.4[b]	70.0[a]	13.0[b]	3.1[b]
Wistar	6.3	28	5.4[b]	58.4[b]	19.5[a]	7.6[a]
Holtzman	6.0	52	16.4[a]	65.7[a]	8.3[c]	1.0[b]
SEMIPURIFIED DIET[e]						
BHE	5.8	35	3.6[c]	68.7[a]	15.6[a]	4.5[a]
Wistar	6.7	27	7.4[b]	61.4[b]	16.7[a]	5.6[a]
Holtzman	6.2	26	12.6[a]	67.2[ab]	11.3[b]	1.2[b]

[a] From Lakshmanan and Marshall.[12]
[b] BHE and Wistar = 7–12 animals per group; Holtzman = 3.
[c] Within a set of means, those that do not have the same superscript are significantly different from one another ($P < 0.05$ or < 0.01); analysis of protein components by moving-boundary electrophoresis.
[d] Stock diet, Animal Foundation diet.
[e] Semipurified diet, no eggs.

TABLE 8 Serum Components of Selected Groups of Fed and Fasted Male and Female Rats of Two Strains: Stock Diet[a]

	Number of Rats	Albumin	Na	K	Ca	Cholesterol
BHE MALES						
Fed	19	3.3 ±0.09	135 ±2	5.8 ±0.21	11.0 ±0.45	102 ±11
Fasted	11	3.1 ±0.18	147 ±3	6.9 ±0.54	9.9 ±0.20	97 ±14
BHE FEMALES						
Fed	19	3.2 ±0.13	141 ±2	5.4 ±0.14	10.9 ±0.42	97 ±3
Fasted	9	3.3 ±0.18	145 ±2	6.0 ±0.16	9.7 ±0.15	95 ±8
WISTAR MALES						
	23	3.4 ±0.03	146 ±1	5.9 ±0.14	10.2 ±0.14	83 ±2
	9	3.5 ±0.03	144 ±3	5.9 ±0.15	9.4 ±0.14	75 ±1
WISTAR FEMALES						
	21	3.6 ±0.03	143 ±2	5.2 ±0.15	10.5 ±0.16	99 ±4
	10	3.6 ±0.03	147 ±3	5.2 ±0.15	9.8 ±0.08	92 ±4

[a]Stock diet, Purina Laboratory Chow. Food removed overnight for approximately 17 hr from rats designated as fasted. All components analyzed by Technicon Autoanalyzer methods.

MARY W. MARSHALL,
ANNA M. ALLEN DURAND,
AND MILDRED ADAMS

TABLE 9 Free Fatty Acids, Glucose, Cholesterol, and Triglycerides in Serum of 300-Day-Old Fed or Fasted Male and Female Rats of Two Strains: Stock Diet[a]

	Males		Females	
	BHE	Wistar	BHE	Wistar
FREE FATTY ACIDS (mEq/l)				
Group 1				
Fed	0.24 ± 0.03 (22)	0.21 ± 0.02 (22)	0.27 ± 0.03 (23)	0.32 ± 0.03 (21)
Fasted	0.43 ± 0.05 (11)	0.43 ± 0.04 (9)	0.42 ± 0.03 (9)	0.51 ± 0.04 (10)
GLUCOSE (mg/100 ml)[b]				
Fed	158 ± 18 (4)	168 ± 16 (12)	133 ± 8 (4)	148 ± 7 (7)
Fasted	136 ± 6 (5)	123 ± 11 (9)	117 ± 3 (3)	119 ± 2 (10)
CHOLESTEROL (mg/100 ml)				
Group 2				
Fed	121 ± 15 (7)	99 ± 4 (8)	117 ± 3 (8)	119 ± 3 (8)
TRIGLYCERIDES (mg/ml)				
Fed	0.712 ± 0.13 (9)	0.624 ± 0.17 (8)	1.34 ± 0.27 (9)	1.19 ± 0.20 (8)

[a]Stock diet, Purina Laboratory Chow. Food removed overnight for approximately 17 hr from rats designated as fasted. Glucose and cholesterol analyzed by Technicon Auto-analyzer. Free fatty acids analyzed by a modification of the method of Novak[26]; triglycerides, by method of Carlson.[27]

[b]Larger numbers of values obtained by another method are available for glucose in BHE rats; averages are similar to those presented here.

small but significant effect of the presence of hydronephrosis on serum cholesterol level. This finding was undoubtedly due to the coexistence of hydronephrosis and nephrosis in some of the same kidneys in this line of rats. It should be pointed out that hydronephrosis is frequently seen in female rats in families where nephrosis occurs and is occasionally seen in the male rats in the same families.

The occurrence of hydronephrosis as a precursor of nephrosis has been ruled out, but the possibility of a difference in expressivity or penetrance of

TABLE 10 Serum Cholesterol Values of Rat Strains without and with Nephrosis: Stock Diet[a]

Strain[b]	Number of Rats	Nephrosis (Histological Evaluation) Score	Serum Cholesterol[c] (mg/100 ml)	
			$\overline{X}$	S.E.
MALES				
Wistar	43	0	101	4
Wistar	87	0	93	2
S-D	20	0–2.0	175	12
BHE	95	0–3.0	124	4
BHE	50	0	111	3
BHE	40	0.5–3.0	135	8
BHE-W (F_2)	74	0	90	2
BHE-W (F_2)	34	0.5–3.0	144	13
FEMALES				
Wistar	37	0	102	4
Wistar	98	0	104	2
S-D	18	0–2.0	117	7
BHE	97	0–2.0	120	2
BHE	88	0	118	3
BHE	10	0–2.0	129	6
BHE-W (F_2)	85	0	104	2
BHE-W (F_2)	11	0–3.0	157	28

[a] Stock diet, Purina Laboratory Chow.
[b] S-D, Sprague-Dawley; BHE-W, BHE rats without and with nephrosis × Wistar rats. All rats were 295–310 days of age.
[c] Cholesterol analyzed by same method by Technicon Autoanalyzer.

the gene for nephrosis resulting in hydronephrosis in some rats and nephrosis in others has not been eliminated. Further studies are under way to determine the genetic correlation between nephrosis and serum cholesterol and the heritability of these characteristics. Table 12 shows that different types of diets, though low in fat and cholesterol, may produce variable effects on serum cholesterol, depending upon the genetic makeup of the rats. Results of a study with rats fed either a stock diet or two levels of defatted egg or lactalbumin showed that effects due to type and level of protein depended to a large extent on inheritance.

Within the BHE strain, rats are found that have high cholesterol but little or no nephrosis (see Table 2). Preliminary results from studies on descendants of these rats showed differences in their requirements for biotin. Whether a different pathway for cholesterol synthesis exists in these rats remains to be seen.

TABLE 11 Effect of Type of Stock Diet, Sex, Heredity, Nephrosis, and Hydronephrosis on Serum Cholesterol Level of 300-Day-Old Rats

Source of Variation	BHE Colony[a]			Inbred line 1[b]			Inbred lines 1 and 2[c]		
	Number of Rats	Cholesterol	P	Number of Rats	Cholesterol	P	Number of Rats	Cholesterol	P
Total	895			742			698		
Line									
1							314	136 ± 9	< 0.01
2							384	215 ± 7	
Diet									
Animal Foundation	226	132 ± 6		355	140 ± 3				
Purina Lab. Chow	669	112 ± 3	< 0.01	387	117 ± 4	< 0.01			
Sex									
Male	454	134 ± 3[d]		340	129 ± 3		332	185 ± 8	
Female	441	110 ± 4	< 0.01	402	130 ± 3	NS	366	166 ± 9	< 0.05
Nephrosis									
With	141	160 ± 5		154	142 ± 4		396	217 ± 9	
Without	754	112 ± 3	< 0.01	588	115 ± 2	< 0.01	302	134 ± 8	< 0.01
Hydronephrosis									
With	243	122 ± 4		104	127 ± 4		222	198 ± 3	
Without	652	122 ± 3	NS	639	119 ± 2	< 0.05	476	186 ± 7	NS

[a] BHE rats of 10 generations, 3 of 10 fed Animal Foundation diet.
[b] Inbred line-1 rats of 17 generations.
[c] Both inbred lines fed Animal Foundation diet for nine generations. Generation effects and all interaction effects will be reported in detail upon completion of analysis based on histological evaluation (BHE rats). Separate analyses were done for effects of nephrosis and hydronephrosis.
[d] Based on analysis for hydronephrosis.

TABLE 12 Effect of Nephrosis on Serum Cholesterol of Two Lines of Inbred Rats Fed Stock or Semipurified Low-Fat, Low-Cholesterol Diet[a]

Diet[b]	Number of Rats	Sex	Line	X̄ (Nephrosis)	Ȳ (Cholesterol)	b	a	F	P_b
Stock	20	Male	1	1.90	170	52	72	34.81	< 0.01
Stock	20	Male	2	2.20	315	55	194	2.36	NS
Stock	17	Female	1	0.74	138	68	89	62.82	< 0.01
Stock	15	Female	2	1.67	263	82	127	2.36	NS
24% DFE	18	Male	1	0.61	135	18	124	3.59	NS
24% DFE	4	Male	2	1.50	214	-35	266	2.58	NS
24% Lact.	19	Male	2	2.42	257	104	5	13.48	< 0.01
47% DFE	16	Male	1	1.44	172	34	123	10.23	< 0.01
47% DFE	20	Male	2	2.35	451	206	-32	5.99	< 0.05
47% Lact.	14	Male	2	3.50	339	112	53	5.07	< 0.05
24% DFE	12	Female	1	0.58	160	41	136	7.70	< 0.05
24% DFE	25	Female	2	1.92	419	125	179	5.60	< 0.05
24% Lact.	12	Female	2	1.50	223	59	135	5.30	< 0.05
47% DFE	14	Female	1	0.82	148	27	126	8.85	< 0.05
47% DFE	4	Female	2	1.72	292	142	44	4.54	NS
47% Lact.	8	Female	2	2.37	277	129	-28	9.83	< 0.05

[a] Linear regression analysis: b = regression coefficient; a = y intercept; $F = \dfrac{\text{regression mean square}}{\text{error mean square}}$; P_b = p values for the regression coefficients.

[b] DFE = defatted whole egg; Lact. = lactalbumin supplied in purified diets containing 43 percent sucrose, 5 percent corn oil, 4 percent salt mixture, and adequate vitamins.

MARY W. MARSHALL,
ANNA M. ALLEN DURAND,
AND MILDRED ADAMS

It has already been pointed out that a difference in biotin requirements was seen in the previous two inbred lines.[9] Full-sib inbreeding of rats that were descendants of a fat BHE male produced families that were, at 300 days of age, composed of individuals that were either fat, nephrotic, or hydronephrotic or had a combination of one or more of these defects. Some of these rats showed differences in their biotin requirements; others showed various degrees of pancreatic tissue atrophy or granularity. Studies to evaluate the nature of the changes in the pancreas and of insulin secretion are continuing.

Effect of Different Types of Dietary Carbohydrate and Fat on Lipid Metabolism of Rat Strains

Earlier studies in the laboratory showed that BHE rats (Table 13) fed purified diets containing sucrose, 15 percent corn oil, 30 percent lactalbumin, and adequate minerals and vitamins until they were 400 days old had higher levels of body fat and serum cholesterol than those fed the same diets with cornstarch (or corresponding diets containing hydro-

TABLE 13 Serum Cholesterol and Carcass Fat of 400-Day-Old BHE Male and Female Rats Fed Stock and Purified Diets with Different Kinds of Fat[a]

Diet[b]	Serum Cholesterol (mg/100 ml)	Body Fat, Wet Weight (%)	Digestible Calories Stored as Fat (%)
MALES			
Stock−AF (ground)	175 ± 9	22.3	3.9
Stock + CO	237 ± 31	25.0	4.7
Sucrose + CO	356 ± 41	33.2	7.0
Cornstarch + CO	172 ± 10	25.3	5.0
Stock + HVO	219 ± 28	23.1	4.3
Sucrose + HVO	190 ± 27	29.8	5.9
Cornstarch + HVO	146 ± 9	26.3	5.1
FEMALES			
Stock−AF	146	21.5	2.8
Sucrose−3% CO	303	38.6	6.5
Sucrose−15% CO	−	38.1	6.7

[a]Data from Marshall et al.[11]
[b]AF = Animal Foundation; CO = corn oil; HVO = hydrogenated vegetable oil; 15 percent fat in diets. Twelve or more male rats per group; 5–7 females.

405

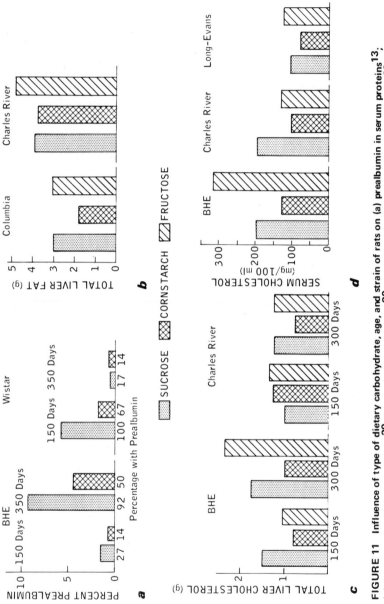

FIGURE 11 Influence of type of dietary carbohydrate, age, and strain of rats on (a) prealbumin in serum proteins[13]; (b) total liver fat (150-day-old rats)[20]; (c) total liver cholesterol[20]; and (d) serum cholesterol (300-day-old rats).[20] Semi-purified diet plus 25 percent whole egg; type of carbohydrate as indicated.

genated vegetable fat instead of corn oil).[11] At that time, the results were in contrast to many reported in the literature for younger rats fed for a much shorter time. Similar responses have been obtained many times since. Diets containing various types of carbohydrate in otherwise identical diets and fed for 150 to 350 days had a different influence on blood prealbumin (Figure 11a), liver fat (Figure 11b), liver cholesterol (Figure 11c), and serum cholesterol (Figure 11d), depending upon age and strain of rat and type of carbohydrate. Food intake or efficiency of utilization could not account for these manifold effects (Figure 12).[17,20] Age and type of dietary fat in diets

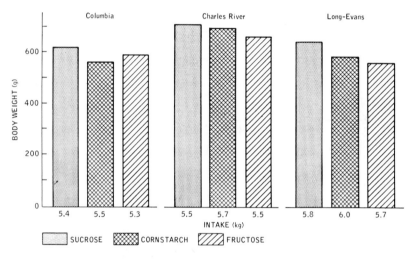

FIGURE 12 Influence of type of carbohydrate on body weight and food intake of three strains of rats at 350 days of age. (From Adams, reference 20.)

containing sucrose influenced levels of serum cholesterol in BHE rats (Table 14).[21] With some types of carbohydrate, liver cholesterol of BHE rats increased with age (Figure 13), while that of Wistar rats decreased.[16]

Table 15 shows that the chief causes of death were respiratory disease in the Wistar strain and kidney disease in the BHE strain. Age at death, however, was mediated by the type of carbohydrate in the diet.[18] Paradoxically, during the histological evaluation of the hearts from several thousand rats without and with nephrosis or high serum cholesterol, or both, little coronary artery disease or related pathology was seen. Further evaluation will be made of this finding.

TABLE 14   Influence of Age and Type of Dietary Fat on Serum Cholesterol Levels of Male BHE Rats 250 and 350 Days of Age[a]

	Serum Cholesterol	
	---	---
Fat in SPE Diet[b]	250 Days (mg/100 ml)	350 Days (mg/100 ml)
HVO	145	359
Lard	121	417
Corn Oil	163	330
Butter	232	315
Blend	250	271
Safflower	199	333

[a]From Adams, reference 21.
[b]Semipurified diet containing 25 percent cooked dried whole egg with sucrose.

% CHANGE

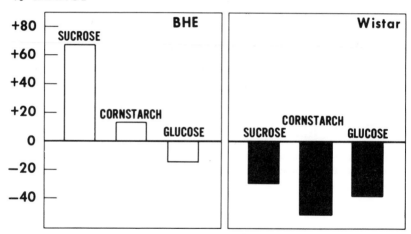

FIGURE 13   Percent change in liver cholesterol of male BHE and Wistar rats with age and type of carbohydrate (at 350 days, based on results for 150 days). (From Taylor *et al.*, reference 16.)

Comment

The variety of responses produced in different strains of rats under the same dietary conditions and the variety of responses produced with different diets in the same strain indicate a number of complex

MARY W. MARSHALL,
ANNA M. ALLEN DURAND,
AND MILDRED ADAMS

TABLE 15 Influence of Strain of Rat and Type of Carbohydrate on Age at Death and Cause of Death of Male BHE and Wistar Rats[a]

Carbohydrate	Age at Death (days)	Kidney Disease (%)	Respiratory Disease (%)
BHE			
Sucrose	444	71	26
Cornstarch	595	77	15
Glucose	543	60	20
WISTAR			
Sucrose	583	6	63
Cornstarch	636	14	71
Glucose	565	0	62

[a]Adapted from Durand *et al.*, reference 18.

genetic–diet interactions that cannot be explained at present. It is obvious that various metabolic patterns exist between strains and between individual rats within strains that permit both acute and chronic adaptations. Although no detailed studies of hormone status have been completed, preliminary evidence* suggests that a strain difference exists in the response to glucose injected intravenously. Abnormal glucose-tolerance curves were obtained when Wistar rats were excited; such curves were not observed in 150-day-old BHE rats. Beyond 150 days, however, a delayed insulin response was occasionally observed. The well-known involvement of insulin, as well as hormones of the thyroid, adrenal, and pituitary in lipogenesis may explain some of these responses.

We need to determine the type of genetic defect we are dealing with, i.e., whether the primary defect is in the kidney or in one of the other organs or whether an abnormal control mechanism influenced by imbalances in hormone secretion is responsible. Other explanations are possible: for example, the presence of a trace mineral imbalance resulting in different requirements for chromium[28]; a deficiency in vitamin E resulting in the breakdown of cell structure and consequent leakage (kidneys of some rats showed defects described in E-deficient rats)[29]; presence of a fatty gene; or diabetes commencing in middle life associated with hyperlipemia and kidney disease. The role of the kidney in the disposal of insulin in rats has been investigated. Less insulin was required by diabetic rats when nephrosis accompanied the diabetes.[30] Rats of Zucker and Zucker,[31,32] described as "fatties" resulting from a mutant gene transmitted as a simple recessive, exhibited hyperlipemia and nephrosis.

*Personal communication with Dr. Carolyn Berdanier.

Sprague-Dawley rats developed spontaneous nephrosis,[33] and a colony of Sprague-Dawley rats[34] has been described as containing "hypo" and "hyper" responders to cholesterol feeding. Slonaker-Addis,[35] Gunn,[36] and Wistar rats[37] have shown the hydronephrotic trait and high protein excretion.[38] Osborne and Mendel rats were reported to gain more body fat on certain types of diets than a strain of black rats.[39] Many of these changes were not seen in very young rats.

Human beings with Type II hypercholesterolemia show some tendency to kidney disease,[3] and an increasing number of carbohydrate-sensitive individuals are being identified.[4] The BHE rat and other strains we have identified provide models for continuing studies that may supply information of benefit to unique individuals in the human population whose dietary requirements may be influenced by their heredity.

References

1. Williams, R. J. 1956. Biochemical individuality: The basis for the genetotropic concept. John Wiley & Sons, Inc., New York.
2. Williams, R. J., and R. B. Pelton. 1966. Individuality in nutrition: Effects of vitamin A deficient and other deficient diets on experimental animals. Proc. Nat. Acad. Sci., U.S. 55:126.
3. Fredrickson, D. S., R. I. Levy, and R. S. Lees. 1967. Fat transport in lipoproteins—an integrated approach to mechanisms and disorders. New Eng. J. Med. 276.
4. Kuo, P. T. 1968. Current metabolic-genetic interrelationship in human atherosclerosis. Ann. Int. Med. 68:449.
5. Callison, E. C., M. Fisher, and E. Orent-Keiles. 1952. Diet as a factor in the production of degenerative changes in tissues of the rat. (Abstr) Federation Proc. 11.
6. Marshall, M. W., and H. E. Hildebrand. 1963. Differences in rat strain response to three diets of different composition. J. Nutr. 79:227.
7. Adams, M. 1964. Diet as a factor in length of life and in structure and composition of tissues of the rat with aging. USDA Home Econ. Res. Rep. No. 24. U.S. Department of Agriculture, Washington, D.C.
8. Durand, A. M. A., M. Fisher, and M. Adams. 1964. Histology in rats as influenced by age and diet. I. Renal and cardiovascular systems. Arch. Pathol. 77:268.
9. Marshall, M. W., and R. P. Lehmann. 1967. Influence of heredity on response of inbred rats to diet. I. Differences in body size, food intakes, incidence of spontaneous kidney defects, kidney weights, urine pH and protein. Metabolism 16:763.
10. Brown, M. L. 1963. Effect of a low dietary level of three types of fat on reproductive performance and tissue lipid content of the vitamin-B$_6$-deficient female rat. J. Nutr. 79:124.
11. Marshall, M. W., H. E. Hildebrand, J. L. Dupont, and M. Womack. 1959. Effect of dietary fats and carbohydrates on digestibility of nitrogen and energy supply, and on growth, body composition and serum cholesterol of rats. J. Nutr. 69:371.
12. Lakshmanan, F. L., and M. W. Marshall. 1966. Influence of diet and heredity on the serum protein components of the rat. Proc. Soc. Exp. Biol. Med. 122:535.

13. Lakshmanan, F. L., E. M. Schuster, and M. Adams. 1967. Effect of dietary carbohydrate on the serum protein components of two strains of rats. J. Nutr. 93:117.

14. Marshall, M. W., B. P. Smith, A. W. Munson, and R. P. Lehmann. 1969. Prediction of carcass fat from body measurements made on live rats differing in age, sex and strain. Brit. J. Nutr. 23:353.

15. Marshall, M. W., B. P. Smith, and R. P. Lehmann. 1969. Dietary response of two genetically different lines of inbred rats: Lipids in serum and liver. Proc. Soc. Exp. Biol. Med. 131:1271.

16. Taylor, D. D., E. S. Conway, E. M. Schuster, and M. Adams. 1967. Influence of dietary carbohydrates on liver content and on serum lipids in relation to age and strain of rat. J. Nutr. 91:275.

17. Lakshmanan, F. L., and M. Adams. 1965. Effect of age and dietary fat on serum protein components of the rat. J. Nutr. 86:337.

18. Durand, A. M. A., M. Fisher, and M. Adams. 1968. The influence of type of dietary carbohydrate. Effect on histological findings in two strains of rats. Arch. Pathol. 85:318.

19. Chang, M. L. W., E. M. Schuster, J. A. Lee, C. Snodgrass, and D. A. Benton. 1968. Effect of diet, dietary regimens and strain differences on some enzyme activities in rat tissues. J. Nutr. 96:368.

20. Adams, M. 1968. Nutritional involvement of carbohydrates. Proceedings of the Symposium on Recent Trends in Nutritional Chemistry, 156th Annual Meeting of the American Chemical Society, September 1968. Abstract No. 89, American Chemical Society, Washington, D.C.

21. Adams, M. 1968. Body composition, liver lipid and serum cholesterol of rats fed low or high cholesterol diets containing different types of fat. Federation Proc. 27. (Abstr).

22. Cole, V. V., and B. K. Harned. 1938. Diabetic traits in a strain of rats. Endocrinology 23:318.

23. Cole, V. V., B. K. Harned, and C. E. Keeler. 1941. Inheritance of glucose tolerance. Endocrinology 28:25.

24. Mayer, J. 1955. The physiological basis of obesity and leanness. Part 1, Nutr. Abstr. Rev. 25:597.

25. Marshall, M. W., B. P. Smith, and R. P. Lehmann. Influence of strain, age and sex on body composition of rats. III. Interrelationships between total body fat, organ weights and serum cholesterol. Manuscript in preparation.

26. Novak, M. 1965. Colorimetric ultramicro method for the determination of free fatty acids. J. Lipid Res. 6:431.

27. Carlson, L. A. 1963. Determination of serum triglycerides. J. Atheroscler. Res. 3:334.

28. Mertz, W. 1967. Biological role of chromium. Federation Proc. 26:186.

29. Martin, A. J. P., and T. Moore. 1939. Some effects of prolonged vitamin E deficiency in the rat. J. Hygiene 39:643.

30. Ricketts, H. T., H. L. Wildberger, and L. Regut. 1963. The role of the kidney in the disposal of insulin in rats. Diabetes 12:155.

31. Zucker, T. F., and L. M. Zucker. 1962. Hereditary obesity in the rat associated with high serum fat and cholesterol. Proc. Soc. Exp. Biol. Med. 110:165.

32. Zucker, L. M. 1965. Hereditary obesity in the rat associated with hyperlipemia. Ann. N.Y. Acad. Sci. 131:447.

33. Berg, B. N. 1965. Spontaneous nephrosis, with proteinuria, hyperglobulinemia, and hypercholesterolemia in the rat. Proc. Soc. Exp. Biol. Med. 119:417.

411

34. Saito, S., and L. C. Fillios. 1964. Certain aspects of cholesterol metabolism and hepatic protein synthesis in the rat. Amer. J. Physiol. 207:1287.

35. Sellers, A. L., S. Rosenfeld, and N. B. Friedman. 1960. Spontaneous hydronephrosis in the rat. Proc. Soc. Exp. Biol. Med. 104:512.

36. Lozzio, B. B., A. I. Chernoff, E. R. Machado, and C. B. Lozzio. 1967. Hereditary renal disease in a mutant strain of rats. Science 156:1742.

37. Astarabadi, T., and E. T. Bell. 1962. Spontaneous hydronephrosis in albino rats. Nature 195:392.

38. Perry, S. W. 1965. Proteinuria in the Wistar rat. J. Pathol. Bacteriol. 89:729.

39. Schemmel, R., O. Mickelsen, G. Jersey, and S. Wegrzyn. 1968. Response of different strains of rats to a high fat ration. (Abstr. 1906), Federation Proc. 27:555.

DISCUSSION

STEVEN WEISBROTH: Please comment on the high blood glucose (130 milligrams percent) in fasted BHE rats on stock diets. Is this normal?

MRS. MARSHALL: It is typical of the BHE rat. The blood glucose of the Wistar rat is also higher than expected at this age, but we feel it is high for a different reason. The Wistar rat is an excitable rat, and some of the variations that we find have been attributed to this characteristic.

DR. ERICHSEN: The second question reads: What were the method and support material for electrophoresis?

MRS. MARSHALL: It is moving boundary electrophoresis. I can talk with Dr. Weisbroth about this method.

H. KAPPEL: Could you explain the meaning of BHE and its origin, please?

MRS. MARSHALL: This is just a name associated with our organization at the time this strain was started. This was the Bureau of Home Economics, really. The Department of Agriculture has been reorganized many times, and many years ago when our organization was called the Bureau of Home Economics, a Yale strain (really an inbred strain originally from Yale) was brought from Columbia University and crossed with a hooded strain that was brought from Pennsylvania State College. Various types of rats were obtained from these crosses. There were gray, gray and white, black, black and white, albino, some brown, and brown and white rats. The odd colors were selected out, and only the black, black and white, and albino rats were retained. We have lost the black rat that was one of the components of that strain. We have now just the albino and the black and white, or the piebald type rat.

C. E. HUNT: On effects of carbohydrate, what were the protein and fat contents?

MARY W. MARSHALL,
ANNA M. ALLEN DURAND,
AND MILDRED ADAMS

MRS. MARSHALL: The protein content was
30 percent, which was 20 percent lactalbumin and 10 percent casein. The
total fat content was 17 percent. Part of that was from the egg fat. Six percent was fat blend, which was typical of the American type diet from the
1955 survey. It was a blend of fats added to the diet at the level of 6 percent.

C. E. HUNT: Were diets isocaloric?

MRS. MARSHALL: Yes, they were.

HOMOGENEOUS AND HETEROGENEOUS ANIMALS FOR THE ENDOCRINOLOGICAL SEARCH

N. Simionescu

The utilization of genetically homogeneous and heterogeneous animals has been widely useful for endocrinology. Many experiments performed on pure strains have pointed out that the endocrine physiology is subject to genetic variations, indicating that the concept of "normality" must include a qualitative and quantitative extension of normal limits. Several investigators have found strain differences in the structure and function of the endocrine glands, especially in the pituitary (see references 8, 14, 16, 22, 25, 29, 45, and 47), the thyroid (see references 7, 8, 13, 15, 17, 18, 23, 27, 28, 31, 40, and 41), the pancreas,[3] the adrenals (see references 1, 2, 4, 5, 6, 12, 26, 35, 36, 39, and 43), the ovaries (see references 10, 16, 24, 25, 34, 46, and 47), and the testicles (see references 8, 20, 33, and 42). Such differences give the investigator the opportunity to select a strain suitable for a particular research purpose, but simultaneously emphasize the need for caution in generalizing results among animal strains and species. Taking into account that the inbred strains are only a part of their respective species and that the human being is a complex heterozygote, we try in this paper to extend the endocrinological information on a broader scale of biological variability by including genetically heterogeneous animals as well as homogeneous ones.

Materials and Methods

Two groups of observations were made:

1. Observations on strain differences in hormonal and behavioral sensitivity were made under 102 sets of experimental conditions, including endocrine in-

414

sufficiency, compensatory hyperplasia, chronic inflammation, autoimmune diseases, shock-induced neoplasia, and organ transplants. The observations were made on the thyroid and the adrenals of mice, rats, guinea pigs, rabbits, and hamsters.

2. Special observations were made on genetic variability in adrenal structure, function, reactivity, and pathology in homogeneous and heterogeneous strains of mice. Studies have been made on 220 mice of the following strains: A_2G inbred (A_2G:i), A_2G noninbred in closed colony (A_2G:ni), RAP inbred (RAP:i), RAP noninbred (RAP:ni), C_3H inbred (C_3H), A inbred (A), $C_{57}BL$ inbred ($C_{57}BL$), and an unknown randomly bred strain (US:ni).*

The experimental conditions included ACTH, SH, or estradiol administration, contralateral adrenalectomy, hypophysectomy, pinealectomy, ovariectomy, orchidectomy, adrenal enucleation, adrenal deglomerulation, periadrenal lipectomy, adrenal transplants (into somatic tissues or into the hepatic portal system), induced adrenal tumors, and hemorrhagic shock.

The investigative methodology was guided by the general premise that on the continuous chain of the genotype–phenotype–dramatype the strain peculiarities may be evident at various levels of investigation of the endocrine glands. These differences were detected as being even greater as the investigation delved deeper and drew nearer to the genetically induced metabolic stereotype of the adrenal cell. Consequently, we recorded significant adrenal variations from the macroscopic level of investigation to the biochemical level.

The techniques used were related to the intracellular (Figure 1) and were directed mainly to those morphophysiological characteristics that many investigators have shown to be genetically induced (see references 4-6, 26, 35, 36, and 43; also, Figure 2).

Results

Macroscopic peculiarities As Table 1 shows, there are significant strain differences in adrenal weights relative to body weight, by sex and age, in mice. For instance, the values 18-19, which are normal for the US:ni strain, indicate significant adrenal hypertrophy for A_2G:i mice. These differences are maintained and are sometimes marked during ontogenic de-

*For convenience, some signs were added to the usual abbreviations.

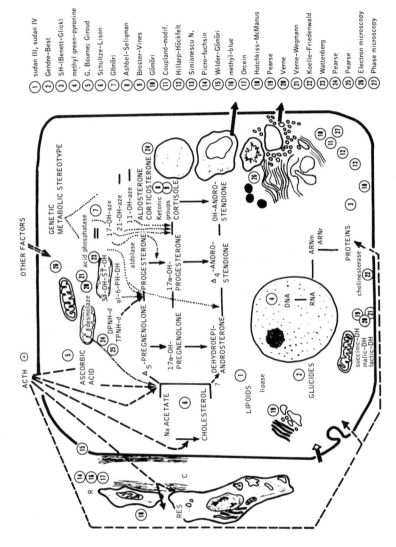

FIGURE 1 Techniques used for histochemical and cytochemical examinations.

1. sudan III, sudan IV
2. Gendre-Best
3. SH-(Benett-Glick)
4. methyl green-pyronine
5. G. Bourne; Giroud
6. Schultze-Lison
7. Gömöri
8. Ashbel-Seligman
9. Broster-Vines
10. Gömöri
11. Coupland-modif.
12. Hillarp-Höckfelt
13. Simionescu N.
14. Picro-fuchsin
15. Wilder-Gömöri
16. methyl-blue
17. Orcein
18. Hotchkiss-McManus
19. Pearse
20. Verne
21. Verne-Wegmann
22. Koelle-Friedenwald
23. Wattenberg
24. Pearse
25. Pearse
26. Electron microscopy
27. Phase microscopy

416

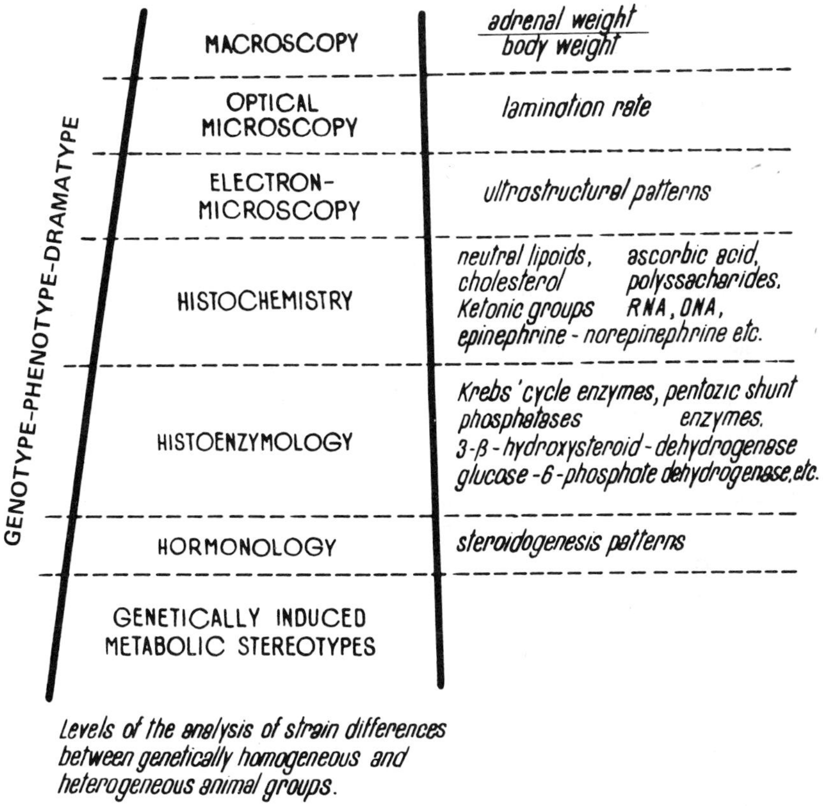

Levels of the analysis of strain differences between genetically homogeneous and heterogeneous animal groups.

FIGURE 2 The methodology of the investigations performed on the adrenal gland in mice.

velopment. The males of highly selected lines show a lower ratio and little change in the ratio of adrenal weight to body weight. These variations can be correlated with changes occurring in other aspects of adrenal structure and function, especially with the postpartum persistence of the X-zone (see below).

Microscopic peculiarities Considering the adrenal as formed by two kinds of glandular tissue, we observed that, in contrast with the relatively smaller amount of medullary tissue in the adrenals of A_2G:i mice, the noninbred US:ni mice have a wide adrenal medulla, especially at maturity.

The X-zone, which is present at birth in all strains of examined mice, degenerates more quickly in males than in females. The X-zone involution is more precocious in inbred than in noninbred mice, the longest persisting X-

TABLE 1 Ontogenic and Sexual Variations in Adrenal Weight

Strain	Age (months)[a]							
	1		2		3		4	
	$\overline{X}$	S.D.	$\overline{X}$	S.D.	$\overline{X}$	S.D.	$\overline{X}$	S.D.
A_2G:i								
Male	15,668 ± 1.8		15,480 ± 2.2		14,696 ± 1.4		14,960 ± 1.6	
Female	15,880 ± 2.0		15,679 ± 1.6		15,134 ± 1.6		15,340 ± 1.8	
C_3H:i								
Male	15,762 ± 1.4		15,214 ± 2.0		14,677 ± 1.8		14,660 ± 2.2	
Female	16,134 ± 1.8		15,892 ± 3.0		15,380 ± 1.5		15,182 ± 1.7	
$C_{57}BL$:i								
Male	16,882 ± 2.6		16,210 ± 1.7		15,612 ± 1.6		15,604 ± 3.1	
Female	17,240 ± 1.4		16,862 ± 1.8		16,142 ± 2.0		15,410 ± 1.9	
US:ni								
Male	19,427 ± 3.4		19,012 ± 4.0		17,122 ± 4.3		17,086 ± 3.3	
Female	18,516 ± 3.1		17,914 ± 3.9		17,044 ± 3.1		16,668 ± 2.4	

[a]$\overline{X}$ = adrenal weight (mg/100 g of body weight); S.D. = standard deviation.

zone occurring in the random animals. The line and the timing of X-zone regression differ significantly between strains (Figure 3).

The lamination rate of the adrenal cortex shows some strain variations. Generally, the heterogeneous strain has a broader fasciculate zone, the enlargement of which is usually due to the concomitant narrowing of the reticulate layer. In experimental conditions some significant changes in the lamination rate have been noted. As shown in Figure 4, the responses of the glomerular layer are more important in the noninbred strains, especially after pinealectomy and ovariectomy. The fasciculate layer is particularly sensitive in the inbred animals, especially the A_2G strain; random breeding decreases the significance of the adrenal sexual layer modulations.

The lamination rate modifications are accompanied by some histochemical changes, each kind of experiment having a more-or-less evident pattern relative to the adrenal content in lipoids, ketonic groups, and ascorbic acid (Figure 4).

As to the histoenzymological strain differences, we must emphasize that in the unknown strain of noninbred mice there is an increased quantity of some Kreb's cycle enzymes (succinic-dehydrogenase and malico-dehydrogenase), as well as an increase in the enzymes directly involved in hormone biosynthesis (glucose-6-phosphate-dehydrogenase and 3-beta-hydroxysteroid-dehydrogenase). ACTH administration stimulates enzymatic activities both in A_2G:i and US:ni

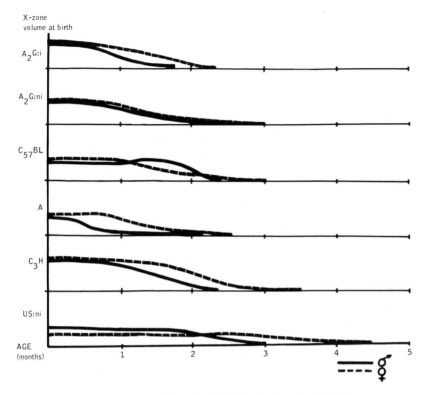

FIGURE 3 Genetic and sex variations in the timing of X-zone involution.

adrenals, the changes occurring in A_2G:ni mice being less intense. A particular enzymatic sensitivity to exogeneous estrogens occurred in the A_2G:i mouse (Table 2).

The cathecolamine content of the adrenal medulla is also subject to strain variations. The chromaphin reaction (Coupland method) and the reaction for epinephrine and norepinephrine (Hillarp-Höckfelt method) are more intensive in $C_{57}BL$, A_2G:ni, and US:ni mice and are strongly decreased in C_3H and A strains. The cathecolamine content of the medullary cells is generally proportional to the acid phosphatase and acid ascorbic quantities found (Table 3).

Strain differences in the adrenal regenerating capacity have been investigated after enucleation and after deglomerulation. In the 16 days following microsurgical removal of the adrenal "nucleus," which is composed of fasciculate, reticulate, and medullary tissues, regeneration is quickest and most marked in the unknown strain. In the inbred animals, the regenerative process is faster in the A and $C_{57}BL$ strains (Figure 5). After removing the glomerular

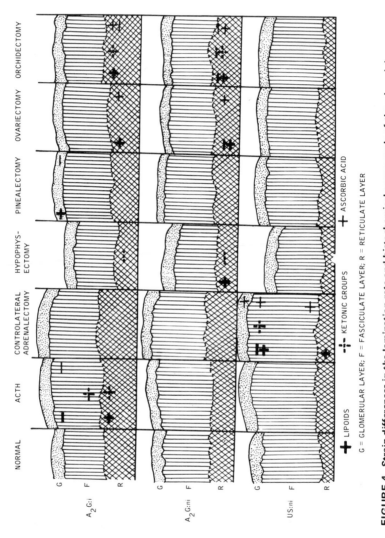

FIGURE 4 Strain differences in the lamination rate and histochemical compounds of the adrenal, in some experimental conditions. (Only the most significant changes are included.)

TABLE 2 Strain Differences in the Adrenal Enzymes of Castrated Mice under Various Stimuli—Females, Three-Month Treatment

Histological Detected Enzymes	Control			ACTH (daily)			Estradiol (0.2 ml daily)		
	A_2G:i	A_2G:ni	US:ni	A_2G:i	A_2G:ni	US:ni	A_2G:i	A_2G:ni	US:ni
DPNH-diaphorase	+	+	++	+	+	++	++	+	++
TPNH-diaphorase	+	+	+	++	+	++	+	+	+
Succinic-dehydrogenase	+	+	++	++	+	+	++	++	+
Malic-dehydrogenase	++	+	+	+	++	+	++	+	++
Lactic-dehydrogenase	+	++	+	++	++	+	++	+	++
Lipase	++	++	+	+	++	+	++	+	+
Cholinesterase	++	+	++	++	++	++	+	+	+
Glucose-6-phosphate-dehydrogenase	+	+	++	++	+	+++	++	+	+
3-β-hydroxysteroid-dehydrogenase	+	+	++	+	+	+++	++	+	++

TABLE 3 Strain Variations in the Adrenal Medulla Cathecol-Amine Content

Histochemical Methods	A_2G:i	A_2G:ni	C_3H	$C_{57}BL$	A	US:ni
Chromaffin reaction	+	++	+	++	+	++
Epinephrine	++	++	+	++	+	++
Norepinephrine	++	+++	+	++	++	++++
Acid phosphatase	+	++	∓	+	+	++
Ascorbic acid	+	+	+	++	±	+

layer (deglomerulation), the centrifugal regeneration that takes place during the following 30 days is particularly important in C_3H and $C_{57}BL$ mice. The lowest capacity for generating a new glomerular zone is observed in the A strain. In general, without being the rule, the randombred animals have a lower ability to remake mineralcorticoid tissues (Figure 6).

The periadrenal fat has a corticoidogenic capacity, and under certain circumstances it can replace a part of the adrenal function. Consequently, when an adrenalectomy with periadrenal lipectomy is performed, the mortality rate is greater, even when the adrenal accessory nodules are preserved. In such experiments we noted that the absolute consanguinity induces a smaller compensatory value of the extra-adrenal adipose tissue. That property is, on the contrary, well-marked in the noninbred mice (Figure 7).

The strain differences in the eosinopenic response to stress were recorded after bleeding in inbred and noninbred mice. From the inbred strains it was observed that in the four hours following the bleeding no significant eosinopenic changes occurred in C_3H mice. During the same period, a slight eosinopenia developed in US:ni mice, while in A_2G and $C_{57}BL$ mice a 20 to 40 percent eosinopenia was counted (Figure 8).

Some obvious strain variations are also present in the viability of transplanted adrenals into the somatic or hepatic portal system. In some strains, like A_2G, the difference between the viability of the intrasomatic and intraportal grafts is very conspicuous compared with the nonsignificant variations noted in $C_{57}BL$ mice. Contrary to the other strains, the intrasomatic grafts performed on A strain mice are more viable than the intraportal ones (Table 4).

At the ultrastructural level, we noted some electronomicroscopic peculiarities of the adrenal cells:

1. In C_3H mice, a great volume of the smooth endoplasmic reticulum vacuoles (store forms), which are often surrounded by amorphous cytoplasmic mass (Figure 9).

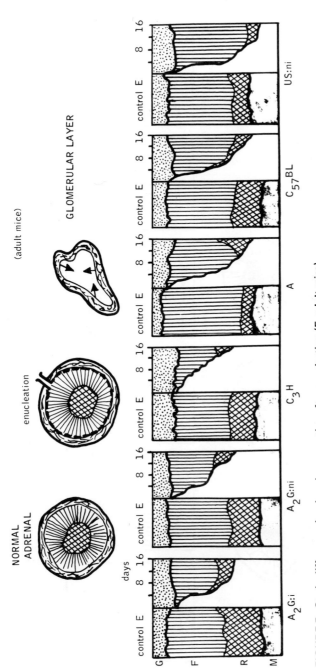

FIGURE 5 Strain differences in adrenal regeneration after enucleation (E, adult mice).

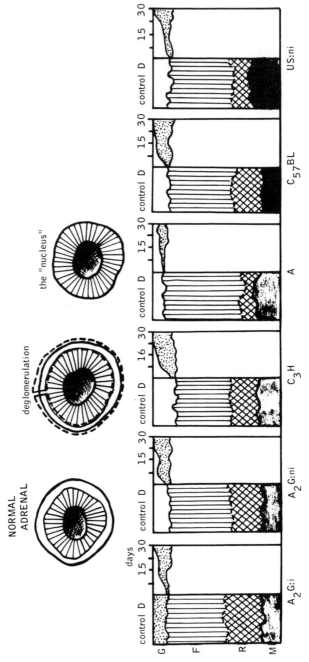

FIGURE 6 Strain differences of adrenal regeneration after deglomerulation (D, adult mice).

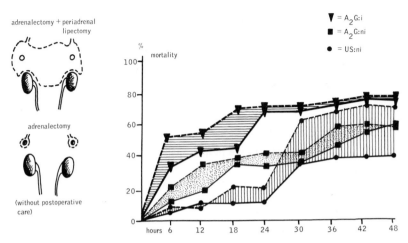

FIGURE 7 Strain differences in the compensatory value of the periadrenal adipose tissue.

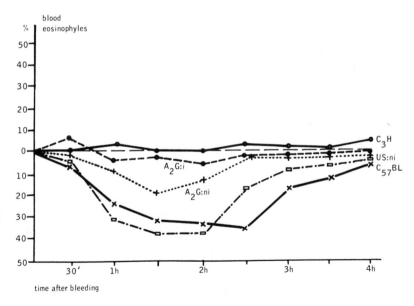

FIGURE 8 Strain differences in the eosinopenia after bleeding.

TABLE 4 Strain Differences in the Adrenal Autograft Viability after Twenty-Six Days

Autografts Sites	A_2G:i (%)	A_2G:ni (%)	C_3H (%)	A (%)	C_{57} (%)	US:ni (%)
Somatic tissues	71 ± 3.6	60 ± 5.9	83 ± 6.6	79 ± 5.2	91 ± 4.1	54 ± 9.4
Hepatic portal system	86 ± 2.3	84 ± 8.1	94 ± 6.1	68 ± 4.4	92 ± 3.6	61 ± 8.9

2. In A_2G mice, especially in the inbred ones, the great number of mitochondria have vesiculous or cylindric internal cristae. The A_2G strain of lower selection is characterized by a more polymorphous ultrastructural pattern of adrenal cells with many intermediate forms between the spongious and oxyphilic cells (Figure 10).

3. In the A strain mice, which possess elaborate cortisole, there is a rich smooth endoplasmic reticulum and macromitochondria.

4. In the US:ni mice, there is a predominance of quickly depleting structures, the microvesicles (Figure 11).

Discussion

The strain variations induced by high selection on the adrenal structure, function, and reactivity prove the ways in which the genetic pattern scale of species homogeneity and variability can become apparent at different levels of investigation. At present for some strains the morphofunctional differences can be correlated with some strain peculiarities of the metabolic stereotype of adrenal cell hormonosynthesis. The A/Cam mice, having a complete enzymatic spectrum, secrete cortisole. Other strains, such as CBA/FaCam, have no 17-hydroxylase or 11-hydroxylase and thus do not produce cortisole (Figure 12). The presence or absence of 11-hydroxylasing activity seems to be of great importance for the final pattern of corticosteroid biosynthesis.[4] It goes without saying how unprofitable and erroneous it is to perform experiments looking for the cortisole on CBA/FaCam mice and those looking for eosinopenia after stress in C_3H mice. Taking into account that hormonogenesis is one of the two basic cellular functions the enzymatic control of which is genetically induced, it is quite clear that if we want to search that part of the genetic factor in the endocrinological experiments or tests, we must not investigate so much the energy metabolism as the hormonogenesis and the mitotic capacity, because these have an evident genetic effect.

Because of the strain differences, the experiment is always, in fact, a par-

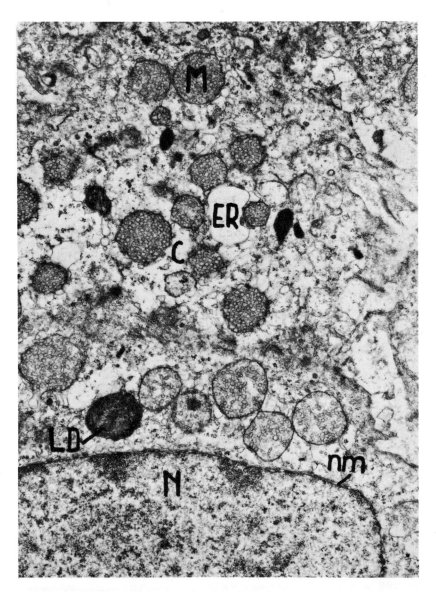

FIGURE 9 Adrenal cell electromicrograph in C₃H mice. Note the relatively scarce mito-chondria (M) surrounded by cytoplasmic ground matrix (C) and containing vesicular in-ternal cristae (IC); rare vesicles of the endoplasmic reticulum (ER), nucleus (N), lipoid droplet (LD), nuclear membrane (nm). (X22,000)

427

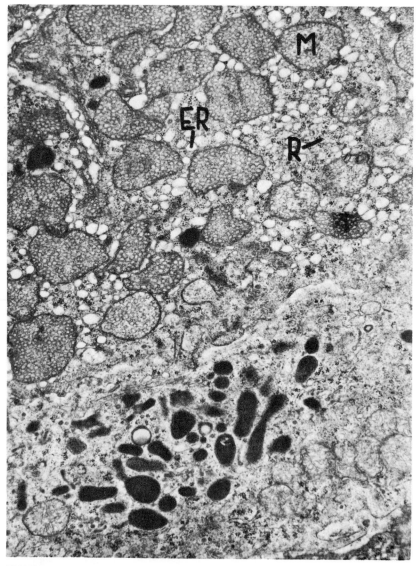

FIGURE 10 Adrenal cell electromicrograph in A₂G:i mice. Note the polimorphous mitochondria (M), well-marked endoplasmic reticulum (ER), and relatively rich free ribosomes (R). (X22,000)

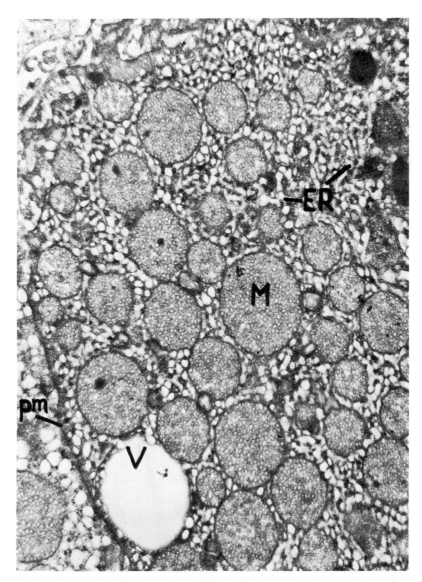

FIGURE 11   Adrenal cell electromicrograph in a US:ni mouse. Note the frequent mitochondria (M) close to the well-developed microvesicular smooth endoplasmic reticulum
(ER); vacuole (V), plasmic membrane (pm). (×22,000)

TABLE 5 Some Adrenal Biologic Characteristics of "Sensitive" and "Nonsensitive" Strains of Mice

Adrenal Peculiarities	"Sensitive Strains"	"Nonsensitive Strains"
Relatively big size	female, male, left, right, $C_{57}BL$	
Relatively small size	DBA/2; RIII	
Accessory adrenals	$C_{57}BL$	
Thin glomerular layer	Peru	
X-zone: precocious involution	female, male; A; $C_{57}BL$; BALB/c	CBA; DBA/1
X-zone: broad	C_3H; BALB/c × C_3H	
Hormonogenesis: cortisole	A; A/Cam	
Hormonogenesis: no cortisole	CBA/FaCam	
Responsiveness to cortisole	C_{57}; C_{57} ×	129/Rr; DBA/1; BALB/c
Eosinopenia after stress	$C_{57}BL$; A_2G	C_3H; US:ni
Hypertrophy, post-estrogens	BABL/c; BALB/cGa; C_{57} × C_3H	C_3H=atrophy
Hyperplasia+microadenomatosis		
–after neonatal gonadectomy	DBA/2; BALB/c × C_3 H female	DE/J
–on old age	NH; A_2G; C_3H	
–after SH+ACTH	A_2G	
Amiloidosis	DE	
Tumors		
–after neonatal gonadectomy	CE/J; BALB/CJ	DE/J
–after SH+ACTH+estrogens	A_2G	
–spontaneous	BALB/c; BALB/cHuDi; CE/WyDi	
Hypercorticism	$C_{57}BL/6J$-ob; NZO	
Teratogenesis after cortisole	C_3H/JKt; $C_{57}BL/6JKt$; DBA/1JKt; A/JKt; CBA/Cag; CBA/JKt; BT/Kt	A_2G

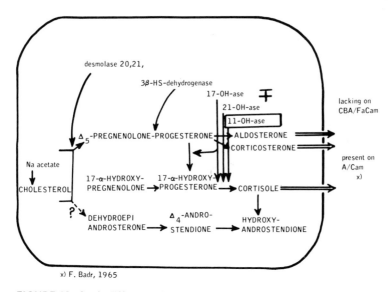

x) F. Badr, 1965

FIGURE 12 Strain differences in the adrenal hormonogenesis.

ticular case. Consequently, in evaluating the significance of the usual responses of the endocrine system during various experiments, one must investigate and determine which part of each dramatype is due to the phenotype and which is due to the genotype. By critical analysis of reactivity components various characters could be distinguished. Thus we might be surprised to discover that what is normal for one strain could be abnormal for another.

Considering our own observations, together with what other investigators presented, one can obtain some practical indications for choosing more-or-less appropriate sensitive strains of mice for an adequate experiment (Table 5). To keep the biological background of a strain or species as real as possible, especially for endocrinological investigations, it is quite proper to work on both genetically homogeneous and heterogeneous animals. Among these, the occurrence of significant adrenal strain differences is only one example of genetic variability.

Conclusion

The strain differences observed in adrenal structure, function, and reactivity show that the genetic pattern of strain and species homogeneity and variability can obviously appear at different levels of laboratory investigation.

The strain variations can be more evident when the investigation is deeper and nearer to the endocrine metabolic stereotype, which is genetically induced and specific for the respective animal group.

The strain characteristics offer the opportunity to select the most suitable strain for a particular endocrine research purpose, taking into account the endocrine peculiarities of the strain, either on the whole or in individual cases.

The findings show that to keep the strain and species biological background as real as possible and to facilitate the application of experimental results to man, a complex heterozygote, it is appropriate to work with both genetically homogeneous and heterogeneous animals.

Taking into account that consanguinity and mutations may change some endocrine characters, there are some prospects of creating new lines having particular endocrine reactivities that will suit the needs of the search for health in this field.

References

1. Ader, R., B. Friedman, and L. J. Grota. 1967. "Emotionality" and adrenal cortical function: effects of strain, test and the 24-hour corticosterone rhythm. Anim. Behav. 15:37.
2. Allen, C., and J. W. Kendall. 1967. Maturation of the circadian rhythm of plasma corticosterone in the rat. Endocrinology 80:926.
3. Argyris, B. F. 1959. Mechanism of insulin resistance in the KL strain of mice: Inactivation of insulin, Endocrinology 64:400.
4. Badr, F. M. and S. G. Spickett. 1965. Genetic variations in the biosynthesis of corticosteroids in *Mus musculus*. Nature 205:1088.
5. Badr, F. M., and S. G. Spickett. 1965. Genetic variations in adrenal weight relative to body weight in mice. Acta Endocrinol. Suppl. 50:92.
6. Badr, F. M., I. G. M. Shire, and S. G. Spickett. 1968. Genetic variation in adrenal weight: Strain differences in the development of the adrenal glands of mice. Acta Endocrinol. 191:58.
7. Barry, G., and E. L. Kennaway. 1937. The structure of the thyroid in mice of different strains. Amer. J. Cancer 29:522.
8. Bartke, A. 1965. The response of two types of dwarf mice to growth hormone, thyrotropin and thyroxine. Gen. Comp. Endocrinol. 5:48.
9. Beatty, R. A., and K. N. Sharma. 1960. Strain differences in spermatozoa from eight inbred strains of mice. Proc. Roy. Soc. Edinburgh. B 68:25.
10. Claringbold, P. J., and J. D. Biggers. 1955. The response of inbred mice to oestrogens. J. Endocrinol. 12:9.
11. Courrier, R. 1958. La nature des besoins en animaux pour les recherches endocrinologiques. ICLA Symp. 1958 "Living animal material for biological research, MRC Lab., Carshalton, Surrey, England.
12. Daughaday, W. 1941. A comparison of the X-zone the adrenal cortex in two inbred strains of mice, Cancer Res. 1:883.
13. Dawson, W. D. 1967. Comparative thyroid activity in two species of *Peromyscus*. Gen. Comp. Endocrinol. 8:1267.

14. Schott, J. 1965. Etude comparative de la cytologie et de l'ultrastructure de l'hypophyse distale de trois espèces d'Amphibiens Anoures: *Rana temporaria* L., *Bufo vulgaris* Laur., *Xenopus laevis* D., Gen. Comp. Endocrinol. 5:631.

15. Eayrs, J. T., and E. D. Williams. 1966. Diet and strain of animal as factors regulating thyroid activity. J. Endocrinol. 34:277.

16. Elefteriou, B. E., and R. C. Church. 1967. Effects of repeated exposure to aggression and defeat on plasma and pituitary levels of luteinizing hormone in $C_{57}BL/6J$ mice. Gen. Comp. Endocrinol. 9:263.

17. Gagliardi, N. C., P. A. Kitos, and I. A. Weir. 1964. Genetic control of thyroid activity in PHH and PHL mice. Genetics 50:249.

18. Giroud, A., and M. Martinet. 1954. Influence de la souche de rat sur l'apparition des caractères thyroxiniennes. Arch. France Pediat. 11:168.

19. Glick, B., and L. J. Dreesen. 1967. The influence of selecting for large and small bursa size on adrenal, spleen and thymus weights. Poultry Sci. 46:396.

20. Huff, S. D., and K. B. Eik-Nes. 1965. Testicular 17-alfa-hidroxylase activity in inbred and hybrid mice with different beta-glucuronidase activity. Steroids, Suppl. 2:97.

21. Kalter, M. 1954. The inheritance of susceptibility to the teratogenic action of cortisone in mice. Genetics 39:185.

22. Kwa, H. G., and F. Verhofstad. 1967. Prolactin levels in the plasma of female $(C_{57}BL/6 \times CBA) F_1$ mice. J. Endocrinol. 38:81.

23. Levy, R. P. 1965. Effect of species differences of mice on the bioassay of thyrotropin. Endocrinology 76:890.

24. Lin, T. P., and D. W. Bailey. 1965. Differences between two inbred strains of mice in ovulatory response to repeated administration of gonadotrophins. J. Reprod. Fertil. 10:252.

25. McLaren, A. 1967. Factors affecting the variations in response of mice to gonadotrophic hormones. J. Endocrinol. 37:147.

26. Meckler, R. J., and R. L. Collins. 1965. Histology and weight of the mouse adrenal: A diallel genetic study. J. Endocrinol. 31:95.

27. Mendoza, L. A., M. Hamburgh, and H. Fuld. 1967. Differences in thyroid activity in several inbred strains of mice. Anat. Rec. 158:275.

28. Mosier, H. D., and C. P. Richter. 1967. Histologic and physiologic comparisons of the thyroid gland of the wild and domesticated Norway rat. Anat. Rec. 158:263.

29. Nandi, S. 1961. Differential responsiveness of A and C_3H mouse mammary tissues to somatotropin containing hormone combinations. Proc. Soc. Exp. Biol. Med. 108:1.

30. Oortmerssen, G. A., and J. A. Beardmore. 1967. An age factor affecting variance of a behavioural character in F_1 hybrids between inbred lines of the house mouse. Experientia 23:328.

31. Premachandra, B. N. 1965. A study of the effect of reserpine of thyroid secretion in several mammalian species. J. Endocrinol. 33:397.

32. Roos, T. B. 1967. Steroid synthesis in embryonic and fetal rat adrenal tissue. Endocrinology 81:716.

33. Sabourdy, M. 1967. L'animal de laboratoire dans la recherche biologique et médicale. Presse Univ. de France, Paris.

34. Sackler, A. M., and A. S. Weltman. 1967. Metabolic and endocrine differences between the mutation whirler and normal female mice. J. Exp. Zool. 164:133.

35. Shire, I. G. M. 1965. Genetic variations in the structure and development of the mouse adrenal cortex. Acta Endocrinol. Suppl. 50:91.

36. Shire, I. G. M. 1965. Genetic variations in the degeneration of the X-zone of female mice. J. Endocrinol. 33:II.

37. Simionescu, N. 1967. Surgical experimental models in endocrinology. p. 471–493. *In* M. Conalty [ed.] Husbandry of laboratory animals, Academic Press, New York - London.

38. Solomon, G. F., Y. C. Merigan, and S. Levine. 1967. Variations in adrenal cortical hormones within physiologic ranges stress and interferon production in mice. Proc. Soc. Exp. Biol. Med. 126:74.

39. Stewart, J., and S. G. Spickett. 1965. Genetic variation in water and electrolyte metabolism in mice. Acta Endocrinol. Suppl. 50:94.

40. Van Heyningen, H. E. 1961. Differences in thyroid function of several strains of mice. Proc. Soc. Exp. Biol. Med. 106:37.

41. Von Schilling, V., H. Frohberg, and H. Oettel. 1967. On the incidence of naturally occurring thyroid carcinoma in the Sprague-Dawley rat. Indust. Med. Surg. 36:678.

42. Wakasugi, N., T. Tonita, and K. Kondo. 1967. Differences of fertility in reciprocal crosses between inbred strains of mice: DDK, KK and NC. J. Reprod. Fert. 13:41.

43. Westenberg, I. A., H. A. Beru, and E. B. Barnawell. 1957. Strain differences in the response of the mouse adrenal to oestrogen. Acta Endocrinol. 25:70.

44. Whitehouse, B. J., and G. P. Vinson. 1967. Pathways of corticosteroid biosynthesis in duck adrenal glands. Gen. Comp. Endocrinol. 9:161.

45. Wolff, G. L. 1965. Hereditary obesity and hormone deficiencies in yellow dwarf mice. Amer. J. Physiol. 209:633.

46. Zarrow, M. X., V. A. Devenberg, and W. D. Kaberer. 1965. Strain differences in the endocrine basis of maternal nest-building in the rabbit. J. Reprod. Fert. 10:397.

47. Young, C. W., and I. E. Legates. 1965. Genetic, phenotypic and maternal interrelationships of growth in mice. Genetics 52:563.

DISCUSSION

DR. LANE-PETTER: To what extent are the strain differences you have referred to influenced by the state of health?

DR. SIMIONESCU: Do you mean to what extent infection could influence the changes that are mentioned here? This condition occurred in one animal colony where we noted that all of the animals used for the experimental research were considered healthy. From a practical point of view, I have no evidence that the infection has any influence on this difference.

DEFINING THE ROLE OF THE GUINEA PIG IN CANCER RESEARCH: A NEW MODEL FOR LEUKEMIA AND CANCER IMMUNOLOGY STUDIES

Stanley R. Opler

The necessity for using experimental animals in cancer research has stimulated investigators to seek animal model systems that might be more illuminating than those now used in studying the broad field of oncogenesis and immunology of specific neoplasms. Avian and murine systems have long been utilized as models for viral oncogenesis, as well as leukemia and mammary tumor research. The ready availability of inbred strains of rats and mice has resulted in their becoming the standard experimental animals in tumor investigation. With specific reference to studies of leukemogenesis, the pioneering attempts at elucidating the role of viruses, chemical carcinogens, and radiation have been carried out in these animals. Details of these studies have been reviewed elsewhere.[1,6,7,12] In view of the presence of viral particles in all mice studied, their vertical transmission, and variance of some of the leukemias and tumors observed from those in humans, interest has developed in finding alternative animal systems. Our own early studies were concerned with a search for another model of acute leukemia that might bear a greater resemblance to human leukemia than mice do.[13] In addition, we were interested in finding a genetically uniform animal for cancer immunology studies, since in many experiments where noninbred animals were used, tumors were rejected as in any homograft. These studies of transplanted tumors were not experiments in cancer research but, rather, observations of transplantation rejection.

Viruses are among the fastest acting known carcinogens in animals and, therefore, it is natural that models of viral oncogenesis have been sought that might directly relate to the human cancer problem. No definite proof of this relationship in humans yet exists, but studies of the viral etiology of human

435

tumors are being very actively pursued in many laboratories. It is certainly of significance that many of the animal tumors investigated to date are of viral origin. All tumors induced by a particular oncogenic virus appear to have the same tumor-specific antigens, which is not the case in chemically induced tumors. In virus-induced animal leukemia, circulating antibody, which damages leukemic cells *in vitro*, is produced, while in chemically induced tumors this does not appear to occur. Transplanted tumors have been the subject of study in most experiments; however, the need to answer the question of whether a host can react against his own tumor has stimulated us to investigate spontaneous tumors in the guinea pig and man. Our studies have shown that the host can respond immunologically by producing specific morphologic patterns associated with longer survival. These changes are observed in lymph nodes that have markedly dilated sinuses containing proliferating macrophages and pyroninophilic lymphocytes. The primary tumor is infiltrated with small lymphocytes. *In vitro* studies show marked reduction of plated cancer cells by host lymphocytes. This study indicates that human breast cancer possesses tumor-specific antigens, as has been shown in animal tumors.

Until quite recently the guinea pig (*Cavia porcellus*, commonly referred to as cavies) had not been used as a model for viral oncogenesis or cancer immunology studies.[16] Historically, a concept had been advanced that the guinea pig exhibited few spontaneous or induced tumors; it was, in fact, generally considered to have such a low incidence of cancer that several investigators pursued studies of guinea pig serum as a tumor inhibitor.[8,10] We have found a pertinent factor that lends credence to this concept: Few people, if any, kept guinea pigs for an appreciable length of time (i.e., until they were at least 4 years of age). On the other hand, these animals have been classically employed in immunology research and in study of diseases relating to humans; among these are scurvy, diphtheria, tuberculosis, endemic and epidemic typhus, Rocky Mountain spotted fever, lymphocytic choriomeningitis, amebiasis, and coccidiomycosis. The guinea pig has been used in animal inoculation studies in order to identify and isolate many pathogenic organisms, reproduce syndromes, and evaluate the pathologic tissue changes characteristic of disease. Moreover, cavies have been extensively employed for work on the biological standardization of vaccines and antiviral agents, as well as in sensitization studies.

Inbreeding of guinea pigs began in 1897 at the Bureau of Animal Husbandry in Bethesda, Maryland, although only 54 animals out of 850 survived a move of the Experiment Station to new quarters. In 1906 extensive experiments were initiated in order to study the effects of inbreeding when continued for successive generations, and several detailed reports of this work were published.[11,25,26] Of the original inbred stock, two families have been maintained (i.e., strains 2 and 13).

In the past, isolated cases of spontaneous and induced leukemia, as well as other neoplasms, have been reported infrequently in the guinea pig. Two cases of leukemia were reported in 1952 in a colony of 51 strain 2 guinea pigs that had been treated with intravenous injections of 1,2,5,6 dibenzanthracene and 20-methylcholanthrene.[9] In 1954, Congdon and Lorenz reported on a series of 10 cases of leukemia in both irradiated and nonirradiated guinea pigs.[4] Although some of these leukemias had been successfully carried by transplantation, it is evident that there was a lack of interest in the guinea pig as a model for cancer research. However, with the finding of a leukemia virus in guinea pigs (Opler virus),[14,15] interest revived in this species as an experimental model for cancer studies (Table 1).

TABLE 1 Animal Leukemia and Lymphoma Virus Research

Date	Investigators	Disease
1908	Ellermann and Bang	Chicken leukemia and lymphoma
1951	Gross	Mouse leukemia
1957	Friend	Mouse leukemia
1959	Moloney	Mouse leukemia
1959	Kaplan and Lieberman	Mouse leukemia[a,b]
1962	Rauscher	Mouse leukemia
1963	Rich	Mouse leukemia
1967	Opler	Guinea pig leukemia

[a]From Progress Against Cancer, 1967, A Report by the National Advisory Cancer Council.
[b]Radiation activates dormant virus.

In our studies there appeared to be a remarkable similarity in the evolution of the acute leukemia carried in our guinea pig colony to that observed in humans, suggesting that further investigation might be relevant to human leukemia.

The true incidence of leukemia in guinea pigs has been obscured by the lack of complete autopsy and hematologic studies in animals dying suddenly of acute disorders. Significant features can be easily overlooked in a cursory examination. Under a policy of carefully examining the bone marrow and peripheral blood, as well as performing autopsies on all dying animals in our colonies, we have detected only rare cases of spontaneous leukemia. The rate of appearance of this disease is so low that it appears comparable to the incidence in humans. However, careful observation will detect the presence of acute leukemia and other spontaneous neoplasms in this species, especially in aged animals.

Our research is concerned with studies of guinea pigs as a model for cancer immunology studies and, specifically, as a model for research on the etiology, pathogenesis, and therapy of acute lymphatic leukemia. In studies of animal or human leukemia it is extremely important to differentiate the cell type as well as to classify it. It would seem that acute lymphatic, acute granulocytic, and chronic lymphatic leukemia behave entirely differently and may, in fact, be unrelated etiologically and pathogenically. The specific type of leukemia seen in the animals used in our work is the acute lymphatic form, which strikingly resembles that disease in children and young adults. It is a fulminating, rapidly fatal disease in humans as well as in guinea pigs, and comparative studies of pathology and hematology reveal marked similarities.[18,19] Studies on chemotherapy and immunotherapy of these animals are now being actively pursued. In preliminary studies, guinea pig leukemia appears to respond to forms of therapy useful in human leukemia.

We have been able to identify a specific virus in tissues from leukemic guinea pigs.[14,15] The presence of this virus has been confirmed by other investigators, and it appears to be uniquely different morphologically and bio-

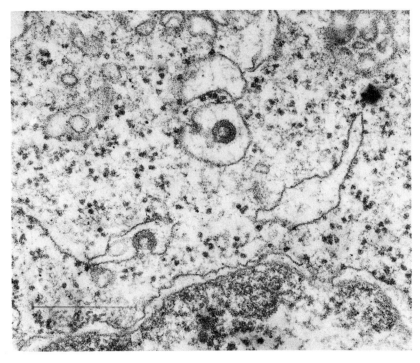

FIGURE 1   Opler virus particles budding from the endoplasmic reticulum. (X 123,000, approximately.)

logically from the murine viruses. Under the electron microscope, the Opler virus measures approximately 100 mμ and buds from the endoplasmic reticulum (Figure 1), in contrast to the murine leukemia viruses, which bud from the outer membrane of cells. C-type particles are seen in the intercellular spaces and in the cytoplasm of cells we have identified as lymphoid cells (Figure 2). The intermediate layer is noted to be less electron dense than the murine viruses. This acute lymphatic leukemia in guinea pigs is transmissible by injection of cell-free material and high-speed plasma pellets, by transplantation of tumor tissue, and by the oral feeding of infected spleen to susceptible animals.[20] Strain 2 guinea pigs are maximally susceptible as newborns and as young animals. Strain 13 animals are resistant. In experiments that show that the immune response is genetically mediated, strain 2 and strain 13 respond differently to various synthetic polypeptides.[2,3] In crossing strain 2 with strain 13, the resulting F_1 hybrids are maximally susceptible to induction of the leukemia with our virus, suggesting that the genetic determinants of strain 2 are dominant.

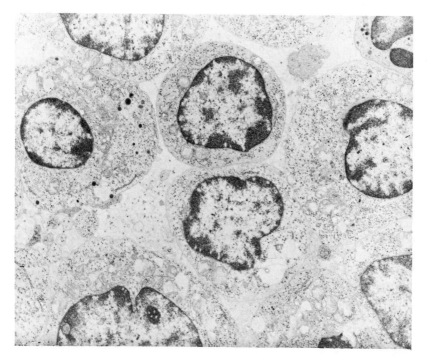

FIGURE 2   Bone marrow: lymphoblasts with virus particles in cytoplasm. ($\times$ 11,000, approximately.)

It is extremely important to use inbred animals in cancer research in order to ensure as much uniformity in the system as possible. Although we have been able to induce this leukemia in outbred guinea pigs by injecting the virus and using various modalities, including immunosuppression or thymectomy, we stress the significance of the use of syngenic animals in cancer immunology research because of the increasing importance of the role of tumor-specific antigens. It is only when tumors are transplanted into genetically similar animals that immune responses affecting growth of cancer can be studied and the presence of tumor-specific antigens identified.

It is of major interest that in this guinea pig model, passenger virions have not been found in nonleukemic or otherwise healthy animals. Nor has it been possible to increase the rate of leukemia significantly by means of radiation. However, in other virus systems, such as the murine, radiation has been shown to potentiate or, in the case of the rad leukemia viruses (i.e., Kaplan and Lieberman), to activate a dormant virus. One possible explanation for this is that radiation damages the immune system in the form of the lymphocyte in these mice, which normally carry passenger viruses, thereby allowing a vertically transmitted or latent virus to become manifest. This would account for the difference in the role of radiation in the murine system as opposed to the guinea pig.

Preliminary biochemical studies indicate that the Opler virus is an RNA virus, incorporating RNA precursors. Tissue-culture experiments are currently in progress. The virus appears to have a cytopathic effect. Replication occurs in the cytoplasm of lymphoid cells and can readily be visualized by observing assembly at a membrane in the majority of lymphoblasts.

We have previously mentioned the rather close resemblance of the leukemia carried in our guinea pig colony to that of acute lymphatic leukemia found in man.[17,23] This experimental disease, which has a relatively short latency period (i.e., from 10 to 40 days, depending upon the dose and mode of transmission), involves practically every organ system (Figures 3 and 4) and is not, therefore, specialized in character as are some of the murine models. As in man, and unlike some of the other animal leukemias, this disease appears to begin in the marrow (Figure 5) and is a true leukemia as opposed to a solid tumor or lymphoma, which secondarily sheds into the bloodstream. The infiltration of the leukemic cells in the guinea pig leukemia is generally the same as that observed in human disease. The clinical course closely parallels that of human leukemia, with the peripheral lymphoblast cell counts ultimately ranging from 100,000 to 250,000 cells per mm^3 (Figure 6). Comparative hematologic experiments show that this model follows closely the changes seen in the human hematopoietic system in cases of acute lymphatic leukemia. This makes it useful in following the effects of therapy, since blood specimens are readily obtainable on a routine basis. Preliminary work has shown that the

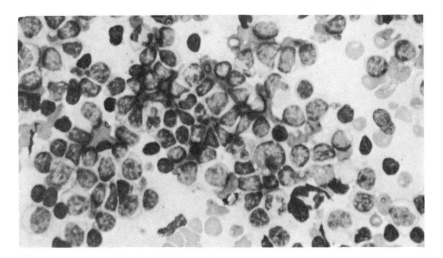

FIGURE 3　Spleen imprint showing lymphoblasts. (X 160, approximately.)

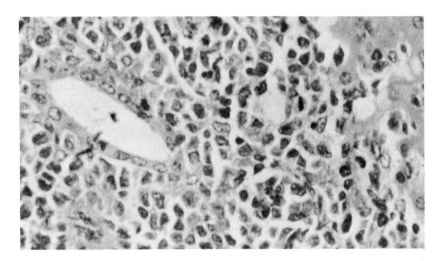

FIGURE 4　Leukemic infiltration of liver. (X 400, approximately.)

guinea pig response is similar to the human response to various therapeutic agents employed in treatment of acute lymphatic leukemia in humans. It has been possible to immunize Hartley guinea pigs and produce antisera that afford some protection against the leukemia in strain 2 animals. We have delayed onset, and prolonged survival, by intradermal inoculation with leukemic cells.

In other studies in our laboratory, the guinea pig has not proven to be re-

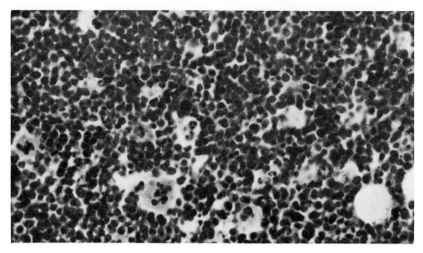

FIGURE 5   Bone marrow smear from leukemic guinea pig. (X 160, approximately.)

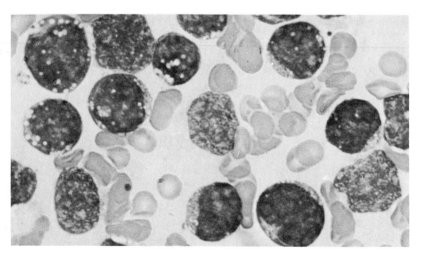

FIGURE 6   Peripheral blood showing lymphoblasts, with count of 200,000 cells/mm^3. (X 400, approximately.)

sistant to the development of a variety of spontaneous solid tumors. If guinea pigs are carefully monitored with blood counts, biopsies, or autopsy studies, or all of these, it is found that both spontaneous and induced neoplasms occur in inbred and outbred animals. Because guinea pigs are used for short-term experiments rather than tumor studies, they have rarely been observed or studied in old age. In addition, there are fewer inbred strains than in other species. In our colony we have observed spontaneous tumors in animals over 4 years of age, including adenocarcinomas of the mammary gland, leiomyomas of the uterus, papillary adenomas of the lung, hepatomas, and a variety of soft-tissue neoplasms. Malignant tumors have an ability to metastasize readily. One possible explanation for the development of spontaneous tumors and metastasis in older animals may be atrophy of the lymphoid system, with impairment of the cell-mediated immune surveillance mechanism. In addition to these spontaneous neoplasms, it has been possible to induce neoplasms with local methylcholanthrene administration, as well as other chemical carcinogens, which may act as immunosuppressants. With the development of tumor immunology and studies of tumor-specific antigens, the guinea pig model may be of value because of the widespread use of guinea pigs in immunology studies in the past and their ability to develop delayed hypersensitivity.

We are currently investigating the role of the reticuloendothelial system in the guinea pig, as well as the role of the lymphocyte in the primary tumor, and the role of immune lymphadenopathy in cancer-survival. This is being carried out in conjunction with studies in the human in which it has been reported that host resistance to breast and gastric cancer was demonstrated, with the presence of lymphocytes in the primary tumor, and hyperplasia of macrophages in the regional lymph nodes was associated with longer survival.[21,22] This appears to be an example of tumor-specific immune reaction. We have observed similar findings in guinea pigs with spontaneous malignant tumors. In a series of survival studies of acute lymphatic leukemia in humans conducted in my laboratory at Stanford, the persistence of normal lymphocytes in the bone marrow of more than 25 percent following therapy was associated with longer survival. Investigations are now under way using the guinea pig as a model for experiments in which antilymphocyte globulin is injected, as in immunosuppressive therapy, to determine if there is an increased incidence of spontaneous tumors as well as increased susceptibility to our virus in strain 13 and outbred guinea pigs. We have observed an increased incidence of tumors in patients given this form of immunosuppression during renal transplantation, and this appears to be true in guinea pigs.

Once a tumor appears, it develops and spreads, according to one theory, because there is increased tolerance to its specific antigens.[24] Alternatively, it seems more likely that the lymphoid system of the tumor host is no longer able to produce an effective cellular immune response. One approach to therapy may be to inject sensitized lymphocytes, to destroy tumor target cells. Chemotherapy may be developed to allow the immune mechanism to function against cancer after removing the majority of malignant cells.

Discussion

What can we learn from the use of these animal models? We are concerned with the following questions.

1. Is leukemia an acute infection? In this animal model leukemia rapidly follows the introduction of the virus in the susceptible host, in a high percentage of cases. In humans the onset of acute leukemia appears to have some of the clinical manifestations of an acute infection.

2. Is there a fundamental difference between irreversible infections and neoplastic disease? So far, biochemical studies have failed to reveal significant metabolic alterations in neoplastic cells that do not appear in non-neoplastic cells. This factor alone has hampered chemotherapy cures, for in attempting to destroy tumor cells it is exceedingly difficult not to damage normal cells. It is also possible that some current forms of cancer therapy may unavoidably damage the cell-mediated immune mechanism. Preservation of this immune mechanism appears to be of the utmost importance to the survival of cancer patients.

3. Does cancer metastasize at its inception with varying growth rates and become manifest at different times in individual hosts due to variations in the efficiency of tumor inhibitory factors that ultimately appear to fail in the majority of patients with disseminated cancer?

4. Can the antigen–antibody response in cancer as well as in infectious disease be harmful, at times, to the host? Immunologic damage has been demonstrated in lymphocytic choriomeningitis infection in animals, including the guinea pig. Damage to the host by circulating antigen–antibody complexes that precipitate out in the kidneys have been demonstrated in serum sickness nephritis.[5] In our present studies in solid guinea pig tumors, humoral antibody appears to be enhancing and to block target cell destruction by immune lymphocytes.

5. With tumor-specific antigens present, can the antibody produced coat the tumor cells and thus prevent the lymphocyte from directly contacting these cancer cells and destroying them?

6. Can sensitized lymphocytes destroy local or metastatic tumor masses?

7. Is it possible that some tumors, including lymphatic leukemia, could result from chronic antigenic stimulation of the host, with the overproduction of lymphocytes?

The answers to these questions can be sought in experiments using animal models in conjunction with critical observation of human cancer.

It is important to realize that many of the major findings in molecular biology were made in bacterial cells and phage systems. Studies are now being pursued in animal cells on the effect of viruses in influencing their genetic mechanism, but in addition, broad studies of the intact animal organism may be necessary for the elucidation of principles of oncogenesis, chemotherapy, and immunotherapy in humans.

Hopefully, in using animal models for studies of tumors, methods may be found for employing the immune mechanism in treatment of cancer in man. The importance of animal experimentation cannot be overestimated, and animal models play an increasingly important role in studies of disease of mankind.

References

1. Andervont, H. B. 1959. Problems concerning the tumor viruses. p. 307–368. *In* F. M. Burnet and W. M. Stanley [ed.] The viruses, Vol. 3, Academic Press, New York.
2. Arquilla, E. R., and J. Finn. 1963. Insulin antibody variations in rabbits and guinea pigs and multiple antigenic determinants on insulin. J. Exp. Med. 118:55–71.
3. Ben-Efraim, S., S. Fuchs, and M. Sela. 1967. Differences in immune response to synthetic antigens in two inbred strains of guinea pigs. Immunology 12:573–581.
4. Congdon, C. C., and E. Lorenz. 1954. Leukemia in guinea pigs. Amer. J. Pathol. 30:337–359.
5. Dixon, F. J., J. J. Vazquez, W. O. Weigle, and C. G. Cochrane. 1958. Pathogenesis of serum sickness. Arch. Pathol. 65:18.
6. Dmochowski, L. 1957. The part played by viruses in the origin of tumours. p. 214–305. *In* R. Raven [ed.] Cancer, Vol. 1, Butterworth's, London.
7. Duran-Reynals, F. A. 1958. Virus induced tumors and the virus theory of cancer. p. 238–292. *In* Homburger [ed.] The physiopathology of cancer, Cassell, London; Harper-Hoeber, New York.
8. Herbut, P. A., W. H. Kraemer, and L. Pillemer. 1958. The effects of components of guinea pig serum on lymphosarcoma $6C_3HED$ in C_3H mice. Blood 13:733–739.
9. Heston, W. E., and M. K. Deringer. 1952. Induction of pulmonary tumors in guinea pigs by intravenous injection of methylcholanthrene and dibenzanthracene. J. Nat. Cancer Inst. 13:705–717.
10. Jameson, E., H. Ainis, and R. M. Ryan. 1958. The inhibition by guinea pig serum on the growth of the Murphy-Sturm lymphosarcoma. Cancer Res. 18:866–868.
11. McPhee, H. C., and O. N. Eaton. 1931. Genetic growth differentiation in guinea pigs. U.S. Dept. Agr. Tech. Bull. 222.
12. Oberling, C., and M. Guerin. 1954. The role of viruses in the production of cancer. Advan. Cancer Res. 2:423–533.
13. Opler, S. R. 1966. Pathology of acute leukemia in an inbred strain of guinea pigs. Proc. Sixth Internatl. Congr. Internatl. Acad. Pathol. (Kyoto), p. 174.
14. Opler, S. R. 1967. Electron microscopy of guinea pig leukemia virus. Proc. Amer. Ass. Cancer Res. 8:52.
15. Opler, S. R. 1967. Observations on a new virus associated with guinea pig leukemia. Preliminary note. J. Nat. Cancer Inst. 38:797–800.

16. Opler, S. R. 1967. Animal model of viral oncogenesis. Nature 215:184.

17. Opler, S. R. 1967. Pathology of cavian viral leukemia. Amer. J. Pathol. 51:1135–1151.

18. Opler, S. R. 1968. New oncogenic virus producing acute lymphatic leukemia in guinea pigs. Proc. Third Internatl. Symp. Comparative Leuk. Res. (Paris), 1967; Bibl. Haemat., (31):81–88 (Karger, Basel/New York).

19. Opler, S. R. 1968. Cavian leukemia: A hematologic and autopsy study. Symp. on Myeloprolif. Disorders of Animals and Man (Richland, Wash.) U.S. Atomic Energy Commission, Washington, D.C.

20. Opler, S. R. 1968. Transmission of viral induced cavian leukemia by the oral route. Oncology 22:273–280.

21. Opler, S. R. 1969. Morphologic expression of host-resistance in humans with breast cancer. Proc. Amer. Ass. Cancer Res. 10:67.

22. Opler, S. R. 1969. Immunological relevance of histological alterations of the reticulo-endothelial system associated with long-term survival in mammary carcinoma. Federation Proc. 28:751.

23. Opler, S. R. 1969. Viral induced acute lymphatic leukemia in guinea pigs resembling human leukemia. Haematologia 3:157–162.

24. Tyler, A. 1962. A developmental-immunogenetic analysis of cancer. p. 533–572. *In* M. J. Brennan [ed.] Biological interactions in normal and neoplastic growths, Little, Brown & Co., Boston, Mass.

25. Wright, S. 1922. The effects of inbreeding and crossbreeding on guinea pigs. I. Decline in vigor. II. Differentiation among inbred families. U.S. Dept. Agr. Bull. 1090.

26. Wright, S., and O. N. Eaton. 1929. The persistence of differentiation among inbred families of guinea pigs. U.S. Dept. Agr. Tech. Bull. 103.

DISCUSSION

TOM WONG: Is there any guinea pig leukemia of the myeloid type?

DR. OPLER: I have not observed any. The leukemia that we see we have studied extensively under the electron microscope, and the cells that are involved in this particular leukemia prove to be lymphoblasts. This is an acute lymphatic leukemia. As far as I know, there has not been an example of anything that I would call myeloid leukemia, although if you do not use the electron microscope and you look at these leukemias only under a light microscope, the cells are so primitive that you really have great difficulty in being sure of the origin. So it would be possible to look at some of these stem cells or blast cells and if you saw some differentiation, you might be tempted to think that you were seeing some myeloid leukemias. But again, I stress the fact that these leukemias, certainly in humans, are all very different. I think this is a model of acute lymphatic leukemia in humans, and in some of the other systems you see a variance, and using the same virus in

446

the same system you get a variety of leukemias, which sometimes leads me to believe that they are not really leukemias.

S. H. WEISBROTH: How did you find and pass the first case?

DR. OPLER: The leukemia had been smoldering probably for a long time. The animals, the strain 2 and strain 13 animals, were housed at NIH, and so all of us who have strain 2 animals—and they are very hard to come by—really trace these animals to NIH. Even the few dealers or breeders that have these animals get them from Wright's colony. These leukemias have been carried by transplantation, but there is also a smoldering incidence of spontaneous leukemia. So, our animals come from NIH, and I would say that the leukemias all come from Condon and Lorenz in NIH. We studied some of these, and having identified the virus, we then concentrated it in plasma pellets, and carried it from there, so that you can now carry it. As I said before, if anyone has strain 2 animals, or any guinea pigs for that matter, and you look at them very carefully, I am sure you will find a spontaneous example of guinea pig leukemia.

M. L. CONALTY: Was the serum asparaginase reduced in the leukemic cavies?

DR. OPLER: We did some preliminary work on that, and although it was not extensive, the answer is that it was not. We also tried to modify the course of the disease by administering it, and this had no effect.

DR. ERICHSEN: Were these leukemias transplantable (a) into the same strain and (b) other strains?

DR. OPLER: These leukemias do best in strain 2. Now, there are some interesting facts. The other inbred strain is strain 13. Strain 13 is resistant to leukemia. However, if you cross strain 2 with strain 13, the F_1 hybrids are maximally susceptible. The immunologists have worked with these strains, and as a matter of fact, they came up with the idea that the immunologic response was genetically mediated because strain 2 and strain 13 saw different parts of the insulin molecule, and in other immunologic studies there were differences. Now, strain 2 apparently has a genetic determinant when it is crossed with strain 13. You can also cross strain 2 with Hartleys or random-bred strains and get F_1 animals that you can carry the leukemia in. If you are interested in immunosuppression and various types of modality, you can, in fact, have this leukemia successfully transplanted into other species.

Our own feeling has been that it is important if one is going to do some cancer work to work with a species in which there is some genetic uniformity, and you can at least experiment in a model that has less unpredictability than in random-bred guinea pigs.

G. PETURSSON: How can one exclude the presence of passenger virus in healthy guinea pigs?

447

DR. OPLER: Actually, this cannot be done. People have been looking at guinea pigs for a long time, and the history of the guinea pig colony that existed from Wright's time on has been studied extensively many times, particularly in an effort to find LCM virus. Some of the early workers found that LCM virus interfered with the leukemia, and then one found prolonged incubation periods and increased survival time. But all we know is that various investigators so far have not found passenger virus.

The other curious, but not necessarily directly related thing, is that this is, I think, the only animal leukemia that does not in fact get potentiated in terms of giving radiation to a nonleukemic animal. If you irradiate some of the other species, and you knock out this lymphoid system—you, in fact, have a so-called dormant passenger virus, which takes over. If you recall the early work of Condon and Lorenz, they were not really interested in leukemia so much as they were in the effects of radiation on various animals. I do not think they were ever really able to show that radiation increased the incidence of leukemia.

The answer to the question is we don't know, but so far a number of rather competent people have looked at it and have not found it. But that is no answer.

N. L. GARLICK: Are you offering material for inoculation to other institutions?

DR. OPLER: Yes, I think we are about at the stage where we would be willing to send this to other laboratories. There are some difficulties. First, these animals are very hard to come by. If you are able to get them, they are over $15. You can keep the virus if you freeze it at minus 90 for 5 months, but in general in order to do some work with it, it is desirable to have these inbred animals. We are hoping to have it in tissue culture, and that would make it more useful for certain types of work.

One of the things we are interested in is studying the leukemia in the intact animal where the immune system has various ways of prevailing on the leukemia. But we do hope to have this virus available for investigators who would like to pursue leukemia studies.

E. BOND: How does this differ from the A-leukemia leukemias in dogs and cats?

DR. OPLER: My experience with dogs and cats is really extremely limited, so I do not think I am qualified to answer that, except on the basis of what I have heard. This leukemia has a relatively short incubation period. It begins in the marrow. It floods the peripheral blood. As far as I know, this is not the case in dogs and cats. I have the impression that the dog and cat leukemias—you will have to excuse my ignorance in that I know more about people than dogs and cats—were more indolent. So all I can say is I am ignorant of dog and cat leukemia. I just do not have the impression that they were fulminating and had the characteristics of this disease.

J. MAISIN: Were you able to extract a cell-specific antigen out of those leukemic cells—an antigen that is different from the viral antigen?

DR. OPLER: This work is preliminary, and I am not sure we have the answer. I know we can protect some of the guinea pigs because we can get antisera in Hartley guinea pigs that protect them. But the immunology needs a bit of work, so I would say we do not know the answer.

SPONTANEOUS TUMORS OF THE SYRIAN HAMSTER: OBSERVATIONS IN A CLOSED BREEDING COLONY AND A REVIEW OF THE LITERATURE

G. L. Van Hoosier, Jr.
H. J. Spjut
J. J. Trentin

The Syrian hamster (*Mesocricetus auratus*) is one of the more frequently used laboratory animals.[44] It was described as a new species, originally *Cricetus auratus*, by Waterhouse in 1839 and introduced into the laboratory in 1930 for research on kala-azar. Subsequently the species has proved especially valuable for studies on a variety of infectious diseases, e.g., tuberculosis, leprosy, and brucellosis, and for tissue transplantation. The Syrian hamster has been reported to have only three histocompatibility loci, in contrast to other animals, including man, with an estimated 12 to 15 or more.[3] The cheek pouch, an invagination of the oral cavity, is a useful site for the transplantation of allografts and xenografts of both normal and neoplastic tissue.

One of the commonest current uses of the species is in experimental carcinogenesis with a variety of physical, chemical, hormonal, and viral agents. Particularly noteworthy are the observations that multiple melanotic lesions develop following a single application of dimethylbenzanthracene, and that kidney tumors develop in males following the administration of estrogens.[51]

The finding that the Syrian hamster is more susceptible to tumor induction by polyoma virus than its natural host, the mouse,[19] has stimulated the use of this laboratory animal to determine the oncogenic potential of a variety of viruses. The hamster has been found to develop tumors following the inoculation of adenoviruses from a variety of species: human,[32,45,46,57,82,83] simian,[47] bovine,[13] canine,[66] and avian.[65] In addition, tumors have been induced with simian virus 40,[18,33] Rous sarcoma virus,[1] and the mouse sarcoma virus.[43]

Information on the incidence of spontaneous tumors is imperative for the

G. L. VAN HOOSIER, JR.,
H. J. SPJUT, AND
J. J. TRENTIN

evaluation of results in experimental oncology. In addition, such data may contribute to comparative oncology and the biology of cancer in general. Reports in the literature on the incidence of spontaneous tumors in the hamster vary considerably. The purposes of this report are (a) to describe eight malignant tumors observed in a closed breeding colony from 1962 to 1967, (b) to review the literature on the numbers and types of tumors reported by other authors, (c) to discuss the presently available data on the etiology of three of the frequently observed tumors, (d) to present possible explanations for the discrepancies in incidence reported by different laboratories, and (e) to compare the incidence and types of tumors found in the hamster with those that occur in other species.

Observations in a Closed Hamster Colony

In 1960 and 1962, breeding stock of noninbred Syrian hamsters was obtained from the colony at the National Institutes of Health and maintained in isolation. Most of the animals are used for testing a variety of human viruses for oncogenic potential. Separate rooms with no cross-circulation of air are used to house different groups of animals, e.g., breeders, adenovirus-inoculated animals, and enterovirus-inoculated animals. Animals are necropsied when dead. Any tissues suspected of being neoplastic are preserved in formalin and examined histologically.

In 1966–1967, approximately 100 animals were tested in collaboration with Dr. John Parker of Microbiological Associates for antibodies to 13 of the commonly occurring rodent viruses. The colony was found essentially free of these agents (G. L. Van Hoosier, Jr., unpublished data) in contrast to most other hamster colonies in which two paramyxoviruses, Simian virus-5 (SV-5), and pneumonia virus of mice (PVM) are enzootic. However, two morphologically distinct agents that are detectable only by electron microscopy are present in our colony[72] and apparently in most other colonies that have been examined.

For classification of tumors, the system published by Stewart et al.[73] will be used (Table 1).

As shown in Table 2 and in Figures 1–8, a total of eight malignant neoplasms were observed in 1,671 animals necropsied at age 188 days or older. Hamsters surviving for shorter time periods were not included because this was the earliest age at which any tumors were observed. The mean incidence for all malignant tumors is 0.5 percent, and the mean age at necropsy for all animals was 347 days. The male:female ratio for all tumors is 3:1, but the higher incidence in males may be more apparent than real. The life-span of the population is illustrated in Figure 9 with the age and sex of animals with malig-

451

TABLE 1 Classification of Tumors of Animals According to Site of Origin and Histology[73]

I.	Tumors of epithelial tissue
	A. Tumors presumably of glandular origin
	B. Tumors of postulated but unconfirmed glandular origin
	C. Tumors of nonglandular epithelial origin
II.	Lymphomas and leukemias
	A. Lymphosarcoma or lymphocytic leukemia, or both, with neoplastic involvement of organs and tissues, or leukemic blood, or both, in most instances
	B. Myelocytic leukemia (including chloroleukemia)
	C. Reticulum-cell sarcoma
	D. Plasma-cell tumor
III.	Tumors of connective tissues
	A. Tumors presumably of connective tissue origin and of specific cell type, e.g., fibrosarcoma
	B. Tumors presumably of connective tissue origin and of nonspecific cell type, e.g., "mixed cell, round cell or sarcoma"
IV.	Tumors of melanin-forming tissue
V.	Tumors of neural tissue
VI.	Tumors composed of mixed tissues
VII.	Tumors not classified elsewhere

nant tumors. Males have a longer life-span than do females in our breeding colony and therefore a greater chance to develop neoplasia. Six of the eight malignant tumors observed were in animals at 12 months of age or older, although only 21 percent of female hamsters alive at 6 months were alive at 12 months, as compared to 52 percent of the males. However, no tumors were observed in an additional 218 females that were mothers of virus-inoculated newborns that were necropsied at a mean age of 561 days.

Review of Literature

The only tumors for which the appropriate reference(s) are not included in the tables are those described above under "Observations in a Closed Breeding Colony." Excluded from the tables are tumors that arose at an unspecified site[12,59] or were metastatic, with the site of the primary tumor unknown or not specified. Also excluded from the tables for malignant tumors are instances in which the total number of animals was not mentioned.[69] Notations to the classification by tissue of origin are included in

TABLE 2 Spontaneous Malignant Neoplasms Observed in a Closed Hamster Colony from 1962 to 1967

Group	Type	Site	Sex	Age (days)	Percent Incidence[a]	Histological Illustration
I Epithelial tissue	Carcinoma	Kidney	♀	717		Figure 1
	Adenocarcinoma	Mesenteric lymph node (metastatic)	♂	456	0.1	Figure 2
II Lymphomas						
Subgroup A	Lymphocytic	Mesenteric lymph node	♂	772		Figure 3
	Lymphocytic	Prefemoral lymph node	♂	391		Not illustrated
Subgroup C	Reticulum cell sarcoma	Posterior cervical lymph nodes	♂	472	0.2	Figure 4
III Connective tissue	Myxoid liposarcoma	Cheek pouch	♂	258		Figures 5 and 6
	Hemangiopericytoma	Cheek pouch	♀	188	0.2	Figure 7
	Undifferentiated sarcoma	Subcutis	♂	661		Figure 8
	All types combined				0.5	

[a]Based on 1,671 hamsters necropsied at or beyond the minimum tumor age (188 days); mean age of necropsy equals 347 days.

453

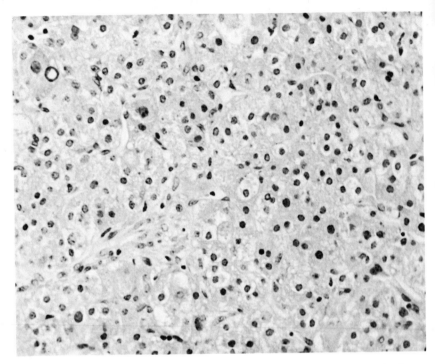

FIGURE 1   Renal nodule from a female hamster. The tumor appears to be a renal cell carcinoma. Hematoxylin-eosin. 190X.

most tables, but reported types are presented according to their frequency of occurrence, not according to the rank order established in Table 1.

BENIGN NEOPLASMS

In understanding the variations in reports on the incidence of tumors in hamsters it is helpful to separate the benign and the malignant neoplasms. The benign neoplasms of epithelial and connective tissue are listed in Table 3 and those of melanin-forming and neural tissue are shown in Table 4, with an indication of their occurrence in relation to all reported benign lesions. The most common is indicated by four plus (++++) and indicates more than 100 cases; common lesions are indicated by three plus (+++) and represent 30 to 100 cases; occasional tumors are indicated by two plus (++) and represent 10 to 30 cases; rare tumors are indicated by one plus (+) and represent fewer than 10 reported cases. The pheochromocytoma listed in Table 4 was recently observed by Dr. Gleiser in an animal from our colony.

The most commonly reported benign neoplasm of the Syrian hamster is

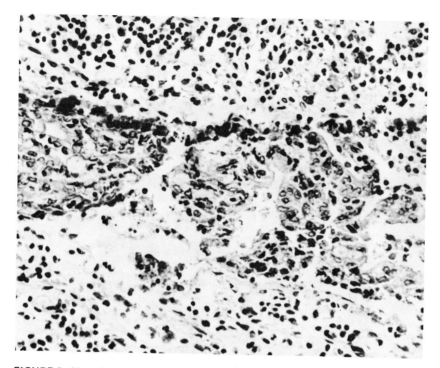

FIGURE 2 Metastic adenocarcinoma in the peripheral sinus of an intraperitoneal abdominal lymph node. The primary site was not found. Hematoxylin-eosin. 320X.

polyps of the intestinal tract. Apparently these are predominately adenomatous rather than inflammatory in nature, but this distinction has not always been made in reports. More than 100 of these were described from a single laboratory and were predominately located in the large intestine, especially the cecum.

Approximately 65 cases of adrenal cortical adenomas have been reported, again with the vast majority of the cases from a single laboratory.

Other commonly reported benign neoplasms include adenomas of the thyroid, papillomas of the forestomach, and hemangiomas of the spleen.

MALIGNANT NEOPLASMS

Thirty-five percent of all reported neoplasms of the Syrian hamster can be classified in the lymphoma and leukemia group II (Table 1). These are presented by subgroup in Table 5; they consist primarily of subgroup II-C, reticulum cell sarcomas, and were either designated as such in the reports or described as histocytic in type. The primary site of these tumors was predominately in one or several lymph nodes, e.g., nodes of the

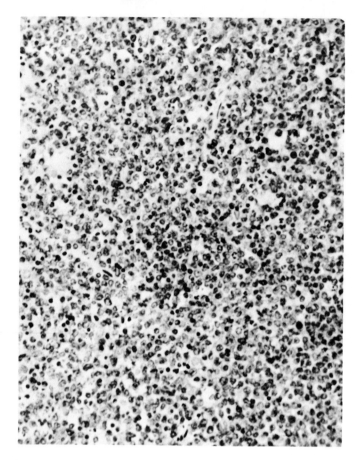

FIGURE 3 Lymphosarcoma of mesenteric lymph nodes in male hamster. The cells are immature lymphocytes. Evidence of phagocytosis is noted. Hematoxylin-eosin. 320X.

gastrointestinal tract or abdominal cavity. Other sites mentioned include the skin, liver, spleen, and cheek pouch. The plasma cell tumors apparently arise predominately if not exclusively in extramedullary sites, e.g., in the subcutis.

The second most frequently reported malignant neoplasm of the hamster is adenocarcinoma of the intestinal tract. The number of reported cases and other tumors of epithelial tissues are listed in Table 6. One of the reports[23] indicates that the colon and cecum are the most common sites of involvement. In addition to the number of cases shown in Table 6, it is possible that the intestines were the site of the primary tumor in cases of metastatic adenocarcinoma observed in lymph nodes of the abdominal cavity with undetermined

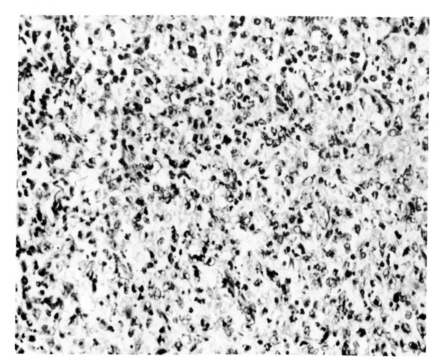

FIGURE 4 Reticulum cell sarcoma in a male hamster. The lesion is of a fairly uniform cell type; nucleoli are prominent. Hematoxylin-eosin. 320X.

sites of origin. The latter group of tumors is not included in the table. The report by Jonas et al.[48] suggests that intestinal adenocarcinoma is a cause of regional enteritis in young animals, but there is some question regarding the neoplastic classification of the lesions.[6] Although a significant number of tumors of the adrenal cortex and two tumors of the parathyroid are listed, their inclusion with malignant tumors is considered tentative since evidence of metastases was not frequently mentioned in the reports cited. It is felt that the benign and malignant tumors of these sites are not easily distinguishable unless metastases have occurred. Ten "hypernephromas of the adrenal" are not included, as it is not clear from the report whether the site of origin was the adrenal gland or the kidney.[2] The other tumors of epithelial tissue have been reported either occasionally (10–20 cases) or rarely (less than 10 cases). The number of tumors of each histological type of a specific organ are shown under the column headed "type." The number of tumors of each type does not always equal the total number of cases; the balance was generally designated only as malignant in the reference.

457

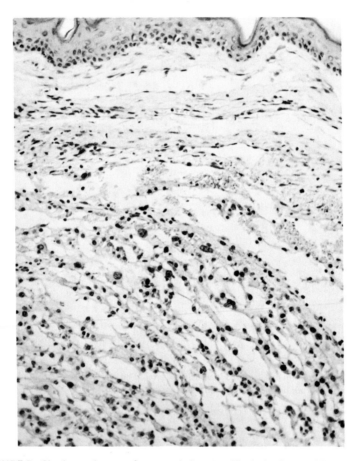

FIGURE 5 Cheek pouch tumor from a male hamster. The lesion is myxoid and poorly circumscribed and does not form a pattern. Hematoxylin-eosin. 200X.

Malignant neoplasms of connective tissue (Group III) and melanin-forming tissue (Group IV) are listed in Table 7. The cheek pouch is the primary site reported for five connective tissue tumors, and an additional tumor at this site has been reported by Busch.[8] The hemangiopericytoma listed is the one described under the first section of this report and is apparently a rare tumor. Two others of this histological type have been reported, but they are not included in the table, either because the site was not reported[51] or the animal had been treated with a carcinogen.[29]

The melanomas are especially noteworthy as they represent the third most frequently reported malignant tumor of the hamster. The back, head, and neck

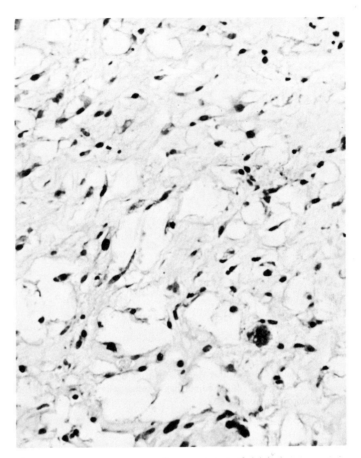

FIGURE 6   At higher magnification some areas of the lesion appear sarcomatous, resembling a liposarcoma. Numerous bizarre nuclei are seen. The spaces associated with a
nuclei suggest fat cells. Hematoxylin-eosin. 320X.

were listed as common primary sites in one report,[25] and the flank organ, a
pigmented area on the lateral aspect of the dorsal side of the abdomen in male
hamsters, was cited in one report.[36]

Table 8 illustrates the frequency, incidence, sex ratio, and mean age of
tumorous animals for the three most frequently reported malignant neoplasms.
The frequency was calculated on the basis of the number of reported cases for
the indicated neoplasm and the total number of malignant tumors listed in
Tables 5, 6, and 7 on 404 cases. The incidence, mean, and range were calculated only from those reports in which the particular neoplasm was reported
and the total number of animals observed was given. The incidence would be

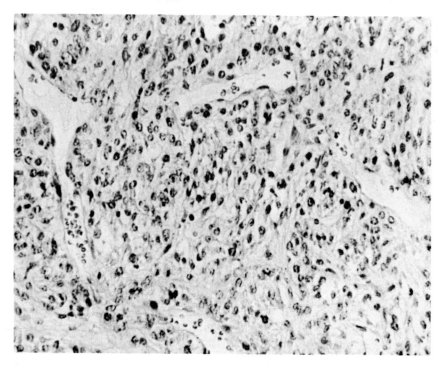

FIGURE 7 A vascular, fairly cellular tumor of the cheek pouch of a female hamster. The perivascular congregation of the somewhat spindled cells resembles the hemangio-pericytoma. Hematoxylin-eosin. 320X.

much less for each of the three types if all animals observed from all reports were included. The mean age of tumorous animals could be calculated only from those animals in which a specific age was included and in the case of adenocarcinoma of the intestines it is not very meaningful as it represents only one animal. The most significant difference in occurrence by sex is the preponderance of melanomas in the male hamster.

Etiology

The hamster type C virus detectable by electron microscopy,[72] but not yet detected by serological procedures, is identical in appearance to viruses in other species, notably the mouse and cat, that have been shown to be etiologically associated with neoplasms of the lymphoma-leukemia group. Our laboratory and others have made extensive efforts to in-

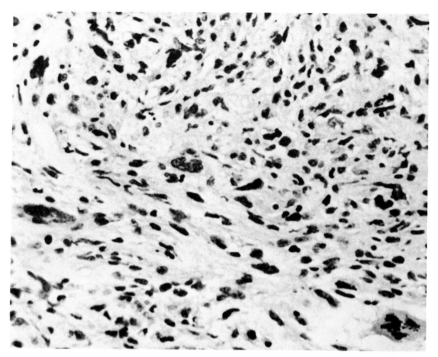

FIGURE 8   Undifferentiated sarcoma which grossly was a subcutaneous mass in a male hamster. Hematoxylin-eosin. 190X.

duce tumors with cell free materials with negative results. Recently, the induction of predominately lymphoid neoplasms has been reported with cell-free materials, and type C viruses were observed in the induced neoplasms.[34] The initial tissue extracts were prepared from papillomas that contained a papova virus,[35] the type associated with warts in a variety of species. Significantly, lymphomas were induced only in a strain of hamsters in which lymphomas or papillomas did not occur spontaneously and transmission experiments were unsuccessful in their own laboratory, where lymphomas are observed in an incidence of approximately 5 percent. It thus appears that viruses are etiologically involved in both papillomas and lymphomas of the hamster, and cell-free transmission is more readily accomplished in animals from colonies in which the natural disease does not occur.

Melanomas in the hamster have also been reported to be transmissible with cell-free extracts.[20]

It is perhaps premature to conclude that three hamster tumors, papilloma, lymphoma, and melanoma are caused by a virus, but recent reports strongly suggest this.

TABLE 3 Benign Tumors of Epithelial and Connective Tissue

Group	Tissue	Tumor Type	Site	Relative Frequency[a]	References
I	Epithelial	Polyps	Intestine	++++	22, 23, 24, 51, 70, 74
			Stomach	+	24
			Uterus	+	24
I	Epithelial	Adenoma	Adrenal cortex	+++	22, 24, 51, 53, 77, 80, 81
			Pancreas	+	24, 51, 80
			Thyroid	+++	51, 53, 81
			Parathyroid	++	24, 51
			Ovary	+	45
I	Epithelial	Adenomatosis	Lung	+	53, 70, 81
I	Epithelial	Papilloma	Forestomach	+++	24, 53, 60, 70, 80, 81
I	Epithelial	Keratoacanthoma	Vagina	+	24, 35, 79
			Skin	+	51
III	Connective	Hemangioma	Spleen	+++	23, 24, 51, 74
			Liver	++	24, 51, 60, 74, 80
			Subcutis	+	51
			Skin	+	51
III	Connective	Lymphangioendothelioma	Mesentery	+	51
III	Connective	Cholangioma	Bile duct	++	53, 70, 77, 80, 81
III	Connective	Thecoma	Ovary	++	22, 24, 51
III	Connective	Fibroma	Uterus	+	74
III	Connective	Leiomyoma	Prostate	+	51
			Uterus	+	51

[a]See page 454 for explanation.

462

G. L. VAN HOOSIER, JR.,
H. J. SPJUT, AND
J. J. TRENTIN

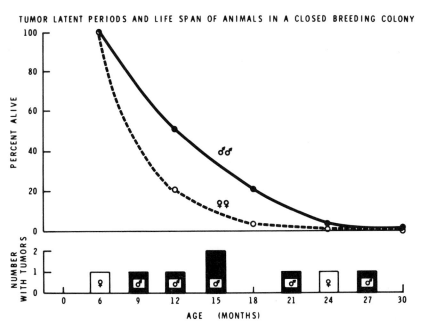

FIGURE 9   Tumor latent periods and life-span of animals in a closed breeding colony.

TABLE 4   Benign Tumors of Melanin-Forming and Neural Tissue

Group	Tissue	Type	Site	Relative Frequency[a]	References
IV	Melanin-forming tissue	Cellular blue nevi	Skin	++	51, 70
V	Neural tissue	Pheochromocytoma	Adrenal	+	
V	Neural tissue	Ganglioneuroma	Adrenal	+	51
V	Neural tissue	Schwannoma	Facial nerve	+	80

[a]See page 454 for explanation.

Factors Contributing to Differences in Incidence Reported by Various Authors

With the possibility that a virus is etiologically involved in two of the most frequently reported malignant tumors and one common benign tumor, the possibility should be considered that the agents are

463

TABLE 5 Malignant Neoplasms of the Hamster: Lymphomas—Group II

Subgroup	Type	Site	Number	References
A	Lymphosarcoma	Lymph nodes	18	16, 21, 38, 42, 74, 78, 79
C	Reticulum cell sarcoma	Lymph nodes	117	7, 16, 22, 24, 32, 41, 42, 60, 74, 75, 77, 78, 79, 80, 81
D	Plasma cell tumor	Extramedullary	8	16, 21, 24, 29, 60, 78

TABLE 6 Malignant Neoplasms of the Hamster: Epithelial Tissues—Group I

Site	Type	Number of Cases	References
SUBGROUP A			
Intestines	Adenocarcinoma	63	16, 22, 23, 24, 51, 74
Adrenal cortex	Carcinoma	26	16, 22, 24, 32, 79, 80
Thyroid	Spindle cell carcinoma–6 Follicular cell carcinoma–5	18	16, 24, 26, 51, 80
Liver and intrahepatic bile duct	Hepatocarcinoma–5 Cholangiocarcinoma–11	16	22, 24, 50, 61, 70, 77, 80
Uterus	Adenocarcinoma–10 Carcinoma cervix–3 Adenoacanthoma–1	13	16, 24, 51, 74, 80
Ovary	Granulosa cell–9	11	10, 16, 63, 79
Kidney	Adenocarcinoma–5 Nephroblastoma–2 Renal cell carcinoma–1	8	16, 22, 24, 31, 51, 80
Cowpers gland	Cystadenocarcinoma	7	24
Pancreas	Carcinoma–3 Adenocarcinoma–2	5	2, 22, 24
Mammary gland	Adenocarcinoma	4	16, 27, 40
Stomach	Adenocarcinoma	3	24
Salivary gland	Undifferentiated carcinoma–1	3	56, 60, 68

TABLE 6 (Continued)

Site	Type	Number of Cases	Reference
Prostate	Adenocarcinoma	2	24
Parathyroid	Carcinoma	2	51
Epididymis	Adenocarcinoma	1	25
Extrahepatic bile duct	Adenocarcinoma	1	24
SUBGROUP B			
Lower respiratory tract	Carcinoma–2 Bronchiogenic adenocarcinoma–1	8	16, 22, 51, 74
SUBGROUP C			
Skin	Squamous cell carcinoma	5	10, 16, 51, 54, 78
	Basal cell carcinoma	3	10, 11, 54
Esophagus	Squamous cell carcinoma	1	9, 10

more prevalent in some colonies than in others and these differences in prevalence are reflected by a difference in incidence of spontaneous tumors.

Some laboratories routinely examine all tissues histologically, while others examine only tissues that appear grossly abnormal. This may contribute more than any other single factor to the variability in different reports, especially for tumors of the adrenal, thyroid, and parathyroid. Differences in incidence of intestinal tract tumors may merely reflect how meticulously the entire tissue is examined.

The differentiation between nodular hyperplasia and adenomas and between adenoma and carcinoma of the adrenal gland may be difficult to make on a histological basis. Tumors of this tissue represent a significant proportion of all tumors in some reports. Thus, the pathologist's attitude toward the interpretation of these lesions can markedly influence the incidence of tumors reported.

The incidences of a variety of types of tumors have been increased in the mouse by inbreeding. Different colonies of hamsters may simulate different strains of inbred mice, as the inbreeding coefficient can become high because of the finite size of the breeding population. It thus seems plausible that different colonies of hamsters may have a different incidence of tumors for genetic reasons.

The age distribution of the population under study can markedly influence

TABLE 7   Malignant Neoplasms of the Syrian Hamster: Connective Tissue—Group III and Melanin-Forming Tissue—Group IV

Site	Type	Number of Cases	References
GROUP III A			
Intestines	Leiomyosarcoma–4 Angiosarcoma–1 Liposarcoma–1	6	22, 23, 24, 32
Cheek pouch	Myxofibrosarcoma–2 Myxoma–1 Myxoid liposarcoma–1 Hemangiopericytoma–1	5	28, 29, 77
Uterus	Leiomyosarcoma	4	24, 51, 80
Subcutis-neck	Fibrosarcoma–1 Leiomyosarcoma–1	2	51, 52
Urinary bladder	Leiomyosarcoma	1	79
Liver-spleen	Angiosarcoma	1	24
Humerus	Chondrosarcoma	1	76
Femur	Osteosarcoma	1	16
Tendon sheath	Giant cell tumor	2	64
GROUP III B			
Subcutis	Sarcoma	4	16, 53, 81
Pleura	Sarcoma	2	16, 77
Peritoneum	Endothelioma–1 Sarcoma–1	2	16, 67
GROUP IV			
Skin	Melanoma	30	5, 22, 24, 25, 36, 39, 51, 79, 80

the results. There were no reports of tumors in hamsters less than 6 months of age, and the majority of the tumors have been observed after 1 year of age.

The possibility of materials tested for carcinogenicity having a potentiating effect on spontaneous tumors should be considered. Several of the reports cited above were on animals from experiments in which tumors in experimental animals were reported as spontaneous.[16,22] It was concluded by the authors that the material under evaluation had no carcinogenic effect.

Several adenoviruses of human, simian, avian, bovine, and canine origin are

G. L. VAN HOOSIER, JR.,
H. J. SPJUT, AND
J. J. TRENTIN

TABLE 8 Most Common Malignant Neoplasms of the Syrian Hamster: Frequency, Incidence, Sex Ratio, and Age

Tumor Type	Frequency[a] (%)	Mean Incidence (%)	Range of Incidence (%)	Sex Ratio (male:female)	Mean Age[b] (months)
Lymphomas	35	1.3	0.2– 9.0[c]	2.6:1[b]	23[c]
Adenocarcinoma of intestines	16	4.8	2.8–13.4	1.4:1	27[d]
Melanomas	7	0.4	0.1– 2.76	10:1	21[e]

[a]Percent of all reported neoplasms.
[b]Calculated on reticulum cell sarcomas (IIC) only.
[c]Calculated on 32 animals.
[d]Only one specific age given.
[e]Calculated on 21 animals.

oncogenic in newborn hamsters. The predominant tumor types are undifferentiated sarcomas and lymphosarcomas. Some of the tumors can be identified definitively as of adenovirus etiology by a variety of criteria, including specific antigenicity. Others of the tumors, however, show little or no antigenicity. Short of a very high incidence of tumors at the site of injection, and demonstration of specific antigenicity, it is difficult to be certain whether some of the tumors, in particular the lymphosarcomas, represent adenovirus-induced tumors or an increased incidence of spontaneous-type tumors in adenovirus-injected animals.[82]

The above reasons and the fact that some authors do not distinguish between benign and malignant tumors in reporting the overall incidence, while other authors have reported primarily, if not exclusively, malignant tumors, appear to be sufficient to explain the variation in the reported incidence of spontaneous tumors in the Syrian hamster.

Incidence and Types of Malignant Tumors in the Hamster as Compared with Other Species

Although the data available leave much to be desired, the incidence of spontaneous malignant neoplasms of the Syrian hamster can be estimated at about 4.0 percent. This is based on 361 tumors observed by various authors in 8,207 hamsters; reports on the balance of the 404 tumors listed in Tables 5, 6, and 7 did not specify the number of animals observed.

In 1960, cancer accounted for 15.6 percent of all human deaths in the United States.[62] Most of the data available on mice are for inbred strains where the incidence is highly variable depending on the strain and the type of tumor. For example, an incidence in excess of 50 percent for lymphoma, hepatoma, and carcinoma of the breast and of the lower respiratory tract can be cited for selected strains.[55] Even in wild mice maintained in captivity, neoplasms of a variety of types are common.[17] If one then includes the domestic fowl in which variations of the leucosis complex may account for as much as 60 percent of all deaths, the human, the mouse, and the chicken appear to be the species most commonly afflicted with malignant neoplasia.

The rat[30,71] and dog[15] are species that occasionally have malignant neoplasms.

The hamster, in common with the guinea pig,[4] rhesus monkey,[49] and rabbit,[37,58] is a species rarely succumbing to neoplastic disease.

In common with the human,[14,62] the gastrointestinal tract and skin is involved frequently in those hamsters that develop tumors. The hematopoietic tissue, in common with many other species, is a common site for neoplasia, although in the hamster these tumors are restricted to cells of the reticular, lymphocytic, and plasmacytic series. Primary tumors of the thymus or neoplasms of the myelocytic series of cells have not been reported. In contrast to some other species, especially the human, malignant tumors of the breast, genital tract, and respiratory tract are rare. In contradistinction to the rat, the incidence of spontaneous pituitary gland tumors is very low in the hamster.

Paradoxically, and perhaps of some basic biological significance, the Syrian hamster is one of the most susceptible animals to the experimental induction of malignant tumors but one of the most resistant animals to spontaneous malignant neoplasms.

Acknowledgments

The authors gratefully acknowledge the assistance of Mrs. D. Boren Ferguson in the preparation of this paper for publication.

The investigation reported in this paper was supported by Public Service Research Grants CA 06941 and K6 CA 14,219.

References

1. Ahlstrom, C. G., and N. Forsby. 1962. Sarcomas in hamsters after injection with Rous chicken tumor material. J. Exp. Med. 115:839–852.
2. Ashbel, R. 1945. Spontaneous transmissible tumours in the Syrian hamster. Nature 155:607.

3. Billingham, R. E., G. H. Sawchuck, and W. K. Silvers. 1960. Studies on the histocompatibility genes of the Syrian hamster. Proc. Nat. Acad. Sci. U.S. 46:1079–1090.

4. Blumenthal, H. T., and J. B. Rogers. 1965. Spontaneous and induced tumors in the Guinea Pig. p. 183–209. *In* W. E. Ribelin and J. R. McCoy [ed.] The pathology of laboratory animals, Charles C Thomas, Springfield, Illinois.

5. Bomirski, A., T. Dominiczak, and L. Nowinska. 1962. Spontaneous transplantable melanoma in the golden hamster (*Mesocricetus auratus*). Unio. Intern. Contra Cancrum ACTA 18:178–180.

6. Boothe, A. D., and N. F. Cheville. 1967. The pathology of proliferative ileitis of the golden Syrian hamster. Pathol. Vet. 4:31–44.

7. Brindley, D. C., and W. G. Banfield. 1961. A contagious tumor of the hamster. J. Nat. Cancer Inst. 26:949–954.

8. Busch, G. 1953. Uber einen ubertragbaren spontantumor beim syrischen Goldhamster. Z. Krebsforsch 59:485–487, cited by Kirkman, H. and F. T. Algard. 1968. Spontaneous and nonviral-induced neoplasms, p. 227–240. *In* R. A. Hoffman, P. F. Robinson, and H. Magalhaes, The golden hamster: Its biology and use in medical research, The Iowa State University Press, Ames, Iowa.

9. Chesterman, F. C. 1962. Pathology of tumours and other lesions in a closed colony of golden hamsters, p. 486. *In* VIII International Cancer Congress, Medgiz Publishing House, Moscow.

10. Chesterman, F. C. 1964. Personal communication, cited by Kirkman, H., and F. T. Algard. 1968. Spontaneous and nonviral-induced neoplasms, p. 227–240. *In* R. A. Hoffman, P. F. Robinson, and H. Magelhaes, The golden hamster: Its biology and use in medical research, The Iowa State University Press, Ames, Iowa.

11. Crabb, E. D., and M. A. Kelsall. 1952. A malignant basaloma transplantable in hamsters. Cancer Res. 12:256.

12. Crabb, E. D., and M. A. Kelsall. 1954. A myxosarcoma, HS-5, transplantable in hamsters. Proc. Amer. Ass. Cancer Res. 1:10.

13. Darbyshire, J. H. 1966. Oncogenicity of bovine adenovirus type 3 in hamsters. Nature 211:102.

14. Dorn, H. F., and S. J. Cutler. 1959. Morbidity from cancer in the United States. Public Health Monograph No. 56. U.S. Government Printing Office, Washington, D.C.

15. Dorn, C. R., D.O.N. Taylor, R. Schneider, H. H. Hibbard, and M. R. Klauber. 1968. Survey of animal neoplasms in Alameda and Contra (COSTA) Counties, California. II. Cancer morbidity in dogs and cats from Alameda County. J. Nat. Cancer Inst. 40:307–318.

16. Dunham, L. J., and K. M. Herrold. 1962. Failure to produce tumors in the hamster cheek pouch by exposure to ingredients of betel quid; histopathologic changes in the pouch and other organs by exposure to known carcinogens. J. Nat. Cancer Inst. 29:1047–1067.

17. Dunn, T. B., and H. B. Andervont. 1963. Histology of some neoplasms and nonneoplastic lesions found in wild mice maintained under laboratory conditions. J. Nat. Cancer Inst. 31:873–902.

18. Eddy, B. E., G. S. Borman, W. H. Berkeley, and R. D. Young. 1961. Tumors induced in hamsters by injection of rhesus monkey kidney cell extracts. Proc. Soc. Exp. Biol. Med. 107:191–197.

19. Eddy, B. E., S. E. Stewart, R. Young, and G. B. Mider. 1958. Neoplasms in hamsters induced by mouse tumor agent passed in tissue culture. J. Nat. Cancer Inst. 20:747–761.

20. Epstein, W. L., K. Fukuyama, M. Benn, A. S. Keston, and R. B. Brandt. 1968. Transmission of a pigmented melanoma in golden hamsters by a cell-free ultrafiltrate. Nature 219:979–980.

21. Finkel, M. P., B. O. Biskis, and C. Farrell. 1968. Osteosarcomas appearing in Syrian hamsters after treatment with extracts of human osteosarcomas. Proc. Nat. Acad. Sci. U.S. 60:1223–1230.

22. Fortner, J. G. 1957. Spontaneous tumors, including gastrointestinal neoplasms and malignant melanomas, in the Syrian hamster. Cancer 10:1153–1156.

23. Fortner, J. G. 1958. An experimental prototype of human colon tumors. AMA Arch. Surg. 77:627–633.

24. Fortner, J. G. 1961. The influence of castration on spontaneous tumorigenesis in the Syrian (golden) hamster. Cancer Res. 21:1491–1498.

25. Fortner, J. G., and A. C. Allen. 1958. Hitherto unreported malignant melanomas in the Syrian hamster: an experimental counterpart of the human malignant melanomas. Cancer Res. 18:98–104.

26. Fortner, J. G., P. A. George, and S. S. Sternberg. 1960. Induced and spontaneous thyroid cancer in the Syrian (golden) hamster. Endocrinology 66:364–376.

27. Fortner, J. G., A. G. Mahy, and R. S. Cotran. 1961. Transplantable tumors of the Syrian (golden) hamster. Part II: Tumors of the hematopoietic tissues, genitourinary organs, mammary glands, and sarcomas. Cancer Res. 21:199–229.

28. Friedell, G. H., B. W. Oatman, and J. D. Sherman. 1960. Report of a spontaneous myxofibrosarcoma of the hamster cheek pouch. Transplant. Bull. 7:97–100.

29. Garcia, H., C. Baroni, and H. Rappaport. 1961. Transplantable tumors of the Syrian golden hamster (*Mesocricetus auratus*). J. Nat. Cancer Inst. 27:1323–1333.

30. Gilbert, C., and J. Gillman. 1958. Spontaneous neoplasms in the albino rat. S. Afr. J. Med. Sci. 23:257–272.

31. Girardi, A. J., M. R. Hilleman, and R. E. Zwickey. 1962. Search for virus in human malignancies. 2. *In vivo* studies. Proc. Soc. Exp. Biol. Med. 111:84–93.

32. Girardi, A. J., M. R. Hilleman, and R. E. Zwickey. 1964. Tests in hamsters for oncogenic quality of ordinary viruses including adenovirus type 7. Proc. Soc. Exp. Biol. Med. 115:1141–1150.

33. Girardi, A. J., B. H. Sweet, V. B. Slotnick, and M. R. Hilleman. 1962. Development of tumors in hamsters inoculated in the neonatal period with vacuolating virus, SV-40. Proc. Soc. Exp. Biol. Med. 109:649–660.

34. Graffi, A., T. Schramm, E. Bender, I. Graffi, K. H. Horn, and D. Bierwolf. 1968. Cell-free transmissible leukoses in Syrian hamsters, probably of viral aetiology. Brit. J. Cancer 22:577–581.

35. Graffi, A., T. Schramm, I. Graffi, D. Bierwolf, and E. Bender. 1968. Virus-associated skin tumors of the Syrian hamster: Preliminary note. J. Nat. Cancer Inst. 40:867–868.

36. Greene, H. S. N. 1958. A spontaneous melanoma in the hamster with a propensity for amelanotic alteration and sarcomatous transformation during transplantation. Cancer Res. 18:422–425.

37. Greene, H. S. N. 1965. Lesions of the spontaneous diseases of the rabbit. p. 330–350. *In* W. E. Ribelin and J. R. McCoy [ed.] The pathology of laboratory animals. Charles C Thomas, Springfield, Illinois.

38. Greene, H. S. N., and E. K. Harvey. 1960. The inhibitory influence of a transplanted hamster lymphoma on metastasis. Cancer Res. 20:1094–1100.

39. Gye, W. E., and L. Foulds. 1939. A note on the production of sarcomata in hamsters by 3:4-benzpyrene. Amer. J. Cancer 35:108.

470

40. Haberman, R. T. Personal communication, cited by B. E. Eddy, S. E. Stewart, R. Young, and G. B. Mider. 1958. Neoplasms in hamsters induced by mouse tumor agent passed in tissue culture. J. Nat. Cancer Inst. 20:747–762.

41. Haemmerli, G., A. Zweidler, and P. Strauli. 1966. Transplantation behavior and cytogenetic characteristics of a spontaneous reticulum cell sarcoma in the golden hamster. Int. J. Cancer 1:599–612.

42. Handler, A. H., R. A. Adams, and S. Farber. 1960. Further studies on the growth of homologous and heterologous lymphoma and leukemia transplants in Syrian hamsters. Acta Unio Intern. Contre Le Cancer 26:1175–1177.

43. Harvey, J. J. 1964. An unidentified virus which causes the rapid production of tumor in mice. Nature 204:1104–1105.

44. Hoffman, R. A., P. F. Robinson, and H. Magalhaes. 1968. The golden hamster: its biology and use in medical research. The Iowa State University Press, Ames, Iowa.

45. Huebner, R. J., M. J. Casey, R. M. Chanock, and K. Schell. 1965. Tumors induced in hamsters by a strain of adenovirus type 3: Sharing of tumor antigens and "neoantigens" with those produced by adenovirus type 7 tumors. Proc. Nat. Acad. Sci. U.S. 54:381–388.

46. Huebner, R. J., W. P. Rowe, and W. T. Lane. 1962. Oncogenic effects in hamsters of human adenovirus types 12 and 18. Proc. Nat. Acad. Sci. U.S. 48:2051–2058.

47. Hull, R. N., I. S. Johnson, C. G. Culbertson, C. B. Reimer, and H. F. Wright. 1965. Oncogenicity of the simian adenoviruses. Science 150:1044–1046.

48. Jonas, A. M., Y. Tomita, and D. S. Wyand. 1965. Enzootic intestinal adenocarcinoma in hamsters. J. Amer. Vet. Med. Ass. 147:1102–1108.

49. Jungherr, E. 1963. Tumors and tumor-like conditions in monkeys. In Epizootiology of cancer in animals. H. E. Whipple [ed.] Ann. N.Y. Acad. Sci. 108:777–792.

50. Kirkman, H. 1950. Different types of tumors observed in treated and in untreated golden hamsters. Anat. Rec. 106:277.

51. Kirkman, H., and F. T. Algard. 1968. Spontaneous and nonviral-induced neoplasms, p. 227–240. In R. A. Hoffman, P. F. Robinson, and H. Magalhaes [ed.] The golden hamster: Its biology and use in medical research, The Iowa State University Press, Ames, Iowa.

52. Klein, M. 1961. A spontaneous metastasizing fibrosarcoma in the golden hamster (Mesocricetus auratus) with some observations on tumor recurrence and metastasis after transplantation. J. Nat. Cancer Inst. 26:1381–1390.

53. Lee, K. Y., B. Toth, and P. Shubik. 1963. Carcinogenic response of the Syrian golden hamster treated at birth with 7,12-dimethylbenz (a) anthracene. Proc. Soc. Exp. Biol. Med. 114:579–582.

54. Lindt, S. 1958. Uber Krankheiten des syrischen Goldhamsters. [Diseases of the Syrian golden hamster (Mesocricetus auratus)]. Schweiz Arch Tierheilk 100:86–97, as cited by Kirkman, H., and F. T. Algard. 1968. Spontaneous and nonviral-induced neoplasms, p. 227–240. In R. A. Hoffman, P. F. Robinson, and H. Magalhaes, The golden hamster: Its biology and use in medical research, The Iowa State University Press, Ames, Iowa.

55. Murphy, E. D. 1966. Characteristic tumors, p. 521–562. In E. L. Green [ed.] Biology of the laboratory mouse, McGraw-Hill, New York.

56. Patterson, W. B. 1963. Duodenal cancer in hamsters induced by a transplantable human cancer of the colon. Unio Internationalis Contra Cancrum Acta 19:640–643.

57. Pereira, M. S., H. G. Pereira, and S. K. R. Clark. 1965. Human adenovirus type 31: A new serotype with oncogenic properties. Lancet 1:21–23.

58. Polson, C. J. 1927. Tumours of the rabbit. J. Pathol. Bacteriol. 30:603–614.

471

59. Popp, I., and T. Predeteanu. 1960. A new spontaneous transplantable tumour of the Syrian golden hamster. Neoplasma 7:363–365.

60. Porta, G. D. 1961. Induction of intestinal, mammary, and ovarian tumors in hamsters with oral administration of 20-methylcholanthrene. Cancer Res. 21:575–579.

61. Porta, G. D., P. Shubik, and V. Scortecci. 1959. The action of N-2-fluorenylacetamide in the Syrian golden hamster. J. Nat. Cancer Inst. 22:463–471.

62. Public Health Service. 1963. Age-adjusted and age specific death rates for malignant neoplasms 1960. USPHS Publ. 1113. U.S. Dept. of Health, Education, and Welfare, Public Health Service, Washington, D.C.

63. Rolle, G. K., and H. A. Charipper. 1949. The effects of advancing age upon the histology of the ovary, uterus and vagina of the female golden hamster (*Cricetus auratus*). Anat. Rec. 105:281–295.

64. Ruffolo, P. R., and H. Kirkman. 1965. Malignant transplantable giant cell tumors of peri-arti-cular connective tissues in Syrian golden hamsters (*Mesocricetus auratus*). Brit. J. Cancer 19:573–580.

65. Sarma, P. S., R. J. Huebner, and W. T. Lane. 1965. Induction of tumors in hamsters with an avian adenovirus (CELO). Science 149:1108.

66. Sarma, P. S., W. Vass, R. J. Huebner, H. Igel, W. T. Lane, and H. C. Turner. 1967. Induction of tumors in hamsters with infectious canine hepatitis virus. Nature 215:293–294.

67. Scott, H. H. 1927. Report on the deaths occurring in the society's gardens during the year 1926. Proc. Zool. Soc. (London), p. 173–198, cited by H. Kirkman and F. T. Algard. 1968. Spontaneous and nonviral-induced neoplasms, p. 227–240. *In* R. A. Hoffman, P. F. Robinson, and H. Magalhaes [ed.] The golden hamster: Its biology and use in medical research, The Iowa State University Press, Ames, Iowa.

68. Sherman, J. D., P. G. Rigby, and R. L. Hackett. 1963. A new spontaneous hamster carcinoma associated with a positive erythroagglutination reaction and anemia. Cancer Res. 23:1689–1693.

69. Shubik, P., G. Pietra, and G. D. Porta. 1960. Studies of skin carcinogenesis in the Syrian golden hamster. Cancer Res. 20:100–105.

70. Shubik, P., G. D. Porta, G. Pietra, L. Tomatis, H. Rappaport, U. Saffiotti, and B. Toth. 1962. Factors determining the neoplastic response induced by carcinogens, p. 285–297. *In* M. H. Brennan and W. L. Simpson [ed.] Biological interactions in normal and neoplastic growth, Little, Brown, and Co., Boston.

71. Snell, K. C. 1965. Spontaneous lesions of the rat. p. 241–302. *In* W. E. Ribelin and J. R. McCoy [ed.] The pathology of laboratory animals, Charles C Thomas, Springfield, Illinois.

72. Stenback, W. A., G. L. Van Hoosier, Jr., and J. J. Trentin. 1966. Virus particles in hamster tumors as revealed by electron microscopy. Proc. Soc. Exp. Biol. Med. 122:1219–1223.

73. Stewart, H. L., K. C. Snell, L. J. Dunham, and S. M. Schylen. 1959. Transplantable and transmissible tumors of animals. Armed Forces Institute of Pathology, Washington, D.C. 378 p.

74. Stewart, S. E., and M. Irwin. 1960. Possible complications in the use of laboratory animals for studies on the viral etiology of human neoplasms. Fourth National Cancer Conference Proceedings, p. 539–557.

75. Strauli, V. P. 1962. Morphogenetische untersuchungen an einem lymphoreticulocytaren sarkom des Goldhamsters. Pathol. Microbiol. 25:301–305.

76. Taylor, D. O. N. 1968. Transplantable chondrosarcoma in the Syrian hamster (*Mesocricetus auratus*). Cancer Res. 28:2051–2055.

G. L. VAN HOOSIER, JR.,
H. J. SPJUT, AND
J. J. TRENTIN

77. Tomatis, L., G. D. Porta, and P. Shubik. 1961. Urinary bladder and liver cell tumors induced in hamsters with o-aminoazotoluene. Cancer Res. 21:1513–1517.

78. Toolan, H. W. 1964. Personal communication, as cited by A. F. Handler. Spontaneous lesions of the hamster. p. 210–240. *In* W. E. Ribelin and J. R. McCoy [ed.] The pathology of laboratory animals, Charles C Thomas, Springfield, Illinois.

79. Toolan, H. W. 1967. Lack of oncogenic effect of the H-viruses for hamsters. Nature 214:1036.

80. Toth, B. 1967. Studies on the incidence, morphology, transplantation and cell-free filtration of malignant lymphomas in the Syrian golden hamster. Cancer Res. 27: 1430–1442.

81. Toth, B., L. Tomatis, and P. Shubik. 1961. Multipotential carcinogenesis with urethan in the Syrian golden hamster. Cancer Res. 21:1537–1541.

82. Trentin, J. J., G. L. Van Hoosier, Jr., and L. Samper. 1968. The oncogenicity of human adenoviruses in hamsters. Proc. Soc. Exp. Biol. Med. 127:683–689.

83. Trentin, J. J., Y. Yabe, and G. Taylor. 1962. The quest for human cancer viruses. Science 137:835–841.

E.L. VAN HOUTEN
H.O. SLOOT 2240
1477494

VII

Primates

PRIMATES IN U.S.
BIOMEDICAL RESEARCH

James A. Shannon

The purpose of this introductory paper is to discuss briefly and generally the use of primates in biological studies as I have seen it develop in the United States. Most of what I say here is contained in two earlier papers. The first of these, written by Willard H. Eyestone in 1966,[1] describes the primate program and the substance of the research that has been made possible by a series of primate centers supported by the National Institutes of Health, beginning in the early sixties. The second is a brief summary relating to the selection of primates for various types of biomedical research. This was written by Leon Schmidt, who has had more experience with one or another type of primates than any other investigator in the United States.

I will discuss quite briefly some aspects of research with nonhuman primates as we have seen it evolve in this country during the past twenty years.

During the late 1940's and early 1950's, although substantial work was done on the organic functioning of the central nervous system and on behavioral problems, the large-scale work with primates was in two areas: in the study of bacterial infections of one sort or another, particularly tuberculosis; and in the study of malaria. The greatest early contribution was to develop in mulatta an animal model system that replicated many aspects of human vivax malaria. It was around this model that an entire national program was developed beginning in the latter years of World War II. This program had a successful conclusion in the early 1950's with the development of a curative drug for this important disease.

Unfortunately, though not as extensively utilized, the mulatta should be almost equally useful in tuberculosis, because of the exquisite sensitivity of this preparation. This opportunity has not been adequately exploited because of an inability to bring together groups of people who are concerned with general chemical concepts of immunology and those who are interested in the ultimate solution of tuberculosis through the development of nonsensitizing solid immunizing agents. Enough preliminary work has been done to demonstrate the capabilities of the model system.

It was these two model situations and their very broad utility, together with an appreciation of the contributions of primates to our understanding of the central nervous system (including behavioral aspects), that led to the belief in the 1950's that there was a need to provide specialized resources so as to permit broader research on nonhuman primates. Facilities available at the time were grossly inadequate.

The National Heart Institute conceived in 1956 and proposed in 1959 the creation of a single national primate center. This was to be large and complex with extensive resources for visiting scientists. The need for primate resources was certainly great, but it seemed unwise to localize such resources in a single geographic site.

The program proposal was modified so as to provide for a series of regionally based centers. Each would have some type of resident staff with a core program and, in addition, there would be resources for visiting scientists with individual programs and for the center to participate in nationally organized research endeavors.

The establishment of these centers was well under way by 1962. In addition, it was apparent that certain universities already had sufficiently broad programs to warrant the availability of adequate primate facilities for their own programs. These were provided for by other construction arrangements in a limited number of institutions.

It is too early to assess the general success of the program. I will say, however, that despite the increasing use of primates during the past ten years, only a beginning use of primates has been made.

The most devastating statistic that I was able to uncover shows that something close to 60 percent of all of the papers describing the use of primates reviewed up to two years ago had made use of only 5 percent of the primates that are available, and 80 percent of the primate species had not been used at all.

The fact that there has been no systematic attempt at the comparative assessment of the net worth of individual primate species that are or could be made available points up deficiencies in our knowledge.

To list a few of the outstanding contributions, one thinks initially of the essential role primates played in the development of understanding of poliomyelitis, and ultimately in the development of both killed and live vaccines.

I mentioned their use in the research on malaria. In retrospect I do not believe the problem of developing curative drugs for vivax malaria could have been solved without the availability of primates.

In tuberculosis, as I noted, the opportunity is great but the general use of primates has been limited.

The baboon, if one can group this rather diverse group into a single category for general discussion, has been found of great utility in two situations. The one relates to the study of congenital malformations, the other to the study of lipid metabolism in relation to the development of atherosclerosis. In this country, there is only one source of supply for such studies, the Southwest Foundation, located in San Antonio, Texas. This center has provided material for a very substantial number of interested scientists.

Much has been done on brain development and the influence of perinatal factors on subsequent deficiencies. Much of this has been done in Puerto Rico by the National Institute of Neurological Diseases and Stroke. These studies show the devastatingly acute effects of short periods of anoxia and the remarkable capacity for the reacquisition of function with time.

Harlow's research at the University of Wisconsin on the development of a normal emotional response to one's environment is now well-known. His more recent work on psychosocial factors in the development of intellectual capability is not as well-known.

Also, the crucial role that the lowly macaque played in the development of the measles vaccine is probably not generally known. This was not written up extensively. For many years the NIH has had a small colony of macaques off the southern shore of Puerto Rico. These animals were kept for several generations in fair isolation from human contact. During the time of the final study of the measles vaccine most of the monkeys were being imported from India, some from Africa. The animals in both situations in the wild or incidental to collection and storage had extensive contact with the human population, and approximately 100 percent of them were immune to the virus that caused measles. In thinning out the colony in Puerto Rico, the thought arose that the isolation there might have produced a nonimmune population for definitive tests of the measles vaccine by the Division of Biologic Standards. This was, in fact, the case, and a serious problem was easily solved.

There is no doubt that primates have made unique contributions to the biomedical sciences, and I would emphasize that a mere beginning has been made.

I would like to mention some numbers that are very impressive to me. In this country in 1967, we used 70,000 primates distributed among about 30 species. Thirty thousand of these were used for research, and about 40,000 in industry.

This seems like a fairly large number, but if we take 30,000, in order to bring it down to a number one can comprehend a little better, and divide it by 100, which is in round numbers the number of U.S. medical schools, we then have 300 monkeys, on the average, per school. Considering the departmental structure of the schools, associated hospitals, and research institutions, there is not a great deal of primate work going on in the United States.

Furthermore, in terms of species, there are only eight species of which more than a thousand animals were used in the entire year. When one goes to a larger number, perhaps 5,000 or more, only three species were used to this extent. An examination of the research performed will show that these animals have been selected for very specific work, but even though the research is costly, little has been done in a systematic fashion to determine whether the species utilized is, indeed, optimal.

References

1. Eyestone, W. H. 1965. Scientific and administrative concepts behind the establishment of primate centers. J. Amer. Med. Ass., 147:1482–1487.
2. Schmidt, L. H. The selection of primate species most suitable for various uses. Paper presented at WHO European Symposium on Use of Non-Human Primates in Medical Research, Lyons, France, December 11–14, 1967. To be published.

VIROLOGICAL STUDY ON
NONHUMAN PRIMATES

S. S. Kalter

According to African folklore, in the beginning there was the Bushman and the baboon. Today there is still the Bushman and the baboon, but, more importantly, there is recognition of a biologic interrelationship between the two. Infectious agents have also evolved through the generations so each animal species frequently developed its own particular counterpart. This evolutionary process, unfortunately, is imperfect because crossing of the species barrier is a common occurrence, with infection and disease, or both, of the new host developing.

Actually, the health of humans and other animals has been recognized as interrelated for a number of years.[1-4,21,42] Several hundred diseases of animals are known to be communicable and they spread with little regard for any species barrier. Awareness and recognition of newer disease entities will unquestionably intensify the current problem. More important is the fact that any attempt to define and characterize a laboratory animal in the search for health introduces a need to develop information relative to the health of animals employed in that search.

Nonhuman primates are one of the more important groups of animals currently used in biomedical research. While use of simians in the laboratory is not new, deliberate use of these animals as models for studies of disease processes, especially as related to man, or in comparative medicine, is relatively recent. It is apparent that many investigators are developing data from experimental animals with totally inadequate background or base-line support. In most instances, little (if any) attempt is made to develop a biologic profile or to characterize the animal of choice. More specifically, experiments are frequently designed that use animals immediately following their purchase from dealers, without taking into consideration the "normal" microbiological flora and fauna (or other biophysiologic condition) of the animals. The entire experimental program may thus be placed in jeopardy by failing to consider the effect or influence of the animals' general health and well-being on the experiment itself. Perhaps equally important is the realization that use of diseased animals or their tissues in the laboratory may seriously endanger the health of the investigator and other laboratory personnel.

It is not our intent to delineate all these parameters but to confine our remarks to virologic problems associated with the use of nonhuman primates in the laboratory. Furthermore, no attempt will be made to review the developing literature in this area other than to refer to a number of recent reviews and experiments. See references 1–4, 9, 10, 12, 16, 21, 22, 29–31, 34, 37, 38, 40–42, 44–47, 56–59, 61, 65, 81, 84, 86, 90. In this presentation an attempt will be made to discuss available virologic information on primates currently employed in various laboratories. An effort will also be made to describe various shortcomings associated with the current practice of handling and capturing these animals and to indicate how these activities may potentially pose additional problems to those inherent to the animal in question.

Nonhuman Primates
Currently Used in Virus Studies

A number of nonhuman primate species, including members of both the prosimian and anthropoid suborders, are in use by various virus laboratories. Generally, New and Old World monkeys as well as certain of the apes, especially chimpanzees, and limited numbers of prosimians are in use. These primates range through the structural and behavioral organization complexities separating the various families in these two suborders. For specific details concerning these primates, a recent publication by Napier and Napier is highly recommended.[77] It is beyond the scope of this report to list all simians. Table 1 includes the majority of primates on which virologic information is already available or under consideration.

The early studies by Enders and his collaborators[19] on developing a cell system for the cultivation of viruses did much to encourage the use of monkeys, especially rhesus and cynomolgus, in the virus laboratory. Inclusion of other species in various virus studies, with possibly a few exceptions, generally resulted from chance or availability of the animal to the laboratory. Planned virus studies on these animals were invariably limited in extent. For example, the majority of viruses recovered from the rhesus and cynomolgus monkeys resulted from investigators' noting a cytopathogenic effect (CPE) in their control tissue culture materials rather than from a planned program searching for viruses.[40–42] The need to maintain such tissue culture controls on African green monkey kidneys as employed in poliomyelitis vaccine production, led Malherbe and his coworkers[65] to the discovery of viruses in kidney cell preparations from these simians.

In contrast to these findings are the studies on the baboon, marmoset and, more recently, the chimpanzee.[16,61,90] Specific programs to ascertain the normal virology (among other biologic parameters) of these animals were developed and explored.

Trapping and Handling

It is unfortunate that procurement of the majority of simians currently utilized in the laboratory is still subject to a system that ignores the fact that many organisms are transmissible from one animal to another. Any attempt, therefore, to define a microbial pattern of an animal is subject to the vagueness and uncertainty associated with an unknown contribution of all those individuals as well as facilities participating in the handling and holding of the animals. As will be described below, sufficient data are now available emphasizing the hazards associated with this interchange of organisms as well as demonstrating the frequency of conversion of various simians following contact with other primates.

Animals to be used by various investigators are generally purchased through commercial sources. A number of institutions using large numbers of animals may, on occasion, have their veterinarians visit the site of capture as well as the "trapper's" holding facility. Unfortunately, these visits do little to change the situation because examinations are cursory, and little is done other than tuberculosis testing. If yellow fever vaccination is required, this may be performed, but the value of this procedure may be questioned in view of recent findings.[55] Animals that are overtly ill or those with diarrhea may be discarded or even sold to other less discriminating purchasers. Recently an attempt has been made to improve conditions, but the basic problem of contamination and exchange of organisms between groups of animals as well as between handlers and animals has been little altered.

Undoubtedly, this situation accounts for many infections, latent or otherwise, that occur. It is to be emphasized, however, that an extensive "normal" virologic flora of various simians is known to exist. These agents may exist overtly or in a latent form, making their presence difficult to delineate. Also of importance to the problem of handling animals is the possible influence of stress of shipping. Recent studies in our laboratory at the Southwest Foundation for Research and Education (SFRE) have shown a potentiation of virus

TABLE 1 Primates Involved in Comparative Virology Studies

Old World Monkeys	Generic Name	New World Monkeys	Generic Name
Gorilla	Gorilla (G. gorilla)	Squirrel	Saimiri (S. sciureus)
Chimpanzee	Pan (P. troglodytes)	Owl	Aotus (A. trivirgatus)
Chimpanzee (pigmy)	Pan (P. paniscus)	Howler	Alouatta (A. belzebul) (type species; others—A. fusca (A. guariba), A. villosa (A. palliata), etc.
Orangutan	Pongo (P. pygmaeus)		
Gibbon	Hylobates (H. Lar) (type species)		
Siamangs	Symphalangus (S. syndactylus)		
Baboons	Papio spp. (P. cynocephalus) group: P. anubis, P. papio, P. ursinus (Chacma)	Spider	Ateles (A. paniscus) (type species; others—A. belzebuth, A. fusciceps, A. geoffroyi)
Hamadryas baboon	Papio (P. hamadryas)	Capuchins	Cebus (C. capucinus) (type species; others—C. albifrons, C. nigrivit-
Mandrill	Mandrillus (M. sphinx)		

Common name	Generic name
Vervet groups-*Mono, Nictitans*, etc.)	*Cercopithecus* (*C. pygerythrus*)
Talapoin (Mangrove) (*Miopithecus*, pigmy guenon)	*Cercopithecus* (*C. talapoin*)
Langur	*Presbytis* (*P. entellus*) (other species not listed)
Patas	*Erythrocebus* (*E. patas patas*)
Mangabeys	*Cercocebus* spp.
Rhesus	*Macaca* (*M. mulatta*)
Cynomolgus (Kara monkey, Crab-eating macaque, *M. irus*)	*Macaca* (*M. fascicularis*)
Bonnet	*Macaca* (*M. radiata*)
Pigtail	*Macaca* (*M. nemestrina*)
Formosan rock	*Macaca* (*M. cyclopis*)
Stumptail	*Macaca* (*M. speciosa*)
Japanese macaque	*Macaca* (*M. fuscata*)
Barbary ape	*Macaca* (*M. sylvana*)

Common name	Generic name
Marmosets	*Callithrix* (Common marmoset, golden, black-tailed, white-fronted, etc.)
Marmosets (pigmy)	*Cebuella* (*C. pygmaea*)
Marmosets (tamarin)	*Saguinus* (*S. tamarin*) (type species; others—*S. nigricollis, S. fusciollis, S. midas*, etc.)
Marmosets (golden lion tamarin)	*Leontideus* (*L. rosalia*)

PROSIMIANS	GENERIC NAME
Tree Shrew	*Tapaia*
Bushbabies	*Galago*
Loris	*Nycticebus*
Lemurs	*Lemur*
Tarsiers	*Tarsius*

shedding following transport of animals from Africa to this institution.[52]

Little difficulty is presented by the mechanics of trapping itself. Many different methods are in use, from large-scale netting of groups or troops of animals to single-cage trapping of individuals. Many trappers are solely dependent upon their own operation for animals. On the other hand, numerous animals are purchased by trappers from "natives" who have obtained them in one way or another and held them for varying periods of time under deplorable conditions. These animals are then kept at the dealer's compound until they are purchased or until a sufficient number is available for shipment. Shipment, generally by air, involves crowding of animals into small cages to conserve space. Flights are not direct but entail a stop enroute for food and watering and change of planes. These stops may also account for exchange of organisms when different animal species are held in the same physical area, common handlers used, or proper sanitary conditions are lacking. Arrival at the importer then requires another holding period before the animals are sent or sold to the investigator. This holding period is frequently erroneously referred to as a quarantine period.

Each investigator's laboratory is another major source of organism interchange, primarily as a result of inadequate housing and space requirements. More often than not, one room serves as a holding place for all the animals used by that laboratory. If there is more than one room, common animal handlers may act as vectors, spreading the different micro-organisms from room to room. This mechanism of trapping and handling animals, while only briefly described, identifies at least 4 opportunities for contamination and organism exchange. Until this system is completely overhauled, there is absolutely no control or guarantee relative to the health of the animal under investigation.

A mechanism patterned along methods now employed by the SFRE in the trapping and handling of their baboons would minimize these sources of contamination.[48,56,57] By collecting and handling only baboons (or only one species), contact with other simians is minimized. Direct shipment of the animals to the San Antonio facility avoids another major source of organism exchange, namely that of commercial trappers as well as holding stations. Use of animal handlers for only one species reduces the interchange of organisms. Careful microbiologic sampling of all personnel and animals involved in the program establishes a comparative baseline profile. This information then allows for intelligent interpretation of data.

One completely uncontrollable area of contamination is that concerned with the animal's habits in the bush. Only too frequently one finds that these animals live in close association with man, even in the most "remote" areas. Primarily vegetarian, many troops of monkeys live close to native villages or isolated farms, raiding the farmers' fields and drinking from the same water supply. Studies designed to ascertain organism exchange under these conditions are extremely difficult. A study on the natives of the area where the animals are trapped is of value, however.

Viruses of Primates

For detection and determination of the virus flora of simians, two methods are currently employed: the serologic method and the virus-isolation method. Others may be available, but data are too scanty for evaluation to be included at this time. For example, electron microscopy of tissue may discern virus-like particles. However, until these particles can be recovered and identified, little is gained.

SIMIAN VIRUS CLASSIFICATION

It is now recognized that many of the viruses recovered from the various animal species studied are indigenous to that animal and are not contaminants (e.g., human virus). Since the original report by Rustigian et al. in 1955[82] on virus-like particles in monkey kidney cell cultures, approximately 60 prototype agents have been described. All of these viruses, with a few possible exceptions, may be considered as counterparts of one or another virus group affecting man.[6] Both deoxyribonucleic acid (DNA) and ribonucleic acid (RNA) viruses are found among the simian viruses examined. Types most frequently encountered are simian adenoviruses and picornaviruses. Others that may be encountered are reoviruses, herpesviruses, myxoviruses, papovaviruses, poxviruses, and a number not yet classified. Virus groups and their simian sources are presented in Table 2.

Current terminology and basis for classification stems primarily from the studies of Hull and his collaborators[40-42] and Malherbe et al.,[65] who introduced the SV (simian virus) and SA (simian agent) series to describe their isolates. These "SV" and "SA" designations have persisted in spite of attempts to integrate the simian viruses into classification schema suggested by other investigators.[39,79] Thus, the use of an "ECMO" (enteric cytopathogenic monkey orphan) nomenclature[38] or an "M" and "C" numerical series[79] have met with only limited acceptance. A new attempt to develop a more practical and less confusing simian virus classification is currently under way.*

SEROLOGIC EVIDENCE FOR VIRUS INFECTION

A number of reports are available concerning the occurrence of "natural" virus infections of simians as detected by the presence

*Ad hoc Working Group for Simian Viruses Subcommittee on Classification, National Institutes of Health: R. L. Heberling, R. N. Hull, S. S. Kalter, H. Malherbe, J. L. Melnick, and J. T. Duff (Executive Secretary).

TABLE 2 Simian Viruses—Their Host and Tissue Sources[a]

ADENOVIRUSES	ENTEROVIRUSES
Rhesus and Cynomolgus Monkeys	Rhesus and Cynomolgus Monkeys (cont'd.)
SV 1-TC	SV42-RS
11-TC	43-RS
15-TC, RS	44-RS
17-TC, RS	45-RS
20-RS	46-RS
23-TC	47-RS
25-TC	48-RS
27-TC	49-RS
30-RS	
31-RS	African Green Monkey: SA5-RS
32-TC, RS	Baboon: A13-RS
33-RS	Marmoset: RS, TS
34-CNS	
36-RS	
38-CNS	UNCLASSIFIED PICORNAVIRUSES
	Rhesus Monkey
African Green Monkey	SV 4-TC
SA 7-TC, RS	28-TC
17-TC	
18-TC	African Green Monkey: SA4-TC
V340-RS, TS, organs	
Squirrel Monkey: RS	REOVIRUSES
Capuchin Monkey; TC, RS	Rhesus Monkey
	SV12-TC
Owl Monkey: RS	59-TC, lung
Cinnamon Ringtail Monkey: RS	African Green Monkey: SA3-TC
Marmoset: RS	Marmoset: RS, TC
	Chimpanzee: TC
ENTEROVIRUSES	
Rhesus and Cynomolgus Monkeys	MYXOVIRUSES
SV 2-TC, RS	
6-TC, RS	A. Rheus Monkey: SV5-TC
16-TC, RS	Cynomolgus Monkey: SV41-TC
18-TC	
19-RS	B. Variety of Primates (Rhesus,
26-CNS	Cynolmolgus, A. C., Chimpanzee, etc.)
35-CNS	
	Foamy virus types 1–7-TC, TS

TABLE 2 (Continued)

HERPESVIRUSES

A. Rhesus and Cynomolgus Monkeys: B. virus-TC, organs
   African Green Monkeys: SA8-TC
   Marmoset, Squirrel, and Owl Monkeys: Marmoset (*Herpes T.*), *H. tamarinus*,
       H. platyrrhinae-TC, organs
   Spider Monkey: SMV, organs

B. African Green Monkey
   SA6-TC, SG
   SGV (Rowe)-TC, SG
   GR2598 (Melnick)-TC

PAPOVAVIRUSES

Rhesus, Cynomolgus, and Patas Monkeys: SV40-TC

African Green Monkey: SA12(?)-TC

POXVIRUSES

Rhesus and Cynomolgus Monkeys
Monkey pox skin lesions
Yaba-like skin lesions

African Green Monkey: Yaba tumors

UNCLASSIFIED VIRUSES

African Green Monkeys
SA10-oral swab
   11-RS
   13-oral swab
   14-TC
Marburg virus (?)

Rhesus, Cynomolgus, Other (?) Monkeys: Simian hemorrhagic disease virus (?)

[a]Source of isolation: TC = tissue culture; RS = rectal swab; CNS = central nervous system;
TS = throat swab; SG = salivary gland.

of specific antibody responses. These studies range from those attempting to determine infections due to a specific organism to broad surveys detecting antibody to numerous viruses of human and simian origin. Interpretation of data obtained from serologic studies may at times be difficult because of duration of demonstrable antibody, nonspecific reactions, the presence of inhibitors as well as antigenic relationships to other organisms. Suitable controls generally provide some assistance in the final interpretation.

Baboons (*Papio* sp.) and African green monkeys (*Cercopithecus aethiops*) were both found to have a high incidence of yellow fever antibody, whereas the galago (*Galago senegalensis*) did not.[94] Baboons in other geographic areas of Africa (Kenya) did not show this incidence of antibody.[51,55]

New World monkeys also show variations in the incidence of antibodies to yellow fever. Evidently, these differences are simply a species resistance, certain monkeys being more susceptible than others. For example, the spider (*Ateles*) and capuchin (*Cebus*) monkeys are more resistant than the marmoset (*Saguinus*) and the howler (*Alouatta*) monkeys. *Aotes* (owl monkey) is highly susceptible, but its living habits minimize contact with the mosquito vector, with a consequent reduction in infectivity rates.[23]

Other surveys of simian sera have included a number for determining the presence of antibody to various representative arboviruses. Previous studies done with baboon sera[51] failed to find HI antibody to the following viruses: Chikungunya Chick L; Eastern encephalitis, New Jersey; Western encephalitis, Fleming; Louping I11 DXLIV; Sindbis AR 1055; Semliki RI-1; California H & R; WE 38873; Cache Valley Like A9171b; Bunyamwera RI-1; Guaroa J-C2; West Nile AR248; Yellow Fever Asibi; Jap. B. G8924; Langat TP21; St. Louis encephalitis Fla P-15; MVE 11A; Marituba BeAn 15; and Dengue II Tr1751, with the exception of one animal with yellow fever antibody. Additional studies in this laboratory[59] have found CF antibody in sera obtained from chimpanzees, baboons, and rhesus monkeys residing in the United States to Western encephalitis (WE) virus. One baboon also had antibody to St. Louis encephalitis (SLE) virus.

It is interesting to note that other studies performed in this laboratory have resulted in the isolation of WE virus from baboons, as well as from mosquitoes in the area.[43] Marmoset sera tested for antibody to WE and EE were negative.[16] Lack of WE and EE antibody in marmoset sera has been confirmed in this laboratory (unpublished data). Chimpanzee sera (U.S.-born and wild-born), gorilla, baboon, orangutan, rhesus, and African green monkeys were tested by Harrison *et al.*[28] for antibody to Chikungunya and related arboviruses. Twenty-six percent of the sera were found to have SN antibody. Included among the positives were 4 of 5 U.S.-born chimpanzees. The significance of this finding is not clear other than to assume antigenic overlapping with other group A viruses. More recently *Erythrocebus patas*, *Cercopithecus aethiops tantalus*, and *C. mona* monkeys were also found to have SN antibody to Chikungunya virus. Two of 12 *C. mona* monkeys were found to have antibody to Pongola virus. It will be recalled that previous studies[59] reported finding WE antibody in chimpanzees of the same colony, as mentioned above.

Measles antibody is rather prevalent in macaques,[18] but varies in extent among different colonies and troops.[59,72,88] Other nonhuman primates demonstrated extensive diversity in measles antibody, not only among the species examined but in the different holding facilities as well.[18,59] For example, one colony of chimpanzees demonstrated a 25.8 percent incidence of measles antibody, whereas another group held in another colony had an incidence of 12.2 percent. Similar variations were found among other primate colonies and primates. Bhatt *et al.*[9] reported 12 of 47 urban rhesus monkeys to have measles HI antibody, while 170 bonnet and 195 langur monkeys of forrestial habitat were completely negative. Similar positive findings were reported by Shah and Southwich[86] on sera obtained from rhesus monkeys living close to humans.

Perhaps the greatest concern to those employing monkeys for various investigative purposes is the herpesviruses. Most important of these simian herpesviruses are B virus (*Herpesvirus simiae*) and marmoset, or *herpes T*, virus (*Herpesvirus tamarinus, Herpesvirus platyrrhinae*). A number of simian herpesviruses, recovered from both Old and New World monkeys, have been reported, but their exact relationship to one another is still not clear.[17,35,66,70,71,83] It would appear that the strains isolated from Old World monkeys are closely related to B virus, whereas the New World monkey isolates are also antigenically related to each other. A number of serum surveys have indicated that this group of viruses is widespread among primates, but the full extent of species susceptibility as well as distribution is unknown.[13,16,27,59,62,63] Questions have been raised regarding interpretation of these serologic findings with regard to antigenic relationships. Finding antibody to one or another of the herpesviruses may be interpreted as indicating infection with an innocuous strain rather than indicating the presence of a virulent form. This philosophy may, however, lead to difficulties inasmuch as herpesviruses are notorious for producing a mild infection in one animal species and a highly lethal infection in another.

Also of importance with regard to the herpesviruses has been a possible association of herpes-like particles (for example, EBV) with malignancy, especially Burkitt lymphoma.[20] Sera from rhesus monkeys, baboons, and chimpanzees failed to show evidence of infection when tested against infected Burkitt cells.[32] On the other hand, two of four baboons experimentally infected with tumor cells did develop low levels of antibody. In contrast to these findings is the report of widespread CF antibody to another cell line derived from Burkitt tumors in chimpanzees, baboons, cynomolgus, rhesus, and African green monkeys.[24] In one study, 50 percent of rhesus monkeys tested within 1–4 days after capture had CF antibody to herpes-like virus antigen. A much smaller percentage of the same animals had CF antibody to herpes simplex virus.[26] As indicated above, the significance of finding herpesvirus antibody is still not clear.

Because of the possible association of herpes-like viruses to infectious mononucleosis,[33] Gerber et al.[25] recently attempted, unsuccessfully, to transmit IM to rhesus monkeys as well as to marmosets. In these studies animals free of detectable antibody were used.

With regard to viruses and tumors, a known oncogenic virus that has received considerable attention is SV40. This virus is included among the papovaviruses and occurs with considerable frequency among various simians, especially rhesus and cynomolgus monkeys. A number of serologic studies have conclusively demonstrated that SV40 is primarily associated with Asian monkeys rather than African animals, although antibody has been demonstrated in a small number of African simians.[59,73,91] SV40 tumor (T) antigen has been reported to occur in rhesus monkeys naturally infected with this virus.[85] The significance of this finding is not clear, especially as the highest incidence occurred in younger animals. Recently Shah et al.[87] were readily able to infect nonimmune rhesus with SV40. Both viral and T antibodies were demonstrable in the majority of inoculated animals. Inoculation of three fetuses at about 90 days' gestation failed to produce detectable tumors. Surprisingly little information is available pertaining to the incidence of SV40 in New World monkeys.

Other studies involving tumorigenic viruses, and nonhuman primates have been associated with adenoviruses, both human and simian, and with Rous sarcoma virus (RSV). For example, Morgan[74] surveyed a number of different animal sera, including baboons (African-born as well as U.S.-born), chimpanzees, and African green monkeys, without detecting any evidence of RSV antibody. In a preliminary study done in this laboratory, sera from representative nonhuman primates were examined for the presence of antibody to adenovirus type 12 and SA7. Many of the animals tested demonstrated neutralizing antibody to both viruses, but none had any evidence of T antibody (unpublished data). The ability to produce tumors in various nonhuman primates—macaques, marmosets, and baboons—with RSV has been clearly demonstrated.[15,50,76,95]

At least seven serotypes of foamy viruses are encountered in preparations of monkey kidney cell cultures, some with greater frequency than others. The role these viruses may play in infection or disease of the host animal is obscure. Their nuisance value to investigators, however, is immeasurable. Stiles[91] in testing rhesus and African green monkeys for CF antibody to foamy virus types 1–3 found the rhesus to be involved primarily with type 1, and the African green with types 2 and 3. Preliminary studies under way in this laboratory failed to demonstrate any significant differences in rhesus monkey sera for the three types of foamy virus employed (Table 3). Insufficient data are available at this time for comparisons between the rhesus and the other primates. It is important, however, in comparing data to take the type of serologic tests used by the different investigators into consideration.

Recently a group of DNA viruses of small particle size has been described[8]

TABLE 3   Foamy Virus Antibody (SN) in Various Primate Sera

Sera	Foamy Virus 1	Foamy Virus 2	Foamy Virus 3
Human			
Africa	–	–	1/21
SFRE	–	–	1/20
Chimpanzee			
Lab 1	–	–	3/28
Lab 2	0/9	0/10	–
Orangutan			
Lab 1	–	–	5/19
Baboon			
African	–	–	6/45
Domestic	–	8/21	16/20
Africa (1963)	–	–	5/24
Rhesus			
Lab 3 (1965)	7/14	8/14	8/16
Vervet			
SFRE	–	–	5/18

as associated with adenoviruses (adeno-associated viruses—AAV). These virions are apparently defective as they cannot replicate in the absence of multiplying adenoviruses. Several serotypes are recognized as associated with adenoviruses derived from different animal sources. Little is known regarding their pathogenic capability, and mention is made at this time only as an indication of their occurrence. For example, Rapoza and Atchison[80] and Mayor and Ito[68] found antibody to types 1 and 4 AAV in rhesus and African green monkey sera. Blacklow et al.[11] showed types 1, 2, and 3 antibody in captive rhesus monkeys but not in the wild. They, as well as Mayor and Ito,[68] failed to find type 4 antibody in human sera. These data suggest AAV types 1–3 are human types, whereas type 4 is simian in origin.

Serologic surveys of simian sera, primarily concerned with defining and characterizing the animal's past antigenic experience, have been extremely limited. Shah and Southwick[86] found serum from one free-living rhesus of 47 tested to have antibody to poliovirus type 2, but not to types 1 and 3. These investigators also reported one serum of 52 to have antibody to Japanese B encephalitis virus; one of 46 with antibody to respiratory syncytial virus; one of 52 with parainfluenza type 1 antibody; of 49 tested for parainfluenza type 2 antibody there was one that was positive, but 14 of 52 sera were positive for type 3 parainfluenza virus. When sera from other free-living rhesus were tested for *Herpesvirus simiae* and SV40 antibody, seven of 47 and 12 of 40 were found to be positive, respectively. In other studies on laboratory housed animals, the following results were found for parainfluenza type 3 anti-

493

body: two of 19 patas, 12 of 20 African green, and 10 of 13 positives among *C. mona*, *C. nictitans*, *M. leucophaeus*, *C. erythrogaster*, and *C. torquatus*. Tests of the same sera for antibody to respiratory syncytial virus were completely negative.

In similar studies on forest monkeys, including the bonnet (*M. radiata*) and langur (*Presbytis entellus*) as well as urban rhesus monkeys, Bhatt et al.[9] found all three species to have antibody to echovirus types 3, 7, 11, 12, and 19. Antibody to all three reovirus types, especially type 1, was present, predominately in the bonnet. In addition, it was found that one rhesus serum reacted with human adenovirus type 2 when types 1 through 7 were tested. None was positive when tested for antibody to rhinovirus (type CV_{30}). All three species of simians were found to have a small but significant number of seropositives when tested against human influenza (A_2/Japan 305/57) and the simian myxovirus SV5. Similarly, all three species were found to have antibody to 2 simian adenoviruses, SV32 and SV33, but only bonnet sera (3 of 54) were found to react with a simian enterovirus (SV49).

Perhaps the most extensive serologic study determining primate contact with antigens of human and simian origin is currently under way at SFRE (unpublished data).[49,51,59,62] In these studies sera obtained from numerous different primates: man (*Homo sapiens*), gorilla (*Gorilla gorilla*), chimpanzee (*Pan* sp.), orangutan (*Pongo pygmaeus*), gibbon (*Hylobates agilis*), baboon (*Papio* spp., *P. hamadryus*), gelada (*Thecopithecus gelada*), rhesus (*Macaca mulatta*), Formosan rock macaque (*M. cyclopis*), stumptail (*M. speciosa*), vervet or African green monkey (*Cercopithecus aethiops*), patas (*Erythrocebus patas*), talapoin (*C. talapoin*), crab-eating cynomolgus (*M. fascicularis*, *M. irus*), marmoset (*Sanguinus* spp.), howler (*Alouatta* spp.), capuchin (*Cebus* sp.), spider (*Ateles* sp.), squirrel (*Saimiri* sp.), slender loris (*Loris tardigradus*), woolly (*Lagothrix*), owl (*Aotus trivirgatus*), bonnet (*M. radiata*), pigtail (*M. nemestrina*), Japanese macaque (*M. fuscata*), and a prosimian (*Galago senegalensis*) are under investigation.

These sera were obtained through the courtesy of numerous cooperating primate laboratories throughout the world. (See acknowledgments.) Briefly, these serum specimens were obtained as part of a World Health Organization (WHO) and National Institutes of Health (NIH) effort to establish a virus reference center, but more importantly, as a study on comparative medicine.[53,54] It should be emphasized that these sera, in many instances, cannot be construed as representing the antibody response of the animal in question as it exists in its native habitat. As indicated above, except for those baboons specifically studied at the time of capture, all other animals have had multiple exposures to other animals, including man, and represent captive animals. It is, therefore, practically impossible to attribute an antibody response as found in an animal's serum to be caused specifically by a virus indigenous to that particular species or to result from an infection as a consequence of contact with another animal occurring after capture.

One conceivable generalization reflects the capability of all primates studied to respond to one or another of the viral antigens with antibody production. Accordingly, antibody has been found in primate sera to all major virus families examined: arboviruses, adenoviruses, herpesviruses, myxoviruses, picornaviruses, poxviruses, reoviruses, and numerous unclassified viruses. Tables 4–11 provide information comparing the incidence of antibody to the different viruses among dissimilar species of primates as well as the possible influence of the facility supplying the animals. Arbovirus infection as represented by antibody studies with three viruses indigenous to the United States, i.e., EE, WE, and SLE, rarely occurs. These animals apparently can develop an infection, but whether these agents are potentially dangerous remains to be determined. A small number of animals (baboons, African greens, chimpanzees) maintained in colonies exposed to the mosquito vector have demonstrated antibody. As mentioned previously, WE virus has been recovered from baboons.[43]

Adenovirus infection occurs very frequently, but a complete analysis regarding specific types—human and/or simian—is needed. Certain adenovirus types are known not to occur naturally in particular areas of the world. SV39 is not found in baboons as tested immediately upon capture. This virus probably does not occur in African simians (Hull, personal communication), and the data accumulated in these studies corroborate this finding. African nonhuman primates that do have this virus probably acquire it as a result of contact with an SV39 shedding animal. Evidently, there is a great need for more elaborate studies with the adenoviruses. It appears, for example, that the CF test that detects the adenovirus group antigen may frequently be negative when used with a positive serum as determined by testing with another adenovirus antigen in another serologic system, e.g., serum neutralization. It is also of interest to note that more adenovirus group antigen seropositives are observed among the higher primates than among the monkeys, except for the gorilla.* A new simian adenovirus, V340, found to be highly pathogenic for African greens and baboons, evidently causes widespread infection. Antibody to this virus is frequently encountered in most primate sera.[59]

Herpesvirus antibody surveillance studies have not been concluded and are in need of expansion to include antigens other than *Herpesvirus hominis* (herpes simplex). The data do suggest that these antibodies are generally found in animals with a long history of contact with humans. In view of the above-mentioned recent association of herpes-like virus particles to various human

---

*It should be noted that serologic findings on the gorilla appear to be somewhat different than other primates. Attempts to demonstrate antibody in sera from 14 gorillas have, in general, been negative for most virus antigens employed.[60]

TABLE 4 Representative Findings upon Testing Primate Sera for Antibody to Arboviruses

Sera	Virus WEE[a]	EEE[a]	SLE[a]
Human			
African	0/32	0/32	1/32
SFRE	0/63	0/63	0/63
1967 Recruits	0/25	0/25	0/25
Gorilla			
Lab 1 (1966)	0/14	0/14	0/14
Lab 1 (1967)	0/14	0/14	0/14
Chimpanzee			
SFRE (preshipment)	0/17	0/17	0/17
(1)	0/18	0/18	0/18
(2)	0/14	0/14	0/14
(4)	0/14	0/14	0/16
Lab 1 (1963)	2/26	0/26	0/21
(lab-born)	0/29	0/29	0/29

Sera	Virus WEE[a]	EEE[a]	SLE[a]
Baboons			
Africa–1963	0/47	0/47	0/25
1964	0/34	0/34	0/16
1966 Site 1	0/3	0/3	0/4
Site 2	0/16	0/16	0/17
Site 3	0/13	0/13	0/19
1968 Site 1	0/18	0/18	1/18
Site 2	0/20	0/20	0/20
Site 3	0/10	0/10	0/10
Site 4	0/23	6/22	4/21
Site 5	0/24	0/24	2/25
SFRE–African	2/30	0/30	0/13
Domestic	1/22	0/22	1/22
Lab 4	0/3	0/3	0/3

Sera	Virus WEE[a]	EEE[a]	SLE[a]
Patas			
Lab 4 (1)	0/22	0/22	0/22
(2)	0/5	0/5	0/5
Lab 5	0/24	0/24	0/24
Cynamolgus			
Lab 5 (1)	0/19	0/19	0/19
(2)	2/23	1/23	1/23
Lab 9	0/3	0/3	0/3
Irus			
Lab 19	0/12	0/12	0/12
Formosan Rock Macaque			
Lab 16	0/25	0/25	0/25
Stumptail			
Lab 4	0/13	0/11	0/11

Lab 4	0/29	0/13	0/30
	0/3	0/3	0/3
	0/8	0/8	0/8
Lab 7	0/24	0/24	0/24
Lab 15	0/1	0/1	0/1
Orangutan			
Lab 1 (1963)	0/18	0/18	0/24
(1966)	0/28	0/28	0/28
(1967)	0/28	0/28	0/28
Gibbon			
Lab 1	0/9	0/9	0/8
Lab 4	0/8	0/8	0/8
Lab 7	0/9	0/9	0/9

SFRE	0/22	0/22	6/22
Lab 5	0/20	0/20	0/20
Lab 11 (1)	0/6	0/6	0/6
(2)	0/24	0/24	0/24
Rhesus			
SFRE	1/25	1/25	2/19
Lab 3 (1965)	0/12	0/12	4/16
3 (1967)	0/17	0/17	0/17
Lab 4 (1)	0/19	0/19	1/20
(2)	0/3	0/3	0/3
Lab 9	0/25	0/25	0/25
Lab 10	0/25	1/25	0/25
Lab 19	2/12	2/12	0/12

Lab 21	0/3	0/3	0/3
Howler			
Lab 6	0/3	0/3	0/3
Japanese Macaque[b]			
Lab 17 (adult)	12/23	11/23	0/23
(young)	5/20	5/20	1/20
Spider			
Lab 21	0/7	0/7	0/7
Marmosets			
Lab 8	0/25	0/25	0/25
Lab 11	0/22	0/22	0/22
Lab 14	0/12	0/12	0/12

[a]Complement-fixation.
[b]Needs further study; probably nonspecific.

497

TABLE 5 Representative Findings upon Testing Primate Sera for Antibody to Adenoviruses

Sera	Adeno Group[a]	SA7[b]	V340[b]	SV39[b]	Adeno 12[b]
Human					
Africans	16/25	13/25	9/34	1/34	5/35
SFRE	27/63	21/44	12/22	–	2/23
1967 Recruits	17/25	–	–	–	–
Gorilla					
Lab 1 (1966)	0/14	–	–	–	–
(1967)	–	–	0/11	0/11	–
Chimpanzee					
SFRE (preshipment)	8/16	1/4	–	–	–
(1)	7/18	3/17	5/17	0/18	–
(2)	9/18	–	–	–	–
(4)	14/15	–	–	–	–
Lab 1 (1963)	20/32	16/54	7/51	2/52	–
(1966)	32/34	–	–	–	–
(lab-born)	21/29	–	–	3/20	–
(1967)	28/69	11/21	9/22	–	0/26
Lab 2	16/18	1/6	0/8	0/20	9/23
	13/18	–	–	–	–
Lab 4	27/29	12/23	3/11	–	3/11
	1/3	–	–	–	–
	1/8	–	–	–	–
Lab 7	18/24	–	–	–	–
Lab 15	1/6	–	–	–	–
Orangutan					
Lab 1 (1963)	5/10	0/19	0/21	1/21	0/23
(1966)	8/28	–	–	–	–
(1967)	1/28	–	–	–	–
Gibbons					
Lab 1	2/9	–	0/6	0/6	0/4
Lab 4	0/8	–	–	–	–
Lab 7	0/9	–	–	–	–
Patas					
Lab 4 (1)	3/24	0/24	3/7	–	0/14
(2)	0/5	0/5	–	–	–
Lab 5	0/24	0/23	–	–	–
Cynomolgus					
Lab 5 (1)	1/18	0/21	9/22	–	–
(2)	9/24	–	–	–	–
Lab 9	0/3	0/2	–	–	–
Irus					
Lab 19	1/13	–	–	–	–
Formosan Rock Macaque					
Lab 16	1/25	0/24	–	–	–

TABLE 5 (Continued)

Sera	Adeno Group[a]	SA7[b]	V340[b]	SV39[b]	Adeno 12[b]
Stumptail					
Lab 4 (1)	0/2	0/2	–	–	–
(2)	0/9	0/9	–	–	–
Lab 9	0/1	0/0	–	–	–
Lab 15	0/2	0/2	–	–	–
Baboons					
Africa 1963	0/27	7/25	3/23	0/24	0/23
1964	1/25	18/25	5/24	0/24	0/24
1966 Site 1	0/5	–	–	–	–
Site 2	0/15	–	10/25	0/25	–
Site 3	0/12	–	9/25	0/25	–
1968 Site 1	0/17	–	–	–	–
Site 2	0/20	–	–	–	–
Site 3	0/10	–	–	–	–
Site 4	0/24	–	–	–	–
Site 5	0/25	–	–	–	–
SFRE-Africans	0/28	5/25	4/24	3/23	0/23
Domestic	1/17	15/25	5/17	14/23	0/24
Lab 4	0/3	–	–	–	–
Lab 5	0/15	–	–	–	0/20
Lab 15	1/12	–	–	–	–
Gelada	0/10	–	–	–	0/10
Vervets					
SFRE	5/65	9/45	17/26	0/26	0/26
Lab 5	1/20	–	–	–	–
Lab 11	3/6	–	–	–	–
	4/24	–	–	–	–
Rhesus					
SFRE	1/14	2/23	3/18	8/19	1/19
Lab 3 (1965)	3/19	–	–	–	–
(1967)	1/17	–	–	–	–
Lab 4	1/21	–	–	–	–
	1/3	–	–	–	–
Lab 9	2/25	–	–	–	–
Lab 10	14/24	–	–	–	–
Lab 19	1/12	–	–	–	–
Talapoin-Lab 11	1/21	3/20	–	–	–
White Faced					
Lab 6	0/0	–	0/12	–	–
	0/0	–	0/24	–	–
Howler					
Lab 6	0/3	0/25	–	–	–
Japanese Macaque					
Lab 17 (adult)	1/23	–	–	–	–
(young)	1/24	–	–	–	–

TABLE 5 (Continued)

Sera	Adeno Group[a]	SA7[b]	V340[b]	SV39[b]	Adeno 12[b]
Spider					
Lab 6	0/0	–	0/14	–	–
Marmosets					
Lab 8	0/24	–	–	–	–
Lab 11	0/15	0/22	–	–	–
Lab 14	1/4	0/13	–	–	–

[a]Complement-fixation.
[b]Serum neutralization.

TABLE 6 Representative Findings upon Testing Primate Sera for Antibody to Herpesviruses

Sera	Herpes[a] Simplex	Sera	Herpes[a] Simplex
Humans		*Gibbons*	
Africans	0/32	Lab 1	0/9
SFRE	3/61	Lab 4	0/8
1967 Recruits	7/25	Lab 7	0/9
Gorilla		*Baboons*	
Lab 1 (1966)	0/14	Africa (1963)	0/47
(1967)	0/14	(1964)	0/34
Chimpanzee		(1966 Site 1)	0/3
SFRE (preshipment)	0/15	Site 2	0/16
(1)	0/18	Site 3	0/13
(2)	0/14	(1968 Site 1)	0/16
(4)	4/14	Site 2	–
Lab 1 (1963)	1/26	Site 3	0/11
(lab-born)	0/28	Site 4	2/10
(1966)	0/43	Site 5	0/21
(1967)	0/69	SFRE African	0/30
Lab 2 (1)	0/14	Domestic	5/22
(2)	0/27	Lab 4	0/3
Lab 4 (1)	0/28	Lab 5	0/21
(2)	0/2	Lab 15	0/2
(3)	2/8	Gelada: Lab 4	0/10
Lab 7	0/24	*Vervets*	
Lab 15	0/1	SFRE	0/22
Orangutan		Lab 5	0/20
Lab 1 (1963)	0/18	Lab 11 (1)	0/6
(1966)	0/28	(2)	0/22
(1967)	0/28		

TABLE 6 (Continued)

Sera	Herpes[a] Simplex	Sera	Herpes[a] Simplex
Rhesus		*Formosan Rock Macaque*	
SFRE	0/25	Lab 16	0/25
Lab 3 (1965)	0/12	*Stumptailed*	
(1967)	0/17	Lab 4 (2)	0/2
Lab 4 (1)	0/22	(3)	0/6
(2)	0/3	*Talapoin*: Lab 11	0/8
Lab 9	2/25	*White Face*: Lab 21	0/3
Lab 10	0/25	*Howler*: Lab 6	0/3
Lab 19	–	*Japanese Macaque*	
Patas		Lab 17 (adult)	0/3
Lab 4 (1)	0/22	(young)	0/6
(2)	0/5	*Spider*: Lab 21	0/7
Lab 5	0/24	*Marmosets*	
Cynomolgus		Lab 8	0/25
Lab 5 (1)	1/17	Lab 11	0/22
(2)	0/15	Lab 14	0/13
Lab 9	0/3		
Irus			
Lab 19	2/10		

[a]Complement-fixation.

TABLE 7 Representative Findings upon Testing Primate Sera for HI Antibody to Myxoviruses

Sera	Inf. A	Inf. A (PR8)	Inf. A₁ (FM1)	Inf. A₂ (Jap 2)	Inf. B (Lee)	Measles	Mumps	Para 1	Para 2	Para 3	RS	SV41	SV5
Human													
Africans	19/24	1/24	5/24	3/24	3/24	14/30	11/28	8/26	1/26	5/26	0/25	8/19	11/23
SFRE	25/62	30/40	31/40	38/40	17/40	34/42	12/47	16/43	17/43	18/43	0/63	15/22	9/23
1967 Recruits	15/25	4/24	18/24	22/24	5/42	8/25	12/25	10/24	4/24	6/24	3/25	–	–
Gorilla													
Lab 1 (1966)	0/14	0/3	0/3	1/3	0/3	1/14	0/14	0/6	0/6	0/6	0/14	–	–
Lab 1 (1967)	0/14	0/9	0/9	1/9	0/9	0/13	1/13	0/11	0/11	3/11	0/14	–	–
Chimpanzee													
SFRE (preshipment)	0/12	0/13	0/13	0/13	0/13	0/17	0/17	0/13	0/13	0/13	11/16	–	–
(1)	0/13	4/12	0/12	2/12	4/12	0/18	0/18	0/12	0/12	4/12	9/18	–	–
(2)	0/15	0/13	0/13	0/13	0/13	0/18	0/18	0/13	0/13	0/13	5/17	–	–
(3)	–	0/16	0/16	0/16	0/16	0/16	2/16	0/16	0/16	4/16	–	–	–
(4)	0/12	0/16	0/16	1/16	0/16	0/16	0/16	0/16	0/16	3/16	0/5	–	–
Lab 1 (1963)	0/17	2/31	0/31	1/31	0/31	11/43	0/13	1/13	0/13	6/13	20/27	4/52	13/49
(lab-born)	0/29	0/25	0/25	0/25	0/25	5/28	2/27	1/25	0/25	4/25	4/29	–	–
(1966)	0/44	0/27	0/27	1/27	0/27	6/42	2/36	0/27	0/27	11/27	4/44	–	–
(1967)	0/68	0/62	0/62	3/62	0/62	0/69	4/64	3/66	1/66	3/66	8/69	–	–
Lab 2 (1)	0/11	6/25	2/25	22/25	5/25	9/23	0/19	0/19	0/19	4/19	0/19	3/24	6/23
(2)	0/20	1/28	0/28	11/28	1/28	0/28	–	–	–	–	0/28	–	–
Lab 4 (1)	0/27	2/25	2/25	6/25	2/25	9/30	22/30	5/25	3/25	16/25	5/29	0/11	1/11
(2)	1/2	0/3	0/3	0/3	0/3	0/3	1/3	0/3	1/3	0/3	0/3	–	–
(3)	0/8	0/8	0/8	2/8	0/8	0/7	3/8	1/8	0/8	3/8	3/8	–	–
Lab 7	0/21	0/24	0/24	0/24	0/24	0/24	1/24	0/21	0/21	6/21	0/24	–	–
Lab 15	1/4	0/7	0/7	7/7	0/7	0/3	5/7	1/7	1/7	4/7	4/6	–	–
Orangutan													
Lab 1 (1963)	0/23	4/22	4/22	10/22	4/22	4/23	0/19	0/17	0/17	0/17	0/6	8/22	4/21
(1966)	0/28	0/17	0/17	6/17	0/17	3/28	0/26	0/20	0/20	0/20	0/28	–	–
(1967)	0/28	0/25	0/25	2/25	0/25	0/28	0/27	0/24	0/24	4/24	1/28	–	–
Gibbon													
Lab 1	0/5	2/9	2/9	2/9	2/9	2/9	0/4	0/4	0/4	0/4	0/9	0/5	1/4

1966 Site 1	0/3	0/4	0/4	1/4	—	0/5	0/5	0/4	0/4	0/4	0/4	—	—
Site 2	1/12	0/16	0/16	6/16	0/6	0/24	0/17	0/16	0/16	0/16	0/15	—	—
Site 3	0/10	1/14	0/14	1/14	0/14	0/17	1/17	0/16	0/16	0/16	0/12	—	—
1968 Site 1	1/11	0/25	0/25	1/25	0/25	0/25	0/25	1/25	1/25	24/25	0/17	—	—
Site 2	0/14	0/20	0/20	2/20	1/20	0/20	3/20	1/20	0/20	20/20	0/20	—	—
Site 3	0/8	1/11	1/11	0/11	0/11	0/11	1/11	0/11	0/11	11/11	0/10	—	—
Site 4	0/17	22/24	0/24	2/24	1/24	0/25	1/25	0/24	2/24	23/24	0/24	—	—
Site 5	0/22	24/25	0/25	5/25	2/25	0/25	2/25	1/25	0/25	25/25	0/25	—	—
SFRE-African	0/13	0/16	0/16	0/16	0/16	29/103	2/15	0/16	0/16	10/16	0/18	1/24	13/24
Domestic	3/19	0/19	0/19	0/19	0/19	1/44	0/21	0/19	0/19	0/19	0/3	0/24	15/24
Lab 4	0/3	0/3	0/3	0/3	0/3	0/3	0/3	0/3	0/3	0/3	0/15	—	—
Lab 5	0/15	0/7	0/7	0/7	0/7	11/24	1/23	0/18	0/17	5/18	0/2	—	—
Lab 15	1/1	0/2	0/2	1/2	0/2	0/2	1/2	0/2	0/2	0/2	0/10	—	—
Gelada Lab 4	0/10	0/8	0/8	0/8	0/8	0/10	0/9	0/8	0/8	2/8	1/65	—	—
African green SFRE	0/22	17/18	14/18	15/18	12/18	1/36	4/25	0/25	0/25	1/25	0/20	1/26	6/25
Lab 5	0/20	0/15	0/15	1/15	0/15	11/23	3/23	0/21	18/21	18/21	0/6	—	—
Lab 11 (1)	0/6	6/20	0/20	0/20	0/20	1/21	2/6	1/21	0/20	4/20	1/24	—	—
(2)	0/23	0/18	1/18	0/18	0/18	0/24	5/24	1/24	2/24	6/24	0/14	—	—
Rhesus SFRE	0/19	0/28	0/28	0/28	0/28	21/27	1/15	0/15	0/15	15/15	0/2	0/17	16/18
Lab 3 (1965)	0/14	5/19	2/19	1/19	1/19	1/19	2/14	0/14	0/14	12/14	0/17	—	—
(1967)	0/17	0/17	0/17	0/17	0/17	13/17	0/17	0/16	6/17	15/16	0/21	—	—
Lab 4 (1)	0/21	0/24	0/24	0/24	0/24	9/25	5/25	0/24	0/24	23/24	0/3	—	—
(2)	0/2	0/3	0/3	0/3	0/3	2/3	0/3	0/3	0/3	0/3	0/25	—	—
Lab 9	0/19	0/27	0/27	0/27	0/27	7/26	5/26	1/27	3/25	7/27	0/24	—	—
Lab 10	0/12	0/23	0/23	0/23	0/23	—	7/23	0/25	0/11	9/25	0/12	—	—
Lab 19	0/7	—	—	—	—	0/12	0/12	0/11	0/11	10/11	—	—	—
Patas Lab 4 (1)	0/24	0/23	0/23	0/23	0/23	4/22	6/25	0/23	0/23	17/23	0/24	—	—
(2)	0/5	0/5	0/5	0/5	0/5	1/5	0/5	0/5	0/5	0/5	0/5	—	—
Lab 5	0/24	1/17	0/17	0/17	0/17	2/24	3/23	0/20	19/23	3/20	0/24	—	—
Cynomolgus Lab 5 (1)	0/18	0/16	0/16	0/16	0/16	20/24	11/24	0/19	0/20	13/19	0/18	—	—
(2)	0/15	—	—	—	—	0/24	0/13	0/23	0/23	20/23	0/24	—	—
Lab 9	0/2	0/4	0/4	0/4	0/4	0/4	0/4	0/4	0/4	0/3	0/3	—	—

TABLE 7 (Continued)

Sera	Inf. A	Inf. A (PR8)	Inf. A₁ (FM1)	Inf. A₂ (Jap 2)	Inf. B (Lee)	Measles	Mumps	Para 1	Para 2	Para 3	RS	SV41	SV5
Irus													
Lab 19	0/5	–	–	–	–	0/13	1/24	1/13	2/13	0/13	0/13	–	–
Formosan Rock Macaque													
Lab 16	0/22	0/24	0/24	0/24	0/24	1/25	6/24	0/24	0/24	4/24	2/25	–	–
Stumptail													
Lab 4 (3)	0/6	0/9	0/9	0/9	0/9	0/9	1/9	0/9	0/9	4/9	0/9	–	–
Talapoin													
Lab 11	0/7	6/20	0/20	0/20	0/20	1/21	18/20	1/21	0/20	4/20	0/21	–	–
White Face													
Lab 21	–	0/5	0/5	0/5	0/5	0/1	0/5	0/5	0/5	1/5	–	–	–
Howler													
Lab 6	0/3	0/6	0/6	1/6	0/6	0/40	2/5	0/6	0/6	1/6	0/3	–	–
Japanese Macaque													
Lab 17 (adult)	–	–	–	–	–	0/24	5/24	0/24	0/24	12/24	0/24	–	–
(young)	–	–	–	–	–	0/24	5/24	0/24	4/24	10/24	0/23	–	–
Spider													
Lab 21	–	0/8	0/8	0/8	0/8	0/3	4/8	0/8	0/8	7/8	–	–	–
Marmosets													
Lab 8	0/22	0/24	0/24	8/24	0/24	0/43	0/24	0/24	0/24	0/24	0/24	–	–
Lab 11	0/15	0/24	0/24	3/24	1/24	0/24	–	0/24	0/24	0/24	0/15	–	–
Lab 14	–	0/15	0/15	15/15ᵃ	15/15ᵃ	–	5/15	0/15	0/15	4/15	–	–	–

[a]Needs further study; probably nonspecific.

504

TABLE 8 Representative Findings upon Testing Primate Sera for SN Antibody to Picornaviruses

Sera	Cox. A9	Cox. A20	Cox. B1	Cox. B2	Cox. B3	Cox. B4	Cox. B5	Cox. B6	Echo 1	Echo 3	Echo 6	Echo 7	Echo 11	Echo 12	Echo 13	Polio 1	Polio 2	Polio 3	SV4	SV16	SV19	SV45	SV49	A13
Humans																								
Africans	15/33	0/27	9/33	18/35	19/33	12/14	1/16	4/14	12/35	10/27	7/11	10/34	7/27	10/33	3/27	27/34	32/35	23/35	1/34	1/26	0/44	0/30	3/28	0/10
SFRE	14/22	0/58	7/23	9/20	8/22	9/21	2/22	4/21	2/22	24/58	4/19	22/57	12/58	36/63	11/58	16/22	21/23	8/9	0/22	6/56	5/23	0/42	5/23	0/23
1967 Recruits	—	0/25	—	—	—	—	—	—	—	0/25	—	8/25	1/25	1/25	0/25	0/25	0/25	—	—	0/25	0/25	0/25	0/25	—
Gorilla																								
Lab 1 (1966)	—	0/14	—	—	—	—	—	—	—	0/14	—	1/14	0/14	0/14	0/14	0/14	—	—	—	1/14	—	0/14	—	—
(1967)	2/11	4/13	0/10	0/11	0/11	0/11	0/11	0/11	0/11	0/13	1/11	2/13	0/13	1/13	0/13	0/13	10/11	9/11	8/11	—	0/10	—	0/13	—
Chimpanzee																								
SFRE (pre-ship)	—	0/17	—	—	—	—	—	—	—	11/17	—	13/17	11/17	3/17	1/17	1/17	—	—	—	3/17	—	0/17	—	—
(1)	4/18	0/17	0/18	0/18	—	1/17	0/17	1/17	0/17	6/17	4/18	9/16	5/17	3/16	1/17	4/18	3/19	3/18	—	1/16	—	0/18	—	—
(2)	—	0/17	—	—	—	—	—	—	—	12/17	—	12/17	3/17	3/17	1/17	—	—	—	—	3/17	—	0/18	—	—
(4)	—	4/15	—	—	—	—	—	—	—	7/16	—	10/16	10/16	1/16	3/15	—	—	—	—	2/15	—	0/18	—	—
Lab 1 (1963)	7/52	0/16	0/52	3/52	0/52	0/52	2/52	3/50	1/51	2/16	10/20	0/25	2/16	12/56	0/16	4/52	16/52	9/25	0/51	0/13	8/51	3/28	9/52	0/52
(lab-born)	—	0/29	—	—	—	—	—	—	—	12/29	—	25/27	3/29	22/27	1/29	—	—	—	—	1/26	—	0/27	—	—
(1966)	—	0/38	—	—	—	—	—	—	—	27/38	—	7/30	9/38	8/41	4/38	—	—	—	—	0/24	—	0/38	—	—
(1967)	6/22	38/64	0/22	1/21	1/21	0/19	0/19	0/21	0/21	29/65	6/25	29/65	23/65	13/65	10/65	14/25	10/25	13/25	—	0/62	0/69	0/21	0/28	—
Lab 2 (1)	4/23	0/23	1/24	1/21	1/24	0/19	0/19	0/15	1/24	17/23	1/4	17/23	3/23	1/23	0/23	1/24	1/24	5/24	0/23	1/23	8/22	1/30	2/24	8/18
(2)	—	2/27	—	—	—	—	—	—	—	17/27	—	7/29	2/27	2/27	0/27	0/27	—	—	—	0/28	—	0/28	—	—
Lab 4 (1)	1/9	0/27	0/11	1/11	1/10	0/9	0/9	0/9	2/11	4/27	—	—	—	—	—	0/11	3/11	4/11	2/11	1/30	4/11	3/11	3/11	1/11
(2)	—	0/3	—	—	—	—	—	0/9	—	0/3	0/3	0/3	0/3	0/8	0/3	0/8	0/3	0/3	—	0/3	—	0/7	0/7	—
(3)	—	0/8	—	—	—	—	—	—	—	8/22	12/25	10/24	2/22	2/24	1/22	1/25	23/25	23/25	0/20	2/8	—	0/24	—	—
Lab 7	—	0/22	—	—	—	—	—	—	—	7/7	—	0/6	—	—	0/7	—	—	—	—	1/24	—	—	—	—
Lab 15	—	0/7	—	—	—	—	—	—	—	—	—	6/7	—	—	—	—	—	—	—	1/1	—	1/6	—	—
Orangutan																								
Lab 1 (1963)	0/22	0/17	0/22	2/22	0/22	0/22	3/22	0/18	0/22	0/17	—	0/19	0/17	4/22	0/17	0/17	0/22	1/22	1/20	0/20	1/22	0/19	1/21	0/6
(1966)	—	0/21	—	—	—	—	—	—	—	1/21	—	16/25	0/21	5/28	0/21	0/21	—	—	—	3/20	—	0/21	—	—
(1967)	—	7/23	—	—	—	—	—	—	—	0/26	—	11/27	0/26	0/27	0/26	0/26	—	—	—	0/20	—	1/28	—	—
Gibbon																								
Lab 1	0/3	0/8	—	—	—	—	—	—	0/6	0/8	0/4	0/9	0/8	3/9	0/8	0/8	0/7	0/7	0/6	0/8	0/5	0/9	0/4	0/6
Lab 4	—	0/7	—	—	—	—	—	—	—	0/7	—	0/8	0/7	0/8	0/7	0/7	—	—	—	1/8	—	0/7	—	—
Lab 7	—	0/9	—	—	—	—	—	—	—	4/9	—	2/9	2/9	2/9	0/9	0/9	—	—	—	—	—	0/9	—	—

TABLE 8 (Continued)

Sera	Cox. A9	Cox. A20	Cox. B1	Cox. B2	Cox. B3	Cox. B4	Cox. B5	Cox. B6	Echo 1	Echo 3	Echo 6	Echo 7	Echo 11	Echo 12	Echo 13	Polio 1	Polio 2	Polio 3	SV4	SV16	SV19	SV45	SV49	A13
Baboon																								
Africa (1963)	0/42	2/30	7/26	1/26	0/53	0/45	0/48	0/16	2/22	11/30	0/39	26/34	7/30	17/35	1/30	0/26	0/26	8/23	1/25	1/30	0/24	3/30	8/24	0/23
(1964)	0/24	2/39	7/24	0/24	0/20	0/22	1/19	0/22	0/24	37/41	0/11	35/41	11/39	25/42	10/39	0/24	0/24	0/23	2/24	0/8	2/23	3/25	9/22	0/24
(1966 Site 1)	–	0/5	–	–	–	–	–	–	–	0/5	–	2/5	0/5	0/5	0/5	–	–	–	–	0/5	0/5	–	–	–
Site 2	0/25	1/17	0/25	0/25	0/25	0/25	0/25	0/25	0/24	1/17	0/25	2/17	1/17	2/15	2/17	0/25	0/25	0/25	–	2/24	0/24	0/17	–	–
Site 3	0/25	0/17	0/25	2/25	0/25	0/24	1/25	0/26	0/25	0/17	0/25	1/17	0/17	4/17	0/17	1/25	1/25	0/24	–	1/17	0/17	1/25	–	–
(1968 Site 1)	0/25	5/22	–	0/25	0/25	0/25	0/25	0/25	0/25	6/23	0/25	22/25	5/21	2/25	0/22	0/22	0/37	0/37	–	11/22	1/25	–	–	–
Site 2	0/20	6/17	–	0/20	1/20	0/20	1/19	0/20	0/20	4/18	0/20	19/19	8/18	0/19	5/18	0/11	0/23	0/23	–	9/17	0/20	–	–	–
Site 3	0/10	4/10	–	0/11	0/11	0/11	0/12	0/11	0/11	4/11	0/24	11/11	5/11	0/11	4/11	0/24	0/11	0/11	–	4/11	1/10	–	–	–
Site 4	0/23	1/25	–	1/23	0/19	0/24	0/24	0/24	0/24	1/25	0/24	11/22	0/25	0/22	1/25	1/25	0/24	0/24	–	0/23	0/25	–	–	–
Site 5	–	2/21	–	1/25	0/25	1/24	2/24	0/24	0/24	1/21	2/23	6/9	1/21	0/8	1/21	0/24	0/24	0/25	–	0/10	0/25	–	–	–
SFRE (African)	0/24	0/15	0/24	2/24	0/23	1/24	2/24	0/20	2/23	2/30	0/7	0/63	1/15	1/99	0/15	0/23	0/25	0/24	–	5/15	0/17	0/17	11/24	2/24
(Domestic)	0/25	0/21	5/25	1/25	0/24	0/23	4/24	0/24	1/24	19/21	0/17	0/30	2/21	1/66	2/21	0/25	0/24	0/24	0/24	3/19	0/21	0/21	1/23	11/21
Lab 4	–	0/3	–	–	–	–	–	–	–	–	0/3	0/3	0/3	0/3	0/3	–	–	–	1/24	0/3	0/3	–	–	–
Lab 5	–	1/23	–	–	–	–	–	–	–	2/23	–	8/24	1/23	0/24	1/23	–	–	–	–	1/20	1/24	–	–	–
Lab 15	–	0/2	–	–	–	–	–	–	–	2/2	–	0/2	1/2	0/2	0/2	–	–	–	–	1/2	0/2	–	–	–
Gelada–Lab 4	–	0/9	–	–	–	–	–	–	–	0/9	–	3/9	0/9	0/9	0/9	–	–	–	–	0/10	0/10	–	1/9	0/9
African green																								
SFRE	–	0/19	3/26	1/26	0/26	0/25	1/26	0/24	1/26	15/32	0/48	1/20	3/19	9/33	1/19	0/26	1/26	–	–	0/19	2/26	0/19	6/26	0/26
Lab 5	–	6/23	–	–	–	–	–	–	–	10/23	–	7/23	8/23	7/23	3/23	–	–	–	4/26	2/22	1/23	1/23	–	–
Lab 11 (1)	–	1/6	–	–	–	–	–	–	–	3/3	–	0/6	0/6	–	–	–	–	–	–	–	–	–	–	–
(2)	–	1/24	–	–	–	–	–	–	–	24/24	–	10/24	3/24	2/24	3/24	–	–	–	–	3/24	0/24	0/24	–	–
Rhesus																								
SFRE	0/14	0/25	0/17	3/18	0/21	0/18	0/18	0/16	2/19	0/25	0/15	4/27	0/25	1/25	1/25	1/19	0/18	0/14	–	0/26	7/18	1/26	6/26	–
Lab 3 (1965)	–	0/19	–	–	0/17	0/11	0/15	0/13	–	1/19	–	3/19	3/19	12/19	2/19	–	–	–	–	5/19	–	2/19	4/18	0/16
(1967)	–	0/16	–	–	–	–	–	–	–	2/16	–	0/17	1/16	0/17	0/16	–	–	–	–	4/17	–	2/17	–	–
Lab 4 (1)	–	0/27	–	–	–	–	–	–	–	0/27	–	6/22	0/27	0/22	0/27	–	–	–	–	2/25	–	1/24	–	–
(2)	–	0/3	–	–	–	–	–	–	–	0/3	–	1/3	0/3	0/3	0/3	–	–	–	–	0/3	–	0/3	–	–
Lab 9	–	0/27	–	–	–	–	–	–	–	0/27	–	4/27	0/27	0/27	3/27	–	–	–	–	3/27	–	3/26	–	–
Lab 10	–	0/22																						

506

Primate / Lab																
(2)	—	—	—	—	—	—	—	—	—	—	1/5	0/5	0/5	0/5	0/5	0/5	0/5	0/5	0/5
Lab 5	—	0/5 8/24	—	—	—	—	—	—	—	—	8/24	6/24	7/24	7/24	2/24	0/23	1/24
Cynomolgus																
Lab 5 (1)	0/25	6/24	0/25	0/25	0/25	0/18	0/12	0/11	11/24	1/25	12/24	5/24	6/24	3/24	1/25 3/25 0/25	3/22	1/24
(2)	—	—	—	—	—	—	—	—	—	—	—	—	—	—	—	1/24
Irus																
Lab 9	—	0/4	—	—	—	—	—	0/4	—	0/4	0/4	0/4	0/4	—	0/4	0/4
Lab 19	—	—	—	—	—	—	—	—	—	—	—	—	—	—	—	0/13
Formosan Rock Macaque																
Lab 16	—	0/25	—	—	—	—	—	0/25	—	1/25	0/25	0/25	0/25	—	0/23	—
Stumptail																
Lab 4 (3)	—	1/9	—	—	—	—	—	4/9	—	5/9	2/9	0/9	3/9	—	3/9	1/9
Lab 15	—	0/11	—	—	—	—	—	2/2	—	2/2	9/11	2/2	0/11	—	—	0/2
Talapoin																
Lab 11	—	3/21	—	—	—	—	—	18/21	—	13/10	2/21	0/20	0/21	—	7/20	2/21
White Face																
Lab 21	—	0/1	—	—	—	—	—	0/5	—	0/9	0/1	0/9	0/1	—	—	0/1
Howler																
Lab 6	—	0/26	—	—	—	—	—	0/26	0/25	0/39	0/26	0/39	0/26	—	0/40	0/24
Japanese Macaque																
Lab 17 (adult)	—	—	—	—	—	—	—	—	—	—	—	—	—	—	—	1/24
(young)	—	—	—	—	—	—	—	—	—	—	—	—	—	—	—	7/24
Spider																
Lab 6	—	0/4	—	—	—	—	—	0/4	2/14	0/4	0/4	0/14	0/4	0/4	1/14	0/6
Marmoset																
Lab 8	—	0/46	—	—	—	—	—	0/46	—	0/46	1/46	0/46	0/45	—	1/44	0/43
Lab 11	—	2/22	—	—	—	—	—	0/22	—	1/22	1/22	1/22	1/22	—	—	—
Lab 14	—	0/15	—	—	—	—	—	0/15	—	0/12	0/15	0/12	0/15	—	0/14	1/24

507

TABLE 9 Representative Findings upon Testing Primate Sera for Antibody to Poxviruses

Sera	Vaccinia[a]	Monkey Pox[a]
Human		
Africans	3/23	13/24
SFRE	5/52	8/57
1967 Recruits	3/25	8/25
Gorilla		
Lab 1 (1966)	0/12	0/11
Lab 1 (1967)	0/13	0/13
Chimpanzee		
SFRE (preshipment)	0/14	4/14
(1)	0/17	
(2)	0/18	
(3)	5/16	

Sera	Vaccinia[a]	Monkey Pox[a]
Baboon		
Africa–1963	0/18	0/16
1964	0/10	1/10
1966 Site 1		
Site 2	0/24	0/24
Site 3		
1968 Site 1	0/24	0/21
Site 2	0/18	1/17
Site 3	0/11	0/10
Site 4	0/25	5/20
Site 5	0/25	10/22
SFRE-African	0/13	0/12

Sera	Vaccinia[a]	Monkey Pox[a]
Patas		
Lab 4	0/20	0/15
Lab 5	0/5	0/5
	0/18	0/20
Cynomolgus		
Lab 5	0/17	0/20
Lab 9	0/24	0/24
	0/4	0/4
Irus		
Lab 19	1/13	2/13
Formosan Rock Macaque		

Chimpanzee (continued)

Species / Lab		
(lab-born)	0/23	0/14
(1967)	0/60	9/69
Lab 2	0/6	0/6
Lab 4	0/29	5/30
	0/3	0/3
	1/8	1/8
Lab 7		2/24
Lab 15	1/7	1/7
Orangutan		
Lab 1 (1963)	0/12	0/12
(1966)	0/16	
(1967)	0/27	1/28
Gibbon		
Lab 1	0/4	
Lab 4	0/8	0/8
Lab 7		

Species / Lab		
Lab 15	0/2	0/2
Gelada	0/9	1/9
Vervet		
SFRE	2/24	1/24
Lab 5	0/20	0/20
Lab 11	0/21	0/6
Rhesus		
SFRE	1/15	1/15
Lab 3 (1965)	0/10	
(1967)	0/15	
Lab 4 (1)	5/24	7/24
Lab 9	0/3	0/3
Lab 10	8/25	0/24
Lab 19	0/12	0/12

Species / Lab		
Talapoin		
Lab 11	0/21	4/21
White Face		
Lab 18	0/7	0/7
Lab 21	0/1	1/1
Japanese Macaque		
Lab 17 (adult)	0/21	1/21
(young)	0/23	2/23
Spider		
Lab 21	2/6	6/6
Marmoset		
Lab 11	2/20	4/20
Lab 18	0/8	2/8
Lab 14	0/15	0/15

[a]Hemagglutination inhibition.

TABLE 10 Representative Findings upon Testing Primate Sera for Antibody to Reoviruses

Sera	Reo 1[a]	Reo 2[a]	Reo 3[a]	SV59[a]	SV12[a]
Human					
African	16/33	16/33	3/18	15/33	15/26
SFRE	44/63	45/63	30/55	46/60	11/56
1967 Recruits	6/25	6/25	11/25	–	4/25
Gorilla					
Lab 1 (1966)	6/14	5/14	11/14	5/14	6/14
(1967)	5/13	6/13	6/11	–	2/10
Chimpanzee					
SFRE (preshipment)	5/17	3/17	15/17	–	15/17
(1)	5/16	5/16	13/16	–	9/15
(2)	6/17	6/17	14/17	–	17/17
(3)	1/15	1/15	10/16	–	4/16
(4)	2/16	1/16	10/15	–	3/15
Lab 1 (1963)	6/17	7/17	5/20	7/17	2/13
(1966)	28/41	34/41	6/30	33/41	19/24
(lab-born)	26/27	26/27	25/27	26/27	19/26
(1967)	35/65	45/65	32/55	–	29/62
Lab 2	15/23	4/23	4/22	4/23	8/23
	9/28	2/28	3/28	–	–
Lab 4	14/29	6/29	23/30	–	25/30
	0/3	0/3	1/3	–	1/3
	5/8	2/8	4/8	–	6/8
Lab 7	8/24	8/24	15/24	–	12/24
Orangutan					
Lab 1 (1963)	3/21	3/21	2/14	3/21	2/19
(1966)	22/28	22/28	19/21	22/28	20/20
(1967)	13/27	19/27	19/27	–	10/20
Gibbon					
Lab 1	4/9	3/9	2/7	3/9	1/8
Lab 4	5/8	1/8	7/8	–	8/8
Lab 7	3/9	1/9	4/9	–	–
Baboon					
Africa (1963)	–	–	25/30	2/30	1/30
(1964)	19/42	12/42	10/35	15/42	0/8
Site 1 (1966)	0/5	0/5	2/5	–	0/5
Site 2	1/15	1/15	10/24	1/15	7/24
Site 3	11/17	6/17	5/17	5/17	6/17
Site 1 (1968)	3/25	1/25	8/22	–	2/22
Site 2	1/19	0/19	9/17	–	1/17
Site 3	0/11	0/11	2/11	–	1/11
Site 4	2/25	0/25	–	–	2/23
Site 5	0/23	0/23	–	–	0/13
SFRE-African	5/15	2/15	6/15	3/15	6/15
Domestic	17/21	14/21	9/21	13/21	8/19

TABLE 10 (Continued)

Sera	Reo 1[a]	Reo 2[a]	Reo 3[a]	SV59[a]	SV12[a]
Baboon (cont'd.)					
Lab 4	0/3	0/3	3/3	–	0/3
Lab 5	0/24	0/24	8/24	–	0/20
Lab 15	0/2	0/2	2/2	–	0/2
Gelada	0/9	0/9	5/9	–	0/10
Vervet					
SFRE	5/19	0/19	9/18	0/19	1/19
Lab 5	7/23	1/23	15/23	–	4/22
Lab 11	2/6	0/6	–	–	–
	15/24	4/24	24/24	–	12/24
Rhesus					
SFRE	20/25	3/23	19/24	3/25	22/26
Lab 3 (1965)	14/18	3/18	3/19	3/18	7/19
(1967)	5/17	0/17	2/15	–	6/17
Lab 4	7/22	3/22	21/26	–	12/25
	0/3	0/3	3/3	–	0/3
Lab 9	8/26	0/26	15/26	–	14/27
Lab 10	0/25	2/25	11/25	–	–
Lab 19	–	–	2/11	–	–
Patas					
Lab 4	13/24	8/24	9/25	–	13/24
	3/5	1/5	4/5	–	3/5
Lab 5	2/24	2/24	16/24	–	2/23
Cynomolgus					
Lab 5	5/24	1/24	12/23	–	2/22
	–	–	3/13	–	–
Lab 9	0/4	0/4	0/3	–	0/4
Irus					
Lab 19	–	–	5/24	–	–
Formosan Rock Macaque					
Lab 16	6/25	0/25	2/25	–	0/25
Stumptail					
Lab 4	0/2	0/2	2/2	–	0/2
	1/9	0/9	5/7	–	1/9
Talapoin					
Lab 11	1/20	0/20	–	–	1/20
White Faced					
Lab 6	0/12	0/12	4/5	–	0/12
	0/25	0/25	14/20	–	0/25
Lab 18	0/5	0/5	3/4	–	–
Howler					
Lab 6	0/39	0/39	3/26	–	0/40
Japanese Macaque					
Lab 17 (adult)	–	–	6/10	–	–
(young)	–	–	18/22	–	–

TABLE 10 (Continued)

Sera	Reo 1[a]	Reo 2[a]	Reo 3[a]	SV59[a]	SV12[a]
Spider					
Lab 6	1/14	0/14	1/6	–	7/14
Lab 21	1/4	1/4	1/3	–	–
Marmoset					
Lab 8	2/46	0/46	7/36	–	–
Lab 11	1/24	0/24	–	–	–
Lab 14	0/15	0/15	0/15	–	0/14
Lab 18	0/8	0/8	3/8	–	–
Galago					
Lab 11	6/24	1/24	–	–	–

[a]Hemagglutination inhibition.

TABLE 11 Representative Findings upon Testing Primate Sera for Antibody to Unclassified Viruses

Sera	Rubella[a]	LCM[b]	Marburg[b]	SHF[b]	Psitt.[b]
Human					
Africans	–	2/32	0/29	0/28	1/32
SFRE	54/54	1/60	0/49	0/57	0/63
1967 Recruits	25/25	0/25	–	–	1/25
Gorilla					
Lab 1 (1966)	8/9	0/14	0/5	–	0/14
(1967)	10/12	0/14	–	0/11	0/14
Chimpanzee					
SFRE (preshipment)	4/2	0/17	2/15	–	1/17
(1)	–	0/18	–	–	0/18
(2)	–	0/18	–	–	0/14
(3)	13/13	–	–	–	–
(4)	14/14	0/16	–	–	0/16
Lab 1 (1963)	0/8	0/21	0/21	–	0/21
(1966)	15/32	1/44	1/34	–	0/44
(lab-born)	1/14	0/29	0/27	–	0/29
(1967)	28/47	0/69	–	0/69	0/69
Lab 2	4/5	2/12	1/4	–	0/11
	1/28	–	14/28	–	0/28
Lab 4	13/20	0/29	0/30	0/23	0/30
	1/3	0/3	–	–	3/3
	2/8	1/8	–	–	0/8
Lab 7	3/21	0/23	–	–	0/24
Lab 15	1/4	0/7	0/1	–	0/1

TABLE 11 (Continued)

Sera	Rubella[a]	LCM[b]	Marburg[b]	SHF[b]	Psitt.[b]
Orangutan					
Lab 1 (1963)	7/15	1/17	0/17	–	3/24
(1966)	4/11	0/28	0/25	0/25	0/28
(1967)	10/28	0/28	–	–	0/28
Gibbon					
Lab 1	–	0/6	–	0/4	0/8
Lab 4	1/8	0/8	0/8	0/8	0/8
Lab 7	–	0/9	–	–	0/9
Patas					
Lab 4	4/19	0/20	–	–	0/22
	1/5	0/5	–	–	0/5
Lab 5	6/23	0/24	2/23	0/23	0/24
Cynomolgus					
Lab 5	4/23	0/19	2/16	0/16	0/19
	0/24	0/24	1/12	–	0/23
Lab 9	0/4	0/4	–	0/2	0/3
Irus Lab 19	1/13	0/13	3/12	–	0/12
Formosan Rock Macaque					
Lab 16	0/17	0/25	4/25	–	0/25
Stumptail					
Lab 4	0/2	0/2	–	–	0/2
	0/9	0/9	–	–	0/9
Talapoin					
Lab 11	0/21	0/21	9/14	0/14	5/20
Baboon					
Africa 1963	2/18	0/25	0/22	0/21	0/21
1964	1/8	16/36	3/7	0/6	0/16
1966 Site 1	0/5	0/4	0/5	0/5	0/4
Site 2	0/16	0/17	2/45	0/40	0/16
Site 3	0/24	0/16	6/33	0/36	0/13
1968 Site 1	0/24	0/18	2/16	0/19	0/18
Site 2	4/17	2/20	0/20	0/17	0/20
Site 3	1/11	1/10	0/11	0/10	1/11
Site 4	0/16	0/24	3/20	0/22	11/22
Site 5	1/24	0/25	4/25	0/25	1/25
SFRE-African	0/14	2/13	0/13	0/13	0/12
Domestic	1/20	3/22	0/18	0/18	0/21
Lab 4	0/3	0/3	–	–	0/3
Lab 5	5/23	0/21	0/18	1/24	0/21
Lab 15	0/2	0/2	0/2	–	0/2
Gelada	0/10	0/10	0/10	–	0/10
Vervet					
SFRE	0/24	5/23	2/19	0/23	0/22
Lab 5	8/22	0/20	7/16	0/21	0/20
Lab 11	0/6	0/6	0/6	0/6	0/6
	0/23	0/24	5/25	–	4/24

TABLE 11 (Continued)

Sera	Rubella[a]	LCM[b]	Marburg[b]	SHF[b]	Psitt.[b]
Rhesus					
SFRE	2/8	1/19	0/18	0/17	0/19
Lab 3 (1965)	3/15	5/16	–	–	0/16
(1967)	1/16	4/17	0/15	–	1/17
Lab 4	13/23	0/20	–	–	0/23
	0/3	0/3	–	–	0/3
Lab 9	10/27	0/27	–	0/27	0/25
Lab 10	1/25	0/25	1/23	–	0/25
Lab 19	0/12	0/12	1/11	–	2/12
Lab 20	0/7	–	0/7	–	–
White Face					
Lab 18	0/3	–	–	–	–
Lab 21	0/1	–	0/4	–	0/3
Japanese Macaque					
Lab 17 (adult)	0/24	0/24	15/22	–	10/23
(young)	0/24	0/24	9/18	–	3/20
Spider					
Lab 21	0/6	–	0/7	–	1/7
Marmoset					
Lab 8	2/12	0/31	0/16	–	0/25
Lab 11	–	0/24	0/14	0/21	0/22
Lab 14	0/40	–	–	–	–
	0/15	1/11	0/3	–	0/13
Lab 18	5/8	–	–	–	–
Galago	1/24	–	0/24	–	–

[a] Hemagglutination inhibition.
[b] Complement fixation.

clinical conditions, an extensive search of simian tissues for similar particles may be in order.

Determinations of myxovirus infections as occurring among primates indicates an extensive, but variable, association with these viruses. The universal occurrence of measles antibody in primates following contact with man, as described by other investigators, has been substantiated. Evidently, measles antibody declines rather rapidly, as CF tests are frequently negative and antibody to this virus may be found only by HI or SN tests. Antibodies to various types and subtypes of influenza are only infrequently found, but simians do respond with antibody production following experimental infection (unpublished data). Furthermore, transmission to cage mates occurs, and infectious virus can be detected for 10 days, and possibly longer, following infection. Serologic evidence of parainfluenza virus infection, especially type 3, is frequently noted, as are mumps and respiratory syncytial virus (RS) antibody. Antibody to RS virus is found in all primates, but the greatest incidence occurs among chimpanzees. It will be recalled that this virus was originally isolated from chimpanzees (chimpanzee coryza agent, CCA) suffering from respiratory disease.[75] Subsequent studies have demonstrated that this RS virus (the name was changed as it inappropriately suggested a chimpanzee origin) is a human agent.[14]

Of equal importance with regard to myxoviruses is the status of SV5. This virus is frequently encountered in preparations of monkey kidney cells. The origin of this virus, however, is unknown. It would appear that SV5 is also of human origin, infecting animals following contact with man.[59,79] Another simian myxovirus, SV41, thought to be distantly related antigenically to SV5, does not show the same antibody pattern. Cursory examination of the serologic responses of various simians suggests that SV41 may be a true simian virus because antibody is found in both wild and captive simians.

Reoviruses have assumed a role of importance among investigators because this virus is frequently encountered in specimens of animal origin. Our original serum survey with reovirus 1 has now been expanded to include reoviruses 2 and 3. Findings here strongly support the extensive presence of antibody to this virus group, especially type 3. However, frequently only type 3 reovirus antibody will be found in a particular group of nonhuman primates. Reoviruses isolated from simians, i.e., SV12 and SV59, are very closely related antigenically to types 1 and 2, respectively. The new types isolated from chimpanzees[81] are in need of study.

Picornaviruses, especially the enteroviruses of human and simian origin, are of importance and of great interest. The majority of both human and simian strains tested evidently cross the species barrier (in both directions), producing infection in all primates. Disease-producing capabilities of these different viruses are, however, still in need of study, although the pathogenesis of polioviruses as well as certain of the other enteroviruses, especially in the chimpanzee, has been reported.[1-4,47] Possible association of coxsackieviruses with paralytic disease of baboons also has been described.[59] Difficulties are encountered when working with a number of the enteroviruses inasmuch as many are only weakly antigenic, either poorly eliciting antibody or perhaps technically difficult to use in standard laboratory serologic procedures. There is not yet evidence for a simian rhinovirus.

Recently, Arita and Henderson[7] have reviewed the literature relative to a possible natural reservoir of smallpox existing in nonhuman primates. This possibility was considered unlikely, although further studies were deemed necessary. Poxvirus studies have recently indicated the presence of antibody among most primates as determined by testing with vaccinia and monkeypox viruses (unpublished data). In this study it was found that the CF test was far less sensitive than the HI procedure for collecting survey data. Accordingly, a substantial number of primates were found to possess HI antibody to both vaccinia and monkeypox antigens. The significance of this finding is not clear, but it would suggest additional investigations into the possibility of a nonhuman primate reservoir. Confirmation of this finding by use of other procedures, e.g., SN tests, are necessary prior to any firm interpretation. It is interesting to speculate regarding the relationship of monkeypox virus to the other true poxviruses in light of these findings. It has also been demonstrated by McConnell and his co-workers that vaccination of animals will prevent outbreaks of monkeypox.[69]

Yaba and Yaba-like viruses are considered with the poxviruses although they are not true members of this group. These viruses emphasize problems associated with mixing simian species. Evidently, the African primates have developed an immunity to these viruses, which is not present in their Asian counterparts. Several outbreaks have recently been reported in primate colonies as a result of the Yaba-like virus. Infection of man with both of these viruses has been recognized.

Recently an outbreak of hemorrhagic disease occurred in a number of monkey colonies in the United States and the Soviet Union. These outbreaks have been devastating, with extremely high mortality rates. An outbreak of this hemorrhagic disease of simians as it occurred in one colony in the United States has been described in detail.[5,78,92,93] Fortunately this virus has shown no ability to produce disease in humans. An attempt to ascertain the incidence of antibody to this virus among various primate populations surprisingly yielded completely negative results, except for the control serum employed (unpublished data). Results such as these are unexpected, and they raise a number of questions that can be answered only by further study.

One of the more disturbing occurrences for investigators employing nonhuman primates was the recent outbreak of disease attributable to an agent referred to as the "Marburg" virus. For details concerning this outbreak a number of publications are available.[64,67,89] A serologic survey of primate sera disclosed that the highest antibody incidence occurred in nonhuman primates of African origin.[58] Only occasionally were Asian simians, and only those known to be in contact with African animals, found to have this antibody. These findings are highly significant but require additional studies on more simian species.

VIRUS ISOLATION AS EVIDENCE FOR INFECTION

The current status regarding isolation of a large number of viruses from various nonhuman primates has been well reviewed.[4,36,39,52] While many of these agents are recoverable from one or another of an animal's body fluids (feces, throat washings), existence of extensive latent virus infections of tissues is also recognized and must be studied more extensively.

Recovery of these viruses from the various simians raises a number of important questions: (1) The source of the organisms: Are they indigenous to that host, or are they contaminants from another animal? (2) Pathogenic qualities: Are these agents capable of causing infection and diseases of the host animal, and if so, under what conditions? (3) Pathogenic capabilities in hosts other than those from which virus was recovered. (4) Extent of tissue localization. (5) Geographic and species distribution. (6) Strain variation: pathogenicity, vaccine potential, mutations, interferon production, antigenic crossing, tumorogenicity, and persistance in host. (7) Existence of animal reservoirs.

Obviously, current information is too limited to offer anything but cursory explanations for most of these questions. For example, the full extent of species susceptibility to the majority of recognized simian viruses is not known. Reports relative to the experimental employment of various simians as models for studies on other animal viruses (especially those derived from human sources) are just beginning to appear in the literature. Difficulties are still encountered when attempting to pinpoint the original source of a virus isolate; RS virus and SV5 are two notable examples of this problem.

With regard to pathogenicity of the simian viruses, again only cursory information is available. Of the 20 simian adenoviruses, serotypes SV1, 11, 15, 17, 23, and 32 have been associated with infections of the upper respiratory tract. SV17 has been found to be the cause of an outbreak of conjunctivitis and rhinorrhea in patas monkeys; SV32 is known to cause a conjunctivitis in rhesus monkeys. A new simian adenovirus, V340, isolated from both African greens and baboons, produces a fatal disease along with a pneumoenteritis. Death occurring in newborn baboons as a result of this virus is probably due to lack of immunity.

Simian picornaviruses are very frequently encountered, especially in new arrivals. It is not known if multiplication of these agents occurs in the intestinal tract and, if so, to what extent. Virus excretion has been found to persist for several months. Of the 15 known simian enterovirus serotypes, none has been demonstrated as the etiologic agent of disease; SV6, 19, and 26 have been found in animals with a bloody diarrhea, but these viruses have been similarly encountered in "normal" animals. Two viruses included among the picornaviruses, SV28 and SA4, are not considered to be enteroviruses because, while they share the general characteristics of enteroviruses, they have never been found in the intestinal tract.

The pathogenic capabilities of the herpesviruses are, unfortunately, a little more familiar to laboratory investigators. Group A herpesviruses resemble *Herpesvirus hominis* (herpes simplex) in their characteristics; Group B resemble the cytomegaloviruses. Most familiar to individuals using simians is *Herpesvirus simiae*, which is frequently recovered from macaques. A closely related virus, SA8, is found in African green monkeys. These two viruses are closely related antigenically to herpes simplex. *Herpes tamarinus* (marmoset virus) is indigenous to the squirrel monkey but was first isolated from the marmoset. Another member of this group is the SMV or spider monkey virus. These viruses are important because of their increased pathogenicity for primate species other than their natural host. *Herpesvirus simiae* is highly lethal for man, but produces a mild localized lesion in the macaque. *Herpesvirus hominis* produces a local disease in man but a highly fatal disease in the owl monkey. A simian cytomegalovirus, SA6, was recovered from African green monkey kidney preparations. Probably other strains and types are in existence. Attempts in our laboratory to produce infection with this virus in baboons failed to elicit overt disease. Antibody to this virus is found in baboon sera.

Two simian viruses are considered as papovaviruses—SV40 and SA12. The only simian virus known to be oncogenic other than certain of the simian adenoviruses is SV40. Because of the type of intranuclear inclusion body it produces, SA12 is included in this group. Apparently neither virus is capable of producing tumors in simians. SA12 is only infrequently recovered.

Of the four recognized types of adeno-associated (AAV) viruses, only type 4 has been found in simians. Furthermore, the evidence suggests that this virus is not found in man. Little is known regarding the pathogenic capability of this virus.

Foamy virus types 1 and 2 have been isolated from macaques and African green monkeys; type 3, only from the African green; type 4, from the squirrel monkey; type 5, from the galago (bushbaby); and types 6 and 7, from chimpanzees. These viruses are primarily adventitious and are not yet associated with any particular disease pattern.

Conclusions and Perspectives

An attempt has been made to demonstrate the ubiquitousness of simian viruses and to develop a concept relating to the potential of a sizable number to cause infection and disease, or both, not only among various species of monkeys but also in man himself. Thus, the importance of obtaining healthy animals as well as maintaining them under carefully controlled conditions is a prime prerequisite for good colony management. Obviously, development of such a system imposes a number of problems on colonies already burdened with many restrictions. However, unless we rapidly come to the realization that research animals must be given very special consideration, as well as precautionary measures taken to maintain high health standards, development of good laboratory animals is simply not in the offing.

We must have a new look at the present methods of trapping and shipping of these research animals. Perhaps most important in this regard is the development of standards and regulations imposing minimum health and hygienic conditions. Training of those involved in the basic concepts of hygiene and disease spread would be of immeasurable assistance. Separation of species and their maintenance as separate species is essential. Use of common animal handlers would obviously invalidate any attempt at the mere physical separation of species. If housing is such that it is impossible to maintain animals separately, then provisions should be made to work with only one species at a time, with sterilization of quarters prior to introduction of the new group of animals.

Quarantine of animals is important but generally of insufficient length. It is now recognized that extensive shedding of virus occurs following shipment of animals and this shedding may persist for 6 months or even longer. Release of animals must be dependent upon adequate clinical supervision supported by extensive laboratory studies. It has been repeatedly pointed out that clinical evaluation alone is inadequate for proper management.

Vaccination offers some promise for the future, but unfortunately at this time only limited experiences are available. Delineation of the important agents of disease must first be made, followed by preparation and evaluation of an adequate vaccination regimen. The problem relating to the role played by adventitious or latent agents in an animal's health is just beginning to emerge. Further study is required relative to these agents, and then the need for vaccines against those found to be important is necessary.

In spite of all these problems, progress is being made. Efforts like the one this volume represents evidence recognition by at least a few individuals of the need for communication among those involved in the use of animals for biomedical research. Similar efforts have been made and more conferences are planned for the immediate future. Both NIH and WHO are fully aware of the problem and are actively taking steps not only to familiarize investigators employing simians with the many problems but, more importantly, to take an active role in assisting these laboratories. As indicated above, a simian virus reference center supported by NIH and WHO is now in existence, offering its services to the scientific community involved in nonhuman primate research.

This program is also concerned with establishment of a simian virus diagnostic capability with all its implications. However, the ultimate goal is to develop an understanding of the exchange and relationship between viruses of man and his fellow primates. A capability is, therefore, provided to investigators involved in the use of nonhuman primates for a more complete and comprehensive interpretation of investigative results. Continuous surveillance of primate sera for the presence of exotic and troublesome viruses informs the scientific community regarding occurrence of these agents in various species primates under study. These data will undoubtedly help improve the health of nonhuman primates when used as biomedical models for human disease, for vaccine production and testing, pharmacological testing, and so on. Establishment of healthier experimental animals minimizes wastage. Cleaner animals will also reduce infections of laboratory personnel. It might be added that this is a continuing program, expanding the list of primates involved as well as their virus contacts.

This facility can do much to familiarize those individuals so engaged with many of the problems and, hopefully, with some solutions.[53,54]

Acknowledgments

This study was funded in part by USPHS grant FR00361-01A1 and WHO grant #Z2/181/27. We are grateful to the following institutions for supplying specimens for our studies: Yerkes Regional Primate Center, Atlanta, Georgia; 6571st Aeromedical Research Laboratory, Holloman A.F.B., New Mexico; School of Aerospace Medicine, Brooks A.F.B., Texas; Delta Regional Primate Research Center, Covington, Louisiana; Institute Merieux, Lyon, France; Gorgas Memorial Institute, Panama; National Institute of Neurological Diseases and Stroke, Bethesda, Maryland; Presbyterian-St. Luke's Hospital, Chicago, Illinois; Animal Farm Division, Fort Detrick, Maryland; Institute of Comparative Biology, Zoological Society of San Diego, San Diego, California; Animal Care Facility, University of California, San Francisco, California; Japan Monkey Center, Aichi, Japan; Division of Comparative Medicine, University of Florida, Gainesville, Florida; National Taiwan University, Taipei, Taiwan; Naval Aerospace Medical Center, Pensacola, Florida; New England Primate Research Center, Southborough, Massachusetts; Oregon Regional Primate Research Center, Beaverton, Oregon; Epidemiology and Research Analysis Section, Foreign Quarantine Program, National Communicable Disease Center, Atlanta, Georgia; Dental Science Institute, University of Texas Dental Branch, Houston, Texas; Laboratory for Experimental Medicine and Surgery in Primates, New York, New York; The Poliomyelitis Research Foundation, Johannesburg, South Africa.

Appreciation is also due the many individuals of the Division of Microbiology and Infectious Diseases, Southwest Foundation for Research and Education, whose combined efforts made this study feasible.

References

1. Animal disease and human health. 1958. Ann. N.Y. Acad. Sci. 70:277–762.
2. Care and diseases of the research monkey. 1960. Ann. N.Y. Acad. Sci. 85:735–992.
3. Comparative virology. 1962. Ann. N.Y. Acad. Sci. 101:327–582.
4. Nonhuman primates in viral research. 1969. Ann. N.Y. Acad. Sci. 162:499–528.
5. Allen, A. M., A. E. Palmer, N. M. Tauraso, and A. Shelokov. 1968. Simian hemor- rhagic fever. II. Studies in pathology, Amer. J. Trop. Med. 17:413–421.
6. Andrewes, C., and H. G. Pereira. Viruses of vertebrates, 2nd Ed., Williams and Wilkins Co., Baltimore, Maryland.
7. Arita, I., and D. A. Henderson. 1969. Smallpox and monkeypox in nonhuman pri- mates. Bull. World Health Org. 39:277–283.
8. Atchison, R., B. C. Casto, and W. M. Hammon. 1966. Electron microscopy of adenovirus-associate virus (AAV) in cell cultures, Virology 29:353–357.
9. Bhatt, P. N., C. D. Brandt, R. A. Weiss, J. P. Fox, and M. F. Shaffer. 1966. Viral infections of monkeys in their natural habitat in Southern India. II. Serological evidence of viral infection. Amer. J. Trop. Med. Hyg. 15:561–566.
10. Bhatt, P. N., M. K. Goverdhan, M. F. Shaffer, D. C. Brandt, and J. P. Fox. 1966. Viral infections of monkeys in their natural habitat in Southern India. I. Some properties of cytopathic agents isolated from bonnet and langur monkeys. Amer. J. Trop. Med. Hyg. 15:551–560.
11. Blacklow, N. R., M. D. Hoggan, and W. P. Rowe. 1968. Serologic evidence for human infection with adenovirus-associated virus. J. N. C. I., 40:319–328.
12. Blinnikov, K. S. 1960. Latent viruses in macacus rhesus monkeys. Prob. Virol. 6:701–704.
13. Burnet, F. M., D. Lush, and A. V. Jackson. 1939. The relationship of herpes and B viruses: Immunological and epidemiological considerations. Aust. J. Exp. Biol. 17:41–51.
14. Chanock, R., B. Roizman, and R. Myers. 1957. Recovery from infants with respira- tory illness of virus related to chimpanzee coryza agent (CCA). I. Isolation, prop- erties, and characterization. Amer. J. Hyg. 66:281–290.
15. Deinhardt, F. 1966. Neoplasms induced by Rous Sarcoma virus in New World monkeys. Nature 210:443.
16. Deinhardt, F., A. W. Holmes, J. Devine, and J. Deinhardt. 1967. Marmosets as laboratory animals. IV. The microbiology of laboratory kept marmosets. Lab. Anim. Care 17:48–70.
17. Emmons, R. W., D. H. Gribble, and E. H. Lennette. 1968. Natural fatal infection of an owl monkey (*Aotus trivigatus*) with herpes T. virus. J. Infect. Dis. 118:153– 159.
18. Enders, J. F. 1940. The etiology of measles. p. 237–267. *In* John E. Gordon *et al.* [ed.] Virus and rickettsial diseases, with special consideration of their public health significance, Symposium held at the Harvard School of Public Health, June 12–17, 1939, Harvard University Press, Cambridge, Massachusetts.
19. Enders, J. F., T. H. Weller, and F. C. Robbins. 1949. Cultivation of Lansing strain of poliomyelitis virus in cultures of various human embryonic tissue. Science 109:85–87.
20. Epstein, M. A., Y. M. Barr, and B. G. Achong. 1965. Studies with Burkitt's lymphona. *In* Methodological approaches to the study of leukemias, Wistar Institute Symposium Monograph No. 4, Philadelphia, Wistar Inst. Press, 1965.

21. Fiennes, R., 1967. Zoonoses of primates, Cornell University Press, Ithaca, New York.

22. Fuentes-Martins, R. A., A. R. Rodriguez, S. S. Kalter, and A. Hellman, 1963. The isolation of enteroviruses from the normal baboon (*Papio doguera*). J. Bacteriol. 85:1045–1050.

23. Galindo, P., and S. Srihongse. 1967. Evidence of recent jungle yellow fever activity in Eastern Panama. Bull World Health Org. 36:151–161.

24. Gerber, P., and S. M. Birch, 1967. Complement-fixing antibodies in sera of human and nonhuman primates to viral antigens derived from Burkitt's lymphoma cells. Proc. Nat. Acad. Sci. U.S. 58:478–484.

25. Gerber, P., and E. N. Rosenblum. 1968. The incidence of complement-fixing antibodies to herpes simplex and herpes-like viruses in man and rhesus monkeys. Proc. Soc. Exp. Biol. Med. 128:541–546.

26. Gerber, P., and E. N. Rosenblum. 1968. The incidence of complement-fixing antibodies to herpes simplex and herpes-like viruses in man and rhesus monkeys. Proc. Soc. Exp. Biol. Med. 128:541–546.

27. Gralla, E. J., S. J. Ciecura, and C. S. Delahunt. 1966. Extended B-virus antibody determinations in a closed monkey colony. Lab. Anim. Care 16:510–514.

28. Harrison, V. R., J. D. Marshall, and N. B. Guilloud. 1967. The presence of antibody to Chikungunya and other serologically related viruses in the sera of sub-human primate imports to the United States. J. Immunol. 98:979–981.

29. Heberling, R. L., and F. S. Cheever. 1964. Some characteristics of the simian enteroviruses. Amer. J. Epidemiol. 81:106–123.

30. Heberling, R., and F. S. Cheever. 1966. A longitudinal study of simian enterovirus excretion. Amer. J. Epidemiol. 83:470–480.

31. Heberling, R. L., and F. S. Cheever. 1960. Enteric viruses of monkeys. Ann. N.Y. Acad. Sci. 85:942–950.

32. Henle, G., and W. Henle. 1967. Immunofluorescence, interference and complement-fixation technics in the detection of the herpes-type virus in Burkitt tumor cell lines. Cancer Res. 27:2442–2446.

33. Henle, G., W. Henle, and V. Diehl. 1968. Relation of Burkitt's tumor-associated herpes-type virus in infectious mononucleosis. Proc. Nat. Acad. Sci. U.S. 59:94–101.

34. Hoffert, W., M. E. Bates, and F. S. Cheever. 1958. Study of enteric viruses of simian origin. Amer. J. Hyg. 68:15–30.

35. Holmes, A. W., R. E. Dedmon, and F. Deinhardt. 1963. Isolation of a new herpes-like virus from South American marmosets. Federation Proc. 22:334.

36. Hsiung, G. D. 1968. Latent virus infections in primate tissues with special reference to simian viruses. Bacteriol. Rev. 32:185–205.

37. Hsiung, G. D., and T. Atoynatan. 1966. Incidence of naturally acquired virus infections of captive monkeys. Amer. J. Epidemiolo. 83:38–47.

38. Hsiung, G. D., and J. L. Melnick. 1958. Orphan viruses of man and animals. Ann. N.Y. Acad. Sci. 70:342–360.

39. Hull, R. N. 1968. The simian viruses. Virology Monographs. Vol. 2, Springer-Verlag, New York.

40. Hull, R., and J. R. Minner. 1957. New viral agents recovered from tissue cultures of monkey kidney cells. II. Problems of isolation and identification. Ann. N.Y. Acad. Sci. 67:413–423.

41. Hull, R., J. R. Minner and C. C. Mascoli. 1958. New viral agents recovered from tissue cultures of monkey kidney cells. III. Recovery of additional agents both from cultures of monkey tissues and directly from tissues and excreta. Amer. J. Hyg. 68:31–44.

42. Hull, R., J. R. Minner, and J. W. Smith. 1956. New viral agents recovered from tissue culture of monkey kidney cells. I. Origin and properties of cytopathogenic agents SV1, SV2, SV4, SV5, SV6, SV11, SV12, and SV15. Amer. J. Hyg. 63:204–215.
43. Hurlbut, H. S., J. Feild, T. E. Vice, and S. S. Kalter. In press. Evidence of arbovirus infections in baboons confined in outdoor cages in Texas.
44. Kalter, S. S. 1960. Animal "orphan" viruses. Bull. World Health Org. 22:319–337.
45. Kalter, S. S. 1964. Enteroviruses in animals other than man. p. 126–159. In Institute on Occupational Diseases Acquired from Animals, January 7–9, 1964. University of Michigan School of Public Health, Ann Arbor, Michigan.
46. Kalter, S. S. 1965. Virus studies on the normal baboon. p. 407–420. In H. Vagtborg [ed.] The baboon in medical research, Proc. of the 1st Symposium on the Baboon and Its Use as an Experimental Animal, 1963. University of Texas Press, San Antonio, Texas.
47. Kalter, S. S. 1966. Picornavirus, p. 207–245. In J. E. Prier [ed.] Basic medical virology, The Williams and Wilkins Co., Baltimore, Maryland.
48. Kalter, S. S. 1968. Baboons. In W. I. B. Beveridge [ed.] Primates in medicine, Vol. 2, Using primates in medical research, Part I. S. Karger, Basel, Switzerland.
49. Kalter, S. S. 1969. Nonhuman primates in viral research, Ann. N.Y. Acad. Sci. 162:499–528.
50. Kalter, S. S., A. K. Eugster, T. E. Vice, C. S. Kim, and I. A. Ratner. 1968. Rous sarcoma in baboons: Development of tumor in an uninoculated animal treated with cortisone. Nature 218:884.
51. Kalter, S. S., R. Fuentes-Martins, R. A. Crandell, A. R. Rodriguez, and A. Hellman. 1964. Virus complement-fixation and neutralization studies on sera from the normal baboon (Papio doguera). J. Bacteriol. 87:744–746.
52. Kalter, S. S., and R. L. Heberling. 1968. Viral flora of tissue sources–simian and human. p. 149–160. In Cell Cultures for Virus Vaccine Production. Natl. Cancer Inst. Mono. No. 29.
53. Kalter, S. S., and R. L. Heberling. 1968. Collaborating center for comparative medicine and simian virus reference center laboratory. Primate Newsletter 7:3–12.
54. Kalter, S. S., and R. L. Heberling. 1969. The study of simian viruses. Work of the WHO collaborating laboratory on comparative medicine: simian viruses. World Health Organization Chronicle 23:112–117.
55. Kalter, S. S., and H. Jeffries-Klitch. 1969. Yellow fever neutralization tests on primate sera. Amer. J. Trop. Med. Hyg. 466–469.
56. Kalter, S. S., R. E. Kuntz, Y. Al-Doory, and A. Katzberg. 1966. Collection of biomedical study materials from baboons in East Africa: Preliminary report. Lab. Anim. Care 16:161–177.
57. Kalter, S. S., R. E. Kuntz, B. J. Myers, A. K. Eugster, A. R. Rodriguez, M. Benke, and G. V. Kalter. 1968. The collection of biomedical specimens from baboons (Papio sp.), Kenya, 1966. Primates 9:123–129.
58. Kalter, S. S., J. J. Ratner, and R. L. Heberling. 1969. Antibodies in primates to the Marburg virus. Proc. Soc. Exp. Biol. Med. 130:10–12.
59. Kalter, S. S., J. Ratner, G. V. Kalter, A. R. Rodriguez, and C. S. Kim. 1967. A survey of primate sera for antibodies to viruses of human and simian origin. Amer. J. Epidemiol. 86:552–568.
60. Kalter, S. S., J. J. Ratner, A. R. Rodriguez, R. L. Heberling, and N. B. Guilloud. 1969. Antibodies to human and simian viruses in the gorilla (Gorilla gorilla). Lab. Anim. Care 19:63–66.

61. Kalter, S. S., J. J. Ratner, A. R. Rodriguez, and G. V. Kalter. 1967. Microbiological parameters of the baboon (*Papio* sp.) virology. p. 757–773 *In* H. Vagtborg [ed.] The baboon in medical research, Vol. II, University of Texas Press, Austin, Texas.

62. Kalter, S. S., A. R. Rodriguez, and J. J. Ratner. 1964. Neutralizing antibodies in baboon serums to measles and B virus (*Herpesvirus simiae*). Bacteriol. Proc. 1964:127.

63. Keeble, S. A. 1960. B. virus infection in monkeys. Ann. N.Y. Acad. Sci. 85:960–969.

64. Kissling, R. E., Q. R. Robinson, F. A. Murphy, and S. G. Whitfield. 1968. Agent of disease contracted from green monkeys. Science 160:888–890.

65. Malherbe, H., and R. Harwin. 1957. Seven viruses isolated from the vervet monkey. Brit. J. Exp. Pathol. 38:539–541.

66. Malherbe, H., R. Harwin, and M. Ulrich. 1963. The cytopathic effects of vervet monkey viruses. S. Afr. Med. J. 37:407–411.

67. Martini, G. A., H. G. Knauff, H. A. Schmidt, G. Mayer, and G. Baltzer. 1968. Uber eine bisher unbekannte, von Affen eingeschleppte infektionskrankheit: Marburg-virus Krankheit, Deutsch. Med. Wochschr. 93:559–571.

68. Mayor, H. D. and M. Ito. 1967. Distribution of antibodies to type 4 adeno-associated satellite virus in simian and human sera. Proc. Soc. Exp. Biol. Med. 126:723–725.

69. McConnell, S., Y. F. Herman, D. E. Mattson, D. L. Huxsoll, C. M. Lang, and R. H. Yager. 1964. Protection of rhesus monkeys against monkeypox by vaccinia virus immunization. Amer. J. Vet. Res. 25:192–195.

70. Melendez, L. V., R.D. Hunt, F. G. Garcia, and B. F. Trum. 1966. A latent herpes-T infection of *Saimiri sciureus* (squirrel monkey). p. 393–397. *In* R. N. T. W. Fiennes [ed.] Some recent developments in comparative medicine, Academic Press, London.

71. Melnick, J. L., M. Midulla, I. Wimberly, J. G. Barrerra-Oro, and B. M. Levy. 1964. A new member of the herpesvirus group isolated from S. American marmosets. J. Immunol. 92:596–601.

72. Meyer, H. M., Jr., B. E. Brooks, R. D. Douglas, and N. G. Rogers. 1962. Ecology of measles in monkeys. Amer. J. Dis. Child. 103:307–313.

73. Meyer, H. M., Jr., H. E. Hopps, N. G. Rogers, B. E. Bivoks, B. C. Bernheim, and W. P. Jones. 1962. Studies on simian virus 40. J. Immunol. 88:796–806.

74. Morgan, H. R. 1967. Antibodies for Rous sarcoma virus (Bryan) in fowl, animal, and human populations of East Africa. II. Antibodies in domestic chickens, wild-fowl, primates, and man in Kenya, and antibodies for Burkitt lymphoma cells in man. J. Nat. Cancer Inst. 39:1229–1234.

75. Morris, J. A., R. E. Blount, Jr., and R. E. Savage. 1956. Recovery of cytopathogenic agent from chimpanzees with coryza. Proc. Soc. Exp. Biol. Med. 92:544–549.

76. Munroe, J. S., and W. F. Windle. 1963. Tumors induced in primates by chicken sarcoma virus. Science 140:1415–1416.

77. Napier, J. R., and P. H. Napier. 1967. A handbook of living primates. Academic Press, New York.

78. Palmer, A. E., A. M. Allen, N. M. Tauraso, and A. Shelokov. 1968. Simian hemor-rhagic fever. I. Clinical and epizootiologic aspects of an outbreak among quarantined monkeys. Amer. J. Trop. Med. Hyg. 17:404–412.

79. Pereira, H. G., R. J. Heubner, H. S. Ginsberg, and J. van der Veen. 1963. A short description of the adenovirus group. Virology 20:613–620.

80. Rapoza, N. D., and R. W. Atchison. 1967. Association of AAV-1 with simian adenoviruses. Nature 215:1186–1187.
81. Rogers, N. G., M. Basnight, C. J. Gibbs, Jr., and D. C. Gajdusek. 1967. Latent viruses in chimpanzees with experimental kuru. Nature 216:446–449.
82. Rustigian, R., P. B. Johnston, and H. Reinhart. 1955. Infection of monkey kidney tissue cultures with virus-like agents. Proc. Soc. Exp. Biol. Med. 88:8–16.
83. Sabin, A. B., and A. M. Wright. 1934. Acute ascending myelitis following a monkey bite with the isolation of a virus capable of reproducing the disease. J. Exp. Med. 59:115–136.
84. Shah, K. V. 1966. Neutralizing antibodies to simian virus (SV40) in human sera from India. Proc. Soc. Exp. Biol. Med. 121:303–307.
85. Shah, K. V., and D. M. Hess. 1968. Presence of antibodies to simian virus 40 (SV40) T antigen in rhesus monkeys infected experimentally or naturally with SV40. Proc. Soc. Exp. Biol. Med. 128:480–485.
86. Shah, K. V., and C. H. Southwick. 1965. Prevalence of antibodies to certain viruses in sera of free-living rhesus and of captive monkeys. Indian J. Med. Res. 53:488–500.
87. Shah, K. V., S. Willard, R. E. Myers, D. M. Hess, and R. DiGiacomo. 1969. Experimental infection of rhesus with simian virus 40. Proc. Soc. Exp. Biol. Med. 130:196–203.
88. Shishido, A. 1966. Natural infection of measles virus in laboratory monkeys. Jap. J. Med. Sci. Biol. 19:221–222.
89. Smith, C. E. G., D. I. H. Simpson, E. T. W. Bowen, and I. Zlotnick. 1967. Fatal human disease from vervet monkeys. Lancet 2:1119–1120.
90. Soike, K. F., F. Coulston, P. Day, R. Deibel and H. Plager. 1967. Viruses of the alimentary tract of chimpanzees. Exp. Molec. Pathol. 7:259–303.
91. Stiles, G. E. 1968. Serologic screening of rhesus and grivet monkeys for SV40 and the foamy viruses. Proc. Soc. Exp. Biol. Med. 127:225–230.
92. Tauraso, N. M., A. Shelokov, A. M. Allen, A. E. Palmer, and C. G. Aulisio. 1968. Two epizootics of simian haemorrhagic fever. Nature 218:876–877.
93. Tauraso, N. M., A. Shelokov, A. E. Palmer, and A. M. Allen. 1968. Simian hemorrhagic fever. III. Isolation and characterization of a viral agent. Amer. J. Trop. Med. 17:422–431.
94. Taylor, R. M., M. A. Haseeb, and T. H. Work. 1955. A regional reconnaisance of yellow fever in Sudan, with special reference to primate hosts. Bull. World Health Org., 12:711–725.
95. Zilber, L. A., B. A. Lapin, and F. I. Adgighytov. 1963. Pathogenicity of Rous sarcoma virus for monkeys. Nature 205:1123-1124.

DISCUSSION

DR. GUILLOUD: Have detailed virological studies been attempted on primates from time of capture through import to show naturally occurring viral agents versus those contracted after capture?

DR. KALTER: Unfortunately, as I indicated, most animals are obtained through commercial sources. The only animal on which there is a complete virologic study, and I should say a complete microbiological study from time of capture, is the baboon. As you know, we have a field station in Kenya in Nairobi, or right outside Nairobi, where these animals are sampled immediately upon capture, and are then followed in longitudinal study. We have information on some of these animals for nearly 4 years. This is the only group of animals about which this can be said. There have been some studies done in India on the rhesus, but I do not think these animals have been followed for an extensive period of time. They have been sampled immediately upon capture, and either released or used in polio research.

All the other primates are generally studied following captivity and for varying periods of time in captivity, so there is no information, such as you ask for, on any animal but the baboon.

I did indicate in the report that we spent a lot of time trying to find what we call remote areas. We set up our cages and we trapped animals, by the time we went to collect our animals, we would find natives poking sticks into into our animal cages and playing with the animals. That is why I brought up that point that "remote areas" is a relative term. It is extremely difficult to find animals that have not been in contact with humans, and obviously with other animals.

DR. ROTH: What are your recommendations for improvement of primate-supply methods, aside from the SDF approach to the problem?

DR. KALTER: We do not have the time to go into an extensive program like that. There are at least three or four committees working on such a proposal. I hope one of them comes into being because it is presently impossible to get clean animals. Our approach is not the answer because it has shortcomings, but eliminating several middlemen is at least helpful because quarantine stations are often pest holes. You simply cannot get a clean animal through under these conditions. It is an important problem, but it is a very difficult problem. I think we have to get stronger and more stringent regulations and that we have to educate those people who are working with the animals in the proper handling from the time of capture to the use of the animal in the laboratory. Education is one of our biggest problems. There isn't any background education in primate handling. We have to start from scratch and teach people how to handle these animals.

DR. ARBRUTYN: Are any laboratories available for characterizing viruses and bacteria from outside sources?

DR. KALTER: Yes; ours.

DR. POPE: Would it be practical or possible to establish breeding colonies as sources?

DR. KALTER: I think this, too, is an important approach that a number of people have taken into consideration and that it can be done. Our efforts in this area is very small, but I do know that other people are trying this. We have roughly 700 baboons under breeding conditions from which we get about 200 to 250 babies a year, which is quite good considering how many fetuses and embryos are destroyed in the process of one or another reproductive program. This I think is an approach for the future, but in terms of the biologic needs (I forget what Dr. Shannon said, but I know it is roughly over 100,000 animals, easily 100,000 primates are used a year), the cost to develop a breeding supply for the future is astronomical.

DR. LANE-PETTER: You showed that after several months in the laboratory, the frequency of virus isolation from primates was diminished. Why?

DR. KALTER: I really do not know, because if we take the same animals and put them through a stress program, they will start shedding virus again. Most of them do not, but some of them will, which indicates a stabilization where the superficial viruses in the gut or throat are now eliminated, and the only foci that contain viruses are deep-rooted. It takes some stress, whatever that means, to make these viruses come forth again. As I said, we have induced this shedding on a number of occasions. We have taken animals, put them in an airplane and flown them around and brought them back, and they are shedding virus again.

We want to do an extensive study along these lines to see what the factors are. As I said, I can't answer that question in detail, other than to point out that we see it happening, and we feel that this is an important thing to be considered when one is doing a long-term chronic experiment requiring a stabilized animal that is not shedding virus. The animal to be used in such an experiment must be kept for a long period of time.

DR. HAYDEN: Is there a vaccine available for the human against herpes B virus?

DR. KALTER: A leading pharmaceutical company has a vaccine that is ready for use. They have been applying to the FDA for distribution, but the last I heard a few months ago, this was not officially released, and until it is released, there is no vaccine.

THE BABOON IN MEDICAL RESEARCH: BASE-LINE STUDIES IN FOURTEEN HUNDRED BABOONS AND PATHOLOGICAL OBSERVATIONS

H. W. Weber, H. D. Brede, C. P. Retief,
F. P. Retief, and E. C. Melby, Jr.

Chacma baboons (*Papio ursinus*) are plentiful in the Western Cape Province of the Republic of South Africa. They cause considerable damage to fruit, vegetable, and grain crops and are therefore declared vermin; farmers destroy them whenever they can. So it is no problem to get baboons for the Stellenbosch-Johns Hopkins Primate Facility at Bellville.

The baboons are caught in self-trapping cages and transported by rail or road to the primate facility.

This colony was originally built to house baboons used in a conjoint transplantation project of the Johns Hopkins and Stellenbosch Universities. It has 30 large open-air cages with a central passageway and temperature-controlled rooms for animals that have been operated on as well as a laboratory. Details of the design are described elsewhere.[1] Expansion of the research program necessitated an extension of the colony with 20 additional covered large cages. These cages are designed so that a trapping cage fits over a sliding trapdoor in one of the walls of the cage. The animal can be herded through the open trapdoor of the large cage into the trapping cage. This facilitates transport to laboratories or to the operating theater. This proved to be very practical, and the same system will be adopted in the original baboon colony. No facilities are provided for breeding purposes because there are no supply problems.

The animals are fed pellets made by Vereeniging Consolidated Mills. At the time of this report the mixture the pellets are made from consists of the following:

Bread crumbs	700 lb
Wheat and bran	150 lb
Degelatinized maize pop	300 lb
Exfoliated maize pop	300 lb
Carcass meal	380 lb
Anti-oxidised fish meal	100 lb
Milk substitute	50 lb
Limestone powder	20 lb
Lucerne meal	200 lb
Bone meal	30 lb
Vit. premix (as for dogs)	2 lb
Mineral premix (as for dogs)	5 lb
(No Mg SO_4)	

Animals housed at the colony vary in size from baby baboons that are bought with their mothers to large males with a body weight of up to 30 kg. Large females weigh between 17 and 19 kg. The usual weight of males is about 16-20 kg, and that of females is between 10 and 12 kg. The percentage of large males among the animals bought for the colony is about 10 percent. This small percentage of trapped large males is owing to the fact that there are only a few large males in each troop, as earlier reported by Hall.[2] The leaders of the troops are generally the first of the troop to trap themselves. The average weight of baboons in the colony used for transplantation experiments is 16.2 ± 5.5 kg. Baboons with this body weight have renal arteries with an inside diameter of 3-3.5 mm, so vascular surgical procedures are technically not overly difficult. (It should be mentioned that our animals are smaller than those reported by Hitchcock.[3]) It might be added that the baboon is not inclined to interfere with sutures or surgical dressings. After arrival at the colony the baboon is placed in quarantine. Blood is drawn for hematological, serological, and biochemical studies. Throat and rectal swabs as well as feces are taken for microbiological investigation. If pathogenic organisms are found, the animal is treated accordingly. After a quarantine period of 2 to 3 weeks and completion of the treatment the baboons are allocated to the various research projects.

Research projects presently carried out include studies in transplantation of various organs, experimental hypertension, organ preservation, and neuroanatomy.

Handling baboons is more difficult than handling dogs, but with only a little experience the laborers and technical assistants are able to catch baboons in the large cages or to inject them intramuscularly without accidents. For time-consuming procedures, such as urea and creatinine clearances or intravenous injections, we found it necessary to administer Sernylan, the dose varying between 0.3 and 1.0 mg/kg of body weight, intramuscularly, depending on the intended procedure. This means that all hematological and biochemical data were obtained under the influence of Sernylan.

Hematology

METHODS

The hematological methods employed at the facility include morphological methods,[4] folate assay,[5] Vitamin B_{12} determination,[6] unsaturated Vitamin B_{12} binding capacity,[7] and separation of Vitamin B_{12} binding proteins.[8]

RESULTS

The hematological data are summarized in Table 1. Erythrocyte, leukocyte, and platelet morphology closely resembled findings in human beings. The total leukocyte count of baboons is slightly higher than in man, and the MCHC is slightly lower. These results are similar to those of de la Peña et al.[9] and differ slightly from those of Moor-Jankowski et al.[10] Moor-Jankowski observed a lymphocytosis of 65 percent with a mean of

TABLE 1 Normal Hematological Data on Baboons

	Males	Females
Hemoglobin	13.5 G% (± 2.3)	12.9 G% (± 1.6)
Hematocrit	41.2% (± 5.6)	40.4% (± 4.4)
MCHC	32.4% (± 1.9)	31.8% (± 2.8)

	Absolute (mm³)	Percentage
Leucocyte count		
Total count	10,300 (± 3,400)	—
Neutrophils	7,000 (± 3,100)	65.9 (± 10.3)
Lymphocytes	2,900 (± 1,000)	29.7 (± 9.5)
Monocytes	240 (± 170)	2.5 (± 1.6)
Eosinophils	90 (± 130)	0.9 (± 1.3)
Basophils	80 (± 70)	0.8 (± 0.9)

Reticulocyte count	0.7% (± 0.5)
Platelet count	336,500/mm (± 117,100)
ESR (West.)	6.2 mm (± 5.3)
Neutrophil lobe count	1.9 lobes/cell (± 0.3)

H. W. WEBER, H. D. BREDE,
C. P. RETIEF, E. P. RETIEF,
AND E. C. MELBY, JR.

8,400 lymphocytes per mm³ of blood. This is more than double the number of lymphocytes, in relative and in absolute figures, found in our material.

The results of studies of hemostasis (Table 2) are within the same range as those published by Hampton et al.,[11] with one difference. Hampton found 336.53 mg/percent of fibrinogen whereas in our material we obtained a mean of 1,126 mg/percent of fibrinogen. This is even higher than that reported by de la Peña[9] who found mean fibrinogen values between 616 and 714 mg/percent in different groups of baboons. Hampton's fibrinogen value is similar to that of man, whereas de la Peña's and our values are considerably higher.

Folate and serum Vitamin B_{12} status are summarized in Tables 3 and 4. Serum folate values are similar to those reported by Huser[12] and are in the same range as in man. The folate values of baboon erythrocytes are much

TABLE 2 Studies of Hemostasis

	Duke Bleeding Time (min)	Lee-White Clotting Time (min)	Prothrombin Index (%)	Partial Thromboplastin Time (sec)	Fibrinogen (mg %)
MEAN:					
Papio ursinus	1.6	9.6	94.8	40.0	1126
RANGE:					
Papio ursinus	1.0–2.0	7.0–12.0	81–108	30–57.0	548–1572
Man	0–5.0	5–12.0	75–110	30–45.0	200– 500

TABLE 3 Folate Status (*L. casei* Activity)

	Folate (ng/ml)	
	Mean	Range
Serum		
P. ursinus	16.0	3.5– 30.0
Man	7.6	3.0– 28.0
Red cells		
P. ursinus	82.8	41.0–194.3
Man	306.0	114.0–853.0
Urine		
P. ursinus	11.5	0– 32.5
Total, µg/24 hr	2.1	0– 65.0
Man	6.8	0.1– 32.8
Total, µg/24 hr	9.5	0.1– 18.0

531

TABLE 4   Serum Vitamin B_{12} Status (Isotope-Dilution Assay)

	Unsaturated B_{12} Binding Capacity			
	Vitamin B_{12} (pg/ml)	Total (pg/ml)	% Beta Globulin- Bound	% Alpha Globulin- Bound
Mean				
P. ursinus	269	2,711	35.6	64.4
Man	384	1,369	73.2	26.8
Range				
P. ursinus	125–646	1,684–3,687	27.9–48.7	51.3–72.8
Man	170–816	525–3,806	85.4–86.3	13.7–34.6

lower than those of human red cells. The serum Vitamin B_{12} level found in our baboons is higher than that reported by Huser. The difference might be caused by the different methods applied. The elevated unsaturated B_{12} binding capacity recorded in our baboons is predominantly due to a marked increase in alpha globulin B_{12}-binding protein.

Biochemistry

METHODS
Blood was drawn from the baboons after an overnight fast. The samples were analyzed for the following:

Hematocrit (International Microcapillary Centrifuge, Model M.B.)
Protein (Reference 13)
Osmolarity (Advanced Instruments Osmometer)
Potassium, sodium (I. L. Flame Photometer)
Chloride (Reference 14)
CO_2 content (Reference 15)
Urea (Reference 16)
Glucose (Hagedorn-Jensen)
Cholesterol (Liebermann Burchard reaction as modified by Pierson, Stern, and Javack)
Creatinin (Reference 17)
Uric acid (Reference 18)
Phosphorus (Reference 19)

H. W. WEBER, H. D. BREDE,
C. P. RETIEF, E. P. RETIEF,
AND E. C. MELBY, JR.

Calcium (Baron's and Bell's EDTA Titration)
Acid and alkaline phosphatase (Reference 20)
Thymol turbidity and thymol flocculation (Reference 21)
Zinc turbidity (Reference 22)
SGPT (Reference 23)
SGOT (Reference 24)
LDR (Reference 25)

RESULTS

Our findings are summarized in Table 5. The mean
values are similar to those in man and in good accord with those of de la Peña
et al.,[9] with only minor discrepancies.

TABLE 5   Biochemical Data

		Mean	S.D.
Hematocrit	%	42	6
Total proteins	g/100 ml	6.4	0.6
Osmolality	mOsm/kg	293	14
Potassium	mEq/l	3.3	0.7
Sodium	mEq/l	145	6
Chloride	mEq/l	100	6
CO_2 content	mMol/l	30.1	4.5
Balance	mEq/l	14	5
Urea	mg/100 ml	48	16
Glucose	mg/100 ml	88	21
Cholesterol	mg/100 ml	100	31
Creatinine	mg/100 ml	1.4	0.5
Uric acid	mg/100 ml	0.4	0.2
Phosphorus	mMol/l	1.7	0.5
Calcium	mEq/l	4.9	0.6
Phosphatase:	Alkaline)	25.5	14.6
	) Bodansky		
	Acid)	8.6	3.7
Bilirubin	B.d. Bergh	N	
	Total: mg/100 ml	0.5	
Thymol turbidity		1	0
Thymol flocculation		0	0
Zinc turbidity		1.4	0.6
SGPT	milliunits (25° C) ml	26	21
SGOT	milliunits (25° C) ml	42	41
LDH	milliunits (25° C) ml	432	301
Urea clearance	ml/min	28.9	21.2
Creatinine clearance	ml/min	60.1	43.7

It is evident from de la Peña's[9] and our values that the baboon has a low serum potassium. This may make the baboon prone to severe electrolyte disturbances whenever potassium is lost, as in enterocolitis. Some of our baboons died suddenly during attacks of enterocolitis. Histological sections of materials from such cases showed hydropic myocardial degeneration and a severe vacuolar nephropathy, which is frequently caused by loss of potassium. However, one must take into account that all the biochemical values in baboons were obtained under the influence of Sernylan.

In our material the blood urea is nearly twice as high as that reported by de la Peña;[9] the cause of this is obscure.

The serum level of uric acid is much lower in baboons than in man. Our findings confirm those of de la Peña.[9]

Our results also confirm de la Peña's[9] observation that acid and alkaline phosphatase levels in the serum are higher than those of man. The age of our baboons varied from juveniles to adults. Statistical subdivision into age groups was not yet possible. These values might, therefore, be influenced by the inclusion of data from still growing animals into the statistics. The SGPT and SGOT levels in our baboons are much higher than those in man, again confirming de la Peña's[9] results.

Serology

METHODS

The following tests were performed: agglutination for heterohemagglutinins, determination of heterolysins in inactivated baboon sera, titration of free hemolysing complement, search for Wassermann antibodies, C-reactive protein reaction, and rheumatoid factor determination.

RESULTS

Heterohemagglutinins against washed sheep erythrocytes occurred in 75 percent of all healthy baboons in a low-titer range. Inactivated baboon serum was used in serial dilutions, and a constant quantity of washed sheep erythrocytes was added. A reading was made of the final dilution that aggregated the sheep cells. Out of 943 untreated baboons, 207 did not have heterohemagglutinins. Serum from six hundred sixty-six baboons agglutinated sheep erythrocytes up to dilution of 1:8, 54 up to 1:16, and 19 from 1:32 and more. The highest titer, a single case, was 1:128. It was possible to absorb the heteroagglutinin with ox red cells, with guinea pig kidney, with subcellular baboon kidney fractions, but not with horse serum. This means that this heteroagglutinin is not a Forssman antibody. According to Landsteiner,[26] baboons do not have Forssman antibodies.

H. W. WEBER, H. D. BREDE,
C. P. RETIEF, E. P. RETIEF,
AND E. C. MELBY, JR.

What is the origin of the heterohemagglutinin in some of the baboons? Since it is lacking in 20 percent of the baboon population, we may be justified in supposing that it is not a genetically preformed substance. It is well known that enterobacterial antigens frequently show serological cross reactions with nonbacterial polysaccharides. Many enterobacterial O antigens show cross reactions with blood-group substances. Baboons are heavily infected with different kinds of Enterobacteriaceae. It therefore seems reasonable to assume that the heterohemagglutinin is induced by Enterobacteriaceae.

Heterohemolysins against sheep or goat red cells can be determined by adding a constant amount of guinea pig complement to the mixture of a dilution series of inactivated baboon serum with a constant amount of sheep or goat washed erythrocytes. It could perhaps be anticipated that heterohemolysins and heterohemagglutinins would be identical antibodies, but this is not true. The heterohemolysins are destroyed at $70°C$, whereas the heteroagglutinins tolerate $70°C$ for 15 minutes. This difference in susceptibility to heat is characteristic for gamma-G and gamma-M antibodies. Gamma-G withstands $70°C$ whereas gamma-M is inactivated at the same temperature. Four hundred six baboons did not have heterohemolysins, 283 had a titer of between 1:2 and 1:8; 112, a titer of 1:16; and 120, a titer of 1:32 or more. Values between 0 and 1:32 were considered as normal. Titers higher than 1:32 are significant and are possibly free-tissue antibodies in the baboon. It was possible to absorb the heterolysins in subcellular baboon kidney fractions, but it was not possible to absorb them by human blood cells.

Nine hundred forty-three normal, healthy untreated baboons had the following complement titers: 1:8 and less, 520; 1:16, 274; 1:32 and more, 149. The highest titer recorded was 1:256.

In baboons, hemolytic whole complement was found to be relatively stable over 12-hour intervals at $0°C$. Methods for assay of baboon whole complement were adapted from those described by Osler and associates.[27] The use of guinea pig complement; Mg^{++}, Ca^{++}, temperature, pH, reaction time, amboceptor, and indicator cell optima was similar.[27]

C-reactive protein antibodies are thermostable up to $70°C$. These antibodies are not specific. They belong to the alpha- and beta-globulins and are significant for cell destruction, especially as anti-DNA factors. C-reactive protein does not react with sheep red cell stromata. This test was never positive.

Rheumatoid factor reacts with the gammaglobulin fixed to the cells as antibody. Only sera that contained appreciable amounts of gamma-G antibody will sensitize cells for agglutination with rheumatoid factor. It seems that gamma-M type globulins are playing an important role in baboon allotransplant rejection reactions, as we have shown.[26] In untreated baboons we have never observed a rheumatoid factor reaction.

Wassermann antibodies may also be regarded as an indication of cell de-

535

struction. Tests were performed in the usual way with inactivated baboon sera and cardiolipin as antigen and also with different subcellular pooled baboon kidney fractions. All these tests were negative.

Bacteriology

Bacteriological specimens were taken from 943 baboons and include specimens of saliva, feces, urine, and skin scrapings. *Mycobacterium tuberculosis* was never found. Table 6 shows the incidence of some bacteria in the oral cavity of apparently healthy baboons. With the exception of five, all these animals were free-living before they joined our colony.

TABLE 6 Incidence of Bacterial Species in the Oral Cavity of Baboons

Bacterial Species	Percentage	Number
Staphylococcus aureus haemolyticus	61.8	576
Streptococci, beta-hemolytic	1.1	11
Enterococci	0.7	7
Escherichia coli	58.3	550
Klebsiella pneumoniae	32.2	304
Proteus	36.4	343
Paracolon	4.3	38
Pseudomonas aeruginosa	3.3	31
Alcaligenes faecalis	1.0	9

Staphylococcus aureus and *Proteus* species were found with a much higher incidence than in Pinkerton's[28] material. The incidence of staphylococci is astonishing, since it is far higher than that noted in humans (± 20 percent) in South Africa. Antibiograms were performed on all those isolated. The highest rate of resistance occurred in the novobiocin and tetracycline groups (Tables 7–10). The antibacterial spectrum of novobiocin resembles that of penicillin G. Therefore, it is not surprising that most of our Gram-negative baboon flora were resistant to novobiocin. Tetracyclines are widely used in South African agriculture. This may be a selecting factor even in the free-living baboon flora.

Pathogenic micro-organisms found in the fecal flora are summarized in Table 11. Nineteen percent of all free-living baboons were carriers of enteropathogenic types of *Escherichia coli*. The most common types of enteropathogenic *E. coli* were O 55, O 119, and O 26. The types O 86, O 111 and O 127 occurred only sporadically.

H. W. WEBER, H. D. BREDE,
C. P. RETIEF, E. P. RETIEF,
AND E. C. MELBY, JR.

TABLE 7   Resistance Pattern of 231 *Staphylococcus Aureus* Isolates from Baboons

Antibiotic	% Resistant Strains
Penicillin (4 O.U.)	30.8
Coxacillin (5 μg)	7.4
Streptomycin (20 μg)	0
Chloramphenicol (30 μg)	4.8
Tetracyclines (50 μg)	13.9
Erythromycin (15 μg)	1.8
Kanamycin (30 μg)	10.0
Neomycin (50 μg)	0.9
Novobiocin (30 μg)	45.7
Gentamycin (10 μg)	0.1
Lincocin (10 μg)	16.5
Diclocil (5 μg)	6.7

TABLE 8   Resistance Pattern of 237 *E. coli* Isolates from Baboons

Antibiotic (μg)	% Resistant Strains
Ampicillin (25)	22.0
Streptomycin (20)	2.5
Chloramphenicol (30)	15.8
Tetracycline (50)	45.0
Erythromycin (15)	10.0
Kanamycin (30)	1.7
Neomycin (50)	1.3
Novobiocin (30)	93.2
Gentamycin (10)	0.9

Salmonella sundsvall was isolated from 52 baboons, not one of which showed signs of clinical disease. *S. sundsvall* has the formula: 6,14,25:z:e,n,x, subgenus 1, group H of the Kauffman White scheme. This type seldom occurs in human beings in the Cape Province of South Africa, but in Canada and Mexico it has been described as the cause of gastroenteritis.[29] These were always sensitive to ampicillin. Ampicillin treatment administered preoperatively was therefore considered advisable to eliminate this potent but latent organism in baboons.

It is well known that monkeys and baboons carry shigellae, mostly without clinical signs. The incidence observed in our colony is similar to that reported

TABLE 9   Resistance Pattern of 106 *Klebsiella pneumoniae* Isolates from Baboons

Antibiotic (μg)	% Resistant Strains
Ampicillin (25)	47.8
Streptomycin (20)	3.8
Chloramphemicol (30)	25.5
Tetracyclines (50)	66.0
Erythromycin (15)	4.9
Kanamycin (30)	0
Neomycin (50)	0
Novobiocin (30)	89.6
Gentamycin (10)	1.0

TABLE 10   Resistance Pattern of 116 *Proteus mirabilis* Isolates from Baboons

Antibiotic (μg)	% Resistant Strains
Ampicillin (25)	6.9
Streptomycin (20)	0.9
Chloramphenicol (30)	1.8
Tetracyclines (50)	90.2
Erythromycin (15)	41.5
Kanamycin (30)	0.9
Neomycin (50)	0.9
Novobiocin (30)	40.2
Gentamycin (10)	2.7

TABLE 11   Cultures of Rectal Swabs from 943 Baboons

Culture	Number of Animals
Pathogenic *E. coli*	180
Salmonella spp.	83
Shigella spp.	78
Coagulase positive	
Staphylococcus aureus	61
Candida albicans	16

TABLE 12 Bacteriological Examination of 210
Gallbladders from Baboons (Post Mortem)

Organism	Number Infected
Salmonellae	
S. typhimurium	8
S. sundsvall	1
Diverse types producing	
gastroenteritis	7
Shigellae	
Sh. flexneri type II	1
Sh. flexneri type V	1
Enteropathogenic E. coli	
Type 026	3
Type 055	3
Type 0111	2
Type 0119	5
B. proteus ⎫	
Pseudomonas ⎬	123
Klebsiella ⎭	
No growth	56

by Abramowa.[30] In our baboons, *Shigella flexneri* type IV was isolated from
43 baboons; *Shigella flexneri* type II, from 25 baboons; and *Shigella flexneri*
type V, from 10 baboons.

Bacteriological findings in gallbladders are summarized in Table 12. It shows
the difference in the incidence of *S. sundsvall* between this flora and the fecal
flora. *S. sundsvall* is possibly not able to cause bacteremia and therefore is
found only rarely in the gallbladder.

About 33.3 percent of all animals were free from parasites, 46.4 percent
carried *Trichuris trichiura*; 13 percent, *trichostrongylus*; 11.9 percent, *Stron-
gyloides stercoralis*; 8.3 percent, *ascaris*; 10.3 percent, *Entamoeba histolytica*;
1.2 percent, *dictocaulus*; and 1.2 percent, paramaecium. *Metastrongylus* seldom occurred, and only in animals from Karoo districts. Some animals had
more than one parasite. Schistosoma was never found because all animals were
trapped in that part of the country that is free from *Bilharziasis*.

Pathology

Autopsies of the internal organs, with the exception
of the brain, were performed on most of the baboons that died at the colony,

the majority of them belonging to one or another experimental series. However, there were some baboons that died from complications of enterocolitis and a very few that died from pneumonia before they had been allocated to a research project.

After more than 600 autopsies, I have the impression that spontaneous diseases other than dysentery are infrequent in our colony.

This report can only touch on morphological manifestations of some spontaneous diseases that were not directly attributable to experimental procedures. From this I make only one exception by discussing cardiac lesions that may have been caused by experimental procedures.

HEART

We did not find coronary sclerosis with myocardial infarcts. Cloudy swelling and hydropic degeneration of the myocardium was frequently observed. In addition to this we noted that approximately 15 percent of the baboons showed inflammatory lesions or focal necroses in the myocardium and endocardium. These lesions were located in the subendocardial layers of the apical regions in both ventricles. In some baboons there was predominantly necrosis with inflammatory reaction. In other baboons

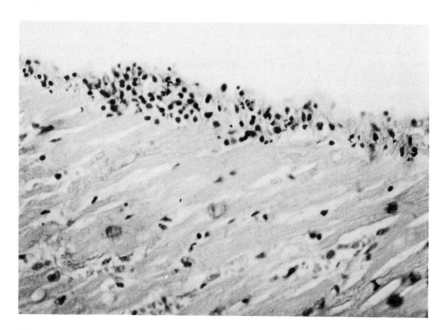

FIGURE 1 Inflammatory infiltration of the parietal endocardium. HE staining, 300×.

H. W. WEBER, H. D. BREDE,
C. P. RETIEF, E. P. RETIEF,
AND E. C. MELBY, JR.

there was inflammatory infiltration without noticeable myocardial necrosis. Frequently the infiltration occupied only the endocardium. The infiltrates consisted of neutrophilic polymorphs and lymphocytes, but occasionally we found infiltrations of eosinophils and plasma cells (Figure 1). The hearts also showed focal fibroelastic endocardial thickening. Rarely we observed subendocardial accumulations of acid and neutral mucopolysaccharides.

These observations were made in baboons dying from dysentery and in those dying from rejection of an allografted kidney, as well as in baboons with postoperative shock.

Endocardial thickening and inflammatory lesions of the parietal endocardium are frequently found in patients of both sexes and all races in Eastern and Southern Africa. Our findings in baboons are not the same as described in man by Connor,[31] and Becker,[32] but they are similar. The cellular infiltration of the parietal endocardium in man was described by Weber.[33] Connor[31] drew attention to the endocardial accumulations of mucopolysaccharides. Becker[32] described the inflammatory fibroelastic endocardial thickening. The cause of these human heart diseases is still obscure. The cardiac lesions in our baboons can be attributed to a variety of factors: Among these, toxins, electrolyte disturbances, and shock have to be considered as possible causes. We have not yet attempted to create experimental conditions that bring about the heart lesions in a higher percentage, but it may well be that the baboon can be used advantageously in the further elucidation of this common heart disease in Africa.

Atherosclerosis, except in very mild degree, was not encountered in our baboons.

LUNG

Bronchopneumonia and, rarely, interstitial pneumonia were similar to the human lesions. Lobar pneumonia was never observed. *Pneumonyssus* was only rarely found in the Chacma baboon, in contradistinction to the Kenian baboon,[34] which is very frequently infested by this parasite. It could be noted that the parasites did not elicit inflammatory reactions in some of the infested animals. Pulmonary embolism was observed in about 4 percent of the baboons that underwent major surgery.

In some arid regions of South Africa the baboons are exposed to dust. The inhaled dust (not yet analyzed mineralogically) is deposited in small cellular granulomas in the alveolar and peribronchial tissue. The birefringent crystals can be easily found within the granulomas. There is only very mild fibrosis in the center of the granulomas (Figure 2). They are all of similar size, not confluent and apparently not leading to a progressive fibrosis. This reaction to dust is quite different from that observed in man.

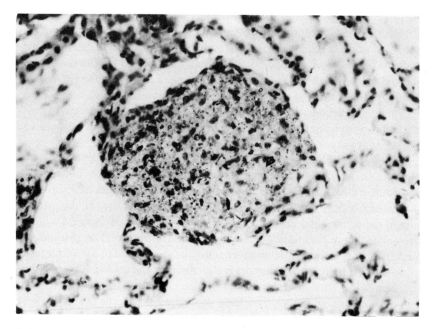

FIGURE 2 Pulmonary dust granuloma. HE staining, 300X.

INTESTINAL TRACT

Diarrhea can be troublesome in a baboon colony. We were unaware of this during the first few months after experimental research commenced at the colony. At this time there was a small turnover of baboons. If a baboon became diarrheic it could easily be isolated. With rising demand it became necessary to house 2 animals per cage and to make use of all cages continuously. Cutting down the demand was difficult as the local surgeons tried to maintain their program and it had been agreed to provide baboons to the team of Professor Barnard in Cape Town. Nevertheless, the turnover of baboons must suit the size of the colony. Only then can intestinal infections be controlled.

The changes we found were similar to human bacillary dysentery involving ulcerative colitis and polymorphonuclear infiltration. In some cases *Shigella flexneri* was found. In other cases the bacteriological findings were negative with regard to enteropathogenic micro-organisms. It has to be added that Shigellae were also found in clinically healthy baboons. A number of baboons with diarrhea died from dehydration and from hypokalemia. Similar difficulties have apparently been encountered by McGill.[35]

KIDNEYS

Interstitial perivascular lymphocytic infiltration was frequently found. These infiltrates were scattered through the kidney and were usually small. The etiology of these infiltrations is obscure. Bacteriological studies were not undertaken because the infiltrations were incidental findings in histological examinations.

Similar infiltrations were found in kidneys that contained precipitates of a crystalline material that we assume to be oxalates. The crystals have a radial structure, are birefringent, and stain with Alcian blue. They are found in the tubular lumen and in the peritubular interstitial tissue. Interstitial crystals elicited a local foreign body reaction with giant cells and mononuclear infiltration.

One case of chronic glomerular nephritis was encountered. This baboon died from hemorrhagic shock within a few hours after a kidney transplant. The grafted kidney showed no signs of incipient rejection. The glomeruli were enlarged. There was proliferation of endothelial cells and a diffuse thickening of glomerular basement membrane. Some glomeruli showed various degrees of fibrosis (Figures 3 and 4). These changes were of such a character and age that they cannot be ascribed to a rejection reaction and have to be considered as existing preoperatively in the donor animal. This animal's complement titer was 1:4; C-reactive protein was negative.

TUMORS

Tumors were observed in only two baboons. In one baboon we found an enlarged spleen that contained a fairly well circumscribed mass of tissue with the histopathological characteristics of a splenic hematoma. It consisted of lymphoid infiltrations and of sinusoids that were filled with blood. This hemangiomatous hematoma replaced large parts of the spleen, the surrounding tissue being within normal limits.

The other tumor was a cystic mass found in the abdominal cavity; it was removed during a kidney allotransplantation. The subsequent autopsy revealed no remnants of the cystic mass and no tumor in other organs either. Histological sections showed a largely necrotic, cystic, and solid tumor that consisted of pleomorphic cellular fibrous tissue and cellular areas where the cells resembled pleomorphic epithelium, in places similar to squamous epithelium (Figure 5). These cells lined cavities varying in size. Around these cell groups accumulations of acid mucopolysaccharides were found. Taking into account the site of the tumor and its histological structure, it was provisionally diagnosed as a peritoneal mesothelioma.

The question arises why there were not more neoplastic lesions in this material. The age of the baboons might be a factor. Many of the baboons in the

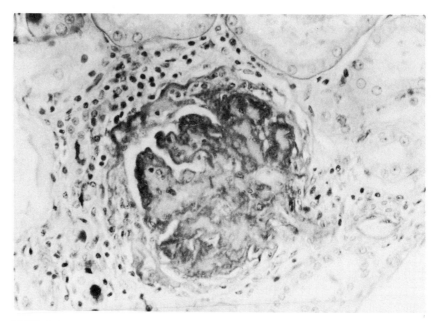

FIGURE 3   Glomerular fibrosis. Alcian blue PAS, 300X.

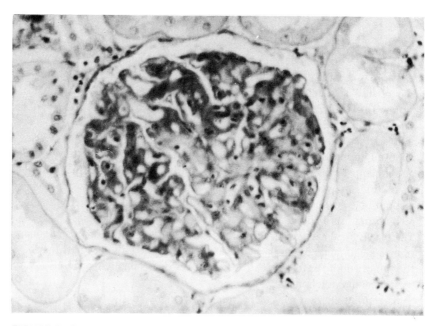

FIGURE 4   Glomerular basement membrane thickening. Alcian blue PAS, 300X.

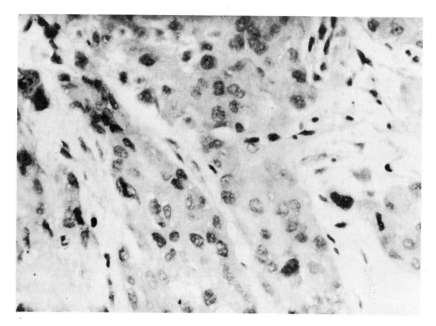

FIGURE 5 Epithelial and fibrous components of peritoneal mesothelioma. HE staining, 300X.

colony are probably of an age at which neoplastic diseases cannot yet be expected in large numbers. Another point is that the pathologist should not expect too many spontaneous diseases in a baboon colony. The baboons represent a healthy population, whereas we are used to a concentration of diseased persons in hospitals.

HEREDITARY FACTORS

A few observations were made that cast some light on inbreeding in wild baboons. Sometimes we found small congenital kidney cysts. These lesions never reached a degree that made them comparable to polycystic kidneys. However, when inquiries were made as to where these baboons came from, it was found that they all were trapped on two farms in the Southern Cape Province. Kidney cysts were never found in baboons from other regions.

Further, when the Stellenbosch-Johns Hopkins transplantation project commenced, baboons from the Table Mountain and from the Cape of Good Hope were used. The average postoperative survival time of kidney allotransplanted baboons was then 22 days.[36] Of this group there is one baboon still surviving after nearly 2 years without any immunosuppressive treatment. When it was realized that inbreeding might influence the graft survival time, a point was made to exchange transplants only between animals from geographically distant groups. The postoperative survival time fell immediately to 9 days, which is the graft survival time reported in the literature. These observations support those of Büttner-Janusch[37] that there is some inbreeding in wild baboons.

Conclusion

The hematological and biochemical results reported herein confirm earlier data with only a few exceptions, e.g., the high urea levels in our baboons. One can, therefore, assume that, by-and-large, valid base lines have now been established. The bacteriological results of our group are slightly different from those reported by Pinkerton et al.[28] The high incidence of *Staphylococcus aureus* in South African material may be due to local conditions.

The pathological findings are complementary to those described by Lapin and Yakovleva.[38] Yet, the body of information on spontaneous diseases in baboons is still very incomplete and gives only a first impression of what has to be expected in the baboon. There is no doubt that infectious diseases of the small and large gut pose a major problem in baboon colonies. It is particularly the newly arrived animal that is in danger of acquiring an intestinal infection. This was noted already by Lapin and Yakovleva.[38] In our colony, pulmonary infections caused minor difficulties as compared with the experience of Lapin and Yakovleva.[38]

The observations on cardiac pathology may be of importance in view of the Southern African endomyocardiopathies. The other findings, however, are too few and too isolated to be viewed in the proper perspective. One fact, however, emerges clearly from the reported base line studies, *viz.*, the baboon is eminently suited as an experimental animal because of its biological similarity to man.

References

1. de Klerk, J. N., J. J. W. van Zyl, H. D. Brede, H. W. Weber, J. A. van Zyl, F. D. du T. van Zijl, A. J. Brink, J. P. Murphy, W. W. Scott, and E. C. Melby. 1968. S. Afr. Med. J. 42:459.
2. Hall, K. R. L. 1965. Experiment and quantification in the study of baboon behaviour in its natural habitat. p. 43–61. *In* H. Vagtborg [ed.] The baboon in medical research, University of Texas Press, Austin, Texas.
3. Hitchcock, C. R., J. C. Kiser, R. L. Telander, E. L. Seljeskog, and J. F. Bascom. 1965. Experiences with extracorporeal cooling and storage of the baboon kidney. p. 567–583. *In* H. Vagtborg [ed.] The baboon in medical research, University of Texas Press, Austin, Texas.

4. Dacie, J. V., and S. M. Lewis. 1963. Practical hematology, 3rd ed. J. & A. Churchill Ltd., London.
5. Herbert, V. 1966. J. Clin. Pathol. 19:12.
6. Lau, K. S., C. Gottlieb, L. R. Wassermann, and V. Herbert. 1965. Blood 26:202.
7. Gottlieb, C., K. S. Lau, L. R. Wassermann, and V. Herbert. 1965. Blood 25:875.
8. Retief, F. P., C. Gottlieb, S. Kochwa, P. W. Pratt, and V. Herbert. 1967. Blood 29:501.
9. de la Peña, A., and J. W. Goldzicher. 1967. Clinical parameters of the normal baboon. p. 379–389. *In* H. Vagtborg [ed.] The baboon in medical research, University of Texas Press, Austin, Texas.
10. Moor-Jankowski, J., H. J. Huser, A. S. Wiener, S. S. Kalter, A. J. Pallotta, and C. B. Guthrie. 1965. Hematology, blood groups, serum isoantigens and preservation of blood of the baboon. p. 363–405. *In* H. Vagtborg [ed.] The baboon in medical research, University of Texas Press, Austin, Texas.
11. Hampton, J. W., and C. Matthews. 1967. Observations on the clotting mechanism in the baboon. p. 659–665. *In* H. Vagtborg [ed.] The baboon in medical research, University of Texas Press, Austin, Texas.
12. Huser, H. J., E. E. Rieber, T. W. Sheehy, and A. R. Berman. 1967. Erythrokinetic studies in the baboon under normal and experimental conditions. p. 391–406. *In* H. Vagtborg [ed.] The baboon in medical research, University of Texas Press, Austin, Texas.
13. Weichselbaum, T. E. 1946. Amer. J. Clin. Pathol. 10:40.
14. Cotlove, E., H. V. Trantham, and R. L. Bowman. 1958. J. Lab. Clin. Med. 51:461.
15. Natelson, S., 1951. Amer. J. Clin. Pathol. 21:1153.
16. Skeggs, L. T. 1957. Amer. J. Clin. Pathol. 28:311.
17. O'Brain, D., and F. A. Ibbott. 1962. Laboratory manual of pediatric micro- and ultramicro biochemical techniques, 3rd ed. Harper and Row, New York.
18. Feichtmeir, T. V., and H. T. Wrenn. 1955. Amer. J. Clin. Pathol. 25:833.
19. Fiske, C. H., and Y. Subbarow. 1925. J. Biol. Chem. 66:375.
20. Shinowara, G. Y., L. M. Jones, and H. L. Reinhardt. 1942. J. Biol. Chem. 142:921.
21. Maclagen, N. F. 1944. Brit. J. Exp. Pathol. 25:234.
22. Kunkel, H. G. 1947. Proc. Soc. Exp. Biol. Med. 66:217.
23. Wroblewski, F., and J. S. La Duc. 1956. Proc. Soc. Exp. Biol. Med. 91:596.
24. Karmen, A. 1955. J. Clin. Invest. 34:131.
25. Kubowitz, F., and S. Ott. 1943. Biochem. Z. 314:94.
26. Brede, H. D. 1968. S. Afr. Med. J. Suppl. 42:22.
27. Osler, A.J., H. J. Randall, B. M. Hill, and Z. Ovary. 1959. J. Exp. Med. 110:311.
28. Pinkerton, M. E., L. H. Boncyk, and J. A. Cline. 1967. Microbiological parameters of the baboon (*Papio* sp.): Bacteriology. p. 717–730. *In* H. Vagtborg [ed.] The baboon in medical research, University of Texas Press, Austin, Texas.
29. Brede, H. D. 1968. S. Afr. Med. J. Suppl. 42:83.
30. Abramova, E. V. 1964. The baboon, an annotated bibliography. The Southwest Foundation for Research and Education, San Antonio, Texas. p. 2.
31. Conner, D. H., K. Somers, M. S. R. Hutt, W. C. Manion, and P. G. D'Arbela. 1967; 1968. Amer. Heart J. 74:687; 75:107.
32. Becker, B. J. P. 1963. Med. Proc. 9:124.
33. Weber, H. W. 1962. Ztschr. Kreislaufforsch. 51:239.
34. Strong, J. P., J. H. Miller, and H. C. McGill. 1967. Naturally occurring parasitic lesions in baboons. p. 503–512. *In* H. Vagtborg [ed.] The baboon in medical research, University of Texas Press, Austin, Texas.

547

35. McGill, H. C., Jr., J. P. Strong, W. P. Newman, and D. A. Eggen. 1967. The baboon in atherosclerosis research. p. 351–363. *In* H. Vagtborg [ed.] The baboon in medical research, University of Texas Press, Austin, Texas.
36. van Zyl, J. A., G. P. Murphy, H. W. Weber, H. D. Brede, J. J. W. van Zyl, P. D. R. van Heerden, F. P. Retief, C. P. Retief, and M. C. Botha. 1968. S. Afr. Med. J. Suppl. 42:6.
37. Büttner-Janusch, J. 1965. Biochemical genetics of baboons in relation to population structure. p. 95–110. *In* H. Vagtborg [ed.] The baboon in medical research, University of Texas Press, Austin, Texas.
38. Lapin, B. A., and L. A. Yakovleva. 1963. Comparative pathology in monkeys. Charles C Thomas Publisher, Springfield, Illinois.

DISCUSSION

DR. MOOR-JANKOWSKI: In your hematological work did you account for the difference between freshly captured animals and animals stablized in captivity, or differences between blood samples obtained under Sernylan anesthesia, nembutal, or with no anesthesia?

DR. WEBER: We did not find any difference between freshly captured animals and animals that were kept for a longer period of time in the colony. All blood samples were taken under Sernylan anesthesia. Baboons have beautifully developed canines, and because they don't particularly like to have their blood drawn, everybody does this using Sernylan. That is what we use, usually 20 milligrams per kilogram.

DR. MOOR-JANKOWSKI: Baboons kept in captivity in America and in Europe are noted to show very little or no gastrointestinal involvements. Your report mentions colitis and dysentery as common in your colony. What could be the reasons for this difference?

DR. WEBER: I am sorry to answer that I do not know. The quality of the colony might be one reason, but we really are still in the dark, and the microbiology department is working full steam to find out why we have these difficulties. But I have to add that after a course of successful treatment the danger of the baboon's getting a second attack of dysentery or diarrhea is greatly reduced.

DR. BARTH: Were your animals bled under sedation or anesthesia.

DR. WEBER: We used Sernylan in all cases where blood had to be drawn.

DR. HAYDEN: Were hematological values obtained from the blood of unanesthetized baboons?

DR. WEBER: No.

H. W. WEBER, H. D. BREDE,
C. P. RETIEF, E. P. RETIEF,
AND E. C. MELBY, JR.

DR. BATRA: Do you feel a 3-week conditioning period is sufficient? What was the incidence of tuberculosis in your colony?

DR. WEBER: Initially when we did not have these difficulties with dysentery, we felt that 3 to 4 weeks of conditioning was sufficient for our purposes. Tuberculosis was never found in the wild baboon.

DR. KRAUS: Were pathogenic amoebae or other protozoa associated with the cases of "dysentery" in the *Chacma* baboons?

DR. WEBER: We found *Entamoeba histolytica* occasionally.

DR. BARTH: Is the colitis not related to amebiasis?

DR. WEBER: Usually not. It looked quite different anatomically, and to find *Entamoeba histolytica* in the stools when they are warm is quite easy. I would not think that the problem is really caused by amebiasis.

DR. WISSLER: Would it be possible that aflatoxin could be causing the endocardial disease. In other words, would they have access to *Aspergillus flavis*, which is well known to stimulate vascular and endocardial proliferation? Would they have access to this mold?

DR. WEBER: I would not think they could get aflotoxin in the western part. That is more a problem in the northern part of South Africa, where it is much warmer. We have a mild cool climate.

DR. FIELDER: Were BUN's run soon after food? High BUN's have been shown to occur after feeding.

DR. WEBER: All the chemical values are from the fasting baboon.

DR. RABSTEIN: Was there a weight gain or loss during the conditioning period?

DR. WEBER: The baboons gained some weight, but of course not much during that short conditioning period.

DNA VIRUSES FROM
SOUTH AMERICAN MONKEYS:
THEIR SIGNIFICANCE
IN THE ESTABLISHMENT
OF PRIMATE COLONIES
FOR BIOMEDICAL RESEARCH

*L. V. Meléndez, M. D. Daniel, R. D. Hunt,
F. G. García, C. E. O. Fraser, T. C. Jones,
and J. Mitus*

The increased use of South American monkeys in biomedical research demands animal species that are well defined as to their health status. However, to define the health status in recently introduced animal species is difficult. A good knowledge of breeding, proper care, nutrition, and microflora is lacking.

During the last 2 to 3 years, our efforts have been mainly devoted to the study of the viral flora of New World nonhuman primates. Several primate species from South America have been studied, and among these, the following species have yielded viruses: *Saimiri sciureus* (squirrel monkey), *Aotus trivirgatus* (owl monkey), *Saguinus oedipus* (cotton-top marmoset), and *Cebus albifrons* (cinnamon ringtail).

A common property of the viruses isolated was that most of them were DNA viruses with the capacity to produce intranuclear inclusions.

The purpose of this paper is to describe studies designed to characterize some of these viral agents.

Outline of Procedures

Most of these viruses were detected as indigenous agents in cell cultures prepared from monkey kidneys; while others were isolated during routine screening of swab samples collected from the natural orifices of these animals.

550

L. V. MELÉNDEZ, M. D. DANIEL,
R. D. HUNT, F. G. GARCÍA, C. E. O. FRASER,
T. C. JONES, AND J. MITUS

The procedures followed to treat these materials have been described previously.[1-3]

Acridine orange staining,[4] thymidine analog treatment,[5,6] and all other physicochemical procedures (ether, heat, filtrations, etc.) followed standard procedures.[7] The ability to develop plaques is briefly mentioned herein for some of the agents (*Herpesvirus T*, *Herpesvirus simplex*, *Herpesvirus saimiri*, and *Cebus* isolate). The plaque procedures have been described previously,[8-11] except for modifications to these methods that are described elsewhere.[12] Electron-microscope procedures, when employed, followed methods formerly described.[13,14]

OWL MONKEY VIRUSES

Isolates were obtained from kidney cultures of two different owl monkeys. Both of these agents were first seen causing a spontaneous alteration in the cell layer, one in 22-day-old culture (owl monkey kidney isolate 203-68 – OMKI 203); the other from 23-day-old culture (OMKI 588).

OMKI 203 and OMKI 588 produced cytopathogenic effect (CPE) in owl monkey kidney cell cultures (OMK), with almost complete destruction in 9 to 10 days after inoculation with undiluted virus.

A titer of 6.0/ml was obtained in these cultures with OMKI 588. No CPE was observed in rabbit kidney cultures (RKP) 10–14 days after inoculation and in squirrel monkey kidney cell line (Hull – LLC-MK4) 20–30 days after inoculation with both isolates.

Intranuclear inclusions similar to those described for herpesviruses were observed in hematoxylin eosin (HE) stained preparations. Polykaryocytes with 4 to 5 nuclei were observed only in OMK cultures inoculated with OMKI 588.

The size of the 2 isolates as determined by filtration was smaller than 220 mμ and greater than 100 mμ. The infectivity of both isolates for OMK cultures was destroyed by ether and heat treatment and by Bromodeoxyuridine.

The intranuclear inclusions fluoresced yellow-green after acridine-orange staining with both agents, indicating its DNA nature. All these common properties strongly suggest that these isolates belong to the herpesviruses group, however, their definite identity among this group of viruses remains to be determined.

Herpesvirus simplex, a known DNA virus for which man is a reservoir host, has also been recognized as the etiological agent of a natural fatal systemic disease in the owl monkey.[3] It is known that this virus can produce a fatal disease in marmoset monkeys[15,16] and owl monkeys after experimental inoculation.[3,17] In our laboratory a herpes simplex strain has been plaque purified, and 2 plaque variants have been obtained: a large plaque virus (LPV) and a small plaque virus (SPV).[18] In preliminary tests, SPV was able to immunize

551

marmoset monkeys against the fatal disease produced by *Herpesvirus simplex*. Studies are in progress to discover if SVP can also protect the owl monkeys against fatal infection.

MARMOSET MONKEY VIRUS

Marmoset kidney cultures (MMK) developed a spontaneous cell layer alteration on the twenty-second day, characterized by the presence of round swollen cells, discretely scattered in small clumps. HE stained preparations revealed the presence of intranuclear inclusions, many of them typical of the herpesvirus group; polykaryocytes were also present in great numbers. Cell culture fluids were collected as marmoset kidney isolate (MKI 125). This isolate had a titer of 4.5 in OMK cultures. Similar CPE was produced in RKP cultures. Staining with acridine-orange indicated its DNA nature. The CPE of MKI 125 was inhibited by ether and heat treatments.

MKI 125 can be considered a member of the herpesvirus group due to the above-mentioned properties. *Herpes simplex* and *Herpesvirus T* antisera did not neutralize the CPE of MKI 125.

CEBUS MONKEY VIRUS

The oral swab material from one cebus monkey of a group of nine yielded a viral isolate—*Cebus* isolate (C.I.). The *in situ* CPE was characterized by clusters of round swollen cells 3 days after inoculation of OMK cultures, when a titer of 6.0 was obtained. HE stained preparations showed intranuclear inclusions. Many of these inclusions had a very dense core and a less dense periphery (Figure 1) very similar to those described for adenoviruses. CPE was also observed in MMK, cebus kidney (CK), squirrel monkey kidney (SMK), human embryo skin and muscle (HESM), and Vero cells. A peculiar feature of this isolate was the presence of cytoplasmic inclusions. These were more frequent and prominent in MMK cultures. These cytoplasmic inclusions were RNA-like, as determined by enzyme treatment and acridine-orange staining. The intranuclear inclusion under acridine-orange showed a yellow-green core (DNA) and a red-orange periphery, while at the same time the reverse situation was also present.

Treatment with thymidine analogs (BDUR and IDUR) did not inhibit the infectivity of C.I. under conditions tested,[5,6] although *Herpes simplex*, a DNA virus, was inhibited.

In other physicochemical properties (heat and ether sensitivity, size as determined by filtration and electron microscopy) and serological properties (complement fixation and hemaglutination), C.I. behaved like a member of the adenovirus group.

C.I. had the ability to produce plaques in cebus monkey kidney (CK) and OMK cell cultures. Antiserum prepared in rabbits and in spider monkeys

L. V. MELÉNDEZ, M. D. DANIEL,
R. D. HUNT, F. G. GARCÍA, C. E. O. FRASER,
T. C. JONES, AND J. MITUS

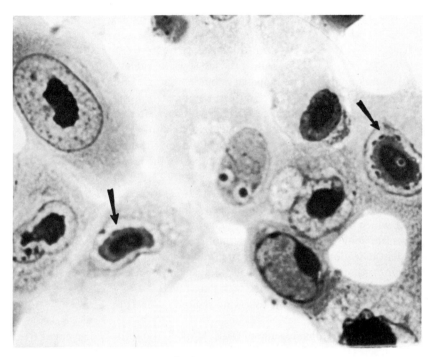

FIGURE 1 MMK cultures 6 days after inoculation with Cebus isolate. Note the nuclear inclusion with very dense core and less dense periphery. X800.

(*Ateles* sp.) had a neutralization index (N.I.) of 3.5 against C.I. The CPE of C.I. was not neutralized by the following herpesvirus antisera: *T*, *simplex*, *B*, and *suis*. Whether C.I. stands as a newer member of the adenovirus group remains to be determined.

SQUIRREL MONKEY VIRUSES

We have found that the squirrel monkey is a reservoir host to several DNA viruses: *Herpesvirus T*,[2,9,19] *Herpesvirus saimiri*[1,20]; and two adenolike viruses (SMAI 96 and SMAI 105).

Herpesvirus T has been reported as a latent agent in this species,[2] and is also known to produce a mild overt disease.[9,19] A detailed description of this virus[10] as well as the characterization of its variants have been described previously.[11] However, mention must be made here of the immunizing capacity of one of these variants—small plaque virus (SPV). In several tests performed in marmoset monkeys and owl monkeys, SPV has conferred immunity against the fatal disease produced by *Herpesvirus T* in these animal species.[21]

553

INDIGENOUS AGENTS OBTAINED FROM
SQUIRREL MONKEY KIDNEY CULTURES

During the last 2 years, kidney cultures from 31 squirrel monkeys were prepared in our laboratory. From these, 21 viral isolates were obtained from 21 different cultures. It must be emphasized that these isolates were obtained between 8 to 23 days after the cultures were prepared and were first detected by the development of a spontaneous cell layer alteration. These isolates have been named as squirrel monkey kidney isolates bearing the number of the monkeys from which the kidney cultures were prepared (SMKI 83, 89, 96, 105, 115, 118, 124, 129, 139, 147, 153, 89-67, 93-67, 95-67, 97-67, 113-67, 114-67, 122-67, 153-67, 166-67, 170-67).

All these isolates have as yet not been fully characterized, except isolate 83,[1] which has been designated *Herpesvirus saimiri.*[20] This is a very interesting agent, and some of its properties not reported previously are described under the section for *Herpesvirus saimiri.* As most of these isolates share with *Herpesvirus saimiri* the property of producing intranuclear inclusions, they have been comparatively studied with this virus.

All these agents share the following common properties:

1. They are cytopathogenic to OMK, MMK, but not for human embryo skin and muscle (HESM), RKP, and rabbit kidney cell line (RKL) cultures, the only exception being SMKI 105, which produces CPE in RKP cultures.

2. All of them develop intranuclear inclusions, with the exception of SMKI 89, 129, 153, and 89-67.

3. The type of intranuclear inclusions are as those described for herpesviruses, with the exception of SMKI 96 and 105, which have adenolike inclusions.

4. Polykaryocytes development. Polykaryocytes were observed only with SMKI 115, 124, 93-67, 95-67, 97-67, 122-67, 153-67, and 166-67.

5. Thirteen of these agents so far studied for thermostability were destroyed at $56°C$ for 30 min. These are 83, 89, 96, 118, 124, 129, 105, 115, 89-67, 95-67, 97-67, 114-67, and 122-67.

6. Ether sensitivity. All the above-mentioned agents except 96 and 105 were found to be destroyed by ether. The other eight isolates are to be tested.

7. *Herpes T* antiserum failed to neutralize the CPE of SMKI 89, 96, 105, 118, 124, 129, 153, 93-67, 95-67, 97-67, 114-67, and 122-67. The other nine have not been tested.

Based on these common properties, most of these agents behave like *Herpesvirus saimiri,* except 96, 105, 129, and 153. Of these four, 96 and 105 behave like adenoviruses, while the other two (129 and 153) remain to be de-

554

L. V. MELÉNDEZ, M. D. DANIEL,
R. D. HUNT, F. G. GARCÍA, C. E. O. FRASER,
T. C. JONES, AND J. MITUS

TABLE 1 Cytopathogenicity of Squirrel Monkey Kidney Isolates and *Herpesvirus saimiri* after Heat and Ether Treatment

Isolate	CPE in	Treatments			Inclusion Type
		None	Heat	Ether	
83[a]	OMK[b]	4.5[d]	0.0	0.0	herpes
89	OMK	5.0	0.0	0.0	none
96	MMK[c]	3.0	0.0	2.5	adeno
105	MMK	4.5	0.0	4.5	adeno
115	OMK	4.0	0.0	0.0	herpes
118	MMK	2.0	0.0	0.0	herpes
124	MMK	1.5	0.0	0.0	herpes
129	MMK	3.5	0.0	0.0	herpes
89–67	OMK	6.0	0.0	0.0	none
95–67	OMK	5.5	0.0	0.0	herpes
97–67	OMK	5.5	0.0	0.0	herpes
114–67	MMK	3.0	0.0	0.0	herpes
122–67	MMK	2.5	0.0	0.0	herpes

[a]*Herpesvirus saimiri.*
[b]Owl monkey kidney cells.
[c]Marmoset monkey kidney cells.
[d]Log 10/ml.

termined. A tabular presentation of most of these properties is provided in Tables 1 and 2.

Herpesvirus saimiri and Malignant Lymphoma in Primates

In a previous communication,[22] we reported the production of a disease resembling reticulum cell sarcoma in marmoset and owl monkeys following the injection of *Herpesvirus saimiri*. Additional studies with this agent in marmosets and the further characterization of the proliferative disorder as acute malignant lymphoma, reticulum cell type are reported here.

A total of 27 marmoset and 3 owl monkeys were inoculated intramuscularly with *Herpesvirus saimiri*. In Study I, six marmosets were inoculated with undiluted virus. In Study II, six marmosets were inoculated with undiluted virus and three with heat-inactivated virus. In Study III, two owl monkeys were inoculated with undiluted virus and one with heat inoculated virus. In

555

TABLE 2 Comparative CPE of Squirrel Monkey Kidney Isolates and *Herpesvirus saimiri* in *In Vitro* Cultures

Isolate	HESM[a] 54[b]	RKP 30	RKL 10	MMK 10	OMK 10
83[c]	–	–	–	+	+
89	–	–	–	+	+
96	–	–	–	+	+
105	–	+	–	+	+
115	nd	–	–	+	+
118	–	–	–	+	+
124	–	–	–	+	+
129	–	–	–	+	+
89-67	–	nd	–	+	+
95-67	–	nd	–	+	+
97-67	–	–	–	nd	+
114-67	–	nd	–	+	nd
122-67	–	–	–	+	nd

[a]HESM = human embryo skin and muscle; RKP = rabbit kidney primary; RKL = rabbit kidney line; MMK = marmoset kidney culture; OMK = owl monkey culture.
[b]Duration of experiment in days.
[c]*Herpesvirus saimiri.*
+ = CPE present.
– = CPE absent.

Study IV, marmosets were inoculated with diluted *Herpesvirus saimiri* as follows: two animals, 10^{-1}; two animals, 10^{-2}; and two animals, 10^{-3}. In Study V, five marmosets were inoculated with virus diluted 10^{-4}. All animals, with the exception of those injected with the heat-inactivated virus, died 13 to 48 days after inoculation. The animals inoculated with undiluted virus (Studies I, II, III) died earlier (13–28 days) than those inoculated with diluted virus (Study IV, 32–41 days; Study V, 41–48 days).

The lesions were, in general, similar in each animal, though the changes were most dramatic in marmosets receiving diluted virus (Studies IV and V). The most consistent finding was extreme enlargement of the spleen and body lymph nodes (Figure 2). The spleens were enlarged from 2 to 5 times, were of firm consistency and generally a deep, red-black color, though on occasion they were mottled with irregularly-shaped, lighter red foci. Lymph nodes were of normal color and firm. The livers, although not appreciably enlarged, were red-brown and had a conspicuous reticular pattern. The renal cortices were swollen and mottled with diffuse areas of tan tissue. The adrenals were also enlarged and contained hemorrhages in their cortices. The bone marrow was grossly normal.

L. V. MELÉNDEZ, M. D. DANIEL,
R. D. HUNT, F. G. GARCÍA, C. E. O. FRASER,
T. C. JONES, AND J. MITUS

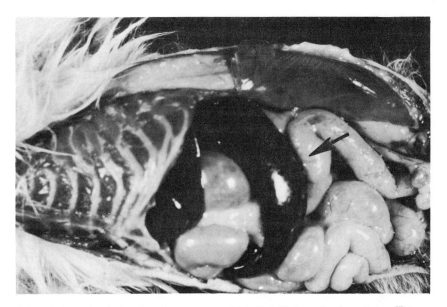

FIGURE 2 Abdominal cavity of a marmoset which died 40 days after inoculation with *Herpesvirus saimiri*: Note the extremely enlarged spleen.

Microscopically, in each animal a cellular infiltrate was observed that was most extensive in the liver, spleen, lymph nodes, thymus, kidney, and adrenal, but was also evident in the lung, mesentery, salivary gland, prostate, testicle, bone marrow, and choroid plexus. The nature of the infiltrate was basically similar to malignant lymphoma of the reticulum cell type as seen in man and lower animals (Figure 3). The infiltrating cells were individually discrete with scant cytoplasm and large, round to oval, pleomorphic, leptochromatic nuclei, which often contained one or more distinct nucleoli. Slight variation in size and occasional indented nuclei were noted. The cytoplasm was eosinophilic, and cellular boundaries were generally indistinct. Mitotic figures were numerous. Some of these cells had undergone necrosis as evidenced by karyorrhexis. The infiltrating cells invaded and replaced the cytoarchitecture of the involved organs. In many lymph nodes the cells invaded the capsule and entered the perinodal tissues.

The features of multiple organ involvement (including nonreticuloendothelial tissues), high mitotic index, invasion and cell type, all support the classification of the syndrome as malignant lymphoma, reticulum cell type. In two animals of Study V, in addition to a large number of reticulum cells, also present were a large number of smaller cells with denser nuclei that were considered to be cells of the lymphocytic series. In these examples, the neoplasm

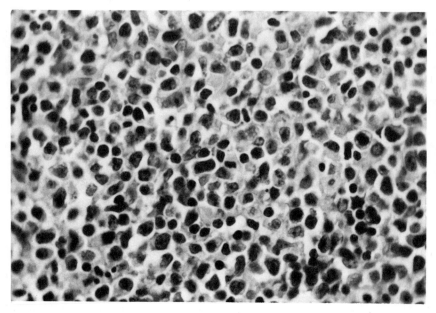

FIGURE 3 Lymph node of a marmoset dying 22 days after inoculation with *H. saimiri.*
The entire node was replaced by reticulum cells. H. E. stain. ×800.

was considered to have differentiated toward malignant lymphoma, lymphocytic or mixed cell type. Another feature of the syndrome was necrosis of mature lymphocytes, manifest in the thymus, lymph nodes, and spleen by extensive karyorrhexis of pre-existing cells.

Peripheral blood studies were performed on animals from Studies II, IV, and V. No unusual features were noted in animals from Study II. In all animals from Study V a moderate increase of white cell count with absolute lymphocytosis was present prior to death (23,400 to 50,100 WBC/mm³; 32–45 percent lymphocytes). Many of these cells were large, with large frequently indented or irregular nuclei with moderately dense chromatin and only occasional nucleoli. The cytoplasm was plentiful and moderately basophilic on Wright's stain. Scanty azurophilic granules were present in some of these cells. These cells were considered to be "atypical lymphocytes," most likely of reactive and not of "neoplastic" origin.

In addition, few larger cells with scanty, strongly basophilic cytoplasm and large nuclei with finer chromatin network and with distinct nucleoli were also seen. This second category most likely represents much more immature and probably neoplastic cells. In Study IV, five animals developed terminal leukocytosis with counts ranging from 2,200 to 107,900/mm³ with absolute lym-

L. V. MELÉNDEZ, M. D. DANIEL,
R. D. HUNT, F. G. GARCÍA, C. E. O. FRASER,
T. C. JONES, AND J. MITUS

phocytosis (76–92 percent lymphocytes). No immature cells were seen. In the remaining animal of Study IV, the total white blood cell count was normal prior to death (7,400/mm^3) with neutropenia (74 neutrophilis/mm^3) and a normal total number of lymphocytes (7,326/mm^3).

Viral isolation was attempted from various tissue (liver, kidney, spleen, lymph node) and oral and anal swabs collected at various periods of time. *Herpesvirus saimiri* was recovered only from liver and kidney of 2 animals that died from Study II. Virus has not been recovered to date from any of the other animals inoculated with live virus.

Discussion

Most of the viral agents mentioned in this report were found as indigenous agents in cultures of tissues derived from New World nonhuman primates. Singularly enough, most of them share two main common properties, though they have been obtained from different animal species. The properties they share are that they are of the DNA type and have the ability to produce intranuclear inclusions. These two common properties are also known to be shared by members of the herpesviruses and adenoviruses, and most of the agents referred to herein belong to either of these two groups, with a greater number in the herpesvirus group.

These efforts to define the viral flora of nonhuman primates have been rewarding. The saimiri species was found to be reservoir host for two herpesviruses, *Herpesvirus T* and *Herpesvirus saimiri*, besides harboring several other agents sharing *Herpesvirus saimiri*-like properties. That this nonhuman primate species could be the reservoir host for more than two different herpesviruses is not unusual, since man, the higher primate, is host to three herpesviruses (*Herpesvirus hominis, Herpesvirus varicella*, and cytomegalovirus). In addition to these, the saimiri species is also host for adenolike viruses (SMKI 96 and SMKI 105).

The cotton-top marmoset is the host for an indigenous viral agent MKI 125. This agent appeared to be a new member of the herpesvirus group based on neutralization tests and CPE behavior.

The *Aotus* species (owl monkey) is also the host for two herpeslike viruses, that so far by their cytopathogenic properties are not related to *Herpesvirus saimiri* and by neutralization are not related to *Herpesvirus T, B*, or *simplex*.

The most rewarding finding in the attempt to characterize these DNA viruses was the isolation of a new oncogenic virus, *Herpesvirus saimiri*. The ability of this agent to produce malignant lymphoma of the reticulum cell type[22] in *Saguinus* and *Aotus* species was remarkable.

With the possible exception of the lesions induced by Yaba virus,[23] which can hardly be considered malignant in that spontaneous regression is the rule, *Herpesvirus saimiri* appears to be the first example of a virus of primate origin that, when injected parenterally in another primate species, is followed by malignancy. Therefore, this represents the first viral-induced malignant lymphoma in primates. The pathological picture is not identical with any particular lymphoma of man, but from our knowledge of lymphomas in numerous animal species in which they have been documented, this would hardly be anticipated. These neoplasms have characteristics unique to each species in which they occur, just as do other forms of neoplasia.

Herpesviruses are considered as etiologic possibilities in some malignancies of man (Burkitt's lymphoma, cervical carcinoma)[24-26] and lower animals (Marek's disease of chickens and frog adenocarcinoma).[27-30] We now have clear evidence that *Herpesvirus saimiri*, a new member of the herpesvirus group, induces a malignant condition in certain species of primates.

That a neoplastic disease could be produced in primates with a primate-derived virus provides a useful model to study oncogenic processes from various angles: etiology, evolution, prevention, and chemotherapy.

Unfortunately, the low antigenic potential of *Herpesvirus saimiri*, as reported earlier,[20] has delayed the typification of the herpesviruses mentioned in this report, particularly those obtained in the saimiri species. Whether any of the other latter agents reported would have malignant properties awaits further study.

The information presented in this report undoubtedly emphasizes the need to define species to be employed in biomedical research. No proper and adequate studies in our quest for health can be expected from recently introduced animal species if these species are not properly defined for this purpose.

Obviously, none of these animal species could be employed as proper and useful models for the study of diseases of man if their own health status remains undefined.

Acknowledgments

We wish to thank Dr. Bernard F. Trum, Director of the New England Regional Primate Research Center, for his encouragement during the course of these studies. The technical assistance of Jill Cadwallader, Margrit Burtscher, Daniel Silva, and Carl Smith is gratefully acknowledged. Thanks are also due to Dr. Robert N. Hull and Dr. J. S. Rhim for providing us with LLC-MK$_4$ and Vero cells, respectively.

This investigation was supported by NIH USPHS Grant No. FR 00168-07.

L. V. MELÉNDEZ, M. D. DANIEL,
R. D. HUNT, F. G. GARCÍA, C. E. O. FRASER,
T. C. JONES, AND J. MITUS

References

1. Meléndez, L. V., M. D. Daniel, R. D. Hunt, and F. G. García. 1968. An apparently new herpesvirus from primary kidney cultures of the squirrel monkey (*Saimiri sciureus*). Lab. Anim. Care 18:374–381.
2. Meléndez, L. V., R. D. Hunt, F. G. García, and B. F. Trum. 1966. Latent herpes T infection in *Saimiri sciureus* (Squirrel monkey). p. 393–397. *In* R. N. T.-W Fiennes [ed.] Some recent developments in comparative medicine, Academic Press, New York & London.
3. Meléndez, L. V., C. Espana, R. D. Hunt, M. D. Daniel, and F. G. García. 1969. Natural herpes simplex infection in the owl monkey (*Aotus trivirgatus*). Lab. Anim. Care 18:38–45.
4. Pollard, M., and T. J. Starr. 1962. Study of intracellular virus with acidrine orange fluorochrome. Prog. Med. Virol. 4:54–59.
5. Smith, K. O., and C. D. Dukes. 1964. Effects of 5-iodo-2' deoxyuridine (IDU) on herpes virus synthesis and survival in infected cells. J. Immunol. 92:550–554.
6. Kissling, R. E., R. Q. Robinson, F. A. Murphy, and S. G. Whitfield. 1968. Agent of disease contracted from green monkeys. Science 160:888–890.
7. Schmidt, N. F. 1964. Tissue culture methods and procedures for diagnostic virology. p. 119–121. *In* E. H. Lennette and N. J. Schmidt [ed.] Diagnostic procedures for viral and rickettsial diseases, American Public Health Association, Inc., New York.
8. Daniel, M. D. and L. V. Meléndez. 1968. Long-term maintenance of cell cultures under agar overlay and development of Herpes T plaques. Proc. Soc. Exp. Bio. Med. 127:919–925.
9. Daniel, M. D., A. Karpas, L. V. Meléndez, N. W. King, and R. D. Hunt. 1967. Isolation of Herpes-T virus from a spontaneous disease in squirrel monkeys (*Saimiri sciureus*). Arch. Ges. Virusforsch. 22:324–331.
10. Daniel, M. D., and L. V. Meléndez. 1968. Herpes T virus plaque assay studies. Arch. Ges. Virusforsch. 25:8–17.
11. Daniel, M. D., and L. V. Meléndez. 1968. Herpes T virus variants: Isolation and characterization. Arch. Ges. Virusforsch. 25:18–29.
12. Daniel, M. D., and L. V. Meléndez. In press. Herpes saimiri. III. Plaque development in multi-agar overlay.
13. Meléndez, L. V., R. D. Hunt, N. W. King, F. G. García, A. A. Like, and E. Miki. 1967. A herpes virus from sand rats (*Psammomys obesus*). Lab. Anim. Care 17:302–309.
14. Brenner, S., and R. W. Horne. 1959. A negative staining method for high resolution electron microscopy of viruses. Biochim. Biophys. Acta 34:103–110.
15. Deinhardt, F., A. W. Holmes, J. Devine, and J. Deinhardt. 1967. Marmosets as laboratory animals. IV. The microbiology of the laboratory kept marmosets. Lab. Anim. Care 17:48–70.
16. Hunt, R. D., L. V. Meléndez, M. D. Daniel, F. G. García, and M. Williamson. In press. Experimental herpes simplex and herpes T infection in marmosets.
17. Katzin, D. S., J. D. Connor, L. A. Wilson, and R. D. Sexton. 1967. Experimental herpes simplex infection in the owl monkey. Proc. Soc. Exp. Biol. Med. 125:391–398.
18. Daniel, M. D., and L. V. Meléndez. In press. Isolation and characterization of herpes simplex plaque variants. Bacteriol. Proc.
19. King, N. W., R. D. Hunt, M. D. Daniel, and L. V. Meléndez. 1967. Overt herpes-T infection in squirrel monkeys (*Saimiri sciureus*). Lab. Anim. Care 17:413–423.

20. Meléndez, L. V., M. D. Daniel, F. G. García, C. E. O. Fraser, R. D. Hunt, and N. W. King. In press. Herpesvirus saimiri. I. Further characterization studies of a new virus from the squirrel monkey. Lab. Anim. Care.

21. Meléndez, L. V., M. D. Daniel, R. D. Hunt, and F. G. García. In press. Herpes-T small plaque (SPV): Immunization against fatal infection in owl and marmoset monkeys.

22. Meléndez, L. V., R. D. Hunt, M. D. Daniel, F. G. García, and C. E. O. Fraser. In press. *Herpes saimiri*. II. An experimentally induced primate disease resembling reticulum cell sarcoma. Lab. Anim. Care.

23. Grace, J. T., Jr., and E. A. Mirand. 1965. Yaba virus infections in humans. Exp. Med. Surg. 23:213–216.

24. Mitchell, J. R., G. R. Anderson, C. A. Bowles, and R. W. Hinz. 1967. Isolation of a virus from Burkitt lymphoma cells. Lancet 1:1358–1359.

25. Epstein, M. A., B. G. Achong, and Y. M. Barr. 1964. Virus particles cultured lymphoblasts from Burkitt's lymphoma. Lancet 1:702–703.

26. Rawls, W. E., W. A. F. Tompkins, M. E. Figueroa, and J. L. Melnick. 1968. Herpesvirus Type 2: Associated with carcinoma of the cervix. Science 161:1255–1256.

27. Biggs, P. M., and L. W. Payne. 1967. Studies on Marek's disease. I. Experimental transmission. J. Nat. Cancer Inst. 39:267–280.

28. Kenzy, S. G., and P. M. Biggs. 1967. Excretion of Marek's disease agent in infected chickens. Vet. Rec. 80:565–568.

29. Lucké, B. 1938. Carcinoma of the leopard frog: Its probable causation by a virus. J. Exp. Med. 68:457–466.

30. Lunger, P. D. 1964. The isolation and morphology of the Lucké frog kidney tumor virus. Virology 24:138–145.

DISCUSSION

DR. SIEBOLD: Did the reticulum cells in the neoplastic lesions caused by the herpesvirus contain intranuclear inclusions.

DR. MELENDEZ: Some of them looked like intranuclear inclusions, but we do not think so. Only in cell culture have we observed the development of intranuclear inclusions.

DR. PETURSSON: Concerning the *Cebus* isolate, is the structure of the virus as seen by negative staining similar to that of adenoviruses?

Have you studied the ultrastructure of the cytoplasmic and nuclear inclusions?

DR. MELENDEZ: Yes, we have studied these in purified preparations, and the electron-microscope studies have revealed herpes virus particles. Now, with the help of Dr. Counselman and Dr. Morgan at Columbia University, we are studying the ultrastructure morphology of these particular viruses. We have already obtained some electron-microscopy pictures. They are, I would say, rather peculiar, particularly in the sense that the only other type of herpesvirus that is suspected of having the ability to produce carcinoma is the frog virus. The ultrastructure and morphology of

L. V. MELÉNDEZ, M. D. DANIEL,
R. D. HUNT, F. G. GARCÍA, C. E. O. FRASER,
T. C. JONES, AND J. MITUS

these particular herpesviruses according to Dr. Morgan is very similar to those of the frog virus.

DR. PETURSSON: Could the *Cebus* isolate possibly contain two types of viruses, one causing nuclear inclusions and the other causing cytoplasmic inclusions?

DR. MELENDEZ: Our answer is yes, it could be possible, but we know it is not the case. We have done electron-microscopy studies with the help of Dr. Arthur Ley at the New England Deaconess Hospital. We do not observe any firal particles in the cytoplasm. We observe only viral particles in the nuclei of the inoculated cells, and these are typical of the adenovirus group.

DR. BOND: Were peripheral blood studies done on the lymphoma?

DR. MELENDEZ: Yes, they were.

DR. MOORE: Have you tried to induce neoplasia in aminal species other than primates? What is the induction period in primates?

DR. MELENDEZ: After intramuscular injection of several viral dilutions, let us say from undiluted virus to 10^{-4}, all the animals die in a period that ranges from 11 to 42 days.

DR. BATRA: Does *Herpes saimiri* affect the bone marrow? The ribs in the illustration (see Figure 2) looked to be affected.

DR. MELENDEZ: Yes, they are.

DR. KRAUS: What routes of inoculation of *Herpes saimiri* have been effective in producing the reticular cell tumors?

DR. MELENDEZ: We have only tested intramuscular injection.

DR. KRAUS: What is the latent period between inoculation and detectable gross and microscopic lesions?

DR. MELENDEZ: Other than the enlargement of the organs that I showed, there is no presentation of a tumor that is recognized microscopically. The presentation of this test goes from 11 to 42 days. We are doing some studies to determine when the first picture of the malignancy appears in these animals through biopsies, but this study is only in progress.

DR. BATRA: Is the blood picture altered in this condition?

DR. MELENDEZ: Yes, it is.

DR. POPE: What are the chances that any of the viruses isolated from the squirrel monkey are transmissible to man?

DR. MELENDEZ: We just have to hope. We do not know.

VIII

Gnotobiotes

NUTRITIONAL AND PHYSIOLOGICAL PARAMETERS INFLUENCING FUNCTION AND METABOLISM IN THE GERMFREE RAT

Bernard S. Wostmann, Julian R. Pleasants, and Bandaru S. Reddy

Introduction

In recent years the importance of experimental animals that are well controlled from both a genetic and an environmental point of view has become more and more obvious. But even the most modern conventional animal husbandry practices may not provide sufficient control for meaningful data to be obtained when the microflora can enter, directly or indirectly, into the experimental equation. At the Lobund Laboratory we have specifically studied the environment-related microbial variable, which we find to influence almost every aspect of function and metabolism of the intact animal.

The consequences of microbially induced variability were well illustrated during our studies of the role of such bioamines as histamine in the host-defense system.[1] Histamine levels in the wall of the small intestine of germfree and conventional rats were determined, since its concentration might reflect the state of adaptive physiological inflammation resulting from direct contact with the intestinal microflora.[2] The results, given in Figure 1, show that while the histamine concentration in the germfree gut wall always approximated 20 μg/g, its concentration in conventional rats varied over the years from 20 to 60 μg, presumably varying with the microbial conditions of the animal house at different periods in time. Obviously, the elimination of the microflora will result in an experimental animal in which such flora-dependent variations have been abolished.

However, with the elimination of the microflora, other defining parameters become more obvious. Our studies on the mobilization of host-defense systems led us to the controlled association of adult germfree rats with *Salmonella*

567

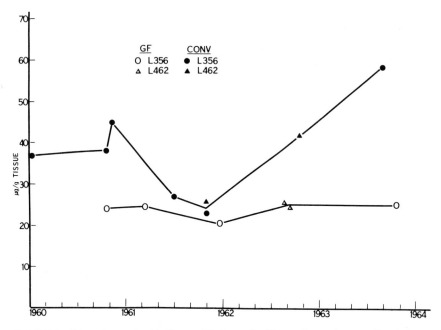

FIGURE 1 Histamine content in ileum of 3–4-month-old germfree and conventional male rats (LOB:(WI)).

typhimurium ND 750. In conventional rats, this association resulted only in transient discomfort, diarrhea, and some loss of weight. In the monoassociated ex-germfree rat, the influence of diet on the sequence of events became obvious. Rats maintained on practical type diet L462[3] consistently showed severe loss of weight, and approximately 25 percent died within 10 days of association. Animals reared on practical type diets L485[4] or 5010-C,* on the other hand, showed little loss of weight, and death never occurred (Figure 2). One major difference between the diets is that the protein fraction in diet L462 is mostly derived from bovine milk (casein, lactalbumin, and whole-milk powder), while the proteins in diets L485 and 5010-C are, for all practical purposes, of vegetable origin. Diet L462 contains approximately 1/3 whole wheat flour and 1/3 yellow corn meal together with 18 percent of bovine milk protein. The major components of diet L485 are corn (59 percent) and soy bean oil meal (30 percent) with no bovine milk proteins added. Since diet L462 has supported excellent reproduction in both germfree and conventional rat and mouse colonies over a number of years, the superior salmonellosis-protection of diets L485 and 5010-C is regarded as a superimposed factor and not as an indirect result

*Ralston Purina Company, St. Louis, Missouri

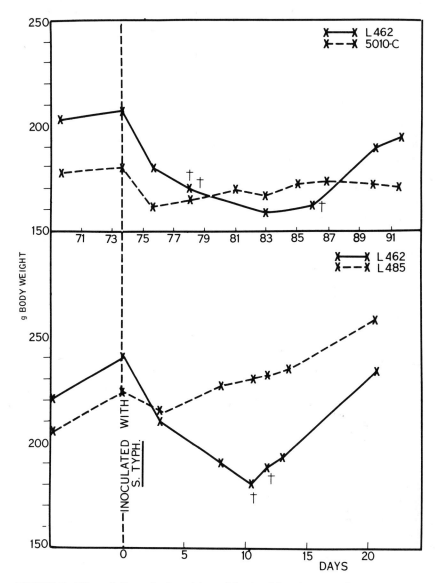

FIGURE 2   Effect of diet on body weight and deaths of female ex-germfree rats
(LOB:(WI)h) mono-associated with *Salmonella typhimurium ND 750*. Upper time scale
indicates actual age.

TABLE 1 Results Obtained from Germfree and Conventional LOB: CFW(SW) Mice Maintained on Solid and Water-Soluble, Low-Molecular-Weight Diets (see text). Number of Animals in Parentheses; Mean and Standard Error Given

	Germfree										Conventional
	L474-E$_{12}$		L462		5010-C		L485		H$_2$O-I[a]	H$_2$O-II[a]	L485
	♂(9)	♀(10)	♂(10)	♀(10)	♂(10)	♀(10)	♂(10)	♀(10)	♂(4)	♂(4)	♂(20)
Body weight (g)	26.8 ±0.5	21.2 ±0.7	28.8 ±0.8	21.7 ±0.3	28.3 ±0.9	22.9 ±0.5	26.8 ±0.4	22.4 ±0.7	–	21.3 ±1.0	24.7 ±0.8
Serum protein (%)	6.1 ±0.1	5.8 ±0.1	6.1 ±0.1	6.0 ±0.2	5.9 ±0.1	5.8 ±0.1	6.0 ±0.1	5.8 ±0.1	5.9 ±0.3	6.2 ±0.1	6.0 ±0.147
WBC × 10^3, total	18.4 ±1.6	19.0 ±1.3	20.0 ±1.7	18.5 ±1.7	8.6 ±0.4	11.0 ±0.6	8.4 ±0.7	8.2 ±0.8	4.6 ±0.7	2.2 ±0.2	9.9 ±0.11
Agranulocytes	15.3 ±1.6	16.8 ±1.3	17.0 ±1.5	16.7 ±1.6	7.0 ±0.6	9.7 ±0.6	6.0 ±0.5	7.3 ±0.7	3.1 ±0.4	1.8 ±0.2	7.5 ±0.2
Granulocytes	3.1 ±0.3	2.2 ±0.4	3.0 ±0.4	1.8 ±0.2	1.6 ±0.3	1.3 ±0.3	2.4 ±0.5	0.9 ±0.1	1.5 ±0.4	0.4 ±0.1	2.4 ±0.2
Weight mesenteric node (mg %)	47 ±3	69 ±7	38 ±4	82 ±5	45 ±5	73 ±5	45 ±6	67 ±6	–	77 ±3	134±11[b]
			42±3[b]								
PFC	29 ±6	–	33 ±5	–	24 ±4	–	32 ±7	–	–	32 ±7	91 ±22[b]

[a]Two independent experiments using diet L479-E$_9$.[14]
[b]Data from earlier experimental series.

BERNARD S. WOSTMANN,
JULIAN R. PLEASANTS,
AND BANDARU S. REDDY

of inferior nutritional quality of diet L462. It would appear that in such simplified experimental systems, differences in more specific protective substances, e.g., in the pacifarin[5] content of the various food components, start to play a role and will have to be controlled.

Diet and Immune Mechanisms

The effect of the composition of supposedly complete diets on function of the lymphoid tissue was further illustrated in a series of experiments that compared data obtained from young adult germfree mice reared on four different solid diets and on a low-molecular-weight, chemically defined, water-soluble, and filter-sterilized formula, which is described in more detail in the next section. Seventy-day-old germfree CFW mice (LOB:CFW (SW)f) that had received diets L462,[3] L474-E$_{12}$,[6] 5010-C, L485,[4] and the water-soluble diet from weaning were studied in terms of serology, histology, and immunology, and the results were compared with those of conventional mice fed diet L485. The results are given in Table 1. No difference was found in total serum proteins between any of the groups. The weights of the mesenteric lymph nodes in all groups were comparable and were lower than those found in conventional mice. The background levels of anti-sheep RBC plaque-forming cells (PFC) were similar in all germfree groups and were substantially lower than in conventional mice.[7]

The major differences among the germfree groups were found in the leukocyte patterns. Total leukocyte counts in mice fed diets L462 and L474-E$_{12}$ were comparable and were approximately twice as high as in the mice reared on diets 5010-C and L485. The latter values were again approximately 2–3 times higher than those found in the mice reared on the water-soluble, low-molecular-weight diet. Lymphocyte values followed this general pattern. Total granulocyte counts of animals fed diets L462 and L474-E$_{12}$ were higher than those of animals maintained on diets 5010-C and L485. In mice fed the water-soluble diet, granulocyte counts tended to be even lower than the values in the germfree animals fed diets 5010-C or L485.

A comparison of the above data with those obtained from conventional CFW mice reared on diet L485 indicates that the leukocyte pattern is affected far more by the character of the diet than by the microbiological status of the animal.

Surprisingly little of the above difference was reflected by other elements of the lymphoid system. In the bone marrow the only significant difference noted among the four solid diet groups was a significantly higher percentage of granulocyte precursors in germfree mice reared on diet L474-E$_{12}$. Granulocyte levels on this diet approached the values found in the conventional animals.

571

Histologically, the germfree small intestine appeared to be similar in mice fed solid diets and water-soluble diets. Lymph nodes were characteristic for the germfree state, with dense nodules predominating and an occasional inactive center visible in the nodule. Some plasma cells were present in the inactive centers, but few, if any, mitotic figures were in evidence. The intermediate and medullary zones of the lymph nodes were sparsely populated with plasma cells and reticulum cells. There were large areas of white pulp in all the spleens examined, but no active centers, few plasma cells, and few mitotic figures were found. Diet, therefore, did not seem to affect the histological appearance of lymph nodes, spleens, or small intestines. Surprisingly, the livers of mice reared on water-soluble diets demonstrated a more than twofold increase in Kupffer cells when compared to animals fed solid diets.

Since the leukocyte pattern of the germfree mice reared on both diets L462 and L474-E$_{12}$ indicated dietary stimulation, and because both of these diets contained substantial amounts of casein and/or lactalbumin, three commercial casein preparations were examined microscopically for the presence of non-viable microbial forms. The results are given in Table 2; they indicate that the

TABLE 2   Bacterial Count on Casein Samples by Direct Microscopic Count (Modified Breed Method)

Casein Source	Number of Individual Bacterial Cells per Gram
Hammersten	6.70×10^8
Vitamin-free casein (GBI)	1.31×10^9
Soluble casein (Na caseinate)	5.34×10^9

incorporation of such material in the diet results in the ingestion of at least 10^9 nonviable micro-organisms, predominantly streptococci, per day with the diet.

The above data indicate that in the absence of a viable microflora, dietary influences are uncovered that appear to play an important role in the functioning of the host-defense systems. Comparison of results of diets L462 and L485 demonstrates that while diet L462 stimulates the production of circulating lymphocytes and granulocytes in mice, it also results in definitely inferior performance in ex-germfree rats monoassociated with *S. typhimurium.* Such effects usually only complicate the evaluation of studies aimed at, e.g., the mobilization of host-defense systems, or of the more detailed functional aspects of the reticuloendothelial system. Control of the dietary variable thus appears to be a first requirement, especially in the otherwise well controlled gnotobiotic system. For this reason, the water-soluble, low-molecular-weight diets first described by Greenstein et al.[8,9] were adapted for use by germfree rats and mice. Since these diets can be sterilized by filtration, they combine the advantage of chemical definition with the virtual absence of macromolecular antigenicity.

Germfree Rats and Mice Reared on Chemically Defined, Low-Molecular-Weight, Water-Soluble, Filter-Sterilized Diets

A general description of these diets, and the sterilization and feeding techniques involved, was given at the third ICLA Symposium.[10] Further improvements in formulation and technique have been published recently.[11] Table 3 gives the latest formulation used for colony production of germfree rodents.

Five successive generations of germfree CFW mice had been maintained on these diets before the colony was lost because of contamination. This fact appears to indicate qualitative completeness of a diet composed only of known, chemically defined components. However definite signs of quantitative deficiency or imbalance in the face of heavy nutritional demands, such as during pregnancy and lactation, still persist. A full review of these data is given by Pleasants et al.[12]

The above diets were used to obtain low stimulation base-line values in the study described in the previous section of this paper. Its low antigenicity is further demonstrated by the immunoelectrophoretic patterns of the serum of CFW mice taken from the reproducing colony mentioned earlier. Figure 3 shows representative immunoelectrophoresis patterns of germfree and conventional mice reared on solid diet L485 and of germfree mice reared on water-soluble, low-molecular-weight formula L479-E_9. The pattern obtained with serum of 65-day-old germfree mouse Number 4 reared on solid diet appears to confirm the earlier observations. Fractions IgG_1 and IgG_2 show very clearly, and IgA is also indicated. The pattern obtained with serum from mouse Number 242 of comparable age reared on the water-soluble diet, however, lacks a sharply delineated IgG fraction, although the presence of some IgA is suggested. In the absence of a demonstrable IgG line, two faint arcs become visible. The faster moving one may well be a vestige of the IgG_2 fraction, while the slower globulin fraction appears in the position where Hurlemann et al.[13] indicate C-reactive protein. Conventional mouse serum demonstrates the usual complement of immune globulins and is especially characterized by the long-drawn-out IgG line in which IgG_1 and IgG_2 are clearly indicated.

Eventually, even in the case of germfree mice reared on these relatively antigen-free diets, a major build up of gamma globulins occurs. One 8-month-

TABLE 3 Composition of Water-Soluble Diet L479-E$_{20}$ Amounts per 100 g of Water-Soluble Solids[a]

Amino Acid–B Vitamin Mix			
L-Phenylalanine	0.90 g	L-Asparagine	1.20 g
L-Leucine	0.80	L-Proline	3.00
L-Methionine	0.85	Glycine	0.50
L-Tryptophan	0.40	L-Serine	1.55
L-Isoleucine	0.50	L-Alanine	0.75
L-Valine	0.70	Na L-glutamate	6.00
L-Lysine HCl	1.25	L-Tyrosine-Ethyl HCl	1.00
L-Arginine HCl	0.75	NaCl·KI mix (125:1)	0.38
L-Threonine	0.50	Potassium acetate	1.06
L-Histidine HCl·H$_2$O	0.55	B vitamins[b]	0.315
Sugar–Mineral Mix			
D-dextrose	72.40 g	MgCl$_2$·6H$_2$O	0.75 g
Ca Fructose (PO$_4$)$_2$	3.2	Ferrous gluconate	0.035
CaCl$_2$·2H$_2$O	0.6	Trace mineral mix[c]	0.035
Oil–Vitamin Mix[d]			
Ethyl linoleate (> 99% pure), mg			1,260
DL-a-tocopherol acetate, mg			26
DL-a-topherol, mg			13
Vitamin K$_1$ (phytyl), mg			1.4
Vitamin A palmitate, IU			1,560
Vitamin D$_3$, IU			36

[a]In comparison to the previously reported diet L479-E$_9$,[14] diet L479-E$_{20}$ contains 2.6 times as much Mg, while the following have been reduced from the L479-E$_9$ levels by the percentages shown in parentheses: tyrosine ethyl ester HCl (50%); Ca (8%); P (41%); Co (90%); Mo (50%).

[b]250 mg choline Cl; 0.5 mg thiamine HCl; 0.75 mg riboflavin; 0.63 mg pyridoxine HCl; 3.75 mg niacin; 25 mg i-inositol; 5 mg calcium pantothenate; 30 mg p-aminobenzoic acid; 0.1 mg biotin; 0.15 mg folic acid; 0.03 mg cyanocobalamin.

[c]26 mg Mn (acetate)$_2$·4H$_2$O; 5.5 mg ZnSO$_4$·H$_2$O; 2.5 mg Cu (acetate)$_2$·H$_2$O; 0.09 mg Co (acetate)$_2$·H$_2$O; 0.3 mg (NH$_4$)$_6$Mo$_6$O$_{24}$·4H$_2$O; 0.011 mg Na$_2$SeO$_3$; 0.48 mg Cr (acetate)$_3$·H$_2$O.

[d]0.15 gm of the oil mixture, absorbed into a small amount of dry glucose, was fed three times weekly to adult mice. Assuming 5 g of solids intake per day, the intake of fat-soluble nutrients per 100 g of water-soluble nutrients would be that listed in the table.

old animal demonstrated a well-defined IgG$_1$ line, and at 11 months of age all mice tested showed the full complement of mouse immunoglobulins, with the possible exception of IgM (Figure 3).

The data suggest that the feeding of the present water-soluble diets results in a minimum of antigenic stimulation. However, a gradual build up of immune globulins with age still occurs. We have now introduced a new filter-sterilization system (Diaflo ultrafiltration membranes) that will eliminate all possible impurities of macromolecular size. Since animal-to-animal contact could be another source of antigenic stimulation, we have started inbred rat and mouse colonies on these chemically defined diets. In this way we hope to obtain total control of microbial, antigenic, and dietary variables of the experimental animal.

However, animals reared under such stringent controls would seem to be far removed from the "normal" living conditions of the conventional rodent. Thus, the question of functional and metabolic normalcy of these animals arises. We have found that growth rate is approximately 10 percent lower than

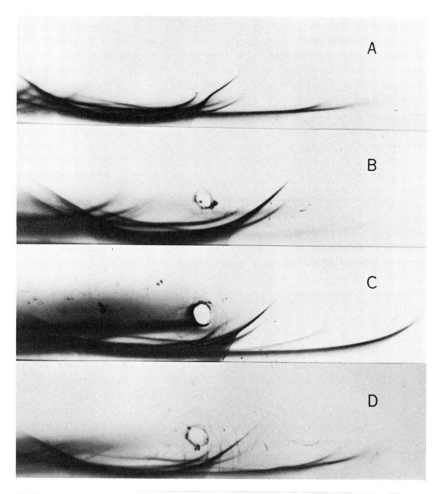

FIGURE 3 Cathodal range of mouse (LOB:CFW(SW)) serum immunoelectrophoresis pattern. A: germfree, 70-day-old, solid diet L485; B: germfree, 70-day-old, water soluble diet; C: conventional, 70-day-old solid diet; D: germfree, 330-day-old, water-soluble diet.

that of comparable germfree rats and mice fed solid diets (Figure 4). Data on life-span are limited, but CFW mice have been maintained on water-soluble diet for almost 2 years in apparent good health.

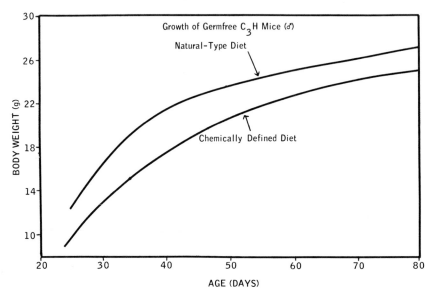

FIGURE 4   Body weight of male germfree mice (C3H/LOB(He Cr)f) fed water-soluble, chemically defined diet L479-E$_{16}$.

A number of studies on function and metabolism of germfree rats (LOB: (W1)h) reared on water-soluble diets have been carried out. Intestinal disaccharidase activities were found to be surprisingly comparable to values found in germfree rats fed steam-sterilized solid diets, even though in the water-soluble formula, glucose and fructose diphosphate were the only carbohydrates, and the animals were thus not exposed to specific disaccharide substrates.[14] Pancreatic enzyme levels also appeared normal in these rats.[15] This is illustrated in Table 4, which shows the pancreatic chymotrypsinogen and chymotrypsin activities at various levels of the intestinal tract. The data also show the higher enzyme concentrations in the lumen of the gut that occur in the absence of an intestinal microflora[14,15] and also suggest a certain amount of protection of the enzyme by the presence of the natural substrate in the solid diet (substrate stabilization). These results and the finding of histologically normal pancreas indicated adequate pancreatic function. This had been in doubt after a study[16] in which a liquid diet supposedly lacking in chromium, and containing Tween-80 had been fed to germfree rats. Chromium

TABLE 4 Effect of Feeding Water-Soluble and Semisynthetic Diets to Germfree and Conventional Rats on Chymotrypsin Activity[a] in Pancreas and Intestinal Contents

	Germfree		Conventional	
	Water-Soluble	Semisynthetic	Water-Soluble	Semisynthetic
Pancreas, units/100 g of BW	896 ± 65[b,c]	1194 ± 104	913 ± 42[d]	1072 ± 28
Small intestine, total units	480 ± 31[d,e]	594 ± 17[e]	246 ± 16[d]	335 ± 20
units/g of contents	259 ± 17[d,e]	306 ± 8.8[d]	123 ± 8[d]	174 ± 10
Cecum, total units	133 ± 9.7[d,e]	189 ± 12.4[e]	8.53 ± 0.92[d]	14.3 ± 1.29
units/g of contents	11.3 ± 0.82[d,e]	17.3 ± 1.14[e]	3.04 ± 0.33[d]	6.5 ± 0.59
Large intestine, total units	11.8 ± 0.74[d,e]	20.1 ± 1.42[e]	5.16 ± 0.36[d]	7.93 ± 0.48
units/g of contents	9.5 ± 0.60[d,e]	14.4 ± 1.01[e]	3.44 ± 0.24[d]	4.96 ± 0.30

[a] Unit of activity is expressed as micromoles of BTEE hydrolyzed per minute.
[b] Averages of 7-11 animals ± SE of mean.
[c] Difference between dietary subgroups within germfree or conventional groups significant, $P < 0.05$.
[d] Differences between dietary subgroups within germfree or conventional groups significant, $P < 0.01$.
[e] Difference from the conventional group fed similar diet significant, $P < 0.01$.

Source: Reddy, B.S., Pleasants, J. R., and Wostmann, B.S.[15]

TABLE 5 Intake, Absorption, Retention, and Plasma Levels of Ca, Mg, and P in Germfree and Conventional 90-Day-Old Male Rats Fed Water-Soluble (L479-E$_9$) and Semipurified Solid (L474-E$_{12}$) Diets; 6–12 Animals per Group; Mean and S.E. Given.

	Intake (mg/day/100 g of BW)		Absorption (%)		Retention (%)		Plasma (mg/100 ml)	
	Germfree	Conventional	Germfree	Conventional	Germfree	Conventional	Germfree	Conventional
Ca H$_2$O-s	29.3[a,b] ±1.0	33.4 ±0.7	50.5[a,b] ±1.3	31.5 ±2.8	46.3[a,b] ±1.6	29.0 ±2.6	11.3 ±0.3	11.0 ±0.2
Solid	32.2 ±0.8	33.4 ±0.9	39.4[b] ±2.6	25.3 ±1.6	35.5[b] ±2.5	23.3 ±1.6	11.1 ±0.3	11.0 ±0.2
Mg H$_2$O-s	2.50[a,b] ±0.08	2.87[a] ±0.06	79.0[a,b] ±1.1	44.9 ±3.0	42.7[b] ±0.9	23.7 ±2.7	2.00 ±0.06	1.96 ±0.06
Solid	4.32 ±0.11	4.46 ±0.14	72.0[b] ±2.1	39.9 ±3.4	44.2[b] ±1.9	27.0 ±3.7	1.89 ±0.05	2.01 ±0.06
P H$_2$O-s	37.8[a,b] ±1.2	43.6[a] ±0.9	72.0[a] ±1.0	74.1[a] ±1.4	32.7[a,b] ±1.1	21.3[a] ±0.5	7.40 ±0.40	7.60 ±0.44
Solid	31.5 ±0.8	32.5 ±1.0	60.6 ±2.0	55.2 ±1.6	48.5[b] ±1.7	35.7 ±1.1	7.56 ±0.24	7.50 ±0.30

[a]Difference between dietary subgroups within germfree or conventional groups significant, $P < 0.05$.
[b]Difference from conventional groups fed similar diet significant, $P < 0.05$.

Source: Reddy, B. S., Pleasants, J. R., and Wostmann, B. S.[20]

578

BERNARD S. WOSTMANN,
JULIAN R. PLEASANTS,
AND BANDARU S. REDDY

enhances insulin activity and prevents overt disturbances of carbohydrate metabolism, while Tween-80 is known to affect certain enzyme activities *in vitro* and *in vivo.* [17,18]

Gustafsson and Norman[19] had found a high incidence of urinary calculi and changes in urinary calcium and phosphate elimination in germfree rats. In our laboratory the use of earlier water-soluble diet formulas had resulted in mild hydronephrosis, with calcification, degeneration, and necrosis of kidney tubules. Recently, we therefore studied the metabolism of Ca, Mg, and P in germfree rats[20] fed formulation L479-Eg.[14] The results are given in Table 5. They indicate that both germfree status and the feeding of water-soluble diet promote the absorption of Ca and Mg, while more P is absorbed by both germfree and conventional rats fed the liquid diet. Surprisingly, these conditions were not reflected in the concentration of these minerals in the plasma. Therefore, although differences in absorption occur resulting from both germfree status and the feeding of liquid diet, these changes in mineral influx appear to be well within the limits of tolerance of the homeostatic mechanisms.

Thus, it appears possible in the gnotobiotic system to control not only the microbiological, but also the dietary variable with its related potential antigenicity. While the present liquid diet reduces performance during periods of very heavy nutritional demand, such as lactation, it appears adequate for normal growth and development. The resulting rodent appears healthy. Its metabolic parameters, as far as they have been studied, appear compatible with a normal functioning within the genotypically determined potential.

The Enlarged Cecum and Its Consequences

The one striking "abnormality" of the germfree rodent is its enlarged cecum. Since in recent years it has become more and more obvious that this enlargement may be related to many of the quantitative differences in function and metabolism observed between germfree and conventional rodents, the phenomenon deserves special attention.

Earlier work had shown that cecal enlargement in the germfree rat starts during the suckling period, and that it is presumably not of dietary origin.[21] Lindstedt *et al.*[22] have shown that large amounts of mucin accumulate in the cecum of the germfree rat, and Gordon has demonstrated the presence of musculoactive[23] and toxic[24] substances. All data indicate that under germfree conditions a net influx of water into the cecum takes place, resulting in a practically isotonic condition of the contents, which under conventional conditions are hypertonic. At this point in time we speculate that interference with the cecal (re)absorption of Na^+, directly or indirectly via Cl^- sequestration,[25] may be the underlying reason. Other data indicate that reduced muscle

TABLE 6 Cardiac Output and Oxygen Consumption of Intact and Cecectomized Germfree and Conventional Male Rats. Age 3–4 Months. 8–13 Animals per Group; Mean and S.E. Given

	Germfree	Germfree Cecectomized	Conventional	Conventional Cecectomized
Body weight g[a]	270	272	277	263
C.O. ml/kg/min	137 ± 7[c]	201 ± 11	203 ± 11	218 ± 5
Body weight g[b]	371	400	417	350
O_2 ml/kg/min	11.2 ± 0.3[d]	13.5 ± 0.3[e]	15.2 ± 0.4	15.3 ± 0.5

[a] Wistar rats: LOB: (WI).
[b] Sprague-Dawley rats: LOB: (SD).
[c] Germfree versus Germfree cecectomized, Conventional, Conventional cecectomized all $P < 0.01$.
[d] Germfree versus Germfree cecectomized, Conventional, Conventional cecectomized all $P < 0.01$.
[e] Germfree cecectomized versus Conventional and versus Conventional cecectomized $P < 0.01$.

tone may be another component of this complex phenomenon.[26,27] Whatever its causes, the enlarged cecum is a very conspicuous aberration. It accumulates abnormal amounts of intestinal material and may be one of the causes of the reduced intestinal-passage time demonstrated in the germfree mouse.[28] Its occurrence may constitute a stressful experience for the animal that could make itself felt in other parameters of function and metabolism.

Morphological studies rather consistently indicated that germfree rats have smaller hearts than conventional controls.[26] Gordon and Wostmann then determined cardiac output and found that germfree rats had only 2/3 of the normal output. Blood volume was likewise reduced to approximately 75 percent of its normal value. Obviously, the germfree rat can function with a much smaller circulatory system per 100 g of body weight.[29] At that time we also studied the influence of the intestinal microflora on cholesterol metabolism in the rat and found an enhancement of the cholesterol–bile acid conversion under the influence of certain flora elements.[30] As a by-product, these studies indicated a lower production of total CO_2 by the germfree animal. Thus a picture seemed to emerge that suggested a lower metabolic rate in the germfree rat.

As a next step, we then determined O_2 consumption and found it approximately 25 percent lower in germfree than in conventional rats. Similar data were obtained by Dr. Levenson and coworkers,[31] while a search of the literature revealed that des Places et al.[32] had not only obtained comparable results a few years earlier, but had also found that in the germfree rat thyroid function as indicated by ^{131}I fixation amounted to approximately 2/3 of the value found in the conventional animal.

It had long been suspected that cecal enlargement might impose an abnormal stress on the germfree rodent. To test this hypothesis, Gordon and Bruckner-Kardoss worked out a technique for surgical removal of the enlarged cecum in the 6-week-old germfree rat.[26,33] Animals thus treated appeared to function normally,[26] and have been kept for 12 months and longer.

Male germfree cecectomized Wistar rats of approximately 100 days of age were used to study the influence of cecectomy on cardiac output and O_2 consumption, and the results were compared with data from intact germfree and conventional rats.[34] The data in Table 6 indicate that cecectomy early in life restored cardiac output values to the "normal" found in the conventional animal. While germfree rats had a 25 percent lower oxygen consumption than comparable conventional animals, the cecectomized germfree animals showed only a slightly but significantly lower oxygen consumption than the conventional controls. The remaining small difference in O_2 consumption may result from oxygen utilization by a viable, actively metabolizing microflora in the conventional rat.

Obviously, the level of general metabolism of the germfree rat is lower than that of its conventional counterpart. This phenomenon appears related, in a yet unexplained way, to the cecal enlargement. Thus, cecal enlargement, whatever its causes, greatly influences physiology and metabolism of the germfree rodent without, however, interfering with such essential functions as growth, reproduction, and longevity.

Conclusion

Gnotobiotic conditions and dietary control constitute experimental refinements that often may prove necessary to obtain satisfactory and meaningful data in biomedical research. When the germfree rodent is used to achieve this definition, it should be kept in mind, however, that this experimental system differs from the conventional system. Data obtained with the intact germfree rodent will have to be interpreted against the background of the consideration that the present germfree rodent, while allowing increased control of experimental variables, constitutes an animal system whose function and metabolism reflect its equilibrium with a different ecosystem.

References

1. Beaver, M., and B. S. Wostmann. 1962. Histamine and 5-hydroxytryptamine in the intestinal tract of germfree animal harbouring one microbial species and conventional animals. Brit. J. Pharmacol. 19:385.
2. Sprinz, H. 1962. Morphological response of intestinal mucosa to enteric bacteria and its implication for sprue and Asiatic cholera. Federation Proc. 21:57.
3. Wostmann, B. S. 1959. Nutrition of the germfree animal. Ann. N.Y. Acad. Sci. 78: 175.
4. Kellogg, T. F., and B. S. Wostmann. 1969. Stock diet for colony production of germfree rats and mice. Lab. Anim. Care. 19:812.
5. Schneider, H. A. 1967. Ecological ectocrines in experimental epidemiology. Science 158:597–603.
6. Wostmann, B. S., and T. F. Kellogg. 1967. Purified starch-casein diet for nutritional research with germfree rats. Lab. Anim. Care. 17:589.
7. Wostmann, B. S., J. R. Pleasants, and P. Bealmear. 1969. p. 287. In E. A. Mirand and N. Back [ed.] Germfree biology. Vol. 3, Advances in experimental medicine and biology, Plenum Publishing Corp., New York.
8. Greenstein, J. P., S. M. Birnbaum, M. Winitz, and M. C. Otey. 1957. Quantitative nutritional studies with water-soluble, chemically defined diets. I. Growth, reproduction and lactation in rats. Arch. Biochem. Biophys. 72:396–416.
9. Birnbaum, S. M., M. E. Greenstein, M. Winitz, and J. P. Greenstein. 1958. Quantitative nutritional studies with water-soluble, chemically defined diets. VI. Growth studies on mice. Arch. Biochem. Biophys. 78:245–247.
10. Wostmann, B. S., J. R. Pleasants, and B. S. Reddy. 1967. Water-soluble, non-antigenic diets, p. 187. In M. L. Conalty [ed.] Husbandry of laboratory animals. Academic Press, London.

11. Pleasants, J. R., B. S. Reddy, and B. S. Wostmann. 1969. p. 307. *In* E. A. Mirand and N. Back [ed.] Germfree biology. Vol. 3, Advances in experimental medicine and biology. Plenum Publishing Corp., New York

12. Pleasants, J. R., B. S. Reddy, and B. S. Wostmann. 1970. Qualitative adequacy of a chemically defined liquid diet for reproducing germfree mice. J. Nutr. 100:498-508.

13. Hurlemann, J., G. J. Thorbecke, and G. M. Hochwald. 1966. The liver as the site of C-reactive protein formation. J. Exp. Med. 123:365-378.

14. Reddy, B. S., J. R. Pleasants, and B. S. Wostmann. 1968. Effect of dietary carbohydrates on intestinal disaccharidases in germfree and conventional rats. J. Nutr. 95:413.

15. Reddy, B. S., J. R. Pleasants, and B. S. Wostmann. 1969. Pancreatic enzymes in germfree and conventional rats fed chemically defined, water-soluble diet free from natural substrates. J. Nutr. 97:327.

16. Geever, E. F., F. S. Daft, and S. M. Levenson. 1965. Pancreatic atrophy and fibrosis in the germfree rat on a chemically defined *diet*. Federation Proc. 24:246.

17. Dawson, A. B., and K. J. Isselbacher. 1960. The estrification of palmitate-1-C14 by homogenates of intestinal mucosa. J. Clin. Invest. 39:150-160.

18. Holt, P. R., H. A. Haessler, and K. J. Isselbacher. 1963. Effects of bile salts on glucose metabolism by slices of hamster small intestine. J. Clin. Invest. 42:777-786.

19. Gustafsson, B. E., and A. Norman. 1962. Urinary calculi in germfree rats. J. Exp. Med. 116:273-284.

20. Reddy, B. S., J. R. Pleasants, and B. S. Wostmann. 1969. Effect of intestinal microflora on calcium, phosphorus and magnesium metabolism in rats. J. Nutr. 99:353.

21. Wostmann, B. S., and E. Bruckner-Kardoss. 1959. Development of cecal distention in germfree baby rats. Amer. J. Physiol. 197:1345.

22. Lindstedt, G., S. Lindstedt, and B. S. Gustafsson. 1965. Mucus in intestinal contents of germfree rats. J. Exp. Med. 121:201-213.

23. Gordon, H. A. 1967. A substance acting on smooth muscle in intestinal contents of germfree animals. Ann. N.Y. Acad. Sci. 147:85-106.

24. Gordon, H. A. 1965. Demonstration of a bioactive substance in caecal contents of germfree animals. Nature 205:571-572; Reduced levels of a bioactive substance in the caecal *content* of gnotobiotic rats monoassociated with *Salmonella typhimurium.* Nature 205:572-573.

25. Asano, T. 1967. Inorganic ions in cecal content of gnotobiotic rats. Proc. Soc. Exp. Biol. Med. 124:424-430.

26. Gordon, H. A., E. Bruckner-Kardoss, T. E. Staley, M. Wagner, and B. S. Wostmann. 1966. Characteristics of the germfree rat. Acta Anat. 64:367-389.

27. Strandberg, K., G. Sedvall, T. Midtvedt, and B. Gustafsson. 1965. Effect of some biologically active amines on the cecum wall of germfree rats. Proc. Soc. Exp. Biol. Med. 121:699-702.

28. Abrams, G. D., and J. E. Bishop. 1967. Effect of the normal microbial flora on gastrointestinal motility. Proc. Soc. Exp. Biol. Med. 126:301-304.

29. Gordon. H. A., B. S. Wostmann, and E. Bruckner-Kardoss. 1963. Effects of microbial flora on cardiac output and other elements of blood circulation. Proc. Soc. Exp. Biol. Med. 114:301-304.

30. Wostmann, B. S., N. L. Wiech, and E. Kung. 1965. Catabolism and elimination of cholesterol in germfree rats. J. Lipid Res. 7:77.

31. Levenson, S. N., D. Kan, M. Lev, and F. S. Doft. 1966. Oxygen consumption, carbon dioxide production and body temperature in germfree rats. Federation Proc. 25:482.

32. Desplaces, A., D. Zagury, and E. Saquet. 1963. [Study of the thyroid function of the germ-free rat.] C. R. Acad. Sci. (Paris). 257:756–758.

33. Bruckner-Kardoss, E., and B. S. Wostmann. 1967. Cecectomy of germfree rats. Lab. Anim. Care. 17:542.

34. Wostmann, B. S., E. Bruckner–Kardoss, and P. L. Knight, Jr. 1968. Cecal enlargement, cardiac output and O_2 consumption in germfree rats. Proc. Soc. Exp. Biol. Med. 128:137.

DISCUSSION

DR. ITURRIAN: Would you comment on the enlarged cecum as related to "possibly different behavioral characteristics" you mentioned.

DR. WOSTMANN: I had not brought this in because of the lack of time, and also because the psychology department had a computer breakdown and I do not have all the data available. There is a difference in behavior and we have been testing it.

What we did was a shock-avoidance test, using a shuttle box. There were 10 observations of 3 germfree groups and seven of another 3 groups of conventional animals. The animals were given 50 trials where they had to escape within 5 minutes after the light came on. The germfree animals escaped before that time in only nine out of 50 cases on the average. The conventional animals did a lot better. These isolated conventionals were littermates of the germfree group. They were conventionalized after weaning, but maintained otherwise under absolutely identical conditions with the germfree group.

The curiosity factor, too, is different. Apparently the germfree animal does not like to move. After this experiment, we took all three groups (two germfree and one conventional), mixed them up as conventional animals, and tested them 6 weeks later. At that time there was no difference in performance whatsoever. On second testing, all three groups had learned equally well apparently, so the germfree groups, even though they did not move so easily, apparently did learn. Right now we are speculating. I talked this over with Dr. Anderson, who is the animal psychologist at Notre Dame, after looking at more recent data than these. We asked ourselves whether we did the right thing by always taking one germfree and one conventional animal, and putting them side by side, or one after the other, in the alleyway or in the shuttle box, because the germfree animal we think might smell the conventional as a foreigner, while the conventional animal might not smell the germfree because of the absence of the characteristic odor. So we may have here a set of data where we overlooked the fact that smell for the one may not mean the same thing as smell for the other, which might invalidate a lot of the things we have done. So, the answer to your question has not yet been found.

DR. GUSTAFSSON: Were the heart weights given
in milligrams percent calculated on the total body weight or body weight
minus the cecal content?

DR. WOSTMANN: The heart weights were calcu-
lated in milligrams percent in both ways. I do not know which one is given
here, but the cecal weights would give you a difference of about 4 or 5 percent of body weight. So whether you do on one basis or the other does not
make too much difference, but all measurements were in milligrams percent.

DR. FOSTER: I have a question related to the
first question: What about the heart weights in cecectomized rats?

DR. WOSTMANN: The heart weights in the
cecectomized rats, as far as we have seen them, are absolutely normal. But
in the cecectomized group we have had only two series, and these were taken
a lot later than the series we have been showing here. The heart weights of
that group are the same as the heart weights of the animals I discuss in my
paper. They do not compare with what I present there because of differences
in animal housing and other characteristics, and because of different food.

DR. FOSTER: Normally, people would think in
terms of increased WBC count as a barometer for infection. I noticed the
very interesting results of different diets on the WBC count in germfree animals. Would you care to comment as to whether you think similar diets
would have the same effect in conventionalized or nongermfree animals?

DR. WOSTMANN: I did not cover this because
we felt we could not investigate both things, and we thought it would be
more interesting right now to check this in the germfree system. I would very
much expect that you would find the same things in conventional animals. As
a matter of fact, we did look at some data we had showing that the levels on
462 diet and CFW mice were high, but we did not do a controlled study. So
we just do not have enough of a basis for comparison between four different
diets. They are a lot higher than you would expect.

UTILIZATION OF DIETS STERILIZED BY GAMMA IRRADIATION FOR GERMFREE AND SPECIFIC-PATHOGEN-FREE LABORATORY ANIMALS

J. S. Paterson and R. Cook

Introduction

It has been suggested that for routine safety control and bioassay, animals will be supplanted by tissue cultures and other *in vitro* systems.[1] However, for many years to come, "vivi-study" will continue to play a vital role in our efforts to unravel the secrets of nature with the objective of bringing benefit to the health and general well-being of both man and his animals.

The production and use of germfree and specific-pathogen-free (SPF) laboratory animals continues to gain momentum in most countries, despite the fact that some members of the older school of medical research workers consider that the results of tests carried out with SPF animals may be misleading when extrapolated to man, who cannot be regarded as "pathogen-free." Our efforts must surely be directed to the task of ensuring that uniform and defined animals become more freely available. Only thus will variation between groups of animals be reduced, and so animal experimentation will, to a greater degree, become more and more qualitative.

In the last decade there has been a considerable increase in the availability of laboratory animals, particularly rats and mice, that merit the designation specific-pathogen-free. Guinea pigs have not been neglected, and SPF colonies have been established at several centers by Davey,[2] Morris,[3] Calhoon and Mathews,[4] Owen and Porter,[5] and by Paterson and Cook.[6] At the Laboratory Animals Centre, Carshalton, Surrey, England, small colonies of SPF rabbits and SPF cats are thriving.[7] Other workers have established groups of animals free of certain agents, particularly parasites. Thus, Greisamer and Gibson[8] have produced ascarid-free dogs, and Hills and MacDonald[9] have produced coccidia-free rabbits.

One of the greatest hazards inherent in any barrier system of producing or using SPF animals is the possibility that infectious agents or parasites may be introduced into breeding quarters or experimental rooms by food or bedding that has been contaminated by the excreta of rodents or birds. This danger may be mitigated to some extent by careful packaging and handling techniques, but clearly the most satisfactory solution is to ensure that the materials are sterile or pasteurized when they pass through the barrier. Various procedures have been described for this purpose, e.g., baking at $600°C$,[10] autoclaving at the relatively low steam pressure of 5 psi for 30 min[2] or at higher pressures for shorter periods, with preliminary and final high-vacuum cycles,[11,12] fumigation with ethylene oxide,[13] and, more recently, the possibility of using microwave energy ovens has been suggested.[14]

Many SPF breeding units do not operate on a scale large enough to justify the capital expenditure involved in the installation of large autoclaves or ethylene oxide chambers, and it was for this reason that in 1961 we turned our attention to the possibility of using diets sterilized by gamma irradiation (Cobalt-60 source), which at that time had become recognized as a safe method of sterilizing disposable medical equipment. We also had in mind that if suitable packaging could be devised, it would greatly simplify the provision of sterile diets to small experimental units engaged in long-term studies with SPF animals.

Gamma rays are emitted by certain radioactive materials and are, in fact, indistinguishable from x rays, which are, of course, generated by an electrical machine. Gamma rays are lethal to micro-organisms, including bacteria, protozoa, and most viruses. They also kill parasites and insects. The unit of radiation dose commonly used is the rad, which is a measure of the energy absorbed by the material through which the radiation passes. One rad is equivalent to 100 ergs of energy per gram. For practical reasons, the multiple unit megarad (Mrad), equivalent to 1 million rads, is a convenient measure to use.

It is important to stress that gamma rays from a Cobalt-60 source do not induce radioactivity in the material being processed, nor do they produce a marked rise in temperature, only $5°C$ at a dose of 2.5 Mrads. Gamma rays have enormous penetrating powers and this ensures that sealed prepacked laboratory animal diets or other materials, even in considerable bulk, may be completely sterilized.

In deciding to investigate the practicability of using whole irradiated diets, we were conscious that although irradiation might not change the physical appearance and texture of the diets, it might nevertheless be impracticable because diets might be rendered unpalatable, or the destruction of nutrients might be so great as to make them useless. Finally, there might be problems associated with induced toxicity or carcinogenesis. We sought first, therefore, to study the breeding performance of comparable groups of breeding rats in our SPF colony, fed *ad libitum* on irradiated and nonirradiated diet PRM. This is a pelleted diet containing a variety of cereal meals, soybean meal, whitefish meal, and dried yeast; it is fortified with a balanced vitamin supplement and trace element mixture (see Table 1).

TABLE 1 Percentage Composition of Diet PRM (Rat and Mouse Diet), and Analysis of Vitamin and Mineral Supplements

Component	Percent
Barley meal	5
Wheat meal	20
Yellow maize meal	10
Oatmeal	20
Fine bran	20
Soybean meal	10
Dried skim milk	7.5
English whitefish meal	5.0
Dried yeast	2.5

Vitamin Supplement

Component	Amount/kg of Diet
Vitamin A	4,000 IU
Vitamin D_3	1,000 IU
Vitamin B_1	2.0 mg
Vitamin B_2	8.0 mg
Nicotinic acid	50.0 mg
Vitamin B_{12}	0.012 mg
Vitamin E (α-tocopherol)	25.0 mg
Vitamin K (Menadione)	10.0 mg
Calcium pantothenate	4.0 mg
Choline chloride	200.0 mg
Folic acid	0.5 mg

Mineral Mix[a]

Component	Percent
Iron sulphate	2.2
Manganese sulphate	2.6
Magnesium oxide	1.5
Calcium iodate	0.03
Sodium chloride	29.95
Dicalcium phosphate	24.3
Chalk	39.29
Copper sulphate	0.1
Cobalt sulphate	0.03

[a]6.7 g of mineral mix is added to each kg of diet.

588

J. S. PATERSON
AND
R. COOK

SPF Studies—Choice of Radiation Dose

The evidence available in 1961, which was later discussed in detail by Ley and Tallentire,[15] indicated that a radiation dose of 2.5 Mrads would be adequate to sterilize cubed or pelleted laboratory animal diets, even though the bacterial population might be as high as 5×10^5 organisms and spores per g. We have used this dose exclusively for our studies on SPF animals, and regular bacteriological examinations of treated diets have shown that complete sterilization is routinely achieved.

In the first trial (November 1961–September 1962), the two groups of rats, fed either an irradiated or a nonirradiated diet, each consisted of 96 monogamous pairs that were mated when approximately 12–14 weeks of age. Detailed records of each litter born were kept, including date of birth, number born, preweaning deaths of young stock, and number weaned. The males were removed from the females 9 months after first mating, and the females were allowed to rear and wean their last litters. Most pairs gave birth to nine litters, some had 10, and two (one in each group) had 12 litters.

The records of the trial were submitted to statistical analysis and it was found that the group fed irradiated diet recorded, on the average:

	Irradiated (%)	Nonirradiated (%)
Smaller preweaning death rate	5.6	6.0
Significantly smaller litters	9.52	9.87
More litters per pair	8.77	8.27
More young weaned per pair	83.5	81.7

The conclusion was that in general there was no marked difference between the two groups.

A second trial (August 1962–September 1963) was conducted in a similar manner, with the following results:

	Irradiated (%)	Nonirradiated (%)
Smaller preweaning death rate	5.25	5.34
Larger litters	9.53	9.42
Significantly fewer litters per pair	8.14	8.57
Fewer young weaned per pair	77.2	79.7

The conclusion, again, was that there was no marked difference between the two groups, and it was decided that our nucleus rat breeding colony of 75 monogamous pairs should now be fed only irradiated diet. However, before

589

committing our production colony to irradiated diet, a more prolonged trial following the normal pattern of breeding pair replacement was carried out in the production colony itself. Beginning in February 1964 and continuing through to January 1965, six monogamous pairs were placed either on irradiated or nonirradiated diet PRM *ad libitum* every second Monday throughout the 12-month period. The males were removed 8 calendar months after first mating, and records were kept as before. In this trial it was found that, although the percentage of young dying was higher in the group fed nonirradiated diet, 7.5 percent compared with 7 percent, they recorded, on the average:

	Irradiated (%)	Nonirradiated (%)
Bigger litters	9.58	9.73
More litters per pair	7.38	7.42
More young per pair weaned	71.0	72.5

None of these differences was significant, and it was again concluded that there was no marked difference between the two groups, so the decision was made to feed only irradiated PRM to both nucleus and production colonies. From April 1, 1963, to March 31, 1968, 125,000 rats were issued to the laboratories for a variety of toxicological and pharmacological studies. None died from intercurrent disease, although they were not maintained under SPF conditions in the laboratories, nor were they fed sterilized diet. At regular monthly intervals during this period, five of the oldest discarded breeding pairs were killed and subjected to a careful examination for parasites and pathogenic bacteria, including mycoplasma. None was found, nor was there any macroscopic evidence of tumor formation or other conditions. The absence of chronic murine pneumonia was demonstrated by the failure of lung suspensions from these aged animals to infect susceptible BALB/C inbred mice by the intranasal route. Breeding nuclei were sent to six research establishments, and after screening, they were judged to be satisfactory for SPF breeding.

A small SPF mouse colony originally founded from a nucleus of gnotobiotic mice of the Webster strain, but gradually switched to our own albino Porton strain by fostering in young obtained by aseptic hysterectomy, has prospered for 6 years on irradiated diet PRM. Production has been better than in our conventional colony, almost certainly because of the absence of worms and the nonoccurrence of infantile diarrhea.

To satisfy a demand for SPF guinea pigs, we have established a hand-reared colony. Guinea pigs may be reared without hand-feeding. We raised guinea pigs obtained by aseptic hysterectomy in solid-floored metal boxes, each measuring 16 in. long, 12 in. wide, and 8 in. high. Each box was individually heated from above by a 250-W energy-regulated infrared heater with a polished metal reflector so placed as to produce a temperature of between 30° and 32°C over approximately one third of the floor area, the remainder being at the ambient temperature of the room. This variation in temperature permitted each guinea pig to choose its own area of comfort and made it possible to put the dishes containing the fluid diet in the cooler part of the box, where evaporation losses were not great.

A spray-dried milk-substitute diet comparable with guinea pig milk was prepared for us by a commercial firm from cow's milk supplemented with protein and fat. This powder, supplemented with vitamins and minerals, was packed in 500-g amounts in glass jars and sterilized by gamma irradiation (2.5 Mrads). The fluid diet was prepared by adding 25 g of the milk substitute to 90 ml of water, plus 10 ml of an aqueous solution containing 0.225 g of potassium acetate and 0.2375 g of magnesium acetate. Two fresh feeds were offered daily. Although the fluid diet was continued for 4 to 5 weeks, a pelleted diet, RGP (see Table 2), and hay, both sterilized by gamma irradiation (2.5 Mrads), were available *ad libitum* from the seventh day on. These were consumed in varying amounts. To ensure that an adequate intake of vitamin C was being maintained, each guinea pig was dosed daily with 10 mg of ascorbic acid in 0.1 ml of distilled water until the twentieth day, by which time sufficient amounts of diet RGP, which contains not less than 1,000 mg of ascorbic acid per kg, were being consumed. The milk-substitute diet was withdrawn gradually between the 28 and 35 days after birth. Of 300 young obtained by aseptic caesarean operation, 264 were successfully raised, an overall survival rate of 88 percent. The colony was stabilized at 140 breeding females in 35 polygamous harems (one male per four females), and for nearly 3 years production has been equal to that achieved by our conventional colony, viz., 15 young per sow per year. Irradiated hay was withdrawn 18 months ago and replaced by hay sterilized by autoclaving at 32 psi (2.25 kg/cm^2) for 4.5 min (136° C/277° F), the sterilizing period being preceded by a high-vacuum cycle and followed by a cycle comprising drying, high vacuum, and flushing of the chamber with sterile air.

We have successfully hand-reared rabbits derived by aseptic hysterectomy on a diet based on cows' colostrum collected within 24 hr of parturition and freeze-dried after homogenization.[6] The resultant powdered colostrum was fortified with minerals, vitamins, and antibiotics and packed in 400 g amounts in glass jars for sterilization by gamma irradiation (2.5 Mrads). The fluid diet was prepared by homogenizing 40 g of the fortified colostral powder in 10 ml of sterile arachis oil and 100 ml of sterile water. The Silverson homogenizer we used is fitted with a fine-mesh stainless-steel sieve to ensure that the final product is free from particles that might block the hole in the nipple of the feeding bottle.

Rabbits feed their young only once in 24 hours, so we adopted the same procedure. On the first day of life, 5 ml of liquid diet was fed, and this was

TABLE 2 Percentage Composition of Diet RGP
(Guinea Pig Diet), and Analysis of Vitamin and
Mineral Supplements

Component	Percent
English fine wheat feed	15
Finely ground oats	12.5
Barley meal	40
Dried grass meal	15
Linseed cake meal	10
Whitefish meal	7.5

Vitamin Supplement

Component	Amount/kg of Diet
Vitamin A	4,000 IU
Vitamin D_3	1,000 IU
Vitamin E (a-tocopherol)	25 mg
Vitamin K (Menadione)	10 mg
Vitamin B_1	2 mg
Vitamin B_2	8 mg
Nicotinic acid	50 mg
Calcium pantothenate	4 mg
Vitamin B_{12}	0.012 mg
Choline chloride	200 mg
Ascorbic acid	1,000 mg
Folic acid	3.0 mg

Mineral Supplement[a]

Component	Percent
Iron sulphate	2.2
Manganese sulphate	2.6
Magnesium oxide	1.5
Sodium chloride	29.95
Dicalcium phosphate	24.3
Chalk	39.29
Copper sulphate	0.1
Calcium iodate	0.03
Cobalt sulphate	0.03

[a]13.3 g of mineral mix is added to each kg of diet.

gradually increased until 40 ml was being taken on the fourteenth day. The fluid diet was continued until the thirty-fifth day, after which it was reduced steadily over the next 7 days, as the animals were weaned to a solid diet of RAF pellets and hay, both sterilized by gamma irradiation (2.5 Mrads). Unfortunately, our first attempt to set up an SPF rabbit unit was unsuccessful. Some 12 weeks after the start, we observed oocysts of *Eimeria magna* in the faeces of more than one animal. However, the project was continued on what might be termed a minimal-disease regime, and for over 2 years, the group has remained free of all respiratory infections and has bred well. As with the SPF guinea pigs, the irradiated hay has been replaced by autoclaved hay. A second attempt to set up an SPF rabbit colony is under way.

Germfree Studies

In our germfree studies, to provide a greater margin of safety, a dose of 4.0 Mrads has been used. Diets and other materials required in the isolators, e.g., bedding and instruments, are packed in fruit-preserving jars. These are placed individually in nylon film bags, which are heat-sealed after excess air has been manually expelled. After irradiation in standard cardboard boxes, the sterile jars are introduced into the germfree isolators either via the dunk tank or the transfer port, whichever method is more appropriate. Small breeding colonies of four strains of inbred mice, Lobund/Webster, CFW, C_3H, and New Zealand Black have been maintained germfree for up to a year on irradiated PRM, while our random-bred albino Porton strain has been similarly maintained for over 5 years. A small colony of breeding rats of the Charles River CD-F (axenic) strain has recently been established and is breeding satisfactorily.

Using the feeding materials and methods previously described for SPF animals, but irradiating the dry-milk replacer diets at 4.0 Mrads, half a dozen guinea pigs and a dozen rabbits have been raised to maturity in the germfree state. These were required for special projects, and to date no attempt has been made to breed these species axenically.

Discussion

From a practical standpoint, I believe the evidence I have been able to present here indicates that it is possible to produce healthy SPF and normal axenic animals on food sterilized by gamma irradiation at 2.5 or 4.0 Mrads. The choice of the dietary constituents or the physical state of the diet (dry or liquid) may be important. Chakhava and Zenkevich[16] failed to rear young conventional guinea pigs on what they termed regular diet sterilized by gamma irradiation at a dose of 3.0 Mrads. Most animals became paralyzed and subsequently died. Affected guinea pigs recovered quickly if fed unsterilized diet. Porter and Lane-Petter[17] found diets sterilized by gamma irradiation suitable for breeding mice, and Festing[18] has reported on reproductive performance of inbred SPF mice fed irradiated diet.

However, little is yet known regarding possible changes in the nutritive value of diets so treated. As far as dry diets are concerned, it would seem that changes are not severe, although it has been reported to us that rats and mice maintained on a gamma-irradiated diet without added synthetic vitamin K in the form of menadione, developed debilitating or even lethal internal hemorrhage. Coates[19] concludes that from the point of view of vitamin stability, gamma irradiation is a satisfactory method of sterilizing diets for germfree and SPF animals. Vitamin losses compared favorably with those reported for diets treated with ethylene oxide or by autoclaving.

It is too early to say that gamma irradiation is one possible solution to the problem of providing sterile laboratory animal diets for gnotobiotic and SPF animals. It is a very convenient though still somewhat costly method of sterilizing diets, and it may well be that if diets sterilized by autoclaving systems incorporating high-vacuum cycles are nutritionally satisfactory, then irradiated diets for laboratory animals will not be a commercially viable enterprise unless the cost factor can be reduced. At the present time, a little over 150 tons of laboratory animal diet are being sterilized annually by gamma irradiation in the Package Irradiation Plant of the U.K. Atomic Energy Authority at their Wantage Laboratory. The facilities are available to both research institutions and commercial firms producing laboratory animal diets. The standard package available weighs 56 lb and contains four 14-lb packs of sterile diet.

Acknowledgments

We would particularly like to thank Mr. Ley and Mr. Crook of the Radiation Branch, Radioisotope Division, U.K. Atomic Energy Authority for all their help and encouragement during the course of these studies, which are continuing.

References

1. Medawar, P. 1969. I.A.T. Manual of Laboratory Practice and Techniques 2nd ed. Crosby Lockwood, London, p.vi.
2. Davey, D. G. 1959. Establishing and maintaining a colony of specific-pathogen-free mice, rats and guinea-pigs. L.A.C. Collected Papers 8:17-34.
3. Morris, J. H. 1963. Management of a specific pathogen-free guinea-pig colony. Lab. Anim. Care 13:96–100.

4. Calhoon, J. R., and P. J. Mathews. 1964. A method for initiating a colony of specific pathogen-free guinea-pigs. Lab. Anim. Care 14:388–394.
5. Owen, D. G., and G. Porter. 1967. The establishment of a colony of specified pathogen-free guinea-pigs. Lab. Anim. 1:151–156.
6. Paterson, J. S., and R. Cook. 1969. Production and use of pathogen-free animals. The I.A.T. Manual of Laboratory Animal Techniques & Practice 2nd ed. Crosby Lockwood, London, p. 420–424.
7. Bleby, J. 1969. Personal communication.
8. Greisemer, R. A., and J. P. Gibson. 1963. The establishment of an ascarid-free beagle dog colony. J. Amer. Vet. Med. Ass. 143:965–967.
9. Hills, D. M., and I. MacDonald. 1956. Hand rearing of rabbits. Nature 178:704–706.
10. Foster, H. L. 1962. The problems of laboratory animal disease. p. 249–259. *In* Establishment and Operation of SPF Colonies. Academic Press, London.
11. Foster, H. L., C. L. Black, and E. S. Pfau. 1964. A pasteurization process for pelleted diets. Lab. Anim. Care 14:373–381.
12. Williams, F. P., R. J. Christie, D. J. Johnson, and R. A. Whitney. 1968. A new autoclave system for sterilizing vitamin-fortified commercial rodent diets with lower nutrient loss. Lab. Anim. Care 18:195–199.
13. Charles, R. T., and A. I. T. Walker. 1964. The use of ethylene oxide for the sterilization of laboratory animal foodstuffs and bedding. J. Anim. Tech. Ass. 15:44–47.
14. Foster, H. L. 1968. The use of microwave sterilization and pasteurization for barrier sustained animal colonies. Lab. Anim. Care 18:356–360.
15. Ley, F. J., and A. Tallentire. 1965. Radiation sterilization—choice of dose. Pharm. J. 195:216–218.
16. Chakava, O. V., M. V. Zenkevich, and T. I. Zazenkina. 1967. Some results with diets designed for germ-free guinea-pigs. Advances in Germ-free Research & Gnotobiology. M. Miyakawa and T. D. Luckey. [ed.] Iliffe, London.
17. Porter, G., and W. Lane-Petter. Observations on autoclaved, fumigated and irradiated diets for breeding mice. Brit. J. Nutr. 19:295–305.
18. Festing, M. 1968. Some aspects of reproductive performance in inbred mice. Lab. Anim. 2:89–100.
19. Coates, M. E., J. E. Ford, M. E. Gregory, and S. Y. Thompson. 1969. Effects of gamma irradiation on the vitamin content of diets for laboratory animals. Lab. Anim. 3:39–49.

DISCUSSION

DR. LATTUADA: What was the highest radiation dosage found that did not sterilize pelleted feed?

DR. PATERSON: Unfortunately I am not a physicist, but I have read some figures on this. I am just quoting from memory now. I think with 2.5 Mrads you might find one organism in 2¼ million tons of animal feeds. So at 4 Mrads, something of the order of 10 million tons can be irradiated. In that you might find one surviving clostridial spore.

At 2.5 megarads, the population reduction factor will be about 10^{15}, so it would have to be an extraordinarily heavily contaminated material that would not in fact be sterilized.

DR. LATTUADA: Were any differences noticed in the sterilization dosage required for 4 percent fat diets compared with 11 percent fat diets?

DR. PATERSON: No. I do not know how the rats and the mice smell, but I can detect a slight difference in smell between the radiated and nonirradiated diets, but whether this is due to any action on the fats producing slight rancidity, I do not know. We have noticed that rats, if they are transferred from nonirradiated diets to irradiated diets do not eat quite so much for about a fortnight, but then the reverse is equally true; if you take rats on irradiated diets and put them on nonirradiated diets, they do not eat quite so much for about a fortnight. So I presume it is a question of acquired taste.

DR. LES: What was the cost per pound of irradiated diet as compared with nonirradiated diet?

DR. PATERSON: There is a little pamphlet put out by the Atomic Energy Commission that tells you. They do not give the cost per pound, but at 2.5 Mrads, the cost per ton is 30 English pounds.

DR. FOSTER: Is that just for the sterilization process?

DR. PATERSON: Yes, for 2.5 Mrads, the cost is 30 pounds, and at 4 Mrads, of course, it is 48 pounds sterling per ton. It is a fairly expensive process. It almost doubles the cost of the diet. That is why I had to say it is expensive, but extremely convenient for handling the diet afterwards, and it is so safe.

DR. FOSTER: Dr. Paterson, I have a question. Is there any problem in sterilizing the entrapped air in the plastic bag? In other words, can you introduce these bags of pelleted feed after being exposed to radiation into a germfree system with complete assurance that the air entrapped is also sterile?

DR. PATERSON: All I can say is that I don't know how many packages we have introduced into germfree isolators, but it must be several hundred, and we have no evidence of contamination. When we do get contaminations, we generally find that we have got a split glove or something. We have no evidence at all that the packages are not sterile. In fact, we have done a tremendous number of bacteriological examinations endeavoring to find bacteria specifically, but I think you would probably agree with me that there is nothing quite so sensitive as a germfree rat. He is a much more capable bacteriologist than I am at finding bugs. It is pretty certain that bacteria and the larger viruses suspended in air would be killed, and of course any parasites would be more easily killed, because those are large.

GNOTOBIOTIC BEAGLES: MAINTENANCE, GROWTH, AND REPRODUCTION

James B. Heneghan, Salvador G. Longoria, and Isidore Cohn, Jr.

Introduction

With each passing year, increased efforts have been directed toward the development of experimental animals that have a defined and controlled parasitic, fungal, and bacterial flora. Animals such as specific-pathogen-free (SPF), germfree (axenic), and gnotobiotic, are initially obtained by Cesarian section, hand-fed, and then maintained within some type of barrier system. In many instances, small colonies of germfree animals, usually rats or mice, are maintained to supply additional breeders for the defined flora colony. Now that reproduction has been obtained in germfree Beagles, a similar approach is possible with those animals.

During the past 6½ years the germfree laboratory at Louisiana State University (L.S.U.) has reared over 200 gnotobiotic dogs to an average age of 5 months. Pure-bred Beagles represented 69 of the 76 puppies delivered into the germfree environment, a survival rate of 91 percent. Two females and one male have been maintained free of contamination for over 3 years; this appears to be the longest time that dogs have been kept germfree. One of these bitches whelped two second-generation litters of four puppies each; these were the first and second times that reproduction had occurred in germfree animals other than rodents, rabbits, or birds. This paper describes the maintenance of germfree Beagles and compares their growth and reproduction with those of conventional Beagles. In addition, the percentage of contaminations, the causes of contamination, and the amount of effort required to rear germfree Beagles are discussed.

597

Historical Background

The first germfree dogs were reared at the Lobund
Laboratory of the University of Notre Dame before 1960. Phillips[9] conducted
a parasitological survey and reported that five of six germfree puppies were
congenitally infected with *Toxocara canis*. One of the five infected puppies
was also contaminated with *Ancylostoma caninum*. Griesemer[2] was the first
to successfully rear gnotobiotic dogs in 1962, and later, in 1963, he[3] described
two methods for the elimination of congenital ascarid parasites. In 1968
Heneghan and colleagues[8] reported the first successful reproduction in germfree dogs. In 1969 Yale[12] described an isolation system that has many engineer-
ing innovations intended to increase environmental control and yet reduce
labor costs of rearing germfree dogs.

Materials and Methods

FACILITIES

The germfree facility in the Department of Surgery
at L.S.U. currently occupies two laboratories: approximately 163 m² are
located in the Medical School (Figure 1), plus an additional 372 m² at the
Magnolia Street Laboratory (Figure 2), which is three blocks from the Medical
School. Both laboratories have high-vacuum steam sterilizers; one laboratory
has an ethylene oxide sterilizing capability. Central isolator air supply systems
consist of a Spencer* turbine and a stand-by blower with automatic switchover.
Electric power to all blowers is "backed-up" by natural-gas-powered auxillary
generators. This system provides from 4 to 8 air changes per hour, depending
upon the size of the isolator and the size of its filters. The laboratory environment is maintained at a temperature of 23° to 27° C and a relative humidity
of 40–60 percent throughout the year by central heating and air conditioning. No attempts are made to further condition the air before it passes into
the isolators, except that a 20-cm-diameter H.E.P.A. filter is used as a pre-
filter in the central isolator air system. In general, this system maintains the
isolator humidity and temperature at 10 percent and 3° C, higher than in the
laboratory.

OBTAINING GNOTOBIOTIC PUPPIES

Healthy, pregnant, pure-bred Beagles were obtained
from commercial breeders who supplied breeding dates for all the bitches.

*Spencer Turbine Co., Hartford, Conn.

JAMES B. HENEGHAN,
SALVADOR G. LONGORIA,
AND ISIDORE COHN, JR.

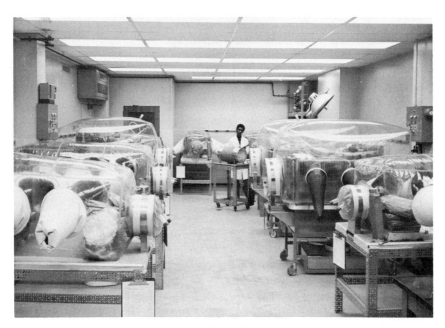

FIGURE 1   Germfree laboratory, L.S.U. Medical School.

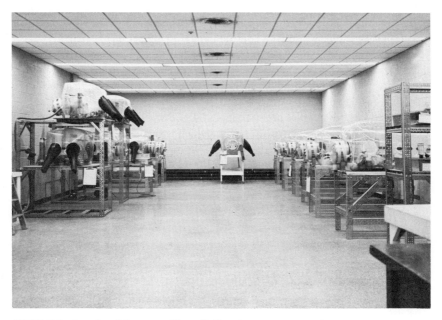

FIGURE 2   Germfree laboratory, Magnolia Street.

However, it was necessary to use additional criteria for scheduling delivery since some suppliers breed their dogs three times over a 5-day period. The degree of fetal calcification observed in x rays of the pregnant dog has been, for us, the most dependable way of scheduling the delivery and reducing prematurity, the most frequent cause of puppy mortality. Other factors, such as a drop in rectal temperature, breast engorgement, soft unformed feces, discharge from the vagina, and nesting behavior were additional aids but, by themselves, were not reliable.

The germfree puppies were delivered by Cesarian section hysterectomy. All precautions required for a germfree delivery by Cesarian section were followed. The entire uterus was excised and passed into a rearing isolator where the puppies were delivered. These techniques have been described previously.[5,7]

HAND FEEDING

Within four to six hours after delivery, the newborn puppies were fed autoclaved Esbilac,[*] a commercially prepared bitch's milk replacement. The Esbilac had been mixed (2 parts water: 1 part milk) in a blendor, placed in 1 liter Square-Pak[†] flasks, and autoclaved for 30 minutes at 121°C. Slight carmelization and coagulation occurred, but this was not a serious problem. Each liter of milk was supplemented with 1.0 cc of Vi-Syneral[‡] B vitamins plus vitamins A, C, D, and E. The daily dosage of vitamins varied between 0.4 cc per kg for newborn puppies to 0.25 cc per kg for weanling puppies. Methionine solution was added to the milk in quantities such that each puppy received approximately 30 mg per day.

The puppies were fed with infant nursing bottles and premature nipples five times a day for the first week, four times a day through the third week, and three times a day throughout the remainder of the hand-feeding period. At each feeding puppies were fed until they refused to suckle.

As soon as the puppies began to lap, at 4 to 5 weeks of age, small portions of solid food were mixed with the milk, which was gradually withdrawn. After weaning, the puppies were maintained on Purina Dog Chow[§] autoclaved for 25 min at 121°C. Water was autoclaved for 60 min at 121°C. A supplement of 2.5 cc of the Vi-Syneral vitamin complex was added to each 2 liters of water (Daily dosage: 0.12 cc of vitamins per kg).

Since our early experience with mongrel dogs had indicated that over 95 percent of all litters were contaminated congenitally with the larvae of the round worm, *Toxocara canis*, all of our Cesarian-derived litters were treated

[*] The Borden Co., New York, New York.
[†] American Sterilizer Co., Erie, Pennsylvania.
[‡] U.S. Vitamin and Pharmaceutical Corp., New York, New York.
[§] Ralston Purina Co., St. Louis, Missouri.

JAMES B. HENEGHAN,
SALVADOR G. LONGORIA,
AND ISIDORE COHN, JR.

FIGURE 3 "Pen-tub" isolator for adult dogs.

with diethylcarbamazine citrate (50 mg per kg) weekly from 2 to 6 weeks of age.

CAGING

Newborn and young puppies were maintained in standard polyvinyl chloride isolators 60 X 60 X 150 cm.* Puppies were placed on cotton towels on the floor of the isolator, which was maintained at 30° C, with a heating pad placed under the isolator. At 4 weeks of age, the puppies were placed in collapsible stainless-steel cages that were assembled in the isolator.

Dogs were placed in pen-tub isolators (Figure 3) when they reached 4 months of age. The cage dimensions of this isolator were approximately 60 cm wide, 120 cm long, and 90 cm high. Two adult Beagles were maintained in each isolator, which was a modification of the standard Partsco Pen-Tub Isolator.† A large stainless-steel tank for storing water replaced the trays, which

*Standard Safety Equipment Co., Palatine, Illinois.
†Partsco, Inc., Columbus, Ohio

had previously formed the top of the cage. It is planned to filter-sterilize water into this storage tank so that a continuous water supply is available for drinking and for washing the cage.

The second major modification was the stainless-steel waste-collection tank located below the isolator and connected to it by a standard 30-cm-diameter plastic transfer sleeve. The floor of the isolator has a 15-cm-diameter hole through which all waste material passed (Figure 4). When the waste tank was full, a large compression rubber stopper was tightened in the opening, and a new tank was connected by using the standard gnotobiotic transfer techniques. In order to ensure sterility of the waste tank, a small amount of water was added to the tank when it was autoclaved. This waste-disposal system has eliminated the transfer of adult dogs to clean isolators every 3 months.

FIGURE 4 "Pen-tub" isolator drain and stopper.

JAMES B. HENEGHAN,
SALVADOR G. LONGORIA,
AND ISIDORE COHN, JR.

STERILIZATION OF SUPPLIES

All the plastic isolation chambers, plastic transfer sleeves used in sterile entries, and most heat-labile materials were sterilized with 2 percent peracetic acid. All heat-stable materials were autoclaved in standard sterilizing cylinders. Liquids were sterilized by autoclaving in sterilizing flasks or by filtration. Special isolators and equipment that were heat and acid labile were sterilized in cylinders with ethylene oxide.

BACTERIOLOGICAL TESTING

Once every 2 weeks and after every sterile entry, bacteriological tests were conducted. Cotton swabs of fresh feces, milk, water, and placenta were placed in screw-cap tubes, passed out of the isolator and inoculated into liquid thioglycholate and brain–heart infusion broth. After a 1-week incubation, blood agar and thioglycholate plates were streaked and incubated anaerobically and aerobically. The details of these techniques have been published previously.[6]

WEIGHING THE DOGS

The weights of the Beagles were determined three times a week for the first 3 weeks, and then once a week until 8 weeks of age. A postmortem scale with an 11 kg capacity was used. After that, the dogs were weighed every 2 weeks using the following procedure. The technician carefully picked up the dogs inside the isolator and determined his weight plus the dogs' weight by standing on a bathroom scale. The dogs' weight was then obtained by subtracting the weight of the technician. Old dogs had to be tranquilized to minimize the chances of torn isolators or gloves.

BREEDING GERMFREE DOGS

In order to obtain successful reproduction, two breeding pairs of Beagles were maintained germfree for 2 years. After reproduction had occurred the first time, only the fecund pair was retained for further breeding. This pair has now been maintained germfree for over 3 years, which is the longest time reported for germfree canines.

Three of the dogs, two females and one male, were littermates born in April 1966; the fourth dog, a male, was born in May 1966. Each breeding pair was reared together in the same isolator. Within 18 months, each female had gone through two normal heat periods with no mating observed. During the first heat period, at approximately 11 months, viable sperm could not be demonstrated in either male. However, 6 months later, during the second heat period, viable sperm were observed in both males, and the failure to breed was attributed to a lack of experience in the males. At this time, artificial insemination was attempted. Although we were able to collect semen successfully

FIGURE 5 First litter of
second-generation germfree
dogs.

from conventional dogs that were experienced studs, all efforts to collect semen from the germfree males failed.

When the third heat period was indicated by vaginal smears, the female in heat was transferred into the isolator with the male that had not been reared with her; fortunately, this male was not her littermate. Three days after the transfer, a mating was observed, followed by another mating one day later. After a 63-day gestation period, the first litter of second-generation germfree dogs was whelped. Approximately 6 months after whelping the first litter, a second mating was observed with the same breeding pair. In both instances, 10 days before the expected whelping date, the male was transferred out of the isolator; this left the pregnant female alone in her pen-tub isolator. When adult dogs were transferred from isolator to isolator they were tranquilized with *Tranvet*,[*] 0.25 to 0.5 mg per kg.

[*]Diamond Laboratories, Des Moines, Iowa.

604

JAMES B. HENEGHAN,
SALVADOR G. LONGORIA,
AND ISIDORE COHN, JR.

Results

SECOND-GENERATION GERMFREE DOGS

The first litter of second-generation germfree dogs was whelped when the bitch was exactly 2 years old. The photograph (Figure 5) was taken at 7 days of age and illustrates that the four male puppies were healthy and were free of congenital defects. The delivery, which occurred on April 7, 1968, was normal, and no assistance was given. Thus, germfree canines were added to the list of germfree animals that have been reared through two generations. This also represented the first time that reproduction had occurred in germfree animals other than rodents, rabbits, or birds.

Six months later, a second litter of second-generation, pure-bred, germfree Beagles was obtained from the same breeding pair. This litter consisted of three males and one female, all of which were healthy and had no congenital defects. Again, the delivery was normal, and no assistance was given.

Now that two litters of second-generation germfree Beagles have been obtained, it has been established that the reproductive system of germfree Beagles is capable of fertilization, implantation, maturation, parturition, and lactation. Also, the onset of the first estrous cycle in germfree dogs (11 months) and the estrual interval (7 months) were normal and were comparable to the same values in conventional Beagles as reported by Smith and Reese.[10]

GROWTH OF SECOND-GENERATION GERMFREE BEAGLES

Growth During Nursing Period

The puppies were not weighed immediately for fear of disturbing the mother. However, when the growth curve was extrapolated backwards (Figure 6), the puppies weighed approximately 300 g at birth. They weighed 375 g at 3 days, 560 g at 1 week, 920 g at 2 weeks, 1,290 g at 3 weeks, and 1,560 g at 4 weeks of age. Thus, adequate nutrition and maternal care promoted a rate of growth that was comparable to conventional Beagles. This rate of growth was twice as fast as the hand-reared, Cesarian-derived, first-generation, germfree puppies during the suckling period, as indicated by their weights at 4 weeks: 0.75 kg for first-generation animals, and 1.5 kg for second-generation animals.

The differences between the first- and second-generation Beagles were emphasized when a litter that was delivered by Cesarian section was compared with the litter that was whelped in the isolator 4 days later. The photographs in Figures 7 and 8 were taken on the same day, when the second-generation puppy was exactly 4 weeks old and the first-generation puppy was 4½ weeks old. Not only was the bitch-nursed puppy approximately twice the size of the hand-fed puppy, but also it was cleaner and more vigorous, additional benefits of maternal care.

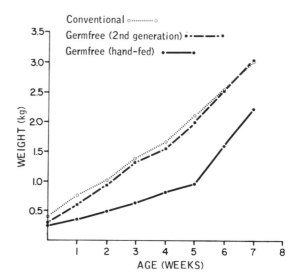

FIGURE 6 Male Beagle growth during nursing period.

FIGURE 7 Second-generation germfree puppy, 4 weeks old.

JAMES B. HENEGHAN,
SALVADOR G. LONGORIA,
AND ISIDORE COHN, JR.

FIGURE 8   First-generation germfree puppy, 4½ weeks old.

Growth During First Eight Months

When the growth of male Beagles was compared during the first 8 months of life (Figure 9), the first-generation germfree dogs gained weight more slowly than both the conventional and second-generation dogs during the hand-feeding and early weaning period, from birth to 2 months. First-generation germfree dogs 2 months old were 1 kg lighter than the other two groups of dogs. However, by 2 months, the rates of growth of all groups (the slope of the curves) were equal, and this 1-kg deficit remained constant for the remaining 6 months. The growth of the second-generation germfree Beagles was the same as that of the conventional Beagles until 5 months, when their growth leveled off, and the second-generation germfree animals weighed 11 kg, the same as the first-generation dogs at 8 months. This characteristic was not significant since only two of the seven second-generation germfree dogs reached an age of 8 months; the others were used for experimentation at 5 or 6 months.

GROWTH OF FIRST-GENERATION GERMFREE BEAGLES

The data in Figure 10 compare the growth of first-generation germfree Beagles with the growth of conventional Beagles as found

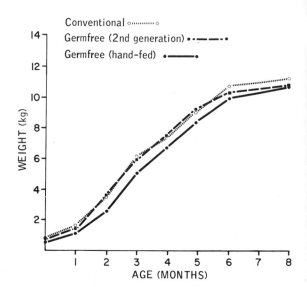

FIGURE 9 Male Beagle growth during first 8 months.

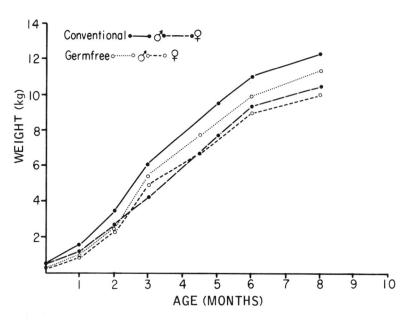

FIGURE 10 Growth of first-generation germfree Beagles.

JAMES B. HENEGHAN,
SALVADOR G. LONGORIA,
AND ISIDORE COHN, JR.

in the literature.[11] From birth to 5 months of age, the germfree Beagle growth curves represent the average weights of 38 males and 31 females; after 5 months, however, experimentation reduced the number of animals remaining to five males and six females at 8 months. For the conventional Beagles, on the other hand, each point on the growth curve represents the average weights of approximately 100 animals. With the exception of the hand-feeding period, the growth rates of the first-generation germfree Beagles were similar to those of the conventional Beagles of the same sex. Neither sex appeared to overcome completely the weight deficit acquired during hand-feeding, and both tended to remain between 0.5 to 1.0 kilograms lighter than their conventional counterparts. However, the female Beagles seemed to be better adapted to the germfree environment than the males since their growth more closely matched that of conventional females.

GENERAL CHARACTERISTICS OF GERMFREE DOGS

Base-line clinical, hematologic, and chemical data are being accumulated on germfree Beagles and will be reported in the future. However, our experience in maintenance and experimentation[1] has revealed that germfree dogs:

- Do not have an enlarged cecum.
- Have formed feces that are slightly softer than conventional.
- Do not have parasites, provided they are treated with diethylcarbamazine citrate and are raised so that reinfection is prevented.
- Are free of congenital parasites if they are second-generation animals.
- Have normal reproductive systems capable of producing offspring.
- Exhibit normal behavior while in the isolators.
- Heal wounds as fast as, and possibly faster than, conventional dogs.
- Reject transplanted tissue in a manner similar to conventional dogs.
- Seem to have faster blood-clotting mechanisms.
- Survive with intestinal strangulation obstruction, whereas conventionals do not.
- Can accumulate copious quantities (over 2 liters) of bile in the peritoneal cavity without developing peritonitis.
- Develop hepatic encephalopathy following portacaval shunt.
- Develop experimental pancreatitis as readily as conventionals.
- Tolerate higher doses of immunosuppressive drugs.

CONTAMINATION IN THE GERMFREE LABORATORY

Since very few investigators have publicized their experiences in this area, and since the successful maintenance of germfree animals requires the reduction of contamination to a minimum, we shall examine in detail the contaminations occurring at L.S.U. while rearing germfree dogs, rats, and mice from 1964 through 1968. During this 5-year period, the laboratory had 54 episodes of contamination: 16 in 1964, 14 in 1965, 13 in 1966, 5 in 1967, and 6 in 1968.

Causes of Contamination

The most frequent causes of contamination (Table 1) were wet filters and holes in isolator gloves. Wet inlet filters, which were moistened by poor air flow and the high relative humidity in the laboratory, represented the most frequent reason for contamination in the early development of the laboratory. Installation of an adequate central isolator air supply system eliminated this problem. However, holes in gloves and holes in the isolators themselves continue to be a problem and are now the most important cause of contamination. The isolator gloves are the weakest link in the barrier system. The only way to minimize this problem is close supervision on the part of the technical personnel; in fact, this approach will tend to reduce all causes of contamination.

TABLE 1 Causes of Contamination 1964 through 1968

Cause	Number Contaminated	Cause	Number Contaminated
Wet filters	13	Poor autoclaving (drain packed)	1
Holes in gloves	13	Cesarian skin contamination	1
Holes in isolator	5	Hole in sterilizing cylinder	1
Holes in air filter	3	Separation of plastic from tub isolator	1
Faulty filter sterilization of liquids	3	Unknown	9
Transportation accidents	2		
Failure in initial isolator sterilization	2		

Total Contaminations 54

Types of Organisms Contaminating Isolators

Sixty-six different types of micro-organisms (Table 2) were isolated from the 54 separate contaminations in the germfree laboratory from 1964 through 1968. *Bacillus* sp. were isolated most frequently (21 times). Because these are spore-formers, one would be inclined to suspect failure in autoclaving or in peracetic acid sterilization. Spore-forming organisms are un-

JAMES B. HENEGHAN,
SALVADOR G. LONGORIA,
AND ISIDORE COHN, JR.

TABLE 2   Types of Organisms Contaminating
Isolators 1964 through 1968

Organism	Number Contaminated
Bacillus species	21
Mold	17
Staphylococcus albus	10
Fungi	7
Enterococci	3
Proteus species	3
Pseudomonas species	2
Staphylococcus aureus	2
Mimapolymorpha	1
Total Contaminations	66

doubtedly the most difficult to destroy. Molds were isolated 17 times and are a reflection of early equipment problems that resulted in wet inlet and outlet filters. *Staphylococcus albus* was identified on 10 occasions and was traced to a hole in one of the isolator gloves. Specific fungi were isolated from seven cultures, and they also were the result of early equipment design problems and wet filters. It should be mentioned that in general many contaminations of *Bacillus* species or *Staphylococcus albus* were monocontaminations. In the case of enterococci, *Proteus*, or *Pseudomonas*, these were not monocontaminations but were usually associated with some other contaminant.

Comparison of Frequency of Contamination for Different Animals

The data in Table 3 illustrate the consecutive number of weeks without a contamination in all the isolators of a given animal

TABLE 3   Contaminations in Maintaining Different Animal Species—Average Consecutive Isolator Weeks Without a Contamination, 1964–1968

Species	Number of Consecutive Weeks without Contamination				
	1964	1965	1966	1967	1968
Dogs	33	42	118	131	122
Rats	25	49	128	237	354
Mice	91	77	81	384	308

species. [Example: 6 isolators maintained germfree for 6 weeks = 36 isolator weeks.] Two conclusions can be drawn from these data. First, the technological proficiency of the germfree laboratory has improved each year, as indicated by the steady rise in the number of consecutive weeks without a contamination. Second, mice are the easiest to maintain germfree; whereas dogs are the hardest.

WORK INVOLVED IN REARING GERMFREE DOGS

The effort (Table 4) required to maintain 4 dogs germfree for 2 years to obtain the first litter of second-generation germfree Beagles was compared to the work necessary to maintain 2 dogs germfree for an additional 8 months to obtain the second litter of second-generation germfree Beagles.

TABLE 4   Second-Generation Germfree Dogs—
Maintenance Required

Maintenance	First Litter[a]	Second Litter[b]
Isolator sterilizations	18	2
Glove entries	4,093	719
Sterile entries	256	39
Milk (l)	30	5
Water (l)	1,671	291
Feed (kg)	616	93

[a]4 Dogs from April 7, 1966 to April 7, 1968.
[b]2 Dogs from April 7, 1968 to December 20, 1968.

ELIMINATION OF IRRITATION OF PERACETIC ACID

Germfree dogs exhibited obvious signs of irritation during a sterile-lock entry when peracetic acid was used as the sterilizing agent. Our laboratory has studied this problem[4] and found that this acid caused an increase in heart rate, blood pressure, and respiratory rate in anesthetized conventional ex-germfree dogs. These effects have been eliminated by rerouting the exit of air from the isolator through the sterilized lock and out the exhaust filter. This provided an adequate method of removing the peracetic acid and its breakdown products.

Summary

Normal reproduction has been obtained for the first and second times in germfree Beagles; this established a high degree of profi-

JAMES B. HENEGHAN,
SALVADOR G. LONGORIA,
AND ISIDORE COHN, JR.

ciency for the current technology. This was the first time that reproduction occurred in large germfree animals and in germfree animals other than rodents, rabbits, or birds.

Germfree Beagles grew at rates comparable to those of conventional Beagles, and they did not have an enlarged cecum. During the hand-feeding period the growth of first generation germfree Beagles was reduced 50 percent. Eight germfree second-generation Beagles grew as well as conventional Beagles during the nursing period and thereafter.

Contamination was most frequently caused by holes in the isolator gloves, the weakest-link in the barrier system. The frequency of contamination in germfree dogs was 3 times, and in germfree rats 2 times, the frequency observed in germfree mice.

One pair of pure-bred Beagles has been maintained germfree for over 3 years, which is the longest time reported for this species. Second-generation germfree Beagles were free of congenital intestinal parasites. The current technology will permit the establishment of germfree Beagle colonies.

The work reported in this paper was supported by USPHS Grant No. RR-00272 from the Animal Resources Branch, Division of Research Facilities and Resources, National Institutes of Health.

References

1. Cohn, I., Jr., J. B. Heneghan, F. C. Nance, G. H. Bornside, J. L. Cain, S. K. Yu, and J. A. Labat. 1967. Germfree surgical research: Current status. Ann. Surg. 166:518–529.
2. Griesemer, R. A. 1963. The gnotobiotic dog. Lab. Anim. Care 13:643–649.
3. Griesemer, R. A. 1963. The establishment of an ascarid-free beagle dog colony. J.A.V.M.A. 143:965–967.
4. Heneghan, J. B., and D. F. Gates. 1966. Effects of peracetic acid as used in gnotobiotics on experimental animals. Lab. Anim. Care 16:96–104.
5. Heneghan, J. B., C. E. Floyd, and I. Cohn, Jr. 1966. Gnotobiotic dogs for surgical research. J. Surg. Res. 6:23–30.
6. Heneghan, J. B. 1968. Contaminations in a germfree laboratory. J. Amer. Ass. Contam. Cont. 1:8–12.
7. Heneghan, J. B., S. G. Longoria, and I. Cohn, Jr. 1968. Maintenance and growth of gnotobiotic beagle dogs. p. 63–67. In M. Miyakawa and T. D. Luckey [ed.] Advances in germfree research and gnotobiology. Proc. International Symposium on Germfree Life Research, Nagoya, Japan, April, 1967. Chemical Rubber Co. Press, Cleveland.
8. Heneghan, J. B., S. G. Longoria, and I. Cohn, Jr. 1969. Reproduction in germfree beagles. p. 367–371. In E. A. Mirand and N. Back [ed.] Germfree biology—experimental and clinical aspects, Vol. 3, Advances in experimental medicine and biology. Plenum Press, New York.

9. Phillips, B. P. 1960. Parasitological survey of Lobund germfree animals. Lobund Reports 3:172–175. (University of Notre Dame Press, Notre Dame, Indiana.)
10. Smith, W. C., and W. C. Reese, Jr. 1968. Characteristics of a beagle colony: I Estrous cycle. Lab. Anim. Care 18:602–606.
11. Spector, W. S. 1956. Handbook of biological data. W. B. Saunders Co., Philadelphia, p. 159.
12. Yale, C. E. 1969. An isolation system for the germfree dog. Lab. Anim. Care 19:103–108.

DISCUSSION

LORELIE MITCHELL: Do germfree dogs develop dental calculi?

DR. HENEGHAN: We have not looked at this ourselves, but we have been sending dental material to a Dr. Lisgarten in Pennsylvania, who has been evaluating some of this material for us. We started this only recently, and we do not have much data on it, but he has indicated that there are—this is in addition to the fact that most of the animals we have sent him are young Beagles, and one would expect to see dramatic evidence in the younger dogs—inflammatory-type responses in the gingiva of these germfree dogs, indicating that plaque formation is probably occurring in the absence of bacteria. We are waiting until we get some older germfree dogs to establish this further.

DR. HAYDEN: Were there any difficulties in keeping the dogs clean in the isolator units?

DR. HENEGHAN: Yes. It is quite difficult to keep the dogs clean in the isolators. As a matter of fact, that is one reason why there were so many isolator sterilizations required to maintain those four animals for 2 years. We had to change them from isolator to isolator just because they get dirty every two and a half to three months. Each pair had to be transferred to a clean isolator.

With this new type of isolator with the tank underneath and being able to clean this out, we could reduce this to possibly once every six months, which is I think the longest I would want an animal to stay in the isolator without the gloves being replaced. So it is hard to keep them clean. I am sure that conventional animals could not exist with the dirt that is in a germfree isolator, but because there are no bacteria there, it does not affect the animal, and we try to keep it as clean as possible.

I think if you looked closely at the bitch in that one picture, you could see that she is quite clean. But it is a problem and it takes our best efforts to keep them as clean as possible.

JAMES B. HENEGHAN,
SALVADOR G. LONGORIA,
AND ISIDORE COHN, JR.

DR. FOSTER: Did I understand you to say that you did not note cecal enlargement in dogs?

DR. HENEGHAN: Yes.

DR. FOSTER: Has anyone theorized why you do not see cecal enlargement in dogs as we do in mice and rats?

DR. HENEGHAN: The only explanation that comes to mind is possibly a different physiologic function of the cecum in rodents as compared to dogs. The cecum is quite a distinct organ in rodents, and may have a function in gastrointestinal physiology, whereas in the dog it is sort of a vestigial organ, something akin to the appendix in man, and therefore may not serve the function that it does in rodents. This may be the reason that it is not enlarged. In all the dogs that we have examined, we have not seen an enlarged cecum.

DR. FOSTER: May I ask you just one more question? Have you ever associated dogs with a microflora, and if so what organisms did you use?

DR. HENEGHAN: Yes, we have associated dogs with *Clostridium*. They have been monoassociated with *Clostridium*, with *E. coli*, with combinations of clostridia and *E. coli*, and we have had a wide variety of organisms contaminating them accidentally, but these are the ones that I can think of offhand that we purposely put in there.

If you are wondering in terms of conventionalizing the dogs after they have been in isolators, we have not found it particularly difficult to conventionalize our germfree dogs, provided we treat them as newborn puppies and give them all the immunizations one would give a newborn puppy, canine distemper, hepatitis, and things like that, and provided we give them a complete flora when we bring them out.

DR. GOLDENSON: Have you found cystic calculi in gnotobiotic beagles?

DR. HENEGHAN: We have not really looked, so I really cannot answer that question.

DEFINING THE
LABORATORY RAT
FOR CARDIOVASCULAR
RESEARCH

I. Albrecht and J. Souhrada

According to *Biological Abstracts*, May–October, 1968, there were 60 publications on the physiology and pathophysiology of the cardiovascular system of the rat published in that period, of which 31 were concerned with hypertension and 14 with ischemic heart disease. Very few, if any, of these studies were carried out under standardized physiological and microbiological conditions in the experimental animals. It is alarming to find that there are striking differences in arterial O_2 saturation between SPF rats and the conventional values published by King[1] and in values of cardiac output determined on such rats in comparison to the data of Wostman.[2]

Skelton,[3] and Masson and Corcoran[4] all reported that male and female rats react in the same manner to adrenal enucleation (pressure rise); Neff and Corell[5] and Jelínek[6] were unable to confirm this, however, and found the pressure rise only in females. Sapierstein *et al.*[7] produced hypertension in rats by administration of 2 percent NaCl as drinking water, but this result was not confirmed by Green *et al.*,[8] van Proosdij-Hartzema and De Jongh,[9] and Jelínek.[10] In view of this lack of confirmation, it is reasonable to doubt whether conventionally reared rats are suitable for research on hypertension and the pathogenesis of ischemic heart disease. The differences in right ventricular weight in Wistar strain rats from three conventional stock farms in Czechoslovakia (labeled *A*, *B*, and *C* are a good illustration of this. In the right half of Figure 1 we can see that the relative weight (mg/100 g of body weight) of SPF rats at age 6–8 months is far less than values from *A* and *C* (columns show average values; circles shows individual values in the group). The difference was not marked in comparison with group *B* in terms of averages, but

I. ALBRECHT
AND
J. SOUHRADA

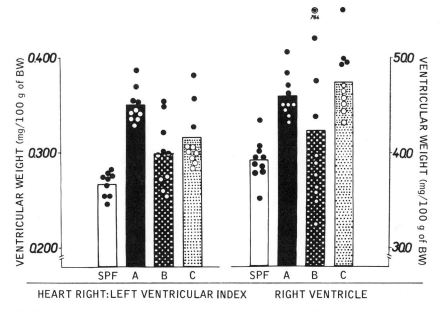

FIGURE 1 Comparison of right ventricular weights and ratios of left:right ventricular weights of SPF and conventional rats from three stock farms in Czechoslovakia.

the interindividual variation in this group was high. The left half of Figure 1 shows the data in terms of the ratio of weights of the right:left ventricles, suggesting that ordinary stock rats have a higher pulmonary resistance than SPF rats in relation to the pulmonary pathology frequently found in the parenchyma.

Lung histology was carried out in 10 rats of group *A* and in 10 SPF animals. The picture seen in the conventionally bred animals can be characterized as a chronic catarrhal bronchitis in 100 percent of the animals, with differences in extent related to age and ecological factors. Figures 2 and 3 compare sections of lungs from SPF rats and conventionally bred animals. The SPF sections showed a strikingly fine bronchial and vascular branching, with an even consistency of the parenchyma, while in the group *A* animals there were massive infiltrations along the bronchi, with local destruction of the bronchial wall and formation of secondary lymphatic follicles. With greater magnification we can see round cell infiltration in the bronchial walls, with a mixture of polymorphic leucocytes. In the mucous membrane there are sites of leucocytic infiltration with purulent foci, and areas of the epithelium show signs of metaplasia; the cells are cubic without flagellae. There are sites of emphysema as well. Figure 4 shows that in the periphery of the tissue there is an inflammatory cellulitis along small bronchi and vessels, extending to the septa, none

617

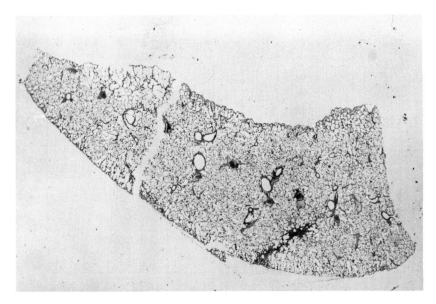

FIGURE 2   Lung section from SPF rat.

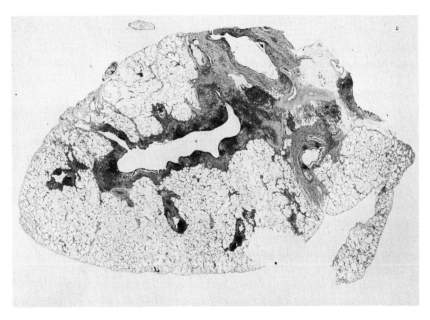

FIGURE 3   Lung section from conventional rat.

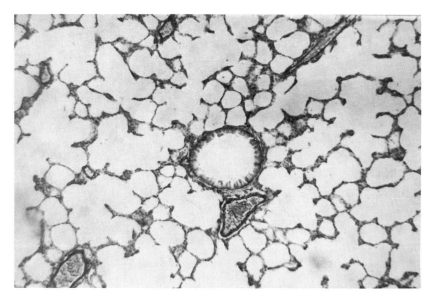

FIGURE 4 Lung section from conventional rat. There is evidence of inflammatory cellulitis along the small bronchi and vessels, extending to the septa.

of which can be seen in sections from SPF animals (Figure 5). This extension of bronchitic and bronchiolitic reaction at the expense of normal parenchyma is associated with an increase in lung weight (Figure 6). Differences can be seen, although they are not all of the same magnitude, between SPF data and A, B, and C data. In groups A and C the above-described picture of peribronchial and perivascular infiltration occurred; in group B there was in addition a definite, though small, incidence of small and medium solitary and multiple abscesses, localized mainly in the peripheral part of the organ. The differences in right:left ventricular index, therefore, may be due to the fact that abscess formation represents a smaller increase in pulmonary resistance that a widespread inflammatory change, with a resultant smaller change in right ventricular weight.

The following questions can therefore be put:

1. How do the SPF versus conventional stock differences—right ventricular weight and pulmonary vasculature—reflect themselves in terms of pulmonary blood flow?
2. Do differences in right heart function also reflect themselves in left heart function, i.e., cardiac output?

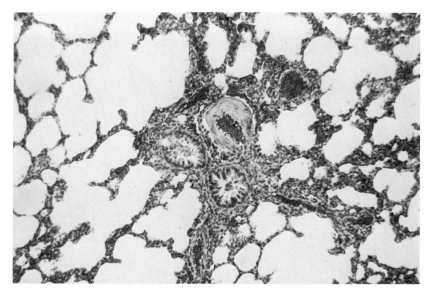

FIGURE 5 Healthy lung section from SPF rat (c.f. Figure 4).

We know that conventionally bred animals have an increased residual lung capacity and that this may be related to the differences in arterial O_2 saturation reported on by King.

We used the heart–lung preparation to study cardiac function. This technique in the rat has already been described by Souhrada.[11] The organs were left *in situ*; venous return was from a reservoir with a water manometer giving the gradient of pressure drop to the right atrium, and thus the flow level. The right and left superior *vena cava* and the azygous vein were tied off. Aortic outflow was through a cannula in the right brachiocephalic trunk. Blood pressure was measured in the left carotid artery with an ELEMA Schönander strain gauge. Aortic flow was stopped beyond the branching of the left subslavian artery after it was tied. A bubble flowmeter was used, and the electrocardiograph served to give heart rate. Values of a variable peripheral resistance were read from a mercury manometer. Cardiac work was calculated as:

$$W \text{ (work)} = \text{flow} \times \text{blood pressure} \times 1.36.$$

These hemodynamic studies were carried out on SPF rats and conventional stock animals from our institute. Pentobarbital anesthesia was used (40 mg/ 100 g of body weight), given intraperitoneally. Arterial blood pressure was measured before opening the thorax. Figure 7 shows that under these conditions SPF rats had higher values. The explanation of this is not clear, but it can be suggested that it is a reflection of a contractility difference rather than one in peripheral vascular resistance. The preparation was then completed with artificial respiration. With a constant peripheral resistance value giving a pressure drop of 140 mm Hg, venous return pressures were varied in steps of 30 mm H_2O from 110 to 290–300. Shortly thereafter, with zero resistance, maximal flow was measured, and then stop flow was carried out by clamping the cannula. Venous pressure was set at 300, and aortic pressure–maximal left ventricular pressure was measured. Figure 8 shows flow curves in the aorta as a function of increasing venous return. Both separate values and average results

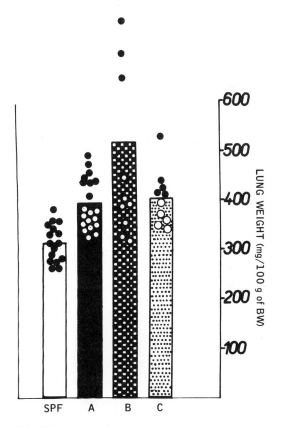

FIGURE 6 Comparative lung weights of conventional and SPF rats. The conventional animals are affected by extended bronchitic and bronchiolitic reaction, which may cause an increase in lung weight.

621

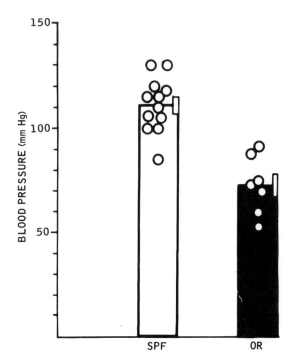

FIGURE 7 Blood pressure values of SPF and conventional rats after administration of pentobarbital anesthesia.

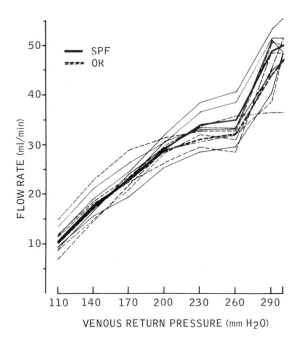

FIGURE 8 Aortic flow of SPF and conventional rats as a function of increasing venous return.

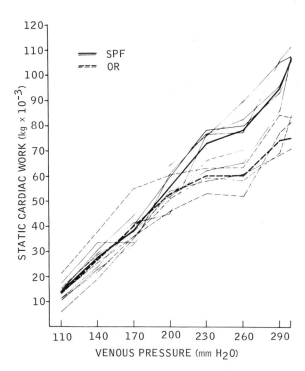

FIGURE 9 Static cardiac work as a function of venous pressure in conventional and SPF rats.

showed no great difference between groups. In other words, right ventricular hypertrophy in the conventional animals was sufficient to secure adequate compensation of function in this regard.

Further results indicated that the higher systemic arterial pressure of the SPF animals was related to cardiac contractility, not to the peripheral bed. Cardiac work calculated from the flow curves shows that with a venous pressure between 230 and 290, the SPF heart carried out more static work than the conventional heart and at 300, these differences are highly significant (Figure 9). With a sudden stop flow, aortic pressure increased, and the functional reserve of the left heart is exhausted. Maximum pressure values under these conditions are a reflection of the maximum ability of the left heart to contract. Figure 10 shows that SPF hearts produced higher maximal left ventricular pressure values. In the terminology of Starling's Law it can be stated that the musculature of the left ventricle in SPF animals can contract more efficiently with the same tension load than that of conventional animals. It can be suggested that this difference is related to the myocardial cells, to O_2 transport and saturation, and to the toxic effects of inflammation. Without doubt, such differences would play a role in studies of hypertension in these

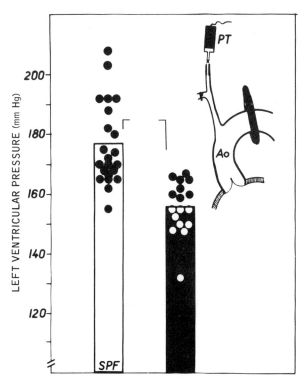

FIGURE 10 Maximal left ventricular pressure values of SPF and conventional rats.

two groups of animals. It would appear that in SPF animals, we are able to study the reaction of the uncomplicated cardiovascular system to a new stimulus rather than the reaction of an already compensated system, the degree of modulation of which is unknown.

References

1. King, T K. C., and D. Bell. 1966. Arterial blood gases in specific pathogen free and bronchitic rats. J. Appl. Physiol. 21:237.
2. Wostmann, B. S., E. Bruckner-Kardoss, and P. L. Knight. 1968. Coecal enlargement, cardiac output and oxygen consumption in germ-free rats. Proc. Soc. Exp. Biol. Med. 128:137.
3. Skelton, F. R. 1956. Adrenal regeneration hypertension and factors influencing its development. A. M. A. Arch. Intern. Med. 98:449.
4. Masson, G. M., and A. C. Corcoran. 1956. Hormonal influences on adrenal regeneration hypertension. Endocrinology 59:201.

5. Neff, A. W., and J. T. Corell. 1957. Influence of sex on adrenal regeneration and accompanying hypertension. Proc. Soc. Exp. Biol. Med. 95:227.
6. Jelínek, J. 1967. Salt intake and sexual differences in sensitivity of rats to adrenal regeneration hypertension. Physiol. Bohemoslov. 16:389.
7. Sapierstein, L. A., W. Brandt, and D. R. Drury. 1950. Production of hypertension in the rat by substituting hypertonic sodium chloride solutions for drinking water. Proc. Soc. Exp. Biol. Med. 73:82.
8. Green, D. M., D. H. Colleman, and M. McCabe. 1948. Mechanisms of desoxycorticosterons action II. Reaction of sodium chloride to fluid exchange. Pressor effects and survival. Amer. J. Physiol. 154:465.
9. van Prosdij-Hartzema, E. G., and D. K. de Jongh. 1955. Investigation on experimental hypertension. I. Observations on the induction of hypertension in rats. Acta Physiol. Pharmacol. Neerl. 4:37.
10. Jelínek, J., M. Kraus, and H. Musilová. 1966. Adaptation of rats of different ages to forced intake of a 2% NaCl solution, without the occurrence of salt hypertension. Physiol. Bohemoslov. 15:137.
11. Souhrada, J. 1967. Size of the heart and cardiac output during experimental polycythaemia. Physiol. Bohemoslov. 16:469.

DISCUSSION

DR. FOSTER: Does your definition of SPF animals coincide at all with the earlier definition I tried to state?

DR. ALBRECHT: Yes, exactly.

DR. FOSTER: In other words, they were gnotobiotes at one time, as opposed to animals that possibly had a few pathogens removed from them.

DR. ALBRECHT: Yes. We have SPF animals, but we tried to understand this program a bit more. As you saw on the graphs, sometimes there is not only the SPF abbreviation, but DOS. We understand this problem to be "defining these rats definite, optimal, and standardized." It is not just a matter of the biological picture. It means the lighting, food, humidity, temperature, all of the things that can play a role in the research of hemodynamics of hypertension and cardiovascular disease.

PARTICIPANTS

DR. IVAN ALBRECHT, Institute of Physiology,
Czechoslovak Academy of Sciences, Prague, Czechoslovakia

DR. GEORGE W. BEADLE, University of Chicago, Chicago, Illinois

DR. ESZTER CHOLNOKY, Laboratory Animals Institute,
Táncsics Milhály Ut 1, Gödöllö, Hungary

DR. ROGER-PAUL DECHAMBRE, Institut Gustave-Roussy,
Villejuif (Val-de-Marnes), France

SIR JOHN C. ECCLES, State University of New York, Buffalo, New York

DR. MICHAEL F. W. FESTING, Medical Research Council,
Laboratory Animals Centre, Carshalton, Surrey, England

DR. MICHAEL W. FOX, Department of Psychology,
Washington University, St. Louis, Missouri

DR. KŌSAKU FUJIWARA, Institute of Medical Science,
University of Tokyo, Tokyo, Japan

DR. CARL T. HANSEN, Laboratory Aids Branch,
National Institutes of Health, Bethesda, Maryland

DR. JAMES B. HENEGHAN, Department of Surgery,
Louisiana State University, School of Medicine, New Orleans, Louisiana

DR. W. BEN ITURRIAN, School of Pharmacy, University of Georgia
Athens, Georgia

DR. VACLAV J. JELINEK (deceased), Research Institute for Pharmacology
and Biochemistry, Prague, Czechoslovakia

DR. S. S. KALTER, Division of Microbiology,
Southwest Foundation for Research and Education, San Antonio, Texas

DR. WILLIAM LANE-PETTER, Houghton, Huntingdon, England

MISS MARY W. MARSHALL, Human Nutrition Research Division,
U.S. Department of Agriculture, Beltsville, Maryland

DR. DOUGLAS H. McKELVIE, Department of Veterinary Medicine,
College of Veterinary Medicine and Biomedical Science,
Colorado State University, Fort Collins, Colorado

DR. L. V. MELÉNDEZ, New England Regional Primate Center,
Harvard Medical School, Southborough, Massachusetts

DR. SOL M. MICHAELSON, Radiation Biology and Biophysics,
University of Rochester School of Medicine and Dentistry,
Rochester, New York

DR. RENÉ MOUTIER, Centre de Sélection et d'Elévage des Animaux de
Laboratoire, Orleans-La Source, France

PROFESSOR OTTO MÜHLBOCK, The Netherlands Cancer Institute,
Amsterdam, The Netherlands

DR. STANLEY R. OPLER, Stanford Research Institute,
Menlo Park, California

DR. JAMES S. PATERSON, Allington Farm, Porton Down,
Salisbury, Wilts., England

DR. DWAYNE SAVAGE, Department of Microbiology,
The University of Texas, Austin, Texas

DR. L. J. SERRANO, Biology Division, Oak Ridge National Laboratory,
Oak Ridge, Tennessee

DR. JAMES A. SHANNON, The Rockefeller University, New York, New York

DR. NICOLAI SIMIONESCU, The Rockefeller University,
New York, New York

DR. HERBERT K. STRASSER, Kelkhaim/Taunus,
Federal Republic of Germany

DR. NEIL B. TODD, Carnivore Genetics Research Center,
Newtonville, Massachusetts

DR. GERALD L. VAN HOOSIER, Department of Veterinary Pathology,
Washington State University, Pullman, Washington

PROFESSOR HORST W. WEBER, Department of Pathology,
Roswell Park Memorial Institute, Buffalo, New York

DR. WOLF H. WEIHE, Biologisches Zentrallaboratorium, Kantonsspital,
Universität Zürich, Zurich, Switzerland

DR. ROGER E. WILSNACK, Huntingdon Research Center, Inc.,
Baltimore, Maryland

DR. GEORGE L. WOLFF, The Institute for Cancer Research, Fox Chase,
Philadelphia, Pennsylvania

DR. BERNARD S. WOSTMANN, The Lobund Laboratory,
University of Notre Dame, Notre Dame, Indiana